AF705213

Stability, Control, and Computation for Time-Delay Systems

Advances in Design and Control

SIAM's Advances in Design and Control series consists of texts and monographs dealing with all areas of design and control and their applications. Topics of interest include shape optimization, multidisciplinary design, trajectory optimization, feedback, and optimal control. The series focuses on the mathematical and computational aspects of engineering design and control that are usable in a wide variety of scientific and engineering disciplines.

Editor-in-Chief

Ralph C. Smith, North Carolina State University

Editorial Board

Athanasios C. Antoulas, Rice University
Siva Banda, Air Force Research Laboratory
John Betts, The Boeing Company (retired)
Stephen L. Campbell, North Carolina State University
Michel C. Delfour, University of Montreal
Fariba Fahroo, Air Force Office of Scientific Research
J. William Helton, University of California, San Diego
Arthur J. Krener, University of California, Davis
Kirsten Morris, University of Waterloo
John Singler, Missouri University of Science and Technology

Series Volumes

Michiels, Wim and Niculescu, Silviu-Iulian, *Stability, Control, and Computation for Time-Delay Systems: An Eigenvalue-Based Approach, Second Edition*

Narang-Siddarth, Anshu and Valasek, John, *Nonlinear Time Scale Systems in Standard and Nonstandard Forms: Analysis and Control*

Bekiaris-Liberis, Nikolaos and Krstic, Miroslav, *Nonlinear Control Under Nonconstant Delays*

Osmolovskii, Nikolai P. and Maurer, Helmut, *Applications to Regular and Bang-Bang Control: Second-Order Necessary and Sufficient Optimality Conditions in Calculus of Variations and Optimal Control*

Biegler, Lorenz T., Campbell, Stephen L., and Mehrmann, Volker, eds., *Control and Optimization with Differential-Algebraic Constraints*

Delfour, M. C. and Zolésio, J.-P., *Shapes and Geometries: Metrics, Analysis, Differential Calculus, and Optimization, Second Edition*

Hovakimyan, Naira and Cao, Chengyu, *$\mathcal{L}_1$ Adaptive Control Theory: Guaranteed Robustness with Fast Adaptation*

Speyer, Jason L. and Jacobson, David H., *Primer on Optimal Control Theory*

Betts, John T., *Practical Methods for Optimal Control and Estimation Using Nonlinear Programming, Second Edition*

Shima, Tal and Rasmussen, Steven, eds., *UAV Cooperative Decision and Control: Challenges and Practical Approaches*

Speyer, Jason L. and Chung, Walter H., *Stochastic Processes, Estimation, and Control*

Krstic, Miroslav and Smyshlyaev, Andrey, *Boundary Control of PDEs: A Course on Backstepping Designs*

Ito, Kazufumi and Kunisch, Karl, *Lagrange Multiplier Approach to Variational Problems and Applications*

Xue, Dingyü, Chen, YangQuan, and Atherton, Derek P., *Linear Feedback Control: Analysis and Design with MATLAB*

Hanson, Floyd B., *Applied Stochastic Processes and Control for Jump-Diffusions: Modeling, Analysis, and Computation*

Michiels, Wim and Niculescu, Silviu-Iulian, *Stability and Stabilization of Time-Delay Systems: An Eigenvalue-Based Approach*

Ioannou, Petros and Fidan, Barış, *Adaptive Control Tutorial*

Bhaya, Amit and Kaszkurewicz, Eugenius, *Control Perspectives on Numerical Algorithms and Matrix Problems*

Robinett III, Rush D., Wilson, David G., Eisler, G. Richard, and Hurtado, John E., *Applied Dynamic Programming for Optimization of Dynamical Systems*

Huang, J., *Nonlinear Output Regulation: Theory and Applications*

Haslinger, J. and Mäkinen, R. A. E., *Introduction to Shape Optimization: Theory, Approximation, and Computation*

Antoulas, Athanasios C., *Approximation of Large-Scale Dynamical Systems*

Gunzburger, Max D., *Perspectives in Flow Control and Optimization*

Delfour, M. C. and Zolésio, J.-P., *Shapes and Geometries: Analysis, Differential Calculus, and Optimization*

Betts, John T., *Practical Methods for Optimal Control Using Nonlinear Programming*

El Ghaoui, Laurent and Niculescu, Silviu-Iulian, eds., *Advances in Linear Matrix Inequality Methods in Control*

Helton, J. William and James, Matthew R., *Extending H^∞ Control to Nonlinear Systems: Control of Nonlinear Systems to Achieve Performance Objectives*

Stability, Control, and Computation for Time-Delay Systems

An Eigenvalue-Based Approach

SECOND EDITION

Wim Michiels
Katholieke Universiteit Leuven
Heverlee, Belgium

Silviu-Iulian Niculescu
Laboratoire des Signaux et Systèmes
Gif-sur-Yvette, France

siam.

Society for Industrial and Applied Mathematics
Philadelphia

Copyright © 2014 by the Society for Industrial and Applied Mathematics

10 9 8 7 6 5 4 3 2 1

All rights reserved. Printed in the United States of America. No part of this book may be reproduced, stored, or transmitted in any manner without the written permission of the publisher. For information, write to the Society for Industrial and Applied Mathematics, 3600 Market Street, 6th Floor, Philadelphia, PA 19104-2688 USA.

Trademarked names may be used in this book without the inclusion of a trademark symbol. These names are used in an editorial context only; no infringement of trademark is intended.

MATLAB is a registered trademark of The MathWorks, Inc. For MATLAB product information, please contact The MathWorks, Inc., 3 Apple Hill Drive, Natick, MA 01760-2098 USA, 508-647-7000, Fax: 508-647-7001, *info@mathworks.com*, *www.mathworks.com*.

Figures 1.9–10, 2.9–10, 7.8–10, and Tables 7.1–2 reprinted with permission from the Institution of Engineering and Technology.
Figures 2.1–8, 3.1–4, 5.3–5, 6.3–5, 13.2–4, and Table 3.1 reprinted with permission from Elsevier.
Figures 5.6, 10.5, and 16.1–2 reprinted with permission from IEEE.
Figures 7.1–7 reprinted with permission from Cambridge University Press.
Figures 8.1–4, 13.7–10, and Tables 8.1–2 reprinted with permission from World Scientific Publishing.
Figures 10.1–3 and Table 10.1 reprinted with permission from Oxford University Press.
Figures 10.6–9, 11.1–2, 11.4, and Table 10.5 reprinted with permission from Taylor and Francis.
Figures 12.1–4 reprinted with permission from Springer Science+Business Media.
Figures 15.1–5 reprinted with permission from AIP.
Figures 16.5–13 reprinted with permission from American Institute of Mathematical Sciences.

Library of Congress Cataloging-in-Publication Data

Michiels, W. (Wim)
[Stability and stabilization of time-delay systems]
Stability, control, and computation for time-delay systems : an eigenvalue-based approach / Wim Michiels, Katholieke Universiteit Leuven, Heverlee, Belgium, Silviu-Iulian Niculescu, Laboratoire des Signaux et Systèmes, Gif-sur-Yvette, France. – Second edition.
pages cm. – (Advances in design and control ; 27)
Revised edition of: Stability and stabilization of time-delay systems.
Includes bibliographical references and index.
ISBN 978-1-611973-62-4
1. Automatic control. 2. Time delay systems. 3. Stability. 4. Eigenvalues. I. Niculescu, Silviu-Iulian. II. Title.
TJ213.M485 2014
629.8'3–dc23 2014029503

siam is a registered trademark.

Contents

Preface to the Second Edition

The interconnection between two (or more) physical systems is always accompanied by *transfer* phenomena (material, energy, information) such as *transport* and *propagation*. Mathematically speaking, transport and propagation phenomena can be represented by *delay elements*. In this way the corresponding overall systems are governed by a special type of differential equations, namely *delay-differential equations*.

Delay-differential equations are also used in modeling various other phenomena coming from biosciences (heredity in population dynamics [202, 223]), chemistry (behaviors in chemical kinetics [367, 334]), physics (lasers [102]), acoustics (wind instruments [102]) or economy (dynamics of business cycles [368]). Further examples in engineering can be found in [359, 290, 130, 102].

As mentioned by El'sgol'ts and Norkin [97] and Răsvan [335], time-delay systems have a long history, and, to the best of our knowledge, the first delay-differential equations are encountered in the work of Bernoulli and Condorcet. However, the theory started to be developed in the second half on the 20th century with the work of the East European Mathematical school: Myshkis [286], Krasovskii [200], and Halanay [145] (to cite only a few), who devoted most of their attention to the extension of the Lyapunov theory to such classes of differential equations. In the 1960s, an increasing interest in the topic appeared also in North America as confirmed by the monographs of Pinney [326] and Bellman and Cooke [23], (the former almost forgotten), with a particular interest in complex-domain approach and related frequency-domain techniques and methods. Subsequently, the theory arrived at some maturity in the 1970s as proven by the publications and monographs devoted to the field during that period. Among them, we mention the pioneering work of Hale [148] (the second edition of the monograph published in 1971), which is one of the most cited reference in the field not only for the fundamental results and approaches, but also for the quality and the clarity of the presentation. For further references and a deeper historical perspective, we refer the reader to [290, 332, 304].

It is important to point out that various references devoted to time-delay systems in engineering existed even before the 1950s as, for example, the papers coauthored by Callender [61, 62] and the editorial of the journal *Engineer* [357], with some contradictory conclusions concerning the effects induced by the delay presence in dynamical systems: sometimes *destabilizing* (mainly by using "huge" gains), and sometimes *stabilizing* (mainly in controlling some oscillatory modes). The explanation of such "dichotomic" behaviors was done case by case, without any attempt at a comprehensive explanation of the situations where stabilizing/destabilizing effects may occur.

Although by now the fundamental results in the theory of functional differential equations are well known and well understood (see, for instance, [23, 148, 151], to cite only a few), the increasing number of applications involving large-scale systems with corresponding complex decision making strategies in which the *delay* (transport, propagation, communication, decision) becomes a "critical" parameter made necessary the development of efficient numerical algorithms and methods for evaluating critical delays and related stability/instability properties. This monograph presents some approaches and techniques in this sense.

Recent approaches in *robust control* opened interesting perspectives and issues in dealing with delays in dynamical systems, where delays are eventually treated as *uncertainty* [130, 290, 32]. Some of them (frequency-sweeping tests, matrix pencil approaches) will be largely discussed in this monograph. Such interpretations of delays as uncertainty were at the origin of an abundant literature in the control area by the end of the 20th century. The corresponding results are expressed in terms of solutions of appropriate Riccati equations [225], and in terms of linear matrix inequalities [32] in connection (or not) with the μ-formalism [318]. An exhaustive overview concerning these approaches in the context of stability analysis can be found in [290].

At the same time, the increasing number of efficient algorithms for dealing with *nonlinear eigenvalue problems* [27, 237, 179] represented another important issue in treating delay systems. As in the finite-dimensional case, the essential properties of time-delay systems (asymptotic behavior, stability, instability, oscillations) are connected with the spectrum location of the corresponding linearized systems. As we shall explain in the following chapters, time-delay systems are infinite-dimensional systems, but with particular spectral properties. Such properties will be explicitly exploited in deriving the main (stability and stabilization) results and related algorithms. Crucial in recent algorithms is the exploitation of some duality in the frequency domain, in the sense that characteristic roots appear as solutions of a finite-dimensional nonlinear eigenvalue problem as well as an infinite-dimensional linear eigenvalue problem. Particular attention will be paid to the distinction between *retarded* and *neutral* systems because, although both belong to the class of time-delay systems, the spectral properties are considerably distinct.

It is important to point out that, excepting the functional differential equations–based representation, there are several ways to *represent* time-delay systems: as evolution equations over infinite-dimensional spaces [25, 84] (infinite-dimensional setting), 2-D (or n-D) systems [215], systems over rings of operators [184], and behavioral-based representations [120]. Throughout the volume, we adopt the functional differential equation based representation, although the connection with ODEs over a function space plays a major role in developing numerical methods. We further assume that the nominal models are completely known. In other words, we do not focus on delay modeling, identification, or identifiability.

Book outline and content

The book is organized in three parts:

(a) Stability analysis of linear time-delay systems.

(b) Stabilization and robust fixed-order control.

(c) Applications.

With Part (a) and Part (b) our intention is to present an analysis of stability, robust stability, and the synthesis of controllers using a *unitary* methodology—the *eigenvalue-based approach*. Without any loss of generality, we mainly concentrate on the following aspects that, to our best knowledge, have not received a full treatment in the literature:

- Sensitivity analysis with respect to delays and to other systems' parameters (continuity of the spectrum with respect to the parameters based on Rouché type theorems and variants, pseudospectra, and related properties);

- Analytical, as well as numerical analysis tools (algorithms for computing characteristic roots, $\mathcal{H}_2$ and $\mathcal{H}_\infty$ norms).

- Design of fixed-order or fixed-structure stabilizing and robust controllers. These approaches, which are recent, even in the context of finite-dimensional systems, are grounded in numerical linear algebra and optimization.

A lot of examples complete the presentation and illustrate the main results proposed in the monograph. Most of the major ideas are explained by using (several) extremely simple, "easy-to-follow" (low-order) examples. Finally, Part (c) of the monograph is devoted to several applications spanning various fields from engineering to biology. All the applications considered start from some generic remarks on the way in which the models are derived, but without any deep discussions on the model derivation and its limitations. The choice of the applications was mainly motivated not only by their impact in engineering, biosciences, and related fields, but also by our own interest in the corresponding topics.

We have made the parts independent of each other as much as possible. However, a number of fundamental results are needed for the whole theoretical development and are presented in Chapter 1.

What's new in the second edition?

The first edition of the book was mainly targeted towards the mathematical control community. The focus was almost exclusively on the stability analysis and stabilization, which encompass only the first step in a control design. With this new edition of the book, first we make the leap from stabilization to the design of robust and optimal controllers, thus enlarging the scope of the book within the control area. Second, by including a lot of new material on numerical methods, we aim at reaching other research communities, in particular the numerical linear algebra and the numerical optimization community. Finally we have extended the number of applications that we address. More specifically, the changes can be summarized as follows:

- In addition to the existing material on the stability analysis and stabilization, we have included several new chapters and sections related to the design of robust controllers, with the emphasis on optimal $\mathcal{H}_2$ and $\mathcal{H}_\infty$ controllers.

- In contrast to the first edition, this new edition contains a *significant part on numerical methods*: (large-scale) eigenvalue computations, level set methods, and structured eigenvalue problems (related to the $\mathcal{H}_\infty$ norm computation), solving generalizations of Lyapunov equations (related to the $\mathcal{H}_2$ norm computation). Throughout the book, an overview of numerical methods for control is proposed, where the details are worked out for the class of time-delay systems. The controller synthesis approaches in the new chapters rely on a direct optimization approach (fixed-order control based on eigenvalue optimization). Eigenvalue optimization and its applications is a relatively young research domain which lies at the intersection of numerical optimization, linear algebra, and control. Appendix A.5 contains, among others, a description of software corresponding to algorithms presented in the book. This software is publicly available.

- While in the first edition most of the material was devoted to delay systems described by retarded functional differential equations, the new edition addresses a larger class of systems represented by delay-differential algebraic equations, covering both retarded and neutral differential equations–based representations. Insights on the existing links between such systems and control feedback problems will be particularly emphasized. We further note that differential algebraic equations represent the standard way to perform modeling in large-scale interconnected systems, such as electronic circuits. In Part I and Part II of this new edition, Section 1.3, Chapter 2, Section 3.5, Chapter 4, Section 5.5, Section 7.4, Section 8.4, and Chapter 9 are completely new.

- New applications have been added to Part III of the book and some of the existing ones have been appropriately updated with new material. To be more precise, some of the first-edition chapters have disappeared, while others have been restructured by including new sections devoted to new topics, and some new chapters have appeared. In particular, the two first edition chapters devoted to output feedback stabilization have been merged into one chapter including the multiple delays case as a section. Furthermore, a new section including some insights on delay effects in networked control systems has been added. The chapter on congestion control algorithms has been completed by a section devoted to Smith predictor–type schemes and related robustness issues (computing the maximal allowable delay uncertainty by using a geometric approach). Next, a new chapter is included on synchronization of delay-coupled nonlinear oscillators including the stability analysis of synchronized equilibria as well as an application to Lorenz oscillators. Finally, the chapter discussing delay models in biosciences now includes an extra section devoted to the stability of biochemical networks. Furthermore, the section dealing with immune dynamics models in leukemia now contains updated material as well as a more elaborated model. Finally, we tried to give a unitary presentation of the delay models in biosciences by using an appropriate multicompartment modeling angle for genetic regulatory networks or a simplified human respiration system model, or for describing immune dynamics in chronic myelogenous leukemia.

Acknowledgments

The idea of writing this monograph appeared a couple of years ago, but, as usual, the *"delay factor"* interfered with the whole process. However, we believe that, finally, the delay had a *"positive"* impact.

We greatly acknowledge the financial support of the Belgian Federal Science Policy Office, through the network IAP VII/19 "Dynamical Systems, Control and Optimization," the KU Leuven Optimization in Engineering Center OPTEC, as well as the Research Council - Flanders (FWO), the KU Leuven Research Council, and the French CNRS (National Center for Scientific Research), through various projects, grants, and fellowships.

Parts of this book have been presented as courses within the framework of the doctoral schools SOCN (Belgium, October–November 2011) and EECI-HYCON (France, March 2013, March 2014). The students' feedback was constructive and helped us to reorganize some material as well as to present results from a different point of view.

We are grateful to ELIZABETH GREENSPAN from SIAM, Philadelphia, for her help and patience during the preparation of the manuscript.

We would like to thank our friends, past and present collaborators, who implicitly or explicitly made a significant contribution to the research results presented in this book. Among them, we mention (in alphabetic order): CHAOUKI T. ABDALLAH (University of New Mexico at Albuquerque, USA), SONDIPON ADHIKARI (University of Bristol,

UK), Mehment Eren Ahsen (University of Texas at Dallas, USA), Dirk Aeyels (Universiteit Gent, Belgium), Catherine Bonnet, Frédéric Mazence (INRIA "DISCO" team and L2S, Supélec at Gif-sur-Yvette, France), Islam Boussaada (IPSA Paris and L2S, Supélec at Gif-sur-Yvette, France), Fabien Chatté (NEOPOST, France), Arben Çela (ESIEE Paris, France), Jie Chen (City University of Hong Kong, China), Joono Cheong (Korea University at Jochiwon, South Korea), Luca di Cicco, Saverio Mascolo (University of Bari, Italy), Michel Dambrine (Université de Valenciennes, France), Denis Dochain (Université Catholique de Louvain, Belgium), Koen Engelborghs (Materialise, Belgium), André R. Fioravanti (Universidade Estadual de Campinas, Brazil), Emilia Fridman (Tel Aviv University, Israel), Peilin Fu (National University at San Diego, USA), Keqin Gu (Southern Illinois University at Edwardsville, USA), Nicola Guglielmi (Università degli Studi dell'Aquila, Italy), Kirk Green (TNO, The Hague, Netherlands), Didier Henrion (LAAS-CNRS, Toulouse, France), Henri Huijberts (Queen Mary University of London, UK), Vladimir L. Kharitonov (Saint Petersburg State University, Russia), Peter S. Kim (University of Sydney, Australia), Doron Levy (University of Maryland at College Park, USA), Xu-Guang Li (Northeastern University at Shenyang, China), Rogelio Lozano (HeuDiaSyC Compiègne, France), Kurt Lust (Flemish Supercomputer Center, Belgium), Karl Meerbergen, Dirk Roose, Erik Steur, Roel Van Beeumen, Stefan Vandewalle, Joris Vanbiervliet (KU Leuven, Belgium), Daniel Melchor Aguilar (IPICyT at San Luis Potosí, Mexico), César Fernando Méndez Barrios (Universidad Autonoma de San Luis Potosí, Mexico), Sabine Mondié (CINVESTAV-IPN at Mexico City, Mexico), Constantin-Irinel Morărescu (Université de Lorraine at Nancy, France), Luc Moreau (Arcelor-Mittal, Belgium), Jaime Moreno (UNAM at Mexico-City, Mexico), Henk Nijmeijer (T.U. Eindhoven, The Netherlands), Sorin Olaru (Supélec at Gif-sur-Yvette, France), Hitay Özbay (Bilkent University at Ankara, Turkey), Thomas Plomteux (BASF, Belgium), Vladmir Răsvan (University of Craiova, Romania), Jean-Pierre Richard (École Centrale de Lille, France), Rodolphe Sepulchre (University of Cambridge, UK), Rifat Sipahi (Northeastern University at Boston, USA), Hakki Ulaş Ünal (Anadolu University, Turkey), Vincent van Assche (Université de Caen, France), Patrick Vansevenant (TVH, Belgium), Koen Verheyden (National Bank Belgium), Erik I. Verriest (Georgia Institute of Technology, USA), Tomáš Vyhlídal, Pavel Zítek (Czech Technical University at Prague, the Czech Republic), Thomas Wagenknecht, who is unfortunately no longer with us, Emmanuel Witrant (Université Joseph Fourier, Grenoble, France), and Zhen Wu (Clear2Play, Belgium).

Special thanks go to Suat Gumussoy (Mathworks, Natick, MA, USA) and Elias Jarlebring (KTH, Stockholm, Sweden).

Last but not least WM would like to thank his wife Barbara and their sons Walt and Georges for the extremely nice time they had together and for their patience in the busy period of finishing the book. He is also thankful to family, friends, and colleagues for the nice time and for all the support. Concerning SIN, there is a special person in his life, Laura, to whom he owes the exceptional support that she gave which helped him overcome overcome all the difficulties, both professional and extraprofessional, in the last 20 years. We dedicate this monograph to all of them, in love and gratitude.

Leefdaal, Belgium, May 2014	Wim Michiels
Gif-sur-Yvette, France, May 2014	Silviu-Iulian Niculescu

Preface to the First Edition

The interconnection between two (or more) physical systems is always accompanied by *transfer* phenomena (material, energy, information), such as *transport* and *propagation*. Mathematically speaking, transport and propagation phenomena can be represented by *delay elements*. In this way the corresponding overall systems are governed by a special type of differential equations, namely *delay-differential equations* (DDEs).

DDEs are also used in modeling various other phenomena coming from biosciences (heredity in population dynamics [202, 223]), chemistry (behaviors in chemical kinetics [367, 334]), or economics (dynamics of business cycles [368]). Further examples in engineering can be found in [359, 290, 130].

As mentioned by El'sgol'ts and Norkin [97] or Răsvan [335], time-delay systems have a long history and, to the best of our knowledge, the first DDEs are encountered in the work of Bernoulli and Condorcet. However, the theory started to be developed in the second half of the 20th century with the work of the East European mathematical school—Myshkis [286], Krasovskii [200], and Halanay [145] (to cite only a few)—who devoted most attention to the extension of the Lyapunov theory to such class of differential equations. In the 1960s, an increasing interest in the topic appeared also in North America as confirmed by the monographs of Pinney [326] and Bellman and Cooke [23], the first one almost forgotten, with a particular interest in the complex-domain approach and related frequency-domain techniques and methods. Next, the theory arrived to some degree of maturity in the 1970s as proven by the publications and the monographs devoted to the field in that period. Among them, we mention the pioneering work of Hale [148] (the second edition of the monograph published in 1971), which is one of the most cited references in the field not only for its fundamental results and approaches, but also for its quality and clarity of presentation. For further references and a deeper historical perspective, we refer the reader to [290, 332, 304].

It is important to point out that various references devoted to time-delay systems in engineering existed even before the 1950s (for example, the papers co-authored by Callender [61, 62] and the editors of the journal *Engineer* [357]), with some contradictory conclusions concerning the effects induced by the delay presence in dynamical systems: sometimes *destabilizing* (mainly by using "huge" gains), and sometimes *stabilizing* (mainly in controlling some oscillatory modes). The explanation of such "dichotomic" behaviors was done case by case, without any attempt at a comprehensive explanation of the situations where stabilizing/destabilizing effects may occur.

Although by now the fundamental results in the theory of functional differential equations (FDEs) are well known and well understood (see, for instance, [23, 148, 151], to cite only a few), the increasing number of applications involving large-scale systems with corresponding complex decision making strategies in which the *delay* (transport, propagation, communication, decision) becomes a "critical" parameter made the development of efficient numerical algorithms and methods for evaluating critical delays and related stability/instability properties necessary. This monograph presents some approaches and techniques in this sense.

Recent approaches in *robust control* opened interesting perspectives and issues in dealing with delays in dynamical systems, where delays are eventually treated as *uncertainty* [130, 290, 32]. Some of them (frequency-sweeping tests, matrix pencil approaches) will be largely discussed in this monograph. Such interpretations of delays as uncertainty are at the origin of an abundant literature in the control area. The corresponding results are expressed in terms of solutions of appropriate Riccati equations [225], linear matrix inequalities [32] in connection or not with the μ-formalism. An exhaustive overview concerning these approaches in the context of stability analysis can be found in [290].

At the same time the increasing number of efficient algorithms for dealing with *nonlinear eigenvalue problems* [237] represented another important issue in treating delay systems. As in the finite-dimensional case, essential properties of time-delay systems (asymptotic behavior, stability, instability, oscillations) are connected with the spectrum location of the corresponding linearized systems. As we shall explain in the following chapters, time-delay systems are infinite-dimensional systems, but with particular spectral properties. Such properties will be explicitly exploited in deriving the main (stability and stabilization) results and related algorithms. In this context, particular attention will be paid to the distinction between *retarded* and *neutral* systems because, although both belong to the class of time-delay systems, their spectral properties are distinct. Most of the approaches presented in this book concern *retarded* delay systems, yet they can be easily extended to the *neutral* case.

It is important to point out that, excepting the FDEs based representation, there are several ways to *represent* time-delay systems—as evolution equations over abstract spaces [25] (infinite-dimensional setting), 2-D systems [215], systems over rings of operators [184], and behavioral based representations [120]. Throughout this volume, we have adopted the FDE based representation. We further assume that the nominal models are completely known. In other words, we do not focus on delay modeling, identification, or identifiability.

Book outline and content

Our intention is to present the stability analysis and synthesis by delayed (state and output) feedback in the linear case by using a *unitary* methodology: the *eigenvalue based approach*. Without any loss of generality we mainly concentrate on the following aspects that, to our best knowledge, have not received a full treatment in the literature:

(a) sensitivity analysis with respect to delays and to other systems' parameters (continuity of the spectrum with respect to the parameters based on Rouché-type theorems and variants, pseudospectra, and related properties);

(b) pole placement strategies in stabilization and (nonlinear) optimization of the spectral abscissa function or robustness indicators. Although such approaches are rather classical in the finite-dimensional case, the extensions to delay systems need some special treatment due to the infinite-dimensional nature of the system. However, the particular spectral properties will be helpful to perform such control strategies in both the retarded and the neutral cases, with some precautions in the latter case.

Many examples complete the presentation and illustrate the main results proposed in the monograph. Most of the major ideas are explained by using (several) extremely simple, easy-to-follow (low-order) examples. Finally, the last part of the monograph is devoted to several applications spanning various fields from engineering to biology. All the applications considered start from some generic remarks on the way in which the models are derived, but without any deep discussions on the model derivation and its limitations. The choice of the applications was mainly explained by their impact in engineering, biosciences, and related fields, but also by our own interest in the corresponding topics.

How to read the book?

The book is organized in three parts:

(a) Stability and robust stability. This part deals with the *analysis* of linear time-delay systems from a stability point of view. It starts with an overview of spectral properties of both retarded and neutral systems. To make the fundamental results apparent, eigenvalue plots are used extensively throughout the text. Then the robustness of stability and related problems are studied using pseudospectra and related quantities such as stability radii. The next three chapters deal with the characterization of stability regions in parameter spaces, both qualitatively (shape of regions, etc.) and quantitatively (explicit computational algorithms). Finally, extensions of the presented results for systems with constant parameters to systems with periodically varying parameters are briefly discussed.

(b) Stabilization and robust stabilization. The second part is devoted to the *synthesis* problems that correspond to the analysis problems treated in the first part, with the focus on stabilization. The first chapter is devoted to an eigenvalue based stabilization approach that is inspired by the classical pole placement method for systems without delay. Next, a numerical case study is presented to illustrate how delays in the control loop affect the stabilizability with state feedback. The following chapter is devoted to the robust stabilization problem, and corresponds to the chapter on pseudospectra presented in the first part. Finally, a new stabilization approach is presented which is based on recently developed methods for nonsmooth optimization.

(c) Applications. A wide class of applications is presented, from congestion analysis in high-performance networks to output feedback stabilization and the analysis of predictor-type controllers, from consensus problems in traffic flows to the stability analysis of various delay models in biosciences. We tried to achieve the right correlation between the theory presented in the first two parts of the monograph and the applications which we consider. In some cases, we present several alternative approaches handling the same stability analysis or control problem.

It is important to point out that we have made the parts independent of each other as much as possible. However, a number of fundamental results are needed for the whole theoretical development; these are presented in the first chapter of the monograph. Since such results can be found in excellent references devoted to the theory of FDEs, we decided to only mention them here, and to pay more attention to some particular approaches and related methodologies that have not received full attention in the literature, such as the *sensitivity analysis* (to cite only one approach).

Acknowledgments

The idea of writing this monograph appeared a couple of years ago, but, as usual, the *delay factor* interfered with the whole process. However, we believe that, finally, the delay had a *positive* impact.

We greatly acknowledge the financial support of the Belgian research projects IAP P5, Dynamical Systems and Control: Computation, Identification and Modeling, and IAP P6, DYSCO (Dynamical Systems, Control and Optimization), funded by the program on Interuniversity Poles of Attraction, initiated by the Belgian State, Prime Minister's Office for Science, Technology and Culture, of the Center of Excellence on Optimization in Engineering of the K.U. Leuven, of the Fund for Scientific Research—Flanders, from which WM received a postdoctoral fellowship in the period of writing of the book, and of the French CNRS (National Center for Scientific Research), which covered in part the traveling between Belgium and France in the period 2002–2006. Starting with 2006, a bilateral French–Flemish collaboration project (Tournesol in 2006, and H. Curien in 2007), entitled *Distributed delays in dynamical systems: Analysis and applications*, helped us to continue and to finish the monograph.

Parts of the book have been presented at the European graduate school FAP (Formation d'Automatique à Paris) in the framework of the CTS (Control Training Site) in Automatic Control in 2005 and 2006. The students' feedback was constructive and helped us to reorganize some material and to present results from a different point of view.

We are grateful to Elizabeth Greenspan from SIAM, Philadelphia, for her help and patience during the preparation of the manuscript.

We would like to thank our friends and past and present collaborators, who implicitly or explicitly made a significant contribution to the research results presented in the book. Among them, we mention Dirk Roose, Tatyana Luzyanina, Stefan Vandewalle, Joris Vanbiervliet, Koen Verheiden (K.U. Leuven, Belgium), Koen Engelborghs (Materialise, Belgium), Patrick Vansevenant (TVH, Belgium), Vincent Van Assche (Université de Picardie, Soissons, France), Kurt Lust (Rijksuniversiteit Groningen, The Netherlands), Thomas Plomteux (BASF, Belgium), Luc Moreau (SIDMAR, Belgium), Dirk Aeyels (Universiteit Gent, Belgium), Rodolphe Sepulchre (Université de Liège, Belgium), Denis Dochain (Université Catholique de Louvain, Belgium), Abhijit Ganguli (Université Libre de Bruxelles, Belgium), Fabien Chatté (NEOPOST, France), Sabine Mondié, Vladimir Kharitonov (CINVESTAV-IPN, Mexico), Jaime Moreno (UNAM, Mexico), Tomáš Vyhlídal (Czech Technical University at Prague, the Czech Republic), Erik I. Verriest (Georgia Institute of Technology, USA), Jean Pierre Richard (Ecole Centrale de Lille, France), Michel Dambrine (Université de Valenciennes, France), Frédéric Mazenc (INRA Montpellier, France), Daniel Melchor Aguilar (IPICYT at San Luis Potosí, Mexico), Chaouki T. Abdallah (University of New Mexico at Albuquerque, USA), Kirk Green (Vrije Universiteit Amsterdam, The Netherlands), Thomas Wagenknecht (University of Manchester, UK), Sondipon Adhikari (University of Bristol, UK), Henk Nijmeijer (T.U. Eindhoven), Henri Huijberts (Queen Mary University of London, UK), Vladmir Răsvan (University of Craiova, Romania), Keqin Gu (Southern Illinois University at Edwardsville, USA), Jie Chen, Peilin Fu (University of California at Riverside, USA), Rifat Sipahi (Northeastern University at Boston, USA), Constantin-Irinel Morărescu (University "Politehnica" at Bucharest, Romania), Qing-Chang Zhong (University of Liverpool, UK), Doron Levy, Peter S. Kim (Stanford University, USA), and Rogelio Lozano (CNRS, Compiègne, France).

Last but not least WM would like to thank Barbara for the extremely nice time they had together and for her patience in the busy period of finishing the book (*mpf, ik zie je graag!*). He is also thankful to his parents, family, and friends for creating a nice atmosphere and for all the support. Concerning SIN, there is a special person in his life, Laura, to whom he owes the exceptional support that she gave to Silviu to overcome all the difficulties both professional and extra-professional in the last fifteen years. We dedicate this monograph to all of them, in love and gratitude.

Leuven, Belgium, January 2007	WIM MICHIELS
Gif-sur-Yvette, France, January 2007	SILVIU-IULIAN NICULESCU

List of Symbols

A^{-1}	inverse of matrix A
$A^{\dagger}$	Moore–Penrose inverse of matrix A
A^{*}	complex conjugate transpose of matrix A
$\det(A)$	determinant of matrix A
$\mathscr{D}(A)$	domain of operator A
A^{-T}	transpose of the inverse matrix of A
$A^{\perp}$	matrix of full column rank whose columns span the orthogonal complement of A
$\mathscr{B}(\vec{\tau},\epsilon)$	open ball of radius $\epsilon \in \mathbb{R}_+$ centered at $\vec{\tau} \in \mathbb{R}_+^m$, $\mathscr{B}(\vec{\tau},\epsilon) := \{\vec{\theta} \in (\mathbb{R})^m : \|\vec{\theta} - \vec{\tau}\| < \epsilon\}$
$\mathscr{B}_+^m$	$\{\vec{r} \in \mathbb{R}_+^m : \|\vec{r}\| = 1\}$
$\mathbb{C}$	set of complex numbers
$\mathbb{C}_-$, $\mathbb{C}_+$	open left half plane, open right half plane
$\mathbb{C}_\sigma$, $\sigma \in \mathbb{R}$	half plane $\{\lambda \in \mathbb{C} : \Re(\lambda) > \sigma\}$
$\operatorname{clos}(Z)$	closure of set Z
$\mathscr{C}([-\tau,\ 0],\ \mathbb{F})$	space of continuous functions from $[-\tau,\ 0]$ to $\mathbb{F}$
$d(\rho,E)$, $\rho \in \mathbb{C}$, $E \subseteq \mathbb{C}$	$\inf_{t \in E} \lvert \rho - t \rvert$
$D(E,F)$, $E,F \subseteq \mathbb{C}$	$\sup_{\rho \in E} d(\rho,F)$
$D_h(E,F)$, $E,F \subseteq \mathbb{C}$	$\max\{D(E,F), D(F,E)\}$, Hausdorff distance
e	$e := \exp(1)$
ϕ	empty set
I	identity matrix (of appropriate dimension)
I_k	identity matrix of size k-by-k
j	imaginary unit, $j = \sqrt{-1}$
$\mathscr{L}_2$	space of square integrable complex functions, e.g., $\mathscr{L}_2([-\tau_{\max}, 0], \mathbb{C}^n) := \{f : [-\tau_{\max}, 0] \to \mathbb{C}^n : \int_{-\tau_{\max}}^{0} \|f(t)\|_2^2 dt < \infty\}$
λ	characteristic root
$\Re(\lambda), \Im(\lambda), \lvert\lambda\rvert$ $\lambda \in \mathbb{C}$	real part, imaginary part, and modulus of λ
$\angle(\lambda)$, $\lambda \in \mathbb{C}$	argument of λ, $\angle(\lambda) \in [0,\ 2\pi)$
$\bar{\lambda}$, $\lambda \in \mathbb{C}$	complex conjugate of λ
$\mathscr{C}(0,1)$	unit circle in the complex plane $\mathbb{C}$
$\mathscr{D}(0,1)$, $(\overline{\mathscr{D}(0,1)})$	(closed) unit disk in the complex plane $\mathbb{C}$
$\Lambda(z) := zM + N$	matrix pencil with indeterminate z $(M, N \in \mathbb{R}^{n \times n})$
$\mu_{\Delta}(A)$	structured singular value of A with respect to the uncertainty set $\mathbf{\Delta}$

$\mathbb{N}$	set of natural numbers (zero is not included)
ω	frequency, imaginary part of eigenvalue
$\mathbb{Q}$	set of rational numbers
$\mathbb{R}$	set of real numbers
$j\mathbb{R}$	imaginary axis
$\mathbb{R}_+$	$\{r \in \mathbb{R} : r \geq 0\}$
$\mathbb{R}^*, \mathbb{R}_+^*$	$\mathbb{R} \setminus \{0\}$, $\mathbb{R}_+ \setminus \{0\}$
$\vec{r} \in \mathbb{R}^m, \vec{n} \in \mathbb{N}^m, \ldots$	short notation for $(r_1, \ldots, r_m)$, $(n_1, \ldots, n_m), \ldots$
range(A)	column space of matrix A
span$(V_1, \ldots, V_m)$	linear space spanned by $V_1, \ldots, V_m$
$\sigma_i(A)$	ith singular value of matrix A, $\sigma_1(A) \geq \sigma_2(A) \ldots \geq \sigma_n(A)$
$\lambda_i(A)$	ith eigenvalue of Hermitian matrix A, $\lambda_1(A) \geq \lambda_2(A) \ldots \geq \lambda_n(A)$
$\sigma(A)$	spectrum of matrix or operator A
$r_\sigma(A)$	spectral radius of A
$r_e(A)$	radius of the essential spectrum of A $\sigma(\Lambda(z)) = \{z \in \mathbb{C} : \det(zM + N) = 0\}$
Tr(A)	trace of matrix A
vec$(A), A = [a_1 \vert \cdots \vert a_q]$	vectorization of A, $\text{vec}(A) = [a_1^T \cdots a_q^T]^T$
mat(a), $a \in \mathbb{C}^{n^2}$	reshaping vector a into a square matrix A such that $a = \text{vec}(A)$
vec$(A_1, A_2, \ldots)$	short notation for $[A_1^T\ A_2^T \cdots]^T$
$\mathbb{Z}$	set of integer numbers
$\bar{E}$	closure of the set E
$\mathscr{P}(E)$	power set of E, set of all subsets
$\lceil \nu \rceil, \nu \in \mathbb{R}$	integer obtained by rounding ν towards infinity
$[\nu], \nu \in \mathbb{R}$	integer obtained by rounding ν towards the nearest integer
$\Vert x \Vert$, $x \in \mathbb{C}^n$	Euclidean norm of x
$\Vert x \Vert_p$, $x \in \mathbb{C}^n$	Hölder p-norm of x
$\Vert A \Vert_p$, $A \in \mathbb{C}^{n \times m}$	induced matrix p-norm
$\Vert A \Vert_Q^2 = \max_i \lambda_i(A^T Q A)$	Q-norm of the matrix A, where $Q = Q^T \in \mathbb{R}^{n \times n}$ an arbitrary positive-definite real matrix
$\Vert \phi \Vert_s$	supremum norm of $\phi \in \mathscr{C}([-\tau, 0], \mathbb{R}^n)$, $\Vert \phi \Vert_s = \sup_{\theta \in [-\tau, 0]} \Vert \phi(\theta) \Vert_2$
$x \geq 0$, $x \in \mathbb{R}^m$	$x_i \geq 0, i = 1, \ldots, m$, where $x = (x_1, \ldots, x_m)$
$\Vert M(j\omega) \Vert_{\mathscr{H}_\infty}$, $\Vert M(j\omega) \Vert_{\mathscr{H}_2}$	$\mathscr{H}_\infty$ norm, $\mathscr{H}_2$ norm of transfer matrix $M(j\omega)$
$\left[\begin{array}{c\|c} A & B \\ \hline C & D \end{array}\right]$	transfer function representation for $T(\lambda) := C(\lambda I - A)^{-1} B + D$
$A \otimes B$, $A \oplus A$	Kronecker product and sum of matrices A and B
deg(P)	degree of the polynomial P

Acronyms

AIMD	additive-increasing multiplicative-decreasing
AQM	active queue management
ARE	algebraic Riccati equation
CML	chronic myelogenous leukemia
DDE	delay-differential equation
DDAE	delay-differential algebraic equation
DNA	deoxyribonucleic acid
FDE	functional differential equation
FIR	finite impulse response
FSA	finite spectrum assignment
GRN	gene regulatory networks
HANSO	hybrid algorithm for nonsmooth optimization
IQC	integral quadratic constraints
IP	internet protocol
LHP	left half plane
LMI	linear matrix inequality
LMS	linear multistep
LTI	linear time invariant
MIMO	multiple input, multiple output
NCS	networked control systems
NFDE	neutral functional differential equation
ODE	ordinary differential equation
PDE	partial differential equation
PID	proportional-integrative-derivative
RHP	right half plane
RNA	ribonucleic acid
RTT	round-trip time
SISO	single input, single output
ssv	structured singular value
TCP	transmission control protocol

Part I

Stability analysis of linear time-delay systems

Chapter 1

Spectral properties of linear time-delay systems

As specified in the prefaces, most reactions of real (physical) systems to external actions and signals never take place instantaneously mainly due to transport and propagation phenomena, and one of the ways to overcome such problems is to include some information on the past in the corresponding model of the systems' dynamics. Such systems are generically called *time-delay systems*.

Roughly speaking, a *time-delay system* is a dynamical system represented by differential equations in some unknown function (and certain of its derivatives) evaluated at arguments which are distributed over some intervals in the past. Differential equations where the right-hand side not only depends on the state variable at the present time but also is a functional evaluated at a solution segment are generally called *functional differential equations* (see, for instance, [148, 151, 78] for the exact definition, and related justifications).

Among the general problems of interest in the theory of time-delay systems, we mention the correct formulation of the initial value problem together with the representation of solutions, and the asymptotic behavior of solutions correlated to the concept of stability.

In this introductory chapter we present several fundamental definitions, properties, and results concerning linear delay-differential equations of retarded and neutral type, with emphasis on their relation with the spectra of appropriately defined operators. Such results, needed in the forthcoming chapters, are presented in a tutorial way. However, some notions and properties essential for the *eigenvalue-based approach* considered in this monograph will receive particular attention and are completed by elementary proofs or by sketched proofs. The presentation closely follows standard references in the field of functional differential equations like [23, 145, 148, 151, 93, 78]. Several relatively simple examples and some discussions concerning the computation of the characteristic roots complete the presentation.

This chapter is organized as follows: Section 1.1 is devoted to linear systems of *retarded* type. The initial value problem, the asymptotic growth rate of solutions, asymptotic stability, and some quantitative and qualitative properties of the spectrum are considered. Section 1.2 introduces the same notions in the context of linear systems of *neutral* type, including an extensive discussion of the delay sensitivity problem. Finally, systems described by linear delay-differential algebraic equations, which are closely related to neutral-type systems, are addressed in Section 1.3. Some notes, together with a list of references, end the chapter.

1.1 ▪ Time-delay systems of retarded type

We discuss spectral properties of linear time-delay systems of retarded type described by the following delay-differential equations (DDEs):

$$\dot{x}(t) = A_0 x(t) + \sum_{i=1}^{m} A_i x(t - \tau_i), \tag{1.1}$$

where $x(t) \in \mathbb{R}^n$ is the state variable at time t, $A_i \in \mathbb{R}^{n\times n}$, $i = 0, 1, \ldots, m$, are real matrices, and $0 < \tau_1 < \tau_2 \cdots < \tau_m$ represent the *time-delays*. The time-delays are *pointwise* or *discrete*, and, hence, they describe the situation where the memory effect is "selective." Such differential equations are sometimes also called differential-difference equations (see, for instance, [23], for some arguments).

Although in this chapter we will not discuss the case of *distributed delays* over some delay intervals, most of the results presented below can be extended to this class of time-delay systems.

1.1.1 ▪ The initial value problem

The initial condition for the time-delay system (1.1) is the function segment

$$\phi \in \mathscr{C}([-\tau_m, 0], \mathbb{R}^n),$$

where $\mathscr{C}([-\tau_m, 0], \mathbb{R}^n)$ is the Banach space of continuous functions mapping the interval $[-\tau_m, 0]$ into $\mathbb{R}^n$ and equipped with the supremum norm, $\|\cdot\|_s$.

Due to the linearity of the mapping $f : \mathscr{C}([-\tau_m, 0], \mathbb{R}^n) \to \mathbb{R}^n$, defined by

$$f(\phi) := A_0 \phi(0) - \sum_{i=1}^{m} A_i \phi(-\tau_i), \tag{1.2}$$

the existence and uniqueness of solutions is guaranteed for all initial conditions.

Furthermore, for a given initial condition, the corresponding solution can be explicitly constructed by using the *method of steps*. More precisely, the function ϕ defined on $[-\tau_m, 0]$ allows us to define the evolution of the delay system (1.1) on the interval $[0, \tau_m]$ as the solution of a standard ODE of the form

$$\dot{\xi}_1(t) = A_0 \xi_1(t) + \sum_{i=1}^{m} A_i \phi(t - \tau_i),$$

with the initial condition $\xi_1(0) = x(0) = \phi(0)$. The next step consists of computing the solution on the interval $[\tau_m, 2\tau_m]$ by using the solution ξ_1 defined on $[0, \tau_m]$. More precisely, the evolution of (1.1) on the interval $[\tau_m, 2\tau_m]$ is given by the solution of the ODE

$$\dot{\xi}_2(t) = A_0 \xi_2(t) + \sum_{i=1}^{m} A_i \xi_1(t - \tau_i),$$

with the initial condition $\xi_2(\tau_m) = x(\tau_m) = \xi_1(\tau_m)$. The solution of the original delay system (1.1) is obtained by considering the collection of all these *"pieces of trajectories"* ξ_k defined on $[(k-1)\tau_m, k\tau_m]$, for all positive integers k. Hence, the computation of the solution of (1.1) at some given time-instant $T > 0$ for a given initial condition ϕ reduces to solving q ODEs, where q satisfies the condition $(q-1)\tau_m \leq T \leq q\tau_m$. This process yields a *unique, globally defined forward solution* of (1.1). It is important to note that this solution becomes *smoother* as the time t is increased: if we start with a continuous initial condition on $[-\tau_m, 0]$, then the corresponding solution of (1.1) becomes continuously differentiable on $(0, \tau_m]$, twice-continuously differentiable on $(\tau_m, 2\tau_m]$, and so on. This also implies that a backward continuation for negative time $t \leq -\tau_m$ requires additional smoothness properties of the initial function ϕ defined on $[-\tau_m, 0]$ (see, e.g., [148] and the references therein).

Although in our case the method of steps guarantees the existence and uniqueness of solutions directly by construction, the existence results for general time-delay systems of retarded type are typically proven using fixed-point theorems (see [23, 145, 290]).

Example 1.1. *Consider the scalar equation*

$$\dot{x}(t) = x(t) - \alpha x(t-\tau), \tag{1.3}$$

where α *and* τ *are real parameters with* $\tau \geq 0$.

If the initial condition ϕ *on* $[-\tau, 0]$ *is given by the constant function* $\phi \equiv 1$, *then the corresponding solution on the interval* $[k\tau, (k+1)\tau]$ *satisfies*

$$x(t) = \alpha^{k+1} + (1-\alpha)\sum_{i=0}^{k} \alpha^i f_i(t) e^{t-i\tau},$$

where the functions f_i *are given by*

$$f_i(t) := \sum_{h=0}^{i} (-1)^h c_{i-h} \frac{t^i}{i!}, \; i = 0, \dots, k,$$

with the coefficients c_i *defined as follows:* $c_0 = 1$, *and*

$$c_i := 1 + i \sum_{h=0}^{i-1} \frac{(i-h)^{h-1}}{h!} \tau^h, \quad i = 1, \dots, k.$$

Indeed, it is easy to see that the solution on the first delay interval $[0, \tau]$ *corresponds to the solution of the initial value problem*

$$\dot{\xi}_1(t) = \xi_1(t) - \alpha, \qquad \xi_1(0) = 1.$$

The remaining steps can be performed by recursion.

A particular case of interest corresponds to $\alpha = 1$, *for which the solution* $x(t)$ *reduces to* $x(t) = 1$ *for all* $t \in [-\tau, \infty)$.

In what follows, let us define

$$x(\phi) : \; t \in [-\tau_m, \infty) \to x(\phi)(t) \in \mathbb{R}^n$$

as the unique forward solution of (1.1) with initial condition $\phi \in \mathscr{C}([-\tau_m, 0], \mathbb{R}^n)$, that is,

$$x(\phi)(\theta) = \phi(\theta) \quad \forall \theta \in [-\tau_m, 0].$$

Then the state at time t is given by the function segment $x_t(\phi) \in \mathscr{C}([-\tau_m, 0], \mathbb{R}^n)$, defined as

$$x_t(\phi)(\theta) = x(\phi)(t+\theta), \quad \theta \in [-\tau_m, 0].$$

Denote by $\mathcal{T}(t)$, $t \geq 0$, the solution operator, mapping the initial data onto the state at time t,

$$(\mathcal{T}(t)\phi)(\theta) = x_t(\phi)(\theta) = x(\phi)(t+\theta), \quad \theta \in [-\tau_m, 0]. \tag{1.4}$$

Sometimes, this solution operator is also called the operator of translation along trajectories [78], since

$$x_{t+t_0}(\phi) = \mathcal{T}(t_0)x_t(\phi), \tag{1.5}$$

for all $t \geq 0$, $t_0 \geq 0$, and $\phi \in \mathscr{C}([-\tau_m, 0], \mathbb{R}^n)$. The solution operator is a strongly continuous semigroup. The semigroup property ($\mathcal{T}(0) = I$, $\mathcal{T}(t+s) = \mathcal{T}(t)\mathcal{T}(s)$, for all $t \geq 0$, $s \geq 0$) follows from the definition and the uniqueness of solutions. A simple proof of the property of the semigroup to be strongly continuous, that is,

$$\forall t \geq 0\ \forall \phi \in \mathscr{C}([-\tau_m, 0], \mathbb{R}^n)\ \lim_{s\to t} \|\mathcal{T}(t)\phi - \mathcal{T}(s)\phi\|_s = 0,$$

can be found in [78].

Let $\mathcal{A}$ be the *infinitesimal generator* of $\mathcal{T}(t)$. This operator satisfies

$$\begin{aligned} \mathcal{D}(\mathcal{A}) = \Bigg\{ & \phi \in \mathscr{C}([-\tau_m, 0], \mathbb{R}^n) : \frac{d\phi}{d\theta} \in \mathscr{C}([-\tau_m, 0], \mathbb{R}^n), \\ & \phi(0) = A_0\phi(0) + \sum_{k=1}^{m} A_k \phi(-\tau_k) \Bigg\}, \\ \mathcal{A}\,\phi = {} & \frac{d\phi}{d\theta}, \end{aligned} \tag{1.6}$$

and allows us to rewrite equation (1.1) as an abstract ODE:

$$\dot{x}_t = \mathcal{A} x_t. \tag{1.7}$$

Summarizing, the same time-delay system can be described in three different ways: by means of a delay (or, more generally, functional) differential equation like (1.1), which is of the form

$$\dot{x}(t) = f(x_t),$$

with f defined as in (1.2), by means of a mapping like (1.5), or as an abstract ODE over an infinite-dimensional function space like (1.7). In what follows, the choice of the delay system representation will be dictated by the problem under consideration. However, in most of the cases and examples throughout the monograph, we will use the DDE-based representation.

1.1.2 ▪ Spectrum: Definitions

The substitution of a sample solution of the form $e^{\lambda t}v$, with $v \in \mathbb{C}^{n\times 1} \setminus \{0\}$, leads us to the *characteristic equation* of (1.1):

$$\det \Delta(\lambda) = 0, \tag{1.8}$$

where

$$\Delta(\lambda) := \lambda I - A_0 - \sum_{i=1}^{m} A_i e^{-\lambda\tau_i}$$

is the *characteristic matrix*. The left-hand side of (1.8) is called the characteristic function. Similarly to the finite-dimensional case, the roots of (1.8) are called the *characteristic roots* of (1.1).

The spectra of the linear operators $\mathscr{A}$, $\mathscr{T}(t)$, and the characteristic roots are related in the following way. The characteristic roots are the eigenvalues of the operator $\mathscr{A}$, which features only a *point spectrum*, that is, $\sigma(\mathscr{A}) = P\sigma(\mathscr{A})$. Furthermore, the algebraic multiplicity of a complex number λ as an eigenvalue of $\mathscr{A}$ is equal to its multiplicity as a zero of the characteristic matrix $\Delta(\lambda)$, while its geometric multiplicity is equal to the dimension of the null space of $\Delta(\lambda)$. An eigenvalue of $\mathscr{A}$ is called *simple* if its algebraic multiplicity is equal to *one* and is called *multiple* otherwise. In the multiple eigenvalue case, we will make a clear distinction between *semisimple* and *non-semisimple* eigenvalues, depending on the relation between the algebraic and geometric multiplicity: a semisimple eigenvalue corresponds to the case where both multiplicities are equal. For $\lambda \in \sigma(\mathscr{A})$, the corresponding eigenfunctions take the form

$$ve^{\lambda\theta}, \quad \theta \in [-\tau_m, 0], \tag{1.9}$$

where $v \in \mathbb{C}^n \setminus \{0\}$ satisfies

$$\Delta(\lambda)v = 0. \tag{1.10}$$

Sometimes, the vector v is called a *right eigenvector* associated to the characteristic root λ. By similarity, we can construct a *left eigenvector* w, which satisfies $w \neq 0$, $w^*\Delta(\lambda) = 0$, and is related to the eigenfunctions of the adjoint of $\mathscr{A}$ (see [148] for more on the construction of the adjoint space).

The spectra of $\mathscr{A}$ and $\mathscr{T}(t)$ are related by

$$\sigma(\mathscr{T}(t)) = \exp(t\sigma(\mathscr{A})) \quad \text{plus possibly } \{0\}. \tag{1.11}$$

Furthermore, if $\lambda \in \sigma(\mathscr{A})$, then $e^{\lambda t} \in P\sigma(\mathscr{T}(t))$, with corresponding eigenfunctions given by (1.9). Conversely, if $z(t) \in \sigma(\mathscr{T}(t))$ and $z(t) \neq 0$, then there exists some $\lambda \in \sigma(\mathscr{A})$ such that $z(t) = e^{\lambda t}$.

Example 1.2. *Recall the scalar system* (1.3), *with* $\alpha = 2$ *and* $\tau = 1/3$. *In Figure* 1.1, *we plot the characteristic roots* λ, *which are the eigenvalues of the operator* $\mathscr{A}$, *as well as the eigenvalues* z *of the operator* $\mathscr{T}(1)$. *These eigenvalues are connected by the relation* $z = e^{\lambda}$.

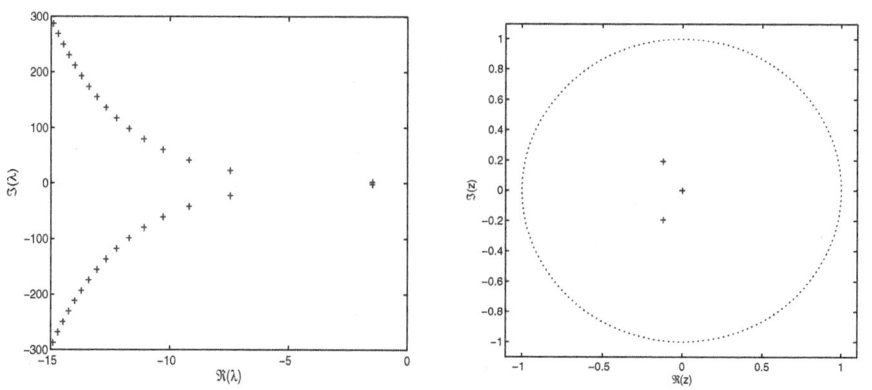

Figure 1.1. *(Left) Characteristic roots of the system* (1.3) *with* $\alpha = 2$ *and* $\tau = 1/3$. *(Right) Eigenvalues of the corresponding solution operator* $\mathscr{T}(1)$.

Example 1.3. *For* $\alpha = 1$, *the characteristic matrix of* (1.3) *becomes*

$$\Delta(\lambda) = \lambda - 1 + e^{-\lambda\tau}.$$

If $\tau = 1$, *then the characteristic root* $\lambda = 0$ *has multiplicity* 2 *since* $\Delta(0) = \Delta'(0) = 0$ *and* $\Delta''(0) \neq 0$. *If* $\tau \neq 1$, *then the root* $\lambda = 0$ *is* simple, *since* $\Delta(0) = 0$, *but* $\Delta'(0) \neq 0$.

Furthermore, the characteristic root at zero is invariant *w.r.t. delay changes, that is,* $\Delta(0)=0$ *for all positive delays. Such an invariance property will prove its interest in the forthcoming chapters.*

1.1.3 ▪ Asymptotic growth rate of solutions and stability

We have the following definitions:

Definition 1.4. *The null solution of* (1.1) *is asymptotically stable*[1] *if and only if*

$$\forall \epsilon > 0\, \exists \delta > 0\, \forall \phi \in \mathscr{C}([-\tau_m, 0], \mathbb{R}^n)\ (\|\phi\|_s < \delta) \Rightarrow (\forall t \geq 0\ \|x_t(\phi)\|_s < \epsilon),$$
$$\forall \phi \in \mathscr{C}([-\tau_m, 0], \mathbb{R}^n)\ \lim_{t\to+\infty} x(\phi)(t) = 0.$$

Definition 1.5. *The null solution of* (1.1) *is exponentially stable if and only if there exist constants* $C > 0$ *and* $\gamma > 0$ *such that*

$$\forall \phi \in \mathscr{C}([-\tau_m, 0], \mathbb{R}^n)\ \|x_t(\phi)\|_s \leq C e^{-\gamma t} \|\phi\|_s.$$

For the case of linear DDEs of retarded type under consideration, exponential stability and asymptotic stability are equivalent. For the relationship between these notions in a more general setting, we refer the reader to the monographs of Halanay [145] and Stépàn [359].

The asymptotic behavior of the solutions of (1.1) and thus, their stability properties, are determined by the spectral radius $r_\sigma(\mathscr{T}(1))$. We have, for instance:

Proposition 1.6. *The null solution of* (1.1) *is exponentially stable if and only if*

$$r_\sigma(\mathscr{T}(1)) < 1,$$

or, equivalently, all characteristic roots of (1.1) *are located in the open left half plane.*

Example 1.7. *The system* (1.3) *with* $\alpha = 2$ *and* $\tau = 1/3$ *is asymptotically stable. As shown in Figure* 1.1, *all characteristic are in the open left half plane. Equivalently, all eigenvalues of the operator* $\mathscr{T}(1)$ *are located inside the unit circle.*

Throughout this monograph, we will mostly restrict our attention to the asymptotic stability notion. However, in some of the applications, we will need other types of stability notions. These will be defined only in the corresponding parts in relation to stability notions presented above.

1.1.4 ▪ Spectrum: Qualitative properties

Although the characteristic function of the time-delay system (1.1) is *transcendental* and has an infinite number of zeros, it has some nice, simple, and interesting properties. Some of these properties, which are useful in the forthcoming chapters, are presented below.

Proposition 1.8. *If there exists a sequence* $\{\lambda_k\}_{k\geq 1}$ *of characteristic roots of* (1.1) *such that*

$$\lim_{k\to\infty} |\lambda_k| \to +\infty,$$

[1]For reasons of conciseness we will often use the less precise formulation "the system (1.1) is asymptotically stable."

then

$$\lim_{k\to\infty} \Re(\lambda_k) \to -\infty.$$

Corollary 1.9. *The following assertions hold:*

(i) *There are only a finite*[2] *number of characteristic roots in any vertical strip of the complex plane, given by*

$$\{\lambda \in \mathbb{C} : \alpha < \Re(\lambda) < \beta\},$$

with $\alpha, \beta \in \mathbb{R}$ *and* $\alpha < \beta$.

(ii) *There exists a number* $\gamma \in \mathbb{R}$ *such that all characteristic roots are confined to the half plane*

$$\{\lambda \in \mathbb{C} : \Re(\lambda) < \gamma\}.$$

This result can be strengthened. The following proposition, which plays an important role in the study of continuity properties of the spectrum, allows us to construct an *envelope curve* around the characteristic roots of (1.1) (see also [23] for further discussions and properties):

Proposition 1.10. *If* λ *is a characteristic root of the system* (1.1), *then it satisfies*

$$|\lambda| \le ||A_0||_2 + \sum_{i=1}^{m} ||A_i||_2 e^{-\Re(\lambda)\tau_i}. \tag{1.12}$$

Proof. The expression $\Delta(\lambda) = 0$ is equivalent to

$$\lambda \in \sigma\left(A_0 + \sum_{i=1}^{m} A_i e^{-\lambda\tau_i}\right).$$

Interpreting the argument of $\sigma(\cdot)$ as a matrix leads to

$$|\lambda| \le \left\| A_0 + \sum_{i=1}^{m} A_i e^{-\lambda\tau_i} \right\|_2,$$

from which (1.12) follows straightforwardly. □

Example 1.11. *For the system* (1.3), *the estimate* (1.12) *becomes*

$$|\lambda| \le 1 + |\alpha| e^{-\tau\Re(\lambda)}.$$

For $\alpha = 2$ *and* $\tau = 1/3$, *the resulting envelope curve* $|\lambda| = 1 + 2e^{-\Re(\lambda)\tau}$ *is depicted in Figure* 1.2.

It is worth mentioning that Proposition 1.10 provides further information concerning the location of the characteristic roots. More precisely, it is easy to see that for any characteristic root λ of (1.8), an upper bound of $\Re(\lambda)$ is given by

$$\Re(\lambda) \le r_0, \quad \text{with } r_0 > 0 \text{ the unique solution of}$$

$$r = ||A_0||_2 + \sum_{i=1}^{m} ||A_i||_2 e^{-r\tau_i}.$$

[2]Multiplicity is taken into account.

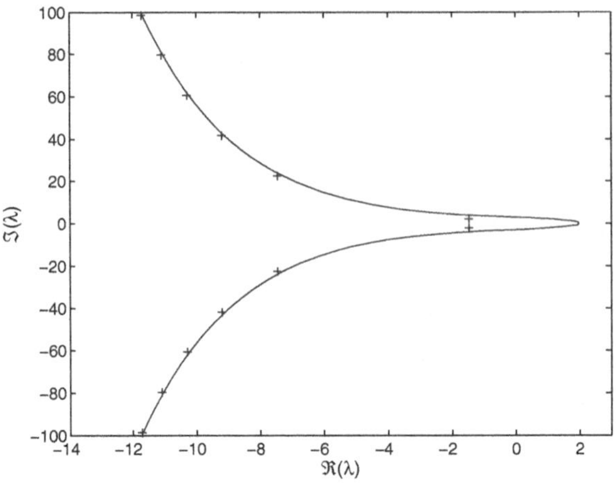

Figure 1.2. *Envelope curve on the characteristic roots for the system* (1.3) *with* $\alpha = 2$ *and* $\tau = 1/3$ *(solid line). The characteristic roots are indicated by* $+$.

Indeed, observe first that the (auxiliary real) function

$$f : \mathbb{R}_+ \mapsto \mathbb{R}, \quad f(x) := x - ||A_0||_2 - \sum_{i=1}^{m} ||A_i||_2 e^{-r\tau_i}$$

is differentiable and strictly increasing. Next, since $f(0) < 0$ and $\lim_{x\to\infty} f(x) = \infty$, it follows straightforwardly that there exists a unique positive root $r_0 > 0$ such that $f(r_0) = 0$. In conclusion, the *rightmost root* λ_r in the complex plane verifies the inequality $\Re(\lambda_r) \leq r_0$. Consider now a root in the imaginary axis, that is, $\lambda_0 = j\omega_0$ (with $\omega_0 \in \mathbb{R}$). The inequality (1.12) provides the following upper bound:

$$|\omega_0| \leq \sum_{i=0}^{m} ||A_i||_2, \tag{1.13}$$

meaning that the interval over the imaginary axis where we can find characteristic roots (if any?!) is *independent* of the delay values. The next result will simply show how the 2 norm can be replaced by any induced matrix Q norm, thus leading to a better estimation of such an interval:

$$||A||_Q^2 = \max_i \lambda_i(A^T QA), \tag{1.14}$$

where $Q = Q^T \in \mathbb{R}^{n\times n}$ is an arbitrary positive-definite real matrix. More precisely, we have the following estimates of the upper bound:

Proposition 1.12. *Let* $Q = Q' > 0 \in \mathbb{R}^{n\times n}$ *be given. If* $\mu \in \mathbb{R}_+$ *is such that*

$$\mu Q > \sum_{i=0}^{m} A_i^T QA_i, \tag{1.15}$$

then any solution of the characteristic equation (1.8) *on the imaginary axis* $\lambda = j\omega$ *satisfies* $|\omega| \leq \sqrt{\mu(m+1)}$.

Proof. With the notation $\tau_0 = 0$, we observe that, for any $i \in \{0,\dots,m\}$, the relation

$$\begin{bmatrix} A_i^T Q A_i & A_i^T e^{j\omega\tau_i} \\ A_i e^{-j\omega\tau_i} & Q^{-1} \end{bmatrix} \geq 0$$

is always verified (Schur complement with respect to the second column and row). Next, we can sum up all the $m+1$ terms leading to

$$\begin{bmatrix} \sum_{i=0}^m A_i^T Q A_i & \sum_{i=0}^m A_i^T e^{j\omega\tau_i} \\ \sum_{i=0}^m A_i e^{-j\omega\tau_i} & (m+1)Q^{-1} \end{bmatrix} \geq 0.$$

Applying again the Schur complement with respect to the second column and row, we obtain

$$\sum_{i=0}^{m} A_i^T Q A_i \geq \frac{1}{m+1}\left(\sum_{i=0}^{N} A_i^T e^{j\omega\tau_i}\right) Q \left(\sum_{i=0}^{m} A_i e^{-j\omega\tau_i}\right), \tag{1.16}$$

and therefore, using (1.15),

$$(m+1)\mu Q > \left(\sum_{i=0}^{m} A_i^T e^{j\omega\tau_i}\right) Q \left(\sum_{i=0}^{m} A_i e^{-j\omega\tau_i}\right). \tag{1.17}$$

Now, since $j\omega$ is a characteristic root, it is an eigenvalue of the complex matrix $\sum_{i=0}^m A_i e^{-j\omega\tau_i}$. Therefore, by considering the eigenvector v of that matrix related to the eigenvalue $j\omega$, we have

$$j\omega v = \left(\sum_{i=0}^{m} A_i e^{-j\omega\tau_i}\right) v. \tag{1.18}$$

Next, by multiplying both sides from the left by $Q^{1/2}$ and by computing the square 2 norm of both vectors, one obtains

$$|\omega|^2 v^* Q v = v^* \left(\sum_{i=0}^{m} A_i^T e^{j\omega\tau_i}\right) Q \left(\sum_{i=0}^{m} A_i e^{-j\omega\tau_i}\right) v. \tag{1.19}$$

Using (1.17), this implies that

$$|\omega|^2 v^* Q v < (m+1)\mu v^* Q v, \tag{1.20}$$

which ends the proof. □

To conclude this section, we address a modal expansion of the solutions of (1.1):

Proposition 1.13 (see [93]). *Consider the solution $x(\phi)$ of system* (1.1) *corresponding to some initial function $\phi \in \mathscr{C}([-\tau_m, 0], \mathbb{R}^n)$. For any $\zeta \in \mathbb{R}$ such that*

$$\det \Delta(\lambda) \neq 0$$

for all $\lambda \in \mathbb{C}$ on the line $\Re(\lambda) = \zeta$, the following asymptotic expansion holds:

$$x(\phi)(t) = \sum_{k=1}^{\ell} p_k(t) e^{\lambda_k t} + o(e^{\zeta t}) \quad \textit{for } t \to \infty,$$

where $\lambda_1,\ldots,\lambda_\ell$ *are the (finitely many) characteristic roots with real part exceeding* ζ*, and* $p_k(t)$*,* $k=1,\ldots,\ell$*, are* $\mathbb{C}^n$*-valued polynomials of degree less than or equal to* m_k-1*, with* m_k *the multiplicity of* λ_k *as a root of the characteristic equation.*

It follows that if all of the characteristic roots are in the open left half plane, then all of the solutions of (1.1) converge to zero exponentially as $t\to+\infty$. This is in accordance with Proposition 1.6.

1.1.5 ▪ Spectrum: Continuity properties

We discuss continuity properties of the characteristic roots and related quantities such as spectral abscissa. As we explicitly address the dependence of the characteristic roots on parameters, we will sometimes write, for instance,

$$\Delta(\lambda;\ \vec{\tau},A_0,\ldots,A_m),$$

instead of $\Delta(\lambda)$, where the two types of arguments, separated by a dot-comma, correspond to variables and parameters, respectively. The short notation $\vec{\tau}$ stands for $(\tau_1,\ldots,\tau_m)$.

Based on Corollary A.2 of the appendix, a corollary of Rouché's theorem, we have the following statement on the continuity of the individual characteristic roots, which says, roughly speaking, that the characteristic roots behave continuously w.r.t. variations of the system matrices and delays:

Proposition 1.14. *Let* λ_0 *be a characteristic root of* (1.1) *with multiplicity k. There exists a constant* $\bar{\epsilon}>0$ *such that for all* $\epsilon>0$ *satisfying* $\epsilon<\bar{\epsilon}$*, there is a number* $\delta>0$ *such that*

$$\Delta(\lambda;\ \vec{\tau}+\delta\vec{\tau},A_0+\delta A_0,\ldots,A_m+\delta A_m),$$

where

$$\begin{array}{ll}\delta\vec{\tau}\in\mathbb{R}^m, & \|\delta\vec{\tau}\|<\delta,\ \vec{\tau}+\delta\vec{\tau}\geq 0,\\ \delta A_k\in\mathbb{R}^{n\times n}, & \|\delta A_k\|_2<\delta,\ k=0,\ldots,m,\end{array}$$

has exactly k zeros[3] *in the disc* $\{\lambda\in\mathbb{C}:\ |\lambda-\lambda_0|<\epsilon\}$.

Next, let the *spectral abscissa function* corresponding to the system (1.1) be defined as follows:

$$c(\vec{\tau};\ A_0,\ldots,A_m):=\sup\{\Re(\lambda):\ \det\Delta(\lambda;\ A_0,\ldots,A_m)=0\}. \tag{1.21}$$

Note from Corollary 1.9 that the spectral abscissa always exists and is finite. Furthermore, there always exist (rightmost) characteristic roots λ such that $\Re(\lambda)=c$. Hence, the supremum operator in (1.21) can be replaced with a maximum, and the exponential stability conditions from Proposition 1.6 can be rephrased as

$$c(\vec{\tau};\ A_0,\ldots,A_m)<0.$$

Regarding the continuity of the spectral abscissa we have the following result:

Theorem 1.15. *The function* $c:\ \mathbb{R}_+^m\times\mathbb{R}^{n\times n\times(m+1)}\to\mathbb{R}$*, defined by*

$$(\vec{\tau},A_0,\ldots,A_m)\mapsto c(\vec{\tau};\ A_0,\ldots,A_m)$$

is continuous.

[3] Multiplicity is taken into account.

Proof. The assertion follows from the combination of Proposition 1.14, Corollary 1.9, and the bound (1.12) on the characteristic roots. For the case where only the delays are varied a detailed proof is given in [85]. □

It is important to point out that the spectral abscissa function is also continuous at these points where the time-delay system reduces to a system without delays. This happens, for instance, at $\vec{\tau} = (0,\dots,0)$, or at parameter values where the system matrices corresponding to the delayed terms vanish.

The following corollary justifies the methods for computing stability regions in parameter spaces, discussed in the subsequent chapters:

Theorem 1.16. *If the matrices $A_0,\dots,A_m$ and the delays $\tau_1,\dots,\tau_m$ are varied, then a loss or acquisition of exponential stability of the null solution of* (1.1) *is associated with characteristic roots on the imaginary axis.*

Example 1.17. *Figure* 1.3 *depicts the rightmost characteristic roots of the system* (1.3) *as a function of the parameter α for a fixed value of τ and vice versa. Notice that the spectral abscissa is continuous, yet not everywhere differentiable. A more elaborate study of continuity properties of the spectral abscissa will be performed in Chapter* 7.

For $\alpha = 0$, the system reduces to

$$\dot{x}(t) = x(t). \tag{1.22}$$

It follows from Proposition 1.14 *and Theorem* 1.15 *that one characteristic root converges to the characteristic root $\lambda = 1$ of* (1.22) *as $\alpha \to 0$, while the real parts of the other characteristic roots move off to $-\infty$. A similar situation occurs as $\tau \to 0+$, where the system reduces to*

$$\dot{x}(t) = (1-\alpha)x(t). \tag{1.23}$$

Note in particular that asymptotic stability for $\tau = 0$ is preserved for small positive values of the delay.

Let us now exploit the continuity results in order to prove the asymptotic stability of (1.3) *for $\tau = 1/3$ and $\alpha = 2$. We fix α and consider τ as a free parameter. The system is obviously*

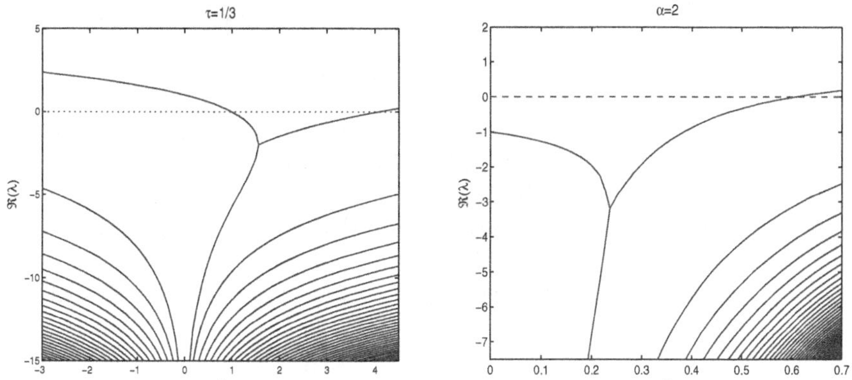

Figure 1.3. *(Left) The rightmost characteristic roots of the system* (1.3) *as a function of the parameter α for $\tau = 1/3$. (Right) The rightmost characteristic roots as a function of the delay-parameter τ for $\alpha = 2$.*

asymptotically stable for $\tau = 0$. Next, by Theorem 1.16 a loss of stability is associated with characteristic roots on the imaginary axis. Substituting $\lambda = j\omega_0$, $\omega_0 > 0$, in the characteristic equation yields

$$j\omega_0 - 1 + 2e^{-j\omega_0\tau} = 0. \tag{1.24}$$

Solving this equation results in $\omega_0 = \sqrt{3}$ and

$$\tau = \frac{\angle(1 + j\sqrt{3}) + 2\pi l}{\sqrt{3}}, \quad l \geq 0.$$

The continuity w.r.t. the delay value implies that the system, stable for $\tau = 0$, remains stable for all delays $\tau \in [0,\ \tau_m)$, where the delay margin *τ_m is given by $\tau_m = (\angle(1 + j\sqrt{3}))/\sqrt{3}$. The remaining step is to compare τ_m with the nominal delay value. In our case, we have $\tau_m > 1/3$, which allows us to conclude that the system is asymptotically stable for $\tau = 1/3$.*

1.2 ▪ Time-delay systems of neutral type

We discuss spectral properties of the neutral equation

$$\frac{d}{dt}\left(x(t) + \sum_{k=1}^{m} H_k x(t - \tau_k)\right) = A_0 x(t) + \sum_{k=1}^{m} A_k x(t - \tau_k), \tag{1.25}$$

where $x(t) \in \mathbb{R}^n$ is the state variable at time t and $0 < \tau_1 < \tau_2 \cdots < \tau_m$ represent the time-delays. In what follows we use the short notation $\vec{\tau} = (\tau_1, \ldots, \tau_m)$.

The reader might be more familiar with the explicit form

$$\dot{x}(t) = A_0 x(t) + \sum_{k=1}^{m} A_k x(t - \tau_k) - \sum_{k=1}^{m} H_k \dot{x}(t - \tau_k), \tag{1.26}$$

which emphasizes that in a system of neutral type the evolution not only depends on the solution at past time-instants but also on the derivative of the solution at past time-instants. In Remark 1.18 we will point out a subtle difference between (1.25) and (1.26).

1.2.1 ▪ The initial value problem

The initial condition for the neutral system (1.25) is the function segment

$$\phi \in \mathscr{C}([-\tau_m,\ 0], \mathbb{R}^n),$$

where $\mathscr{C}([-\tau_m,\ 0],\ \mathbb{R}^n)$ is the Banach space of continuous functions mapping the interval $[-\tau_m,\ 0]$ into $\mathbb{R}^n$ and equipped with the supremum norm, $\|\cdot\|_s$. The forward solution of (1.25) on the interval $[0,\ t_0]$, $t_0 > 0$, with initial condition $x(t) = \phi(t), t \leq 0$, is an absolutely continuous function that satisfies the differential equation (1.25) almost everywhere on the interval $[0,\ t_0]$.

Due to the fact that the map $\mathscr{N} : \mathscr{C}([-\tau_m,\ 0], \mathbb{R}^n) \to \mathbb{R}^n$, defined by

$$\mathscr{N}(\phi) := \phi(0) + \sum_{k=1}^{m} H_k \phi(-\tau_k),$$

is atomic at zero, existence and uniqueness of solutions of (1.25) are guaranteed. Let

$$x(\phi) : \ t \in [-\tau_m,\ \infty) \to x(\phi)(t) \in \mathbb{R}^n$$

be the unique forward solution with initial condition $\phi \in \mathscr{C}([-\tau_m, 0], \mathbb{R}^n)$, that is,

$$x(\phi)(\theta) = \phi(\theta) \quad \forall \theta \in [-\tau_m, 0].$$

Then the state at time t is given by the function segment $x_t(\phi) \in \mathscr{C}([-\tau_m, 0], \mathbb{R}^n)$ defined as

$$x_t(\phi)(\theta) = x(\phi)(t+\theta), \quad \theta \in [-\tau_m, 0].$$

Denote by $\mathscr{T}_N(t)$, $t \geq 0$, the solution operator, mapping initial data onto the state at time t,

$$(\mathscr{T}_N(t)\phi)(\theta) = x_t(\phi)(\theta) = x(\phi)(t+\theta), \quad \theta \in [-\tau_m, 0]. \tag{1.27}$$

This is a strongly continuous semigroup. Let $\mathscr{A}_N$ be its infinitesimal generator, that is,

$$\begin{aligned} \mathscr{D}(\mathscr{A}_N) = \Bigg\{ & \phi \in \mathscr{C}([-\tau_m, 0], \mathbb{R}^n) : \frac{d\phi}{d\theta} \in \mathscr{C}([-\tau_m, 0], \mathbb{R}^n), \\ & \frac{d\phi}{d\theta}(0) + \sum_{k=1}^{m} H_k \frac{d\phi}{d\theta}(-\tau_k) = A_0\phi(0) + \sum_{k=1}^{m} A_k\phi(-\tau_k) \Bigg\}, \\ \mathscr{A}_N\, \phi = {} & \frac{d\phi}{d\theta}. \end{aligned}$$

Remark 1.18. *From* (1.26) *it might be surprising that the existence and uniqueness of solutions of* (1.25) *do not require that the initial function be differentiable almost everywhere. If one would alternatively define the solution on* $[0, t_0]$ *as an absolutely continuous function on the interval, satisfying* (1.26) *almost everywhere, then an additional differentiability requirement on the initial condition would indeed be needed. The key is that* (1.25) *only requires that* $x(t) + \sum_{k=1}^{m} H_k x(t-\tau_k)$ *is differentiable for almost all* $t \geq 0$*, instead of* $\dot{x}(t)$*, which is more restricting.*

The associated delay-difference equation of (1.25) is given by $\mathscr{N}(x_t) = 0$, or

$$x(t) + \sum_{k=1}^{m} H_k x(t-\tau_k) = 0. \tag{1.28}$$

For any initial condition $\phi \in \mathscr{C}_D([-\tau, 0], \mathbb{R}^n)$, where

$$\mathscr{C}_D([-\tau_m, 0], \mathbb{R}^n) = \{\phi \in \mathscr{C}([-\tau_m, 0], \mathbb{R}^n) : \mathscr{N}(\phi) = 0\}$$

is a closed subspace of $\mathscr{C}([-\tau, 0], \mathbb{R}^n)$, a forward solution of (1.28) is uniquely defined. Let

$$y(\phi) : t \in [-\tau_m, \infty) \to y(\phi)(t) \in \mathbb{R}^n$$

be the unique forward solution of (1.28) with initial condition $\phi \in \mathscr{C}_D([-\tau_m, 0], \mathbb{R}^n)$. Then the state at time t is given by the function segment $y_t(\phi) \in \mathscr{C}_D([-\tau_m, 0], \mathbb{R}^n)$,

$$y_t(\phi)(\theta) = y(\phi)(t+\theta), \quad \theta \in [-\tau_m, 0].$$

We denote by $\mathscr{T}_D(t)$ the corresponding solution operator.

1.2.2 • Spectrum: Definitions

The substitution of a sample solution of the form $e^{\lambda t}v$, with $v \in \mathbb{C}^{n\times 1} \setminus \{0\}$, leads us to the *characteristic equation* of (1.25)

$$\det(\Delta_N(\lambda)) = 0,$$

where

$$\Delta_N(\lambda) := \lambda\left(I + \sum_{k=1}^{m} H_k e^{-\lambda\tau_k}\right) - A_0 - \sum_{k=1}^{m} A_k e^{-\lambda\tau_k} \tag{1.29}$$

is the *characteristic matrix*.

The zeros of (1.29) are called the *characteristic roots* of (1.25). The spectra of the linear operators $\mathscr{A}_N$, $\mathscr{T}_N(t)$ and the characteristic roots are related in the following way. The characteristic roots are the eigenvalues of the operator $\mathscr{A}$, which only features a *point spectrum*, that is, $\sigma(\mathscr{A}_N) = P\sigma(\mathscr{A}_N)$. Furthermore, the algebraic multiplicity of a complex number λ as an eigenvalue of $\mathscr{A}_N$ is equal to its multiplicity as a zero of $\Delta_N(\lambda)$, while its geometric multiplicity is equal to the dimension of the null space of $\Delta_N(\lambda)$. For $\lambda \in \sigma(\mathscr{A})$, the corresponding eigenfunctions take the form

$$ve^{\lambda\theta},\ \theta \in [-\tau_m, 0], \tag{1.30}$$

where the eigenvector $v \in \mathbb{C}^n \setminus \{0\}$ satisfies

$$\Delta_N(\lambda)v = 0. \tag{1.31}$$

The spectra of $\mathscr{A}_N$ and $\mathscr{T}_N$ are related by

$$\sigma(\mathscr{T}_N(t)) = \operatorname{clos}(\exp(t\sigma(\mathscr{A}_N))) \ \text{ plus possibly } \{0\}. \tag{1.32}$$

Furthermore, if $\lambda \in \sigma(\mathscr{A}_N)$, then $e^{\lambda t} \in P\sigma(\mathscr{T}_N(t))$, with corresponding eigenfunctions given by (1.30). It is important to mention that $\mathscr{T}_N(t)$ in general features an essential spectrum. This is the part of the spectrum that cannot be removed with a compact perturbation,

$$\sigma_e(\mathscr{T}_N(t)) := \bigcap_{K \text{ compact}} \sigma(\mathscr{T}_N(t) + K).$$

For the system (1.25) the essential spectrum of $\mathscr{T}(t)$ appears as the collection of complex numbers λ for which there are infinitely many eigenvalues in any open disk centered at λ (multiplicity taken into account).

The characteristic equation of the associated delay-difference equation (1.28) is given by $\Delta_D(\lambda) = 0$, where

$$\Delta_D(\lambda) := \left(I + \sum_{k=1}^{m} H_k e^{-\lambda\tau_k}\right). \tag{1.33}$$

The zeros of (1.33) are called the characteristic roots of the delay-difference equation (1.28). These characteristic roots are related with the spectrum of $\mathscr{T}_D(t)$ as follows:

$$\sigma(\mathscr{T}_D(t)) = \operatorname{clos}\left\{e^{\lambda t} :\ \lambda \in \mathbb{C} \text{ and } \Delta_D(\lambda) = 0\right\} \ \text{ plus possibly } \{0\}. \tag{1.34}$$

The spectrum of $\mathscr{T}_D(t)$ only features an essential spectrum, that is,

$$\sigma_e(\mathscr{T}_D(t)) = \sigma(\mathscr{T}_D(t)). \tag{1.35}$$

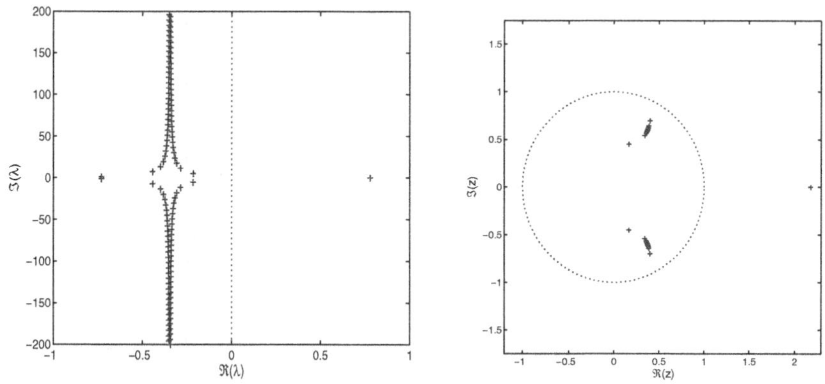

Figure 1.4. *(Left) Characteristic roots of the neutral system* (1.37)–(1.38). *(Right) Eigenvalues of the corresponding operator* $\mathscr{T}_N(1)$.

A very important result in what follows, which connects the spectra of $\mathscr{T}_N(t)$ and $\mathscr{T}_D(t)$, is the following:

$$\sigma_e(\mathscr{T}_N(t)) = \sigma(\mathscr{T}_D(t)). \tag{1.36}$$

Example 1.19. *We consider the neutral system*

$$\frac{d}{dt}\left(x(t) - \frac{3}{4}x(t-\tau_1) + \frac{1}{2}x(t-\tau_2)\right) = \frac{1}{4}x(t) + \frac{3}{4}x(t-\tau_1), \tag{1.37}$$

where

$$\vec{\tau} = (1,2). \tag{1.38}$$

In Figure 1.4 *we plot the characteristic roots* λ*, which are the eigenvalues of the operator* $\mathscr{A}_N$*, as well as the eigenvalues* z *of the operator* $\mathscr{T}_N(1)$*. These are connected via the relation* $z = e^{\lambda}$*. The operator* $\mathscr{A}_N$ *only features a point spectrum; in particular, the characteristic roots are all isolated and of finite multiplicity. The operator* $\mathscr{T}_N(1)$ *features an essential spectrum that corresponds to the accumulation points of the eigenvalues given by*

$$z_e^{\pm} = \frac{3 \pm \sqrt{23}j}{8}. \tag{1.39}$$

In Figure 1.5 *we plot the characteristic roots of the associated delay-difference equation*

$$x(t) = \frac{3}{4}x(t-\tau_1) - \frac{1}{2}x(t-\tau_2). \tag{1.40}$$

The characteristic roots can be computed analytically as follows:

$$1 - \frac{3}{4}e^{-\lambda} + \frac{1}{2}e^{-2\lambda} = 0 \tag{1.41}$$

$$\Leftrightarrow e^{\lambda} = \frac{3 \pm \sqrt{23}j}{8} \tag{1.42}$$

$$\Leftrightarrow \lambda = -\log 2 \pm j\left(\text{atan}\frac{\sqrt{23}}{3} + 2\pi l\right),\ l \in \mathbb{Z}. \tag{1.43}$$

The fact that the right-hand sides of (1.39) *and* (1.42) *are equal is a consequence of* (1.35)–(1.36).

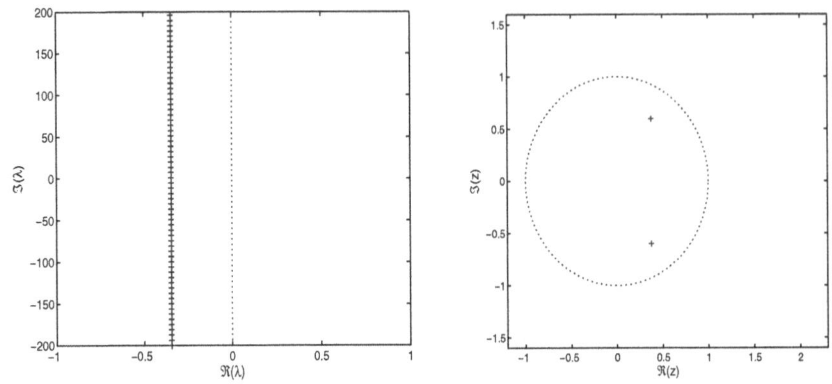

Figure 1.5. *(Left) Characteristic roots of the delay-difference equation* (1.40)–(1.38). *(Right) Eigenvalues of* $\mathscr{T}_D(1)$.

1.2.3 ▪ Asymptotic growth rate of solutions and stability

The definition of stability notions is similar to that for ODEs and DDEs of retarded type:

Definition 1.20. *The null solution of* (1.25) *is asymptotically stable*[4] *if and only if*

$$\forall \epsilon > 0\ \exists \delta > 0\ \forall \phi \in \mathscr{C}([-\tau,\,0],\mathbb{R}^n)\ (\|\phi\|_s < \delta) \Rightarrow (\forall t \geq 0\ \|x_t(\phi)\|_s < \epsilon),$$
$$\forall \phi \in \mathscr{C}([-\tau,\,0],\mathbb{R}^n)\ \lim_{t\to+\infty} x(\phi)(t) = 0.$$

Definition 1.21. *The null solution of* (1.25) *is exponentially stable if and only if there exist constants* $C > 0$ *and* $\gamma > 0$ *such that*

$$\forall \phi \in \mathscr{C}([-\tau,\,0],\mathbb{R}^n)\ \|x_t(\phi)\|_s \leq Ce^{-\gamma t}\|\phi\|_s.$$

It is clear that exponential stability implies asymptotic stability. Contrary to the case of linear delay equations of retarded type, the converse does not hold in general, as illustrated by an example in [393] (see also [50]).

The asymptotic behavior of the solutions of (1.25), and thus their stability properties, is determined by the spectral radius $r_\sigma(\mathscr{T}_N(1))$. In particular we have the following result:

Proposition 1.22. *The null solution of* (1.25) *is exponentially stable if and only if*

$$r_\sigma(\mathscr{T}_N(1)) < 1,$$

or, equivalently, all characteristic roots are located in the open left half plane and bounded away from the imaginary axis.

For the delay-difference equation (1.28), associated with (1.25), stability definitions and their relation with spectral properties are similar. We have, for instance:

Definition 1.23. *The null solution of* (1.28) *is exponentially stable if and only if there exist constants* $C > 0$ *and* $\gamma > 0$ *such that*

$$\forall \phi \in \mathscr{C}_D([-\tau,\,0],\mathbb{R}^n)\ \|y_t(\phi)\|_s \leq Ce^{-\gamma t}\|\phi\|_s.$$

[4] For reasons of conciseness we will often use the less precise formulation "the system (1.25) is asymptotically stable."

Proposition 1.24. *The null solution of* (1.28) *is exponentially stable if and only if*

$$r_\sigma(\mathscr{T}_D(1)) < 1,$$

or, equivalently, all characteristic roots of (1.28) *are located in the open left half plane and bounded away from the imaginary axis.*

Let us now relate the exponential stability of the null solution of (1.28) with the exponential stability of null solution of (1.25). The following result is a direct corollary of (1.36) and Propositions 1.22 and 1.24:

Proposition 1.25. *A necessary condition for the exponential stability of the null solution of the neutral equation* (1.25) *is the exponential stability of the null solution of the delay-difference equation* (1.28).

1.2.4 ▪ Spectrum: Qualitative properties

Instrumental to the study of qualitative features of the spectra related to the neutral equation (1.25), we first study some qualitative features of the spectra related to the delay-difference equation (1.28).

Delay-difference equation

Define the collection of the real parts of all the characteristic roots of (1.28) as

$$Z_D := \{\Re(\lambda) : \det \Delta_D(\lambda) = 0\} \tag{1.44}$$

and let the spectral abscissa c_D be its supremum:

$$c_D := \sup \{\Re(\lambda) : \det \Delta_D(\lambda) = 0\}. \tag{1.45}$$

It is clear that $c_D < \infty$. The following result characterizes the set Z_D:

Proposition 1.26. *Define*

$$\mathscr{I}_D = \left\{ \zeta \in \mathbb{R} : \det\left(I + \sum_{k=1}^{m} H_k e^{-\zeta \tau_k} e^{i\theta_k} \right) = 0 \textit{ for some } \vec{\theta} \in [0,\ 2\pi]^m \right\}.$$

Then the following hold:

- $\bar{Z}_D \subseteq \mathscr{I}_D$.
- *If the delays* $\vec{\tau}$ *are rationally independent,*[5] *then* $\bar{Z}_D = \mathscr{I}_D$, *hence;* $\bar{Z}_D$ *consists of a finite number of intervals. If the delays* $\vec{\tau}$ *are commensurate, then* Z_D *consists of a finite number of points.*

The next proposition (from [13]) describes the typical behavior of the spectrum, characterized by chains of characteristic roots whose real parts are bounded, yet whose imaginary parts tend to infinity:

[5]The m components of $\vec{\tau} = (\tau_1, \ldots, \tau_m)$ are rationally independent if and only if $\sum_{k=1}^{m} n_k \tau_k = 0,\ n_k \in \mathbb{Z}$ implies $n_k = 0\ \forall k = 1, \ldots, m$. For instance, two delays τ_1 and τ_2 are rationally independent if their ratio is an irrational number. See Section A.4 of the appendix for more information on the interdependency of numbers.

Proposition 1.27. *If* $\zeta \in \bar{Z}_D$, *then there exists a sequence of characteristic roots of* (1.28), $\{\lambda_n\}_{n\geq 1}$, *satisfying*

$$\lim_{n\to\infty} \Re(\lambda_n) = \zeta, \quad \lim_{n\to\infty} = \Im(\lambda_n) = \infty.$$

If the delays are commensurate, then the chains of roots can be computed analytically. For $\vec{\tau} = \vec{n}\tau_0$ with $\tau_0 \in \mathbb{R}_+$ and $\vec{n} \in \mathbb{N}^m$, the characteristic equation is given by

$$p(z) := \det\left(I + \sum_{k=1}^{m} H_k z^{-n_k}\right) = 0,$$

where $z = e^{-\lambda\tau_0}$. Note that p becomes a polynomial in z after a multiplication with z^{n_m}. Hence, the characteristic roots are given by

$$\left\{\lambda \in \mathbb{C}:\ \lambda = \frac{-\mathrm{Log}(z_k) + l2\pi j}{\tau_0},\ p(z_k) = 0,\ l \in \mathbb{Z}\right\}.$$

Note that Z_D consists of at most n points, namely

$$\left\{r \in \mathbb{R}:\ r = \frac{\mathrm{Log}(z_k|}{\tau_0},\ p(z_k) = 0\right\},$$

a property in accordance with Proposition 1.26. We refer to Example 1.19 and, in particular, Figure 1.5 for a numerical example. For details about the spectrum of delay-difference equations in the noncommensurate case we refer the reader to [242].

Neutral equation

Equation (1.25) also features chains of characteristic roots, whose position is determined by the associated *delay-difference equation*:

Proposition 1.28. *If* $\zeta \in \bar{Z}_D$, *with* Z_D *defined by* (1.44), *then there is a sequence of characteristic roots* $\{\lambda_n\}_{n\geq 1}$ *of* (1.25), *satisfying*

$$\lim_{n\to\infty} \Re(\lambda_n) = \zeta, \quad \lim_{n\to\infty} = \Im(\lambda_n) = \infty.$$

Proof. The proof follows from the relation (1.36) and Proposition 1.27. □

Intuitively this result is expected, since for $\lambda \neq 0$, the characteristic equation can be written in the form

$$\Delta_D(\lambda) = \frac{1}{\lambda}\left(A_0 + \sum_{k=1}^{m} A_k e^{-\lambda\tau_k}\right).$$

If $|\lambda| \gg \Re(\lambda)$, then the right-hand side is very small. Hence, characteristic roots with a large modulus but small real part are expected to be approximate zeros of $\Delta_D(\lambda)$.

In the half plane $\Re(\lambda) > c_D$ the set of characteristic roots of neutral systems has many properties similar to the retarded case, for instance:

Proposition 1.29. *For any* $\epsilon > 0$, *the system* (1.25) *only has a finite number of characteristic roots in the right half plane*

$$\Re(\lambda) \geq c_D + \epsilon,$$

where c_D *is defined in* (1.45).

Proof. The assertion follows from the fact that $\mathcal{T}_N(t)$ only has point spectrum in the set $|z| \geq r_e(\mathcal{T}_D(1))$, consisting of eigenvalues of finite multiplicity, and from the relation between the spectrum of this operator and the spectrum of $\mathcal{A}_D$. □

Also, here one can derive some envelope curves containing all characteristic roots with $\Re(\lambda) > c_D$, yet as we shall see in the next subsection, this is only useful when taking small perturbations explicitly into account.

To clarify the above results we again refer to Example 1.19. In particular the correspondence between the characteristic roots with large moduli shown in Figures 1.4 and 1.5 is a consequence of Proposition 1.28. In the right half plane there is only one characteristic root, in accordance with Proposition 1.29.

1.2.5 ▪ Spectrum: Continuity properties

The exponential stability of the delay-difference equation (1.28) associated with (1.25) may be sensitive to infinitesimal delay perturbations, which strongly affects the continuity properties of the characteristic roots of (1.25). For this, we first take a closer look at this phenomenon and discuss its relation with the spectral properties of the neutral equation. Next we discuss some continuity properties of the characteristic roots of the neutral equation (1.25) and related quantities such as spectral abscissa.

As we explicitly address the dependence of characteristic roots on parameters, we will write, for instance,

$$\Delta_N(\lambda;\ \vec{\tau}, H_1, \ldots, H_m, A_0, \ldots, A_m),$$

instead of $\Delta_N(\lambda)$, where the two types of arguments (variable and parameters) are separated with a dot-comma.

Delay sensitivity problem of the associated delay-difference equation

It is well known that the spectral radius $r_\sigma(\mathcal{T}_D)$, although continuous in the system matrices H_k, is *not* continuous in the *delays* $\vec{\tau}$ (see e.g., [151, 152, 159]), which carries over to the spectral abscissa

$$c_D(\vec{\tau};\ H_1, \ldots, H_m) := \sup\left\{\Re(\lambda) :\ \det\left(I + \sum_{k=1}^{m} H_k e^{-\lambda\tau_k}\right) = 0\right\}. \tag{1.46}$$

One consequence is that arbitrarily small perturbations on the delays may destroy stability of the delay-difference equation. This has led to the introduction of the concept of *strong stability* in [152]: we say that the null solution of equation (1.28) is strongly exponentially stable if it remains exponentially stable when subjected to small variations in the delays. Theorem 2.2 and Corollary 2.2 of [152] provide the following conditions:

Proposition 1.30. *The null solution of the delay-difference equation* (1.28) *is strongly exponentially stable if and only if* $\gamma_0 < 1$, *where*

$$\gamma_0(H_1, \ldots, H_m) := \max_{\vec{\theta} \in [0,\ 2\pi]^m} r_\sigma\left(\sum_{k=1}^{m} H_k e^{j\theta_k}\right).$$

Furthermore, if $\gamma_0 > 1$, *then equation* (1.28) *is exponentially unstable for rationally independent delays.*

Notice that the quantity γ_0 does not depend on the value of the delays; in other words, exponential stability locally in the delays is equivalent to exponential stability globally in the delays [152].

Even if the delay-difference equation is strongly exponentially stable, it is very useful to have *more precise* information about the position of the real parts of its characteristic roots, and in particular the upper bound (1.46). Due to *lack of continuity* of this quantity w.r.t. the delays we are led from a practical point of view to the smallest upper bound, which is *insensitive* to small delay changes. More precisely, we define this *safe* upper bound $C_D(\vec{\tau})$, which we call the robust spectral abscissa of (1.28), as follows:

Definition 1.31. *Let* $C_D(\vec{\tau};\ H_1,\dots,H_m)\in\mathbb{R}$ *be defined as*

$$C_D(\vec{\tau};\ H_1,\dots,H_m):=\lim_{\epsilon\to 0+} c_D^{\epsilon}(\vec{\tau};\ H_1,\dots,H_m),$$

where

$$c_D^{\epsilon}(\vec{\tau};\ H_1,\dots,H_m)=\sup\{c_D(\vec{\tau}+\delta\vec{\tau};\ H_1,\dots,H_m):\ \delta\vec{\tau}\in\mathbb{R}^m \text{ and } \|\delta\vec{\tau}\|\leq\epsilon\}.$$

Clearly we have $C_D(\vec{\tau};\ H_1,\dots,H_m)\geq c_D(\vec{\tau};\ H_1,\dots,H_m)$, and, as we shall illustrate, the inequality can be *strict*.

In order to present a computational expression for $C_D(\vec{\tau};\ H_1,\dots,H_m)$, we define the function $f:\ \mathbb{R}\to\mathbb{R}_+$ in the following way:

$$f(\alpha;\ \vec{\tau},H_1,\dots,H_m)=\max_{\vec{\theta}\in[0,\,2\pi]^m} r_\sigma\left(\sum_{k=1}^{m} H_k e^{-\alpha\tau_k} e^{j\theta_k}\right). \tag{1.47}$$

This function is *continuous* in both its *argument* α and *parameters* $\vec{\tau}$ and H_k, $k=1,\dots,m$. Notice that $\gamma_0(H_1,\dots,H_m)=f(0;\ \vec{\tau},H_1,\dots,H_m)$. We have the following result [264, Theorem 6]:

Theorem 1.32. *The robust spectral abscissa* $C_D(\vec{\tau};\ H_1,\dots,H_m)$ *is equal to the unique zero of the strictly decreasing function*

$$\alpha\in\mathbb{R}\to f(\alpha;\ \vec{\tau},H_1,\dots,H_m)-1,$$

where f *is defined as in* (1.47). *Furthermore,* $C_D(\vec{\tau};\ H_1,\dots,H_m)$ *is continuous in both the delays* $\vec{\tau}\in\mathbb{R}_+^m$ *and the parameters* $H_1,\dots,H_m$.

Remark 1.33. *As* $f(\alpha;\ \vec{\tau},H_1,\dots,H_m)$ *is strictly decreasing, a (robust) bisection algorithm is appropriate for the computation of* $C_D(\vec{\tau};\ H_1,\dots,H_m)$.

Remark 1.34. *Since* $f(\alpha;\ \vec{\tau},H_1,\dots,H_m)\leq\sum_{k=1}^m\|H_k\|e^{-c\tau_k}$, *an upper bound on* $C_D(\vec{\tau})$ *is given by the unique solution of*

$$\sum_{k=1}^{m}\|H_k\|e^{-c\tau_k}-1=0. \tag{1.48}$$

Accordingly, $\sum_{k=1}^m\|H_k\|<1$ *is a* sufficient *condition for strong exponential stability.*

We mention some special cases where the above expressions become very simple. In the case of one delay ($m=1$) we have $\gamma_0=r_\sigma(H_1)$,

$$f(\alpha;\ \tau_1,H_1)=r_\sigma(H_1)e^{-\alpha\tau_1},\qquad C_D(\tau_1)=\frac{1}{\tau_1}\log r_\sigma(H_1).$$

Next, when the equation is scalar ($n = 1$) but includes multiple delays, we have

$$\gamma_0(H_1,\ldots,H_m) = \sum_{k=1}^{m} |H_k|, \qquad f(\alpha;\, \vec{\tau}, H_1,\ldots,H_m) = \sum_{k=1}^{m} |H_k| e^{-\alpha\tau_k}.$$

Finally, consider the case of nonscalar systems including two delays ($m = 2$). In this case, the strong exponential stability reduces to the analysis of the location of the generalized eigenvalues of some appropriate finite-dimensional regular matrix pencils. More precisely, the following result [114] holds:

Proposition 1.35. *Define the matrices*

$$B_0 = H_1 \otimes H_2^T, \tag{1.49}$$

$$B_1 = H_1 \otimes H_1^T + H_2 \otimes H_2^T - I, \tag{1.50}$$

$$B_2 = H_2 \otimes H_1^T. \tag{1.51}$$

Then $\gamma(H_1, H_2) < 1$ *if and only if* $\rho(H_1) < 1$, *and additionally one of the following equivalent conditions holds:*

(i) $\rho(H_1 + z_2 H_2) < 1 \; \forall z_2 \in \overline{\mathscr{D}(0,1)}$;

(ii) $\rho\left(H_1 + e^{j\theta} H_2\right) < 1 \; \forall \theta \in [0, 2\pi]$;

(iii) $\sigma(H_1 + z_2 H_2) \cap \mathscr{C}(0,1) = \emptyset \; \forall z_2 \in \mathscr{C}(0,1)$ *and* $\rho(H_1 + H_2) < 1$;

(iv) $\rho(H_1 + H_2) < 1$, *and the matrix pencil* $\Lambda(z) := zM + N$ *satisfies the condition* $\sigma(\Lambda) \cap \mathscr{C}(0,1) = \emptyset$, *where*

$$M := \begin{bmatrix} I & 0 \\ 0 & B_2 \end{bmatrix}, \qquad N := \begin{bmatrix} 0 & -I \\ B_0 & B_1 \end{bmatrix}.$$

Proof. The sufficiency of condition (i) is straightforward (see, for instance, [31]). To see the necessity, suppose that $\rho(H_1 + z_2 H_2) \geq 1$ for some $z_2 \in \overline{\mathscr{D}(0,1)}$. Since $\rho(H_1) < 1$, by continuity, there must exist some $z_{2,c} \in \mathbb{C}$, $| z_{2,c} | \leq 1$, such that $\rho(H_1 + z_{2,c} H_2) = 1$. This simply means that the matrix $H_1 + z_{2,c} H_2$ has some eigenvalues (at least one) on $\mathscr{C}(0,1)$, which is impossible. Next, to establish condition (ii), it suffices to note that

$$\max_{z_2 \in \overline{\mathscr{D}(0,1)}} \rho(H_1 + z_2 H_2) = \max_{z_2 \in \mathscr{C}(0,1)} \rho(H_1 + z_2 H_2),$$

which follows by the fact that $H_1 + z_2 H_2$ is analytic in the closed unit disk $\overline{\mathscr{D}(0,1)}$, and hence $\rho(H_1 + z_2 H_2)$ is a *subharmonic function* of z_2 [37]. Next, it is evident that condition (ii) implies (iii). Conversely, condition (iii) also implies (ii). Indeed, suppose otherwise that $\rho\left(H_1 + e^{j\theta_c} H_2\right) \geq 1$ for some $\theta_c \in [0, 2\pi]$. Then, since $\rho(A_1 + A_2) < 1$, it follows from the continuity of $\rho(H_1 + e^{j\theta} H_2)$ that there exists some $\theta \in [0, \theta_c]$ such that $\rho(H_1 + e^{j\theta} H_2) = 1$. This implies that $\sigma(H_1 + e^{j\theta} H_2) \cap \mathscr{C}(0,1) \neq \emptyset$, thus leading to a contradiction. Finally, to establish the equivalence between (iii) and (iv), note first that from the property of the Kronecker product (see, e.g., [124]), it follows that $\sigma(H_1 + z_2 H_2) \cap \mathscr{C}(0,1) = \emptyset$ holds for all $z_2 \in \mathscr{C}(0,1)$ if and only if for all $z_2 \in \mathscr{C}(0,1)$,

$$1 \notin \sigma\left((H_1 + z_2 H_2) \otimes (H_1 + z_2 H_2)^*\right).$$

Next, some simple but tedious algebraic manipulations show that the last property is equivalent to the condition $\sigma(\Lambda) \cap \mathscr{C}(0,1) = \emptyset$, with the matrix pencil Λ defined by $\Lambda(z) = zM + N$ (see, e.g., [114] for the corresponding computations). □

Example 1.36. *Consider the delay-difference equation* (1.40)*. For* $\vec{\tau} = (1,2)$ *its null solution is exponentially stable but not strongly exponentially stable because* $\gamma_0(-3/4, 1/2) = |3/4| + |1/2|$ *is larger than one. Furthermore, we have* $c_D((1,2);\ -3/4, 1/2) \approx -0.3466$*, which is* strictly smaller *than* $C_D((1,2);\ -3/4, 1/2) \approx 0.1616$*. This illustrates the noncontinuity of* $c_D(\vec{\tau};\ -3/4, 1/2)$ *w.r.t.* $\vec{\tau}$*. In the left frame of Figure* 1.6 *we plot the characteristic roots of* (1.40) *for both the nominal delays* $\vec{\tau} = (1,2)$ *(indicated by +) and the perturbed delays* $\vec{\tau} = (0.99, 2)$ *(indicated by* o*). In the right frame we plot the eigenvalues of the operator* $\mathscr{T}_D(1)$*. The dashed curves are described by*

$$\Re(\lambda) = C_D((1,2); -3/4, 1/2)\ |z| = \exp(C_D((1,2);\ -3/4, 1/2)). \tag{1.52}$$

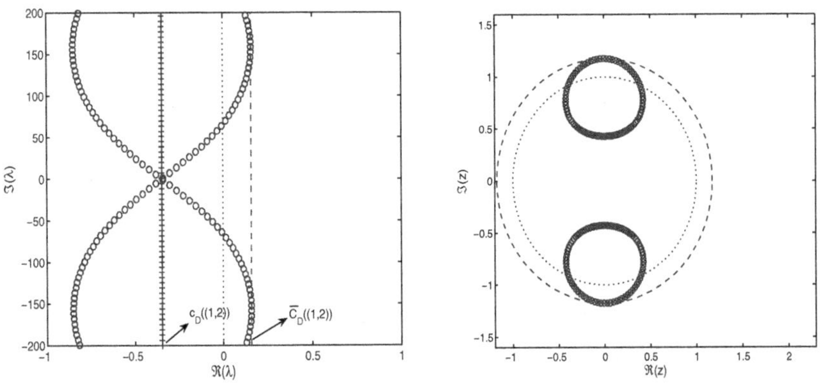

Figure 1.6. *(Left) Characteristic roots of the delay-difference equation* (1.40)*. (Right) Corresponding eigenvalues of* $\mathscr{T}_D(1)$*.*

Relation with the spectrum of the neutral equation

From (1.35)–(1.36) it follows that not only the delay-difference equation (1.28) but *also* the neutral equation (1.25), has characteristic roots with real part *arbitrarily close* to $C_D(\vec{\tau};\ H_1, \ldots, H_m)$ for certain (arbitrarily small) delay perturbations.

Example 1.37. *In the left frame of Figure* 1.7 *we plot the characteristic roots of* (1.37) *for both the nominal delays* $\vec{\tau} = (1,2)$ *(indicated by +) and the perturbed delays* $\vec{\tau} = (0.99, 2)$ *(indicated by* o*). In the right frame we plot the eigenvalues of the operator* $\mathscr{T}_N(1)$*. The dashed curves again correspond to* (1.52)*.*

If $\Re(\lambda) > c_D(\vec{\tau};\ H_1, \ldots, H_m)$, then the matrix $\Delta_D(\lambda;\ \vec{\tau}, H_1, \ldots, H_m)$ is invertible. If, in addition, $\Re(\lambda) > C_D(\vec{\tau};\ H_1, \ldots, H_m)$, then the following estimate holds:

$$\begin{aligned} \|(\Delta_D(\lambda;\ \vec{\tau}, H_1, \ldots, H_m)^{-1})\|_2 &= \left\| \left(I - \sum_{k=1}^{m} H_k e^{-\Re(\lambda)\tau_k} e^{-j\Im(\lambda)\tau_k} \right)^{-1} \right\|_2 \\ &\leq \max_{\vec{\theta} \in [0,\, 2\pi]^m} \left\| \left(I - \sum_{k=1}^{m} H_k e^{-\Re(\lambda)\tau_k} e^{j\theta_k} \right)^{-1} \right\|_2 \end{aligned} . \tag{1.53}$$

The right-hand side of (1.53) is well defined because $f(\Re(\lambda);\ \vec{\tau}, H_1, \ldots, H_m) < 1$ if $\Re(\lambda) >$

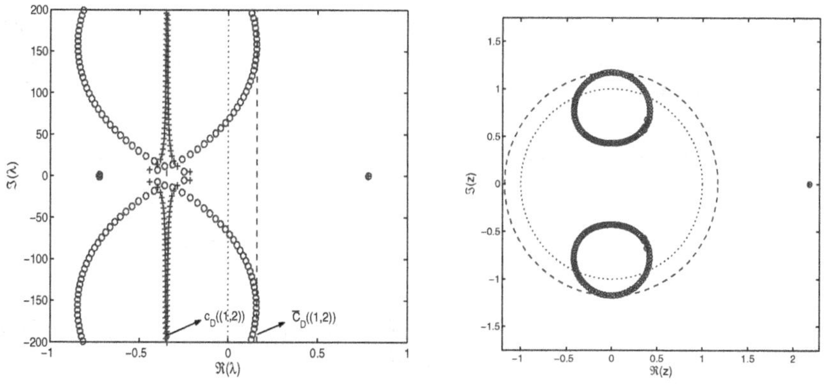

Figure 1.7. *(Left) Characteristic roots of the neutral equation* (1.37). *(Right) Corresponding eigenvalues of* $\mathscr{T}_N(1)$.

$C_D(\vec{\tau};H_1,\ldots,H_m)$. This leads to a lemma which will play a crucial role in the proof of some continuity properties of the spectrum discussed in what follows.

Lemma 1.38. *If λ is a characteristic root of the neutral system* (1.25) *with*

$$\Re(\lambda) > C_D(\vec{\tau};\ H_1,\ldots,H_m),$$

then it satisfies

$$|\lambda| \le b(\Re(\lambda);\ \vec{\tau},H_1,\ldots,H_m,A_0,\ldots,A_m), \tag{1.54}$$

where

$$\begin{aligned} &b(\Re(\lambda);\ \vec{\tau},H_1,\ldots,H_m,A_0,\ldots,A_m) \\ &:= \max_{\vec{\theta}\in[0,\ 2\pi]^m} \left\| \left(I + \sum_{k=1}^{m} H_k e^{-\Re(\lambda)\tau_k} e^{j\theta_k} \right)^{-1} \right\|_2 \\ &\quad \times \left(\|A_0\|_2 + \sum_{i=1}^{m} \|A_k\|_2 e^{-\Re(\lambda)\tau_k} \right). \end{aligned} \tag{1.55}$$

Proof. Because $\Delta_D(\lambda;\ \vec{\tau},H_1,\ldots,H_m)$ is invertible, we can write the characteristic equation in the form

$$\det\left(I - \Delta_D(\lambda;\ \vec{\tau},H_1,\ldots,H_m)^{-1} \left(A_0 + \sum_{k=1}^{m} A_k e^{-\lambda\tau_k} \right) \right) = 0.$$

This equation can be interpreted as

$$\lambda \in \sigma\left(\Delta_D(\lambda;\ \vec{\tau},H_1,\ldots,H_m)^{-1} \left(A_0 + \sum_{k=1}^{m} A_k e^{-\lambda\tau_k} \right) \right),$$

which implies

$$|\lambda| \le \left\| \Delta_D(\lambda;\ \vec{\tau},H_1,\ldots,H_m)^{-1} \left(A_0 + \sum_{k=1}^{m} A_k e^{-\lambda\tau_k} \right) \right\|_2.$$

By further working out this estimate and using (1.53) one arrives at (1.54). $\square$

It is important to mention that for a given $\Re(\lambda) > C_D$, the quantity (1.55) is *continuous in the delays* at the nominal delay values and, hence, not sensitive to infinitesimal delay perturbations.

Example 1.39. *Let us revisit the example* (1.40), *for which the estimate* (1.54)–(1.55) *becomes*

$$|\lambda| \leq \frac{\frac{1}{4} + \frac{3}{4}e^{-\Re(\lambda)\tau_1}}{1 - \frac{3}{4}e^{-\Re(\lambda)\tau_1} - \frac{1}{2}e^{-\Re(\lambda)\tau_2}}. \tag{1.56}$$

With $\vec{\tau} = (1,2)$ *the curve defined by* (1.56) *is shown in bold in Figure* 1.8, *as well as the characteristic roots for the nominal delays* $\vec{\tau}(1,2)$ *and the perturbed delays* $\vec{\tau} = (0.99,2)$. *The left and right frame, corresponds to a different scaling of the real and imaginary axis. Despite of the discontinuity of the function* $\vec{\tau} \to c_D(\vec{\tau}; -3/4, 1/2)$ *at* $\vec{\tau} = (1,2)$, *the envelope curve* (1.56) *is not affected by small delay perturbations.*

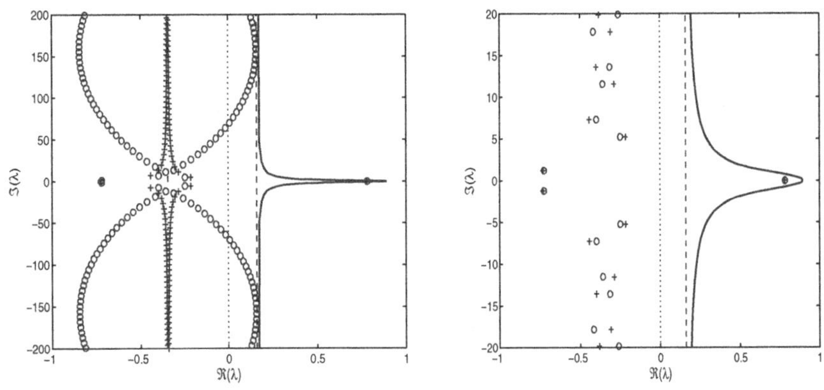

Figure 1.8. *Characteristic roots of the neutral equation* (1.37), *and the curve defined by* (1.56).

Continuity properties and stability switches

Also in the neutral case the individual characteristic roots behave continuously w.r.t. the system's parameters (following from Corollary A.2):

Proposition 1.40. *Let* λ_0 *be a characteristic root of the neutral equation* (1.25) *with multiplicity k. There exists a constant* $\bar{\epsilon} > 0$ *such that for all* $\epsilon > 0$ *satisfying* $\epsilon < \bar{\epsilon}$, *there is a number* $\delta > 0$ *such that*

$$\Delta_N(\lambda;\ \vec{\tau} + \delta\vec{\tau}, H_1 + \delta H_1, \ldots, H_m + \delta H_m, A_0 + \delta A_0, \ldots, A_m + \delta A_m),$$

where

$$\begin{aligned} &\delta\vec{\tau} \in \mathbb{R}^m, && \|\delta\vec{\tau}\| < \delta,\ \vec{\tau} + \delta\vec{\tau} \geq 0, \\ &\delta H_k \in \mathbb{R}^{n\times n}, && \|\delta H_k\|_2 < \delta,\ k = 1, \ldots, m, \\ &\delta A_k \in \mathbb{R}^{n\times n}, && \|\delta A_k\|_2 < \delta,\ k = 0, \ldots, m, \end{aligned}$$

has exactly k zeros[6] *in the disc* $\{\lambda \in \mathbb{C} :\ |\lambda - \lambda_0| < \epsilon\}$.

[6]Multiplicity is taken into account.

Let the *spectral abscissa function* corresponding to the neutral equation (1.25) be defined as follows:

$$\begin{aligned} &c(\vec{\tau};\, H_1,\ldots,H_m,A_1,\ldots,A_m) \\ &\qquad := \sup\{\Re(\lambda):\ \det\Delta_N(\lambda;\, \vec{\tau},H_1,\ldots,H_m,A_0,\ldots,A_m)=0\}. \end{aligned}$$

In contrast to the retarded case, this function is in general not continuous, as we illustrate with an example:

Example 1.41. *The characteristic matrix of the system*

$$\begin{aligned} &\frac{d}{dt}\left(x(t)-\frac{3}{4}x(t-\tau_1)+\frac{1}{2}x(t-\tau_2)\right) \\ &\qquad\qquad = -\left(x(t)-\frac{3}{4}x(t-\tau_1)+\frac{1}{2}x(t-\tau_2)\right) \end{aligned} \tag{1.57}$$

is given by

$$\Delta_N(\lambda)=(\lambda+1)\left(1-\frac{3}{4}e^{-\lambda\tau_1}+\frac{1}{2}e^{-\lambda\tau_2}\right)=(\lambda+1)\Delta_D(\lambda).$$

Hence, the characteristic roots of (1.57) *consists of the characteristic roots of the delay-difference equation* (1.40)*, in addition to a characteristic root at* $\lambda=-1$*. The discontinuity of the function* $\vec{\tau}\mapsto c_D(\vec{\tau};-3/4,1/2)$ *at* $\vec{\tau}=(1,2)$*, discussed in Example* 1.36*, carries over in this case to the function* $\vec{\tau}\mapsto c(\vec{\tau};\ -3/4,1/2,-1,3/4,-1/2)$*.*

Again, the problem can be solved by modifying the spectral abscissa function in such a way that small delay perturbations are taken explicitly into account.

Definition 1.42. *Let the* robust spectral abscissa $C(\vec{\tau})$ *be defined as*

$$C(\vec{\tau};\, H_1,\ldots,H_m,A_0,\ldots,A_m):=\lim_{\epsilon\to 0+} c^{\epsilon}(\vec{\tau}), \tag{1.58}$$

where

$$c^{\epsilon}(\vec{\tau}):=\sup\{c(\vec{\tau}+\delta\vec{\tau}):\ \delta\vec{\tau}\in\mathbb{R}^m \text{ and } \|\delta\vec{\tau}\|\le\epsilon\}.$$

Proposition 1.43. *For every* $\vec{\tau}\in(\mathbb{R}^+)^m$*, we have*

$$C(\vec{\tau})=\max(C_D(\vec{\tau}),c(\vec{\tau})). \tag{1.59}$$

Proof. We distinguish between two cases.

Case 1: $c(\vec{\tau})>C_D(\vec{\tau})$. Take $\epsilon>0$ such that $\epsilon<c(\vec{\tau})-C_D(\vec{\tau})$. From Lemma 1.38 and the fact that the characteristic function is analytic, there are only a finite number of characteristic roots in the half plane

$$\{\lambda\in\mathbb{C}:\ \Re(\lambda)>C_D(\vec{\tau})+\epsilon\}. \tag{1.60}$$

Moreover, this number remains invariant under sufficiently small delay perturbations, provided that for the nominal delays $\vec{\tau}$ there are no characteristic roots with real part equal to $C_D+\epsilon$. By the continuity properties of the individual characteristic roots with respect to the delay parameters, we conclude that $C(\vec{\tau})=c(\vec{\tau})$.

Case 2: $C_D(\vec{\tau}) \geq c(\vec{\tau})$. First, from $c(\vec{\tau}) \geq c_D(\vec{\tau})$ and Lemma 1.28 we have $C(\vec{\tau}) \geq C_D(\vec{\tau})$. Next, for any $\epsilon > 0$ there are no characteristic roots in the right half plane (1.60). By Lemma 1.38 and Rouché's theorem this remains so for sufficiently small delay perturbations. This fact and the property that $C_D(\vec{\tau})$ is independent of $\vec{\tau}$ lead us to the conclusion $C(\vec{\tau}) = C_D(\vec{\tau})$. □

Theorem 1.44. *The function*

$$(\vec{\tau}, H_1, \ldots, H_m, A_0, \ldots, A_m) \mapsto C(\vec{\tau};\ H_1, \ldots, H_m, A_0, \ldots, A_m)$$

is continuous.

Proof. We first prove continuity at a point where

$$c(\vec{\tau};\ H_1, \ldots, H_m, A_0, \ldots, A_m) \geq C_D(\vec{\tau};\ H_1, \ldots, H_m).$$

Fix $\epsilon > 0$. Let $\gamma > 0$ be such that

$$\begin{aligned} b_\gamma(\Re(\lambda)) := \sup\{ & b(\Re(\lambda);\ \vec{\tau} + \delta\vec{\tau}, H_1 + \delta H_1, \ldots, H_m + \delta H_m, \\ & A_0 + \delta A_0, \ldots, A_m + \delta A_m) :\ \delta\vec{\tau} \in \mathbb{R}^m, \|\delta\vec{\tau}\|_2 < \gamma,\ H_k \in \mathbb{R}^{n\times n}, \\ & \|H_k\|_2 < \gamma,\ k = 1, \ldots, m,\ A_k \in \mathbb{R}^{n\times n},\ \|A_k\|_2 < \gamma,\ k = 0, \ldots, m\} \end{aligned}$$

is defined and bounded if $\Re(\lambda) \geq c(\vec{\tau};\ H_1, \ldots, H_m, A_0, \ldots, A_m) + \epsilon$ (see Lemma 1.38 for the definition of b). Define the compact set

$$\Omega := \left\{ \lambda \in \mathbb{C} :\ \Re(\lambda) \geq c(\vec{\tau};\ H_1, \ldots, H_m, A_0, \ldots, A_m) + \epsilon,\ |\lambda| \leq b_\gamma(\Re(\lambda)) \right\}. \quad (1.61)$$

It is clear that on Ω the characteristic matrix

$$\Delta_N(\lambda;\ \vec{\tau}, H_1, \ldots, H_m, A_0, \ldots, A_m)$$

has no zeros. From Rouché's theorem (see Section A.1 of the appendix) it follows that there exists a number γ_2 such that

$$\Delta_N(\lambda;\ \vec{\tau} + \delta\vec{\tau}, H_1 + \delta H_1, \ldots, H_m + \delta H_m, A_0 + \delta A_0, \ldots, A_m + \delta A_m) \quad (1.62)$$

has no zeros in Ω whenever

$$\|\delta\vec{\tau}\|_2 < \gamma_2, \quad \|\delta H_k\| < \gamma_2,\ k = 1, \ldots, m, \quad \|\delta A_k\| < \gamma_2,\ k = 0, \ldots, m. \quad (1.63)$$

If γ_2 is taken smaller than γ, then, by Lemma 1.38, (1.63) also implies that (1.62) has no zeros satisfying

$$\Re(\lambda) \geq c(\vec{\tau};\ H_1, \ldots, H_m, A_0, \ldots, A_m) + \epsilon.$$

Since the above analysis can be repeated for any $\epsilon > 0$ we arrive at

$$\begin{aligned} & \forall \epsilon\ \exists \gamma_2\ (\|\delta\vec{\tau}\|_2 < \gamma_2 \text{ and } \|\delta H_k\| < \gamma_2,\ k = 1, \ldots, m \text{ and } \|\delta A_k\| < \gamma_2,\ k = 0, \ldots, m) \\ & \Rightarrow (\forall \lambda \in \mathbb{C} \text{ with } \Re(\lambda) \geq c(\vec{\tau};\ H_1, \ldots, H_m, A_0, \ldots, A_m) + \epsilon : \\ & \Delta_N(\lambda;\ \vec{\tau} + \delta\vec{\tau}, H_1 + \delta H_1, \ldots, H_m + \delta H_m, A_0 + \delta A_0, \ldots, A_m + \delta A_m) \neq 0). \end{aligned}$$

Roughly speaking, this statement expresses that infinitesimal perturbations cannot lead to a (discontinuous) growth of the spectral abscissa function c. Combining this fact with the continuity of $C_D(\vec{\tau};\ H_1, \ldots, H_m)$ (Theorem 1.32) and the continuity of the individual characteristic roots (Proposition 1.40), we arrive at the continuity of $\beta(\vec{\tau},H_1,\ldots,H_m,A_0,\ldots,A_m)$.

Next, we consider the case where

$$c(\vec{\tau};\ H_1,\ldots,H_m,A_0,\ldots,A_m) < C_D(\vec{\tau};\ H_1,\ldots,H_m).$$

The proof is by contradiction. Since the function C_D is continuous, a violation of the statement of the theorem implies the existence of a number $\epsilon > 0$ such that

$$\begin{aligned}
&\forall \beta > 0\ \exists \delta\vec{\tau} \text{ with } \|\delta\vec{\tau}\|_2 < \beta, \qquad \exists \delta H_k \text{ with } \|\delta H_k\|_2 < \beta, \quad k=1,\ldots,m,\\
&\exists \delta A_k \text{ with } \|\delta A_k\|_2 < \beta, \quad k=0,\ldots,m:\\
&c(\vec{\tau}+\delta\vec{\tau},H_1+\delta H_1,\ldots,H_m+\delta H_m,A_0+\delta A_0,\ldots,A_m+\delta A_m)\\
&> C_D(\vec{\tau};\ H_1,\ldots,H_m)+\epsilon.
\end{aligned} \tag{1.64}$$

Let $\{\beta_n\}_{n\geq 1}$ be a sequence of strictly positive real numbers satisfying $\lim_{n\to\infty}\beta_n = 0$. Expression (1.64) implies the existence of a corresponding sequence $\{\lambda_n\}_{n\geq 1}$ of complex numbers satisfying $\Re(\lambda) \geq C_D(\vec{\tau};\ H_1,\ldots,H_m)+\epsilon$, and sequences of perturbations $\{\delta\vec{\tau}_n\}_{n\geq 1}, \{\delta H_{k,n}\}_{n\geq 1}, \{\delta A_{k,n}\}_{n\geq 1}$ with $\|\delta\vec{\tau}_n\|_2 < \beta_n, \|\delta H_{k,n}\|_2 < \beta_n, \|\delta A_{k,n}\|_2 < \beta_n$, such that

$$\Delta_N(\lambda;\ \vec{\tau}+\delta\vec{\tau}_n,H_1+\delta H_{1,n},\ldots,H_m+\delta H_{m,n},A_0+\delta A_{0,n},\ldots,A_m+\delta A_{m,n}) \tag{1.65}$$

has a zero for $\lambda=\lambda_n$. Since for sufficiently small β_n, all zeros of (1.65) satisfying $\Re(\lambda)\geq C_D(\vec{\tau};\ H_1,\ldots,H_m)+\epsilon$ can be constrained to a compact set (analogous to (1.61)), the sequence $\{\lambda_n\}_{n\geq 1}$ has a converging subsequence with limit $\bar{\lambda}$. It is easy to show, using Rouché-type arguments, that

$$\Delta_N(\bar{\lambda};\ \vec{\tau},H_1,\ldots,H_m,A_0,\ldots,A_m)=0.$$

It follows that

$$\begin{aligned}
&c(\vec{\tau};\ H_1,\ldots,H_m,A_0,\ldots,A_m)\\
&\qquad \geq C_D(\vec{\tau},\ H_1,\ldots,H_m)
\end{aligned}$$

and we arrive at a contradiction. □

The next result lays the theoretical basis for the methods for computing stability regions in parameter spaces, discussed in the next chapters:

Theorem 1.45. *Assume that the delay-difference equation associated with the neutral equation* (1.25) *is strongly exponentially stable, that is,* $\gamma_0(H_1,\ldots,H_m)<1$.

If the matrices $A_0,\ldots,A_m$ *and the delays* $\tau_1,\ldots,\tau_m$ *are varied, then a loss or acquisition of exponential stability of the null solution of* (1.25) *is associated with characteristic roots on the imaginary axis.*

Proof. Let U be an *arbitrary* compact subset of $\mathbb{R}_+^m$, to which the delays are constrained. As $\gamma_0(H_1,\ldots,H_m)=f(0;\ \vec{\tau},H_1,\ldots,H_m)<1$ Theorem 1.32 implies that

$$\max_{\vec{\tau}\in U} C_D(\vec{\tau};\ H_1,\ldots,H_m)<0. \tag{1.66}$$

By Theorem 1.44 it then follows that the spectral abscissa function

$$(\vec{\tau},A_0,\ldots,A_m)\mapsto c(\vec{\tau};\ H_1,\ldots,H_m,A_0,\ldots,A_m)$$

is *continuous* whenever

$$c(\vec{\tau};\, H_1,\ldots,H_m,A_0,\ldots,A_m) \geq \max_{\vec{\tau}\in U} C_D(\vec{\tau};H_1,\ldots,H_m).$$

A loss of exponential stability is thus characterized by $c=0$. Since for any $\epsilon>0$, there are only a finite number of characteristic roots with $\Re(\lambda)\geq C_D(\vec{\tau})$, a situation where $c=0$ corresponds to the presence of characteristic roots on the imaginary axis. □

Remark 1.46. *If the matrices H_k, $k=1,\ldots,m$, are also varied, then the situation is more complex, since γ_0 depends on these matrices, and strong exponential stability of the delay-difference equation may be lost. According to Theorem 1.44, a transition from $\beta<0$ to $\beta>0$ then corresponds to either the case discussed above ($c=0$ for the critical parameters and characteristic roots on the imaginary axis) or $C_D=0$ for the critical parameters. Note that in the latter case the condition $\beta>0$ does not necessarily imply instability, yet instability can always be achieved by applying infinitesimal perturbations to the delays.*

1.3 ▪ Systems described by delay-differential algebraic equations

We consider delay-differential algebraic equations (DDAEs), also called descriptor systems, of the form

$$E\dot{x}(t)=A_0x(t)+\sum_{i=1}^{m}A_ix(t-\tau_i), \tag{1.67}$$

where $x(t)\in\mathbb{R}^n$ and matrix $E\in\mathbb{R}^{n\times n}$ is allowed to be singular. The time-delays $\tau_i, i=1,\ldots,m$, satisfy

$$0<\tau_1<\tau_2<\cdots<\tau_m$$

and the uppercase letters are real-valued matrices of appropriate dimensions.

The motivation for the system description (1.67) in the context of designing controllers lies in its generality in modeling interconnected systems. For instance, the feedback interconnection of the system

$$\begin{cases} \dot{z}(t)=\sum F_iz(t-r_i)+\sum G_iu(t-r_i),\\ y(t)=\sum H_ix(t-r_i)+\sum L_iu(t-r_i) \end{cases} \tag{1.68}$$

and the controller

$$\begin{cases} \dot{z}_c(t)=\sum \hat{F}_iz_c(t-s_i)+\sum \hat{G}_iy(t-s_i),\\ u(t)=\sum \hat{H}_iz_c(t-s_i)+\sum \hat{L}_iy(t-s_i) \end{cases} \tag{1.69}$$

can be directly transformed to (1.67), where

$$x=[z^T\ z_c^T\ u^T\ y^T],\qquad \{\tau_1,\ldots,\tau_m\}=\{r_i\}\cup\{s_i\}.$$

In this way no elimination of inputs and outputs is required, which may not even be possible in the presence of delays [141]. Another favorable property is the linear dependence of the matrices of the closed-loop system on the elements of the matrices of the controller. The increase in the number of equations, on the contrary, is a minor problem in most applications because the delay-difference equations or algebraic constraints are related to inputs and outputs, as illustrated above, and the number of inputs and outputs is usually much smaller than the number of state variables. Finally, we note that also neutral systems can be dealt with in this framework by introducing slack variables. The neutral equation

$$\frac{d}{dt}\left(z(t)+\sum_{i=1}^{m} G_i z(t-\tau_i)\right)=\sum_{i=0}^{m} H_i z(t-\tau_i) \tag{1.70}$$

can be rewritten as

$$\left\{\begin{array}{l} \dot{v}(t)=\sum_{i=0}^{m} H_i z(t-\tau_i), \\ \quad\;\; =-v(t)+z(t)+\sum_{i=1}^{m} G_i z(t-\tau_i), \end{array}\right. \tag{1.71}$$

where v is the slack variable. Clearly (1.71) is of the form (1.67) if we set

$$x(t)=[v(t)^T \; z(t)^T]^T.$$

1.3.1 ▪ The initial value problem

Let matrix E in (1.67) satisfy

$$\operatorname{rank}(E)=n-\nu,$$

with $1 \leq \nu < n$, and let the columns of matrix $U \in \mathbb{R}^{n\times\nu}$, respectively, $V \in \mathbb{R}^{n\times\nu}$, be a (minimal) orthonormal basis for the left, respectively, right, null space of E, which implies

$$U^T E=0, \quad EV=0. \tag{1.72}$$

Throughout the book we make the following assumption.

Assumption 1.47. *Matrix $U^T A_0 V$ is nonsingular.*

Let the columns of $U^{\perp} \in \mathbb{R}^{n\times\nu}$, respectively, $V^{\perp} \in \mathbb{R}^{n\times\nu}$, be a basis for the orthogonal complement of U, respectively, V, such that

$$\mathbf{U}:=\left[U^{\perp} \; U\right], \quad \mathbf{V}:=\left[V^{\perp} \; V\right],$$

are orthogonal matrices. The equations (1.67) can now be separated into coupled delay-differential and delay-difference equations: a premultiplication of (1.67) with $\mathbf{U}^T$ and the substitution

$$x=\mathbf{V}\,[x_1^T \; x_2^T]^T,$$

with $x_1(t)\in\mathbb{R}^{n-\nu}$ and $x_2(t)\in\mathbb{R}^{\nu}$, yield the coupled equations

$$\begin{aligned} E^{(11)}\dot{x}_1(t)&=\sum_{i=0}^{m} A_i^{(11)} x_1(t-\tau_i)+\sum_{i=0}^{m} A_i^{(12)} x_2(t-\tau_i), \\ 0&=A_0^{(22)} x_2(t)+\sum_{i=1}^{m} A_i^{(22)} x_2(t-\tau_i)+\sum_{i=0}^{m} A_i^{(21)} x_1(t-\tau_i), \end{aligned} \tag{1.73}$$

where

$$E^{(11)}=U^{\perp T} E V^{\perp} \tag{1.74}$$

and

$$\begin{aligned} A_i^{(11)}&=U^{\perp T} A_i V^{\perp}, \quad A_i^{(12)}=U^{\perp T} A_i V, \\ A_i^{(21)}&=U^{T} A_i V^{\perp}, \quad\;\; A_i^{(22)}=U^{T} A_i V, \quad i=0,\dots,m. \end{aligned} \tag{1.75}$$

Matrix $E^{(11)}$ is invertible, following from

$$\operatorname{rank}(E^{(11)}) = \operatorname{rank}(\mathbf{U}^T E \mathbf{V}) = \operatorname{rank}(E) = n - \nu.$$

In addition, Assumption 1.47 corresponds to the invertibility of matrix $A_0^{(22)}$.

Equations (1.73) are semi-explicit delay-differential algebraic equations of index 1, because delay-differential equations are obtained by differentiating the second equation. It precludes the occurrence of impulsive solutions [113]. Moreover, the invertibility of $A_0^{(22)}$ prevents the equations from being of advanced type and, hence, noncausal. The avoided potential causality problem is illustrated with the equations

$$\begin{cases} \dot{x}_1(t) = x_1(t) + x_2(t), \\ \quad 0 = 0\, x_2(t) + x_2(t-\tau) + x_1(t), \end{cases}$$

where determining $x_2(t)$ requires the knowledge of $x_1(t+\tau)$. Finally, we note that equation (1.71) satisfies Assumption 1.47.

A forward solution of (1.67) on the interval $[0,\ t_0]$, $t_0 > 0$, with initial condition $x(t) = \phi(t), t \leq 0$, is an absolutely continuous function that satisfies the differential equation (1.67) almost everywhere on the interval $[0,\ t_0]$. In [109] it is shown that for every initial condition $\phi \in X$, where

$$X := \left\{ \phi \in \mathscr{C}([-\tau_m,\ 0],\ \mathbb{R}^n) :\ U^T A_0 \phi(0) + \sum_{i=1}^{m} U^T A_i \phi(-\tau_i) = 0 \right\}, \tag{1.76}$$

and for every $t_0 > 0$, a forward solution $x(\phi)$ exists and is uniquely defined on $[0,\ t_0]$. The constraint in (1.76) expresses that the initial function must satisfy the second equation of (1.73) at $t = 0$. It reduces to an algebraic constraint in the delay free case.

1.3.2 ▪ Properties of the spectrum and connections with stability

In this section the spectral properties of equation (1.67) are discussed. In the technical derivation connections with the neutral equation

$$\begin{cases} E^{(11)} \dot{x}_1(t) = \sum_{i=0}^{m} A_i^{(11)} x_1(t-\tau_i) + \sum_{i=0}^{m} A_i^{(12)} x_2(t-\tau_i), \\ A_0^{(22)} \dot{x}_2(t) = -\sum_{i=1}^{m} A_i^{(22)} \dot{x}_2(t-\tau_i) - \sum_{i=0}^{m} A_i^{(21)} \dot{x}_1(t-\tau_i), \end{cases} \tag{1.77}$$

obtained by differentiating the second equation in (1.73), play an important role.

Exponential stability

Exponential stability of the null solution of (1.67) is defined as follows:

Definition 1.48. *The null solution of* (1.67) *is exponentially stable if and only if there exist constants* $\delta > 0$ *and* $\gamma > 0$ *such that*

$$\forall \phi \in X\ \forall t \geq 0\ \|x(\phi)(t)\| \leq \delta e^{-\gamma t} \|\phi\|_s.$$

Stability conditions can be expressed in terms of the position of the *characteristic roots*, i.e., the roots of the equation

$$\det \Delta(\lambda) = 0, \tag{1.78}$$

where Δ is the characteristic matrix,

$$\Delta(\lambda) := \lambda E - A_0 - \sum_{i=1}^{m} A_i e^{-\lambda\tau_i}.$$

In particular, we have the following result, whose proof can be found in Appendix 1 of [241].

Proposition 1.49. *The null solution of* (1.67) *is exponentially stable if and only if $c < 0$, where c is the* spectral abscissa,

$$c := \sup\{\Re(\lambda) : \det \Delta(\lambda) = 0\}. \tag{1.79}$$

Continuity of the spectral abscissa and strong stability

We discuss the dependence of the spectral abscissa of (1.67) on the delay parameters $\vec{\tau} = (\tau_1, \dots, \tau_m)$. In general, the function

$$\vec{\tau} \in (\mathbb{R}^+)^m \mapsto c(\vec{\tau}) \tag{1.80}$$

is not everywhere continuous, which carries over from the spectral properties of delay difference equations (see, e.g., [13, 242, 266]). In light of this we first outline properties of the function

$$\vec{\tau} \in (\mathbb{R}^+)^m \mapsto c_D(\vec{\tau}) := \sup\{\Re(\lambda) : \det \Delta_D(\lambda; \vec{\tau}) = 0\}, \tag{1.81}$$

with

$$\Delta_D(\lambda; \vec{\tau}) := U^T A_0 V + \sum_{i=1}^{m} U^T A_i V e^{-\lambda\tau_i}. \tag{1.82}$$

Note that (1.82) can be interpreted as the characteristic matrix of the delay-difference equation associated with the neutral equation (1.77).

The property that the function (1.81) is not continuous led in [264] to the smallest upper bound, which is *insensitive* to small delay changes.

Definition 1.50. *For $\vec{\tau} \in (\mathbb{R}^+)^m$, let the robust spectral abscissa $C_D(\vec{\tau}) \in \mathbb{R}$ be defined as*

$$C_D(\vec{\tau}) := \lim_{\epsilon \to 0+} c_D^{\epsilon}(\vec{\tau}),$$

where

$$c_D^{\epsilon}(\vec{\tau}) := \sup\{c_D(\vec{\tau} + \delta\vec{\tau}) : \delta\vec{\tau} \in \mathbb{R}^m \text{ and } \|\delta\vec{\tau}\| \le \epsilon\}.$$

Several properties of this upper bound on c_D are listed below (see [241] for an overview).

Proposition 1.51. *The following assertions hold:*

1. *The function*

$$\vec{\tau} \in (\mathbb{R}^+)^m \mapsto C_D(\vec{\tau})$$

is continuous.

2. *For every* $\vec{\tau} \in (\mathbb{R}^+)^m$, *the quantity* $C_D(\vec{\tau})$ *is equal to the unique zero of the strictly decreasing function*
$$\alpha \in \mathbb{R} \to f(\alpha;\, \vec{\tau}) - 1, \tag{1.83}$$
where $f : \mathbb{R} \to \mathbb{R}^+$ *is defined by*
$$f(\alpha;\, \vec{\tau}) := \max_{\vec{\theta} \in [0,\, 2\pi]^m} r_\sigma \left(\sum_{k=1}^{m} (U^T A_0 V)^{-1} (U^T A_k V) e^{-\alpha \tau_k} e^{j\theta_k} \right). \tag{1.84}$$

3. $C_D(\vec{\tau}) = c_D(\vec{\tau})$ *for rationally independent*[7] $\vec{\tau}$.

4. *For all* $\vec{\tau}_1, \vec{\tau}_2 \in (\mathbb{R}_0^+)^m$, *we have*
$$\operatorname{sign}(C_D(\vec{\tau}_1)) = \operatorname{sign}(C_D(\vec{\tau}_2)) := \Xi. \tag{1.85}$$

5. $\Xi < 0\ (> 0)$ *holds if and only if* $\gamma_0 < 1\ (> 1)$ *holds, where*
$$\gamma_0 := \max_{\vec{\theta} \in [0,\, 2\pi]^m} r_\sigma \left(\sum_{k=1}^{m} (U^T A_0 V)^{-1} (U^T A_k V) e^{j\theta_k} \right). \tag{1.86}$$

For the single delay case, some of the expressions can be simplified.

Corollary 1.52. *If* $m = 1$, *then we have*
$$C_D(\vec{\tau}) = \frac{1}{\tau_1} \log \left\{ r_\sigma((U^T A_0 V)^{-1} (U^T A_1 V)) \right\}$$
and
$$\gamma_0 = r_\sigma((U^T A_0 V)^{-1} (U^T A_1 V)).$$

We now come back to the DDAE (1.67), more precisely, to the properties of the spectral abscissa function (1.79). The lack of continuity of the spectral abscissa function (1.79) leads us again to an upper bound that takes into account the effect of small delay perturbations.

Definition 1.53. *For* $\vec{\tau} \in (\mathbb{R}_0^+)^m$, *let the* robust spectral abscissa $C(\vec{\tau})$ *be defined as*
$$C(\vec{\tau}) := \lim_{\epsilon \to 0+} c^\epsilon(\vec{\tau}), \tag{1.87}$$
where
$$c^\epsilon(\vec{\tau}) := \sup \{ c(\vec{\tau} + \delta \vec{\tau}) :\ \delta \vec{\tau} \in \mathbb{R}^m \text{ and } \|\delta \vec{\tau}\| \le \epsilon \}.$$

The following characterization of the robust spectral abscissa (1.58) constitutes the main result of this section.

[7] The m components of $\vec{\tau} = (\tau_1, \ldots, \tau_m)$ are rationally independent if and only if $\sum_{k=1}^{m} n_k \tau_k = 0$, $n_k \in \mathbb{Z}$, implies $n_k = 0\ \forall k = 1, \ldots, m$. For instance, two delays τ_1 and τ_2 are rationally independent if their ratio is an irrational number.

Proposition 1.54. *The following assertions hold:*

1. *The function*
$$\vec{\tau} \in (\mathbb{R}^+)^m \mapsto C(\vec{\tau}) \tag{1.88}$$
is continuous.

2. *For every $\vec{\tau} \in (\mathbb{R}^+)^m$, we have*
$$C(\vec{\tau}) = \max(C_D(\vec{\tau}), c(\vec{\tau})). \tag{1.89}$$

Proof. The proof is analogous to the proofs of Proposition 1.59 and Theorem 1.44. □

In line with the sensitivity of the spectral abscissa with respect to infinitesimal delay perturbations, which has been resolved by considering the robust spectral abscissa (1.58) instead, we define the concept of strong stability.[8]

Definition 1.55. *The null solution of* (1.67) *is strongly exponentially stable if there exists a number $\hat{\tau} > 0$ such that the null solution of*

$$E\dot{x}(t) = A_0 + \sum_{k=1}^{m} A_k x(t - (\tau_k + \delta\tau_k))$$

is exponentially stable for all $\delta\vec{\tau} \in (\mathbb{R}^+)^m$ satisfying $\|\delta\vec{\tau}\| < \hat{\tau}$ and $\tau_k + \delta\tau_k \geq 0$, $k = 1, \dots, m$.

The following result provides necessary and sufficient conditions.

Theorem 1.56. *The null solution of* (1.67) *is strongly exponentially stable if and only if $C(\vec{\tau}) < 0$, or, equivalently, $c(\vec{\tau}) < 0$ and $\gamma_0 < 1$, where γ_0 is defined by* (1.86).

We illustrate the above results with a numerical example.

Example 1.57. *Consider the system*

$$\begin{bmatrix} 1 & 0 \\ 0 & 0 \end{bmatrix} \dot{x}(t) = \begin{bmatrix} 0 & -\frac{1}{8} \\ -1 & 1 \end{bmatrix} x(t) + \begin{bmatrix} 0 & 0 \\ 0 & -a \end{bmatrix} x(t-\tau_1) \begin{bmatrix} 0 & 0 \\ 0 & \frac{1}{2} \end{bmatrix} x(t-\tau_2), \tag{1.90}$$

where a is a parameter. In Figure 1.9 *we plot the rightmost characteristic roots for $a = 1/4$ and two sets of delay values: $\vec{\tau} = (1,2)$ and $\vec{\tau} = (0.99, 2)$. The dotted line corresponds to the spectral abscissa $c((1,2))$, and the dashed line corresponds to $C_D((1,2))$. In this case the robust spectral abscissa satisfies*

$$C((1,2)) = c((1,2)).$$

Moreover, since the robust spectral abscissa is strictly negative, the null solution of (1.90) *is strongly exponentially stable.*

For parameter $a = 3/4$ the analysis has been repeated, and the results are displayed in Figure 1.10. *In this case we have*

$$C((1,2)) = C_D((1,2)) > 0 > c((1,2)),$$

and thus, although the zero solution is asymptotically stable for $\vec{\tau} = (1,2)$, the stability is not robust with respect to small delay perturbations. This property is consistent with the characteristic roots shown for the perturbed delay values $\vec{\tau} = (0.99, 2)$.

[8] This terminology is borrowed from the theory of neutral delay-differential equations [152, 264].

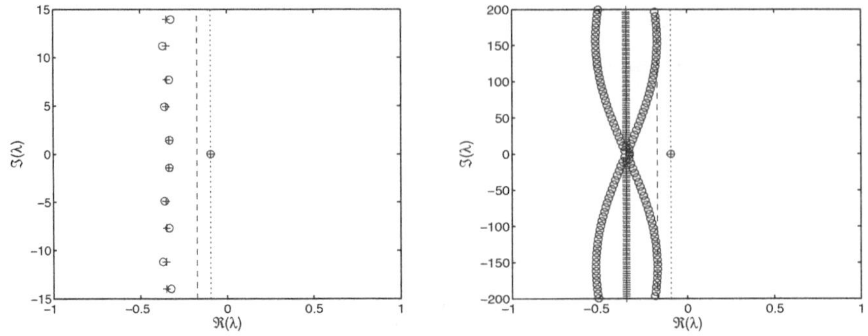

Figure 1.9. *Characteristic roots of* (1.90) *for* $a = 1/4$ *and delays* $\vec{\tau} = (1,2)$ *(pluses) and* $\vec{\tau} = (0.99,2)$ *(circles). The difference between the left and the right panes lies in the scaling of the vertical axis* [241].

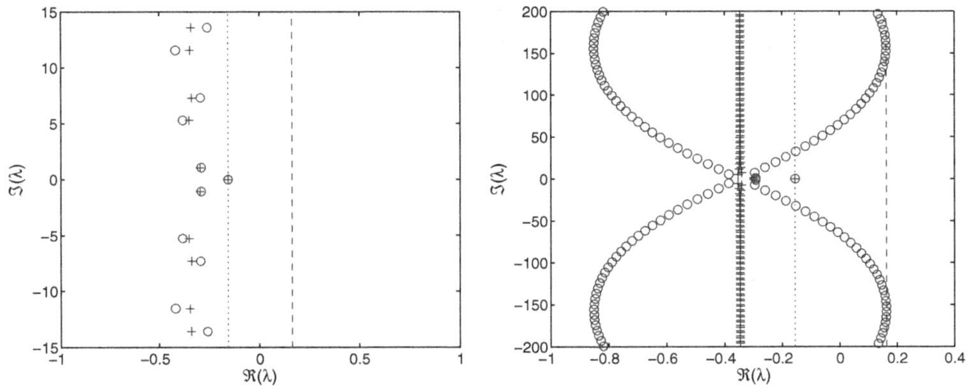

Figure 1.10. *Characteristic roots of* (1.90) *for* $a = 3/4$ *and delays* $\vec{\tau} = (1,2)$ *(indicated by +) and* $\vec{\tau} = (0.99,2)$ *(indicated by* o*)* [241].

1.4 ▪ Notes and references

We introduced some classes of linear time-delay systems and outlined spectral properties, thereby taking a stability analysis point of view and using eigenvalue plots to illustrate the main results. Our goal was not to present a complete theory, but to focus on these properties which play an important role in the rest of the book. Key references for the general theory of functional differential equations, of which the delay equations presented form a special class, are [148, 151, 196, 197, 93]. For a general introduction to infinite-dimensional systems from a system's theory point of view we refer the reader to [84].

The section on retarded systems is based on Cooke's chapter [78], with complements from [148, 151]. The terminology "piece of trajectories" in defining the state notion x_t of a time-delay system was suggested by Krasovskii in [200]. The construction of solutions using the "step-by-step" method follows closely the arguments in [290] (see also [145]). To the best of our knowledge, Bellman was the first to propose such a construction of the solutions in the context of delay-difference equations. For a more general discussion on the method of steps we refer the reader to [97]. Next, concerning Proposition 1.10, it is worth mentioning that other envelopes have been considered in [276] and [404], which could provide less conservative results. Proposition 1.12 can be found in [107], where the result was used for constructing appropriate linear controllers that improve the first delay interval, guaranteeing closed-loop stability under the assumption that the delay-free system can be stabilized by using the same controller. Finally, a different estimation of this upper bound is proposed in [303] by using a different interpretation of quasi-polynomials in terms of positive polynomials. Further remarks and comments on the distribution of zeros of the characteristic function for linear systems of retarded type can be found in Bellman and Cooke [23] (see also some discussions in Kolmanovskii and Myshkis [196]).

Sections 1.2.1–1.2.3 on the neutral equation are based on results of [151, 152, 153, 264, 242, 323], which are applied to (1.25) and further developed by taking into account the specific structure of this equation. The result in Proposition 1.35 is taken from [114]. Indeed, the corresponding characteristic equation is nothing else than the characteristic equation of the Fornasini–Marchesini model, corresponding to some particular class of $2-D$ systems. Next, further connections between the stability of the trivial solutions of the continuous-time delay-difference equations and the corresponding discrete-time delay-difference equations can be found in [358]. The results on delay sensitivity of Section 1.2.4 are based on [264]. To the best of our knowledge, a detailed analysis of continuity properties of the spectrum, in particular of the spectral abscissa, as presented in Section 1.2.5, is not performed in the existing literature, although some of the ideas and results are implicitly present in some papers (for instance in [132]). This clarifies why, in Section 1.2.5, the proofs are fully developed. For further reading on the delay sensitivity problem of stability we refer the reader to [266]. This reference also treats the case where not all delays are independent but may depend on a smaller number of independent parameters, e.g., $(\tau_1, \tau_2, \tau_3) = (r_1, r_2, r_1 + r_2)$.

Finally, the presented theory of delay-differential algebraic equations in Section 1.3 and, in particular, the analysis of the sensitivity of stability with respect to small perturbations, are based on [241, 141] and the references therein. Please note that the simplest class of models described by delay-differential algebraic equations is represented by the so-called *lossless propagation* models [336]. Roughly speaking, *lossless propagation* is understood as the phenomenon associated with long transmission lines for (some) physical signals. In engineering, this problem is strongly related to *electric* and *electronic applications*, e.g., circuit structures consisting of multipoles connected through LC transmission lines (a long list of references may be provided, starting with the pioneering paper of Brayton [40]. We note that the lossless propagation occurs also for *nonelectric* "signals" such as water, steam, or gas flows and pressures. With respect to this, we may cite, for instance, the pioneering (but almost forgotten) papers of Kabakov and Sokolov [182] on steam pipes for combined heat-electricity generation, the waterhammer case, and many others. In all these cases, the mathematical model is described by a *mixed initial* and *boundary value problem* for hyperbolic partial differential equations. Further examples with respect to these topics can be found in Hale and Verduyn Lunel [151] as well as the paper of Halanay and Răsvan [147] (see also the monographs [335] and [290]).

Throughout this chapter we have restricted ourselves to linear time-delay systems with *pointwise* (discrete) delays (although most of the results in this chapter can be generalized to more general classes of functional differential equations), because the remainder of the book is almost exclusively devoted to problems involving this type of delays. If other types of delays are considered in particular places (e.g., in Chapters 12 and 14), then it concerns systems which can be brought into a form with pointwise delays using particular model transformations in the sense of [218, 278] (see also [131]). At such places, the necessary additions to the stability theory of this chapter will be explicitly provided.

Chapter 2

Computation of characteristic roots

2.1 ▪ Introduction

We consider the linear time-delay system of retarded type described by the delay-differential equation

$$\dot{x}(t) = A_0 x(t) + \sum_{k=1}^{m} A_k x(t - \tau_k), \tag{2.1}$$

where $x(t) \in \mathbb{R}^n$ is the state variable at time t, $A_k \in \mathbb{R}^{n \times n}$, $k = 0, 1, \dots, m$, are real matrices, and τ_i, $i = 1, \dots, m$, represent the time-delays.

We recall that the characteristic roots of (2.1) can be found as the solutions of the nonlinear eigenvalue problem

$$\Delta(\lambda)v = 0, \quad \lambda \in \mathbb{C},\ v \in \mathbb{C}^n, v \neq 0, \tag{2.2}$$

where $\Delta(\lambda)$ is the characteristic matrix,

$$\Delta(\lambda) := \lambda I - A_0 - \sum_{k=1}^{m} A_k e^{-\lambda \tau_k}. \tag{2.3}$$

Several numerical approaches for computing characteristic roots are based on the discretization of either the solution operator associated with (2.1) (cf. [41, 42]) or the infinitesimal generator of the solution operator (cf. [43, 59, 42, 392]).

The solution operator $\mathcal{T}(t)$, $t \geq 0$, associated with (2.1), is given by (1.4), and its infinitesimal generator $\mathcal{A}$ by (1.6). Relations between the characteristic roots and the spectrum of the solution operator $\mathcal{T}$ and its infinitesimal generator $\mathcal{A}$ are described in detail in Chapter 1. Summarizing the characteristic roots are the solutions of the eigenvalue problem

$$\mathcal{A}\phi = \lambda\phi, \quad \lambda \in \mathbb{C},\ \phi \in \mathcal{C}([-\tau_{\max}, 0], \mathbb{R}^n),\ \phi \neq 0, \tag{2.4}$$

where $\tau_{\max}$ is the maximal delay. Hence, inferred from the equivalent representation of (2.1) as (1.7), the characteristic roots can be computed as either the solutions of the finite-dimensional nonlinear eigenvalue problem (2.2), or as the solutions of the infinite-dimensional linear eigenvalue problem (2.4). The characteristic roots and the spectrum of the solution operator $\mathcal{T}(t)$, $\sigma(\mathcal{T}(t))$ are related by

$$\det \Delta(\lambda) = 0 \Leftrightarrow \lambda = \frac{1}{t} \ln \mu, \quad \mu \in \sigma(\mathcal{T}(t)) \backslash \{0\} \quad \forall t > 0.$$

The above relations lead to the idea of transferring the problem of computing approximations of characteristic roots to a corresponding eigenvalue problem for a suitable matrix discretization of the solution operator or its infinitesimal generator.

In [101] Engelborghs and Roose proposed a method for computing characteristic roots, based on discretizing the solution operator using a linear multistep (LMS) time-integration method (see, e.g., [22] for a survey on time-integration of DDEs), and implemented it in the software package DDE-BIFTOOL for the bifurcation and stability analysis (cf. [100]). This package contains a heuristic to determine the step-length in order to compute all characteristic roots in a given right half plane [42]. In [391] Verheyden, Luzyanina, and Roose described an improved step-length heuristic to reduce the size of the algebraic eigenvalue problem. This new heuristic has been incorporated in DDE-BIFTOOL v.2.03.

In [43] Breda, Maset, and Vermiglio presented an approach for computing characteristic roots, based on a discretization of the infinitesimal generator via a spectral method. The approach was implemented in the software package TRACE-DDE [46]. This software package does not include an automatic choice of the number of discretization points. In [43] it is suggested to determine the number of discretization points based on the bounds describing the local, asymptotic convergence of the individual characteristic roots; however, this requires an accurate knowledge of error constants. In this chapter, the relation between a spectral discretization method and a rational approximation is used to derive a procedure that can automatically select the number of discretization points as small as possible, but sufficiently large to compute all characteristic roots in a given right half plane. The approach also exploits the property that the error on individual characteristic roots can be removed by Newton corrections on the nonlinear equations (2.2). In this way the role of solving the discretized eigenvalue problem reduces to generating sufficiently good estimates for all characteristic roots in the half plane under consideration, instead of fully accurate solutions.

Throughout the chapter we make the following assumption:

Assumption 2.1. *The delays are ordered, and the maximum delay, τ_m, is equal to one.*

Note that Assumption 2.1 can be made without losing generality: if it is initially not satisfied, one can always resolve this by rescaling the characteristic equation.

The structure of the chapter is as follows. In Section 2.2 we discuss methods for computing characteristic roots of (2.1), which are based on a spectral discretization. The emphasis is on the problem of robustly computing all characteristic roots in a prescribed right half plane. In Section 2.3 extensions towards neutral equations and delay-differential algebraic equations are addressed. In Section 2.4 methods applicable to large-scale problems are outlined. Finally, in Section 2.5 the conclusions are presented.

2.2 ▪ Computing all characteristic roots in a right half plane

2.2.1 ▪ Spectral discretization of the eigenvalue problem

First, we review the spectral discretization of the infinitesimal generator as in [43], we describe its properties, and we outline the algorithm used to compute characteristic roots.

A spectral discretization approach for computing characteristic roots

We consider the spectral discretization of the eigenvalue problem (2.4), using a mesh Ω_N of $N+1$ distinct points in the interval $[-\tau_m, 0]$,

$$\Omega_N = \{\theta_{N,i},\ i = 1,\dots,N+1\}, \tag{2.5}$$

with

$$-\tau_m \leq \theta_{N,1} < \cdots < \theta_{N,N+1} = 0.$$

The discretization is based on replacing the space $\mathscr{C}([-\tau_m, 0], \mathbb{R}^n)$ by the space X_N of discrete functions defined on the grid Ω_N. Specifically, any $\phi \in \mathscr{C}([-\tau_m, 0], \mathbb{R}^n)$ is discretized into the block vector $\mathbf{x} \in X_N$ with components

$$\mathbf{x}_i = \phi(\theta_{N,i}) \in \mathbb{C}^n, \quad i = 1, \ldots, N+1.$$

Let $\mathscr{L}_N \mathbf{x}$, $\mathbf{x} \in X_N$, be the unique $\mathbb{C}^n$-valued interpolating polynomial of degree less than or equal to N, satisfying

$$(\mathscr{L}_N \mathbf{x})(\theta_{N,i}) = \mathbf{x}_i, \quad i = 1, \ldots, N+1.$$

Operator $\mathscr{A}$, given by (1.6), can now be approximated by matrix $\mathbf{A}_N : X_N \rightarrow X_N$, defined by

$$\begin{aligned} (\mathbf{A}_N \mathbf{x})_i &= (\mathscr{L}_N \mathbf{x})'(\theta_{N,i}), \quad i = 1, \ldots, N, \\ (\mathbf{A}_N \mathbf{x})_{N+1} &= A_0 (\mathscr{L}_N \mathbf{x})(0) + \sum_{k=1}^{m} A_k (\mathscr{L}_N \mathbf{x})(-\tau_k). \end{aligned}$$

By using the Lagrange representation of $\mathscr{L}_N \mathbf{x}$, the explicit expression

$$\mathbf{A}_N = \begin{bmatrix} a_{1,1} & \cdots & a_{1,N+1} \\ \vdots & \ddots & \vdots \\ a_{N+1,1} & \cdots & a_{N+1,N+1} \end{bmatrix} \in \mathbb{C}^{n(N+1) \times n(N+1)} \tag{2.6}$$

is obtained, where

$$a_{il} = \begin{cases} \ell_l'(\theta_{N,i}) I_n, & l \in \{1, \ldots, N+1\}, \quad i \in \{1, \ldots, N\}, \\ \sum_{k=0}^{m} A_k \ell_l(-\tau_k) & l \in \{1, \ldots, N+1\}, \quad i = N+1. \end{cases} \tag{2.7}$$

The functions ℓ_l in (2.7) are the Lagrange polynomials relative to Ω_N, i.e., polynomials of degree N satisfying

$$\ell_l(\theta_{N,i}) = \begin{cases} 1, & i = l, \\ 0, & i \neq l, \end{cases} \quad i, l \in \{1, \ldots, N+1\}. \tag{2.8}$$

The discretization allows us to approximate the operator eigenvalue problem (2.4) by the matrix eigenvalue problem

$$\mathbf{A}_N \mathbf{x} = \lambda \mathbf{x}, \quad \lambda \in \mathbb{C}, \quad \mathbf{x} \in \mathbb{C}^{n(N+1)}, \quad \mathbf{x} \neq 0. \tag{2.9}$$

In what follows, the grid points in Ω_N are chosen as

$$\theta_{N,i} = \frac{\tau_m}{2}(\alpha_{N,i} - 1), \quad \alpha_{N,i} = -\cos \frac{\pi i}{N+1}, \quad i = 1, \ldots, N+1, \tag{2.10}$$

for which spectral convergence of eigenvalues of $\mathbf{A}_N$ to corresponding eigenvalues of $\mathscr{A}$, i.e., an approximation error $\mathscr{O}(N^{-N})$, is observed. This can be explained by the fact that grid points (2.10) have the same asymptotic distribution as a grid of Chebyshev extremal points, for which spectral convergence has been proven in [43]. We refer the reader to [377] for the connections between convergence properties and the distribution of grid points. As shown in [177] the use of the grid (2.10) further allows us to rewrite the eigenvalue problem of $\mathbf{A}_N$ as

$$(\Sigma_N - \lambda \Pi_N)\mathbf{c} = 0, \quad \lambda \in \mathbb{C}, \quad \mathbf{c} \in \mathbb{C}^{(N+1)n}, \quad \mathbf{c} \neq 0, \tag{2.11}$$

where

$$\Pi_N = \frac{\tau_m}{4} \begin{bmatrix} \frac{4}{\tau_m} & \frac{4}{\tau_m} & \frac{4}{\tau_m} & \cdots & & \cdots & \frac{4}{\tau_m} \\ 2 & 0 & -1 & & & & \\ & \frac{1}{2} & 0 & -\frac{1}{2} & & & \\ & & \frac{1}{3} & 0 & \ddots & & \\ & & & \frac{1}{4} & \ddots & -\frac{1}{N-2} & \\ & & & & \ddots & 0 & -\frac{1}{N-1} \\ & & & & & \frac{1}{N} & 0 \end{bmatrix} \otimes I_n \tag{2.12}$$

and

$$\Sigma_N = \begin{bmatrix} R_0 & R_1 & \cdots & R_N \\ & I_n & & \\ & & \ddots & \\ & & & I_n \end{bmatrix}, \tag{2.13}$$

with

$$R_i = A_0 + \sum_{k=1}^{m} A_k T_i\left(-2\frac{\tau_k}{\tau_m} + 1\right), \quad i = 0, \ldots, N.$$

Here the functions T_i, $i = 0, \ldots, N$ are the Chebyshev polynomials of the first kind and order i. Because Π_{N_1} and Σ_{N_1} are submatrices of Π_{N_2} and Σ_{N_2} for $N_2 > N_1$, an increase of the number of the discretization points can be dealt with by simply extending the matrices. Note that this property forms the basis of the iterative method described in [177] and outlined in Section 2.4.

Summing up the above results, the characteristic roots of (2.1) appear either as the solutions of the finite-dimensional nonlinear eigenvalue problem (2.2) or as the solutions of the infinite-dimensional linear eigenvalue problem (2.4), which can be discretized into (2.9) or (2.11). Both viewpoints can be combined in a computational scheme. The discretization of the linear infinite-dimensional problem allows us to obtain estimates for all characteristic roots in a region where the approximation is accurate. These estimates can subsequently be corrected by a local method acting on the nonlinear equations (2.2). This brings us to the following algorithm.

ALGORITHM 2.1. Computation of characteristic roots.

1. Fix N and compute the eigenvalues of the pencil (Σ_N, Π_N), defined by (2.12)–(2.13).

2. Correct these approximate characteristic roots by applying Newton's method to equations (1.31).

We conclude with a central result, on which the analysis in the following sections strongly relies. The following proposition provides an interpretation of the approximation of $\mathscr{A}$ by $\mathbf{A}_N$, or, equivalently, by the pencil (Σ_N, Π_N), in terms of an approximation of the exponential functions in the characteristic matrix (2.3).

Proposition 2.2. *For $\lambda \in \mathbb{C}$, let $p_N(\cdot;\,\lambda)$ be the polynomial of degree N satisfying*

$$\begin{cases} p_N(0;\,\lambda)=1, \\ p_N'(\theta_{N,i};\,\lambda)=\lambda p_N(\theta_{N,i};\,\lambda), \quad i=1,\ldots,N. \end{cases} \tag{2.14}$$

Moreover, let

$$\Delta_N(\lambda) := \lambda I - A_0 - \sum_{k=1}^{m} A_k p_N(-\tau_k;\,\lambda). \tag{2.15}$$

Then the following statements are equivalent:

$$1.\ \exists \mathbf{x} \in \mathbb{C}^{(N+1)n},\ \mathbf{x} \neq 0 :\ (\lambda I - \mathbf{A}_N)\mathbf{x} = 0. \tag{2.16}$$

$$2.\ \exists v \in \mathbb{C}^n,\ v \neq 0 :\ \Delta_N(\lambda) v = 0. \tag{2.17}$$

The proof of Proposition 2.2 can be found in Section 3 of [43]. In the appendix of [140] it has been shown that the functions

$$\lambda \mapsto p_N(-\tau_i;\,\lambda), \quad i=1,\ldots,m, \tag{2.18}$$

are proper rational functions of order N, with *common poles* and with the property of uniform convergence on compact sets to the functions

$$\lambda \mapsto e^{-\lambda \tau_i}, \quad i=1,\ldots,m. \tag{2.19}$$

Note that the latter property is expected since conditions (2.14) are mixed interpolation/collocation conditions for the function $[-1,\ 0] \ni t \mapsto e^{\lambda t}$. Hence, the effect of the discretization of $\mathscr{A}$ can be interpreted as the effect of a rational approximation of the characteristic matrix.

2.2.2 ▪ Approach to determine the number of discretization points

We derive a procedure for computing *all* characteristic roots in a given right half plane, i.e., the set

$$\{\lambda \in \mathbb{C} :\ \Re(\lambda) \geq r\}, \tag{2.20}$$

where $r \in \mathbb{R}$. This problem is well posed since the number of characteristic roots of (2.1) in any right half plane is finite; see Chapter 1. When taking a spectral discretization approach combined with local corrections, as described in Section 2.2.1, the problem can be translated into an automatic selection of the number of discretization points, N, in Algorithm 2.1. On the one hand, the value of N should not be chosen too large, since (Σ_N, Π_N) are $n(N+1) \times n(N+1)$ matrices; hence computing all eigenvalues with a general-purpose eigensolver has complexity $\mathcal{O}(nN)^3$. On the other hand, the characteristic roots in the half plane under consideration need to be sufficiently well approximated by eigenvalues of (Σ_N, Π_N) such that Newton's method converges to the desired solutions. The latter imposes a lower bound on the value of N.

The proposed procedure to determine N is based on the connection between a spectral discretization and a rational approximation of exponential functions, described by Proposition 2.2. The main steps can be summarized as follows. First, a region is determined which contains all characteristic roots. Subsequently, the number of discretization points is selected in such a way that the rational approximation is accurate in this region. For clarity of presentation the procedure is explained by means of the problem of computing all characteristic roots in the right half plane

$$\{\lambda \in \mathbb{C} :\ \Re(\lambda) \geq 0\}. \tag{2.21}$$

The general case (2.20) can be treated by a preliminary shifting of the origin of the complex plane, i.e., by the substitution $\lambda \leftarrow \lambda - r$. For sake of conciseness we omit all proofs. These can be found in [409].

Estimate the region containing all characteristic roots in the right half plane

We start with a technical result.

Proposition 2.3. *All characteristic roots of* (2.1) *satisfying* $\Re(\lambda) \geq \xi$, $\xi \in \mathbb{R}$, *belong to the set*

$$\Omega_\xi := \left\{ \bigcup \sigma\left(A_0 + \sum_{k=1}^{m} A_k z_k\right) : z_k \in \mathbb{C}, |z_k| \leq e^{-\xi\tau_k}, k = 1, \ldots, m \right\} \bigcap \{\lambda \in \mathbb{C} : \Re(\lambda) \geq \xi\}. \tag{2.22}$$

The boundary of Ω_ξ *is included in*

$$\Psi_\xi := \left(\bigcup_{\vec{\omega} \in [0, 2\pi)^m} \sigma\left(A_0 + \sum_{k=1}^{m} A_k e^{-\xi\tau_k} e^{j\omega_k}\right)\right) \bigcap \{\lambda \in \mathbb{C} : \Re(\lambda) \geq \xi\}. \tag{2.23}$$

In order to estimate the location of characteristic roots in the right half plane (2.21) we can directly use the set Ψ_0, i.e., Ψ_ξ for $\xi = 0$. However, this estimate may be very conservative if the real part of λ is large, since its derivation is based on the bounds $|e^{-\lambda\tau_k}| \leq 1, k = 1, \ldots, m$. In our implementation we therefore split up the right half plane into two regions

$$\{\lambda \in \mathbb{C} : 0 \leq \Re(\lambda) \leq x\}, \ \{\lambda \in \mathbb{C} : \Re(\lambda) \geq x\}, \ x > 0, \tag{2.24}$$

which allows us to use the estimate Ψ_0 for the first region and the estimate Ψ_x for the second region. This brings us to the following corollary:

Corollary 2.4. *Let* $x > 0$. *The characteristic roots of* (2.1) *in the right half plane* (2.21) *belong to a region whose boundary is included in*

$$\Psi := \Psi_{[0,x]} \bigcup \Psi_x, \tag{2.25}$$

where

$$\Psi_{[0,x]} := \left(\bigcup_{\vec{\omega} \in [0, 2\pi)^m} \sigma\left(A_0 + \sum_{k=1}^{m} A_k e^{j\omega_k}\right)\right) \bigcap \{\lambda \in \mathbb{C} : 0 \leq \Re(\lambda) \leq x\} \tag{2.26}$$

and

$$\Psi_x := \left(\bigcup_{\vec{\omega} \in [0, 2\pi)^m} \sigma\left(A_0 + \sum_{k=1}^{m} A_k e^{-x\tau_k} e^{j\omega_k}\right)\right) \bigcap \{\lambda \in \mathbb{C} : \Re(\lambda) \geq x\}. \tag{2.27}$$

With the following example we illustrate the bounds obtained from Proposition 2.3 and Corollary 2.4. The choice of x in the software will be discussed in Section 2.2.2.

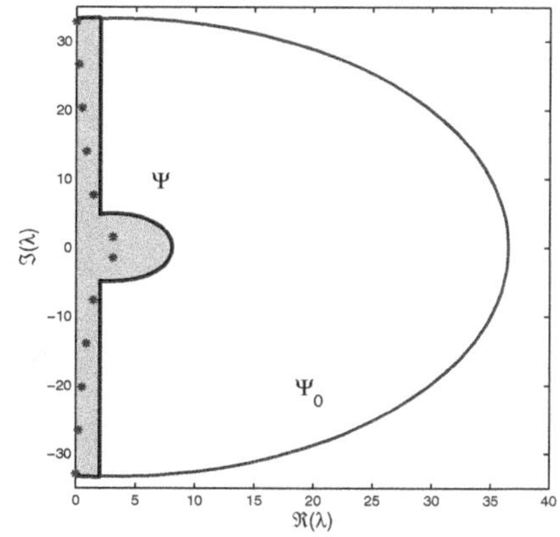

Figure 2.1. *Comparison of* Ψ_0 *and* Ψ [409].

Example 2.5. *Consider the time delay system*

$$\dot{x}(t) = 3.2x(t) - 33.34x(t-1). \tag{2.28}$$

In Figure 2.1 *the area in the right half plane which is bounded by the outer curve is the roots region estimated from* Ψ_0*, while the area bounded by the imaginary axis and the inner curve is the estimated region from* Ψ*, for* $x = 1.93$*. The characteristic roots in the right half plane are indicated with pluses* $(*)$*.*

Automatic selection of the number of discretization points

Let $\varepsilon > 0$ be a given (small) real number. For any integer $N \in \mathbb{N} \setminus \{0\}$, let the set S_N be defined by

$$S_N := \left\{ \lambda \in \mathbb{C} : \max_{t\in[-1,\,0]} \frac{|e^{\lambda t} - p_N(t;\,\lambda)|}{|e^{\lambda t}|} < \varepsilon \right\}, \tag{2.29}$$

where $p_N(\cdot;\,\lambda)$ is the polynomial of degree N, defined by conditions (2.14). As a first property of the set S_N, the following implication holds:

$$\lambda \in S_N \Rightarrow \frac{\left|e^{-\lambda\tau_k} - p_N(-\tau_k;\,\lambda)\right|}{|e^{-\lambda\tau_k}|} < \varepsilon, \quad k = 1,\ldots,m.$$

Hence, by Proposition 2.2 the effect of the discretization of $\mathscr{A}$ into $\mathbf{A}_N$ can be interpreted as the effect of an approximation of the exponential functions in the characteristic matrix, with a relative accuracy better than ε inside S_N. As a second property, the set S_N is independent of the system matrices and the delays, and can be computed beforehand. Its shape is illustrated in Figure 2.2 for $\varepsilon = 0.05$ and various values of N.

The automatic selection of the number of discretization points in Algorithm 2.1 is based on choosing N as small as possible such that the condition

$$\Psi \subseteq S_N \tag{2.30}$$

is satisfied, where Ψ is given as in (2.25). Figure 2.3 explains this principle for system (2.28). In what follows, we give more details about the computation of N.

Because both Ψ and S_N are symmetric with respect to the real axis and we are interested in computing the characteristic roots in the right half plane, we can restrict ourselves to

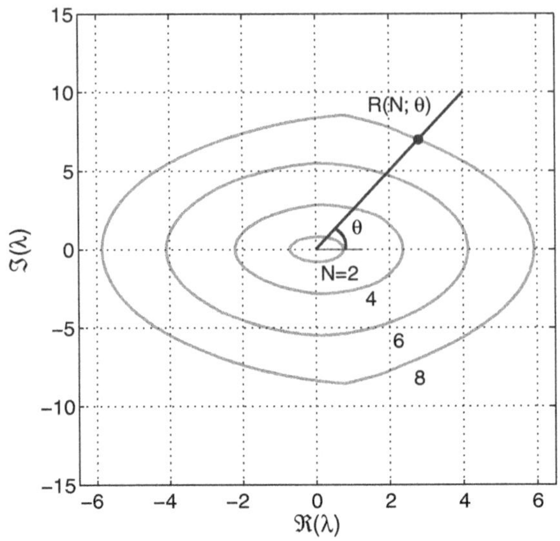

Figure 2.2. *The boundary of the region S_N for different values of N* [409].

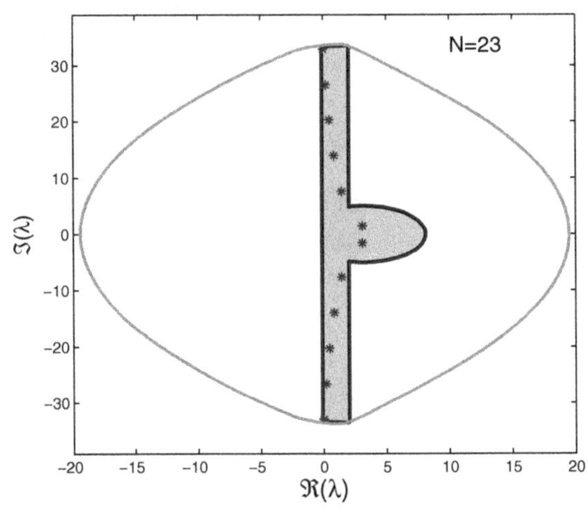

Figure 2.3. *Principle behind the selection of N. The shadowed area is the region bounded by* Ψ. *The region bounded by the outer curve is S_N, where $N = 23$ is the smallest integer for which* (2.30) *holds* [409].

the first quadrant. For $\theta \in [0, \pi/2]$ and $N \in \mathbb{N} \setminus \{0\}$ we define

$$R(N;\, \theta) := \left\{ r > 0 :\ re^{j\theta} \in \partial S_N \right\}, \tag{2.31}$$

where ∂S_N is the boundary of S_N. This quantity is shown in Figure 2.2. As indicated by this figure, the function $N \mapsto R(N;\, \theta)$ has linear growth in N for all values of θ and, accordingly, (2.31) can be accurately approximated as

$$R(N;\, \theta) \approx b(\theta) + a(\theta)N. \tag{2.32}$$

Similarly to the set S_N, the functions a and b do not depend on the system data and can be computed in advance. They are shown in Figure 2.4.

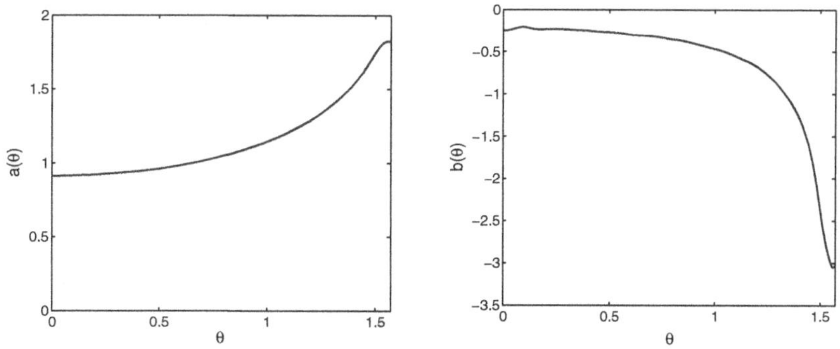

Figure 2.4. *The functions $\theta \to a(\theta)$ and $\theta \to b(\theta)$, corresponding to $\varepsilon = 0.05$* [409].

For any $z \in \mathbb{C}$ satisfying $\angle z \in [0, \pi/2]$, we define N_z as the smallest value of N for which $z \in S_N$. From (2.32) it follows that

$$N_z = \left\lceil \frac{|z| - b(\angle z)}{a(\angle z)} \right\rceil .$$

Finally, we can compute the minimal value of N such that (2.30) holds as

$$N = \max\{N_z : z \in \Psi, \angle z \in [0, \pi/2]\}. \tag{2.33}$$

Implementation aspects and the choice of parameters

In the definition of S_N, (2.29), it is not necessary to choose ε very small. The only requirement is that the approximation of the exponential functions by p_N be sufficiently accurate such that the resulting approximation error of the characteristic roots can be removed by Newton's method, in the second step of Algorithm 2.1. Intensive numerical experiments have led to the value $\varepsilon = 0.05$. Note that for the selected N, ε is an upper bound on the approximation error of the exponential functions holding for the whole region S_N, while the local error improves exponentially when $|\lambda|$ tends to zero.

In order to evaluate (2.33), the set Ψ is discretized by restricting $\vec{\omega}$ in (2.26)–(2.27) to a grid consisting of p equidistant points in every dimension. This leads to $p^m/2$ grid points on $[0, 2\pi)^m$, where the factor $1/2$ stems from the fact that the symmetry of Ψ with respect to the real axis can be exploited. Since for every grid point two eigenvalue problems of size $n \times n$ need to be solved, the computational complexity of determining N is of order

$$\mathcal{O}(p^m n^3). \tag{2.34}$$

In the software the user can set the value of p. The default value is $p = 20$.

Finally, the choice of $\varkappa$ in the partition (2.24), on which the set Ψ is based, depends on the choice of p and Ψ_0. It is computed as

$$\varkappa = \sin\left(\frac{2\pi}{p}\right) \times (\max\{\Re(\lambda) : \lambda \in \Psi_0\}). \tag{2.35}$$

Formula (2.35) is motivated by the following example.

Example 2.6. *We reconsider system* (2.28). *Figure* 2.5 *shows the sets Ψ_0 and Ψ, also displayed in Figure* 2.1, *along with points obtained by discretizing the interval* $[0, 2\pi)$ *in* (2.26)–(2.27)

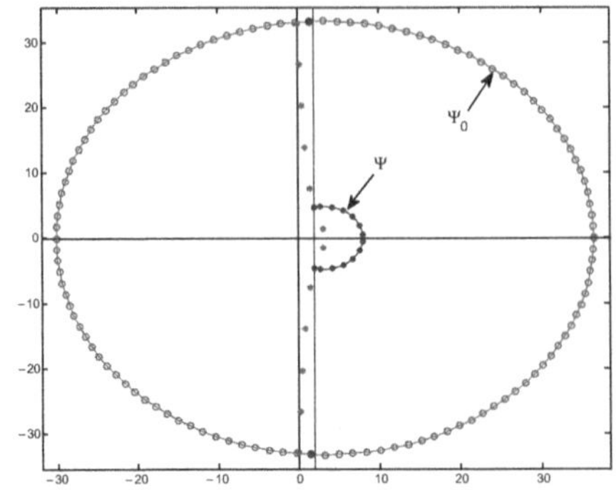

Figure 2.5. *Discretization (sampling) of the sets Ψ_0 and Ψ for system* (2.28). *The bold points are used for the evaluation of* (2.33) [409].

where $p = 120$ is chosen for the purpose of visualization. This discretization is used in evaluating (2.33). *On the one hand, it is preferable to have x small, to have an accurate estimate of the roots location and a good choice of N. On the other hand, if x is taken too small, there is a potential risk that no point of Ψ_0 with real part in the interval $[0,\ x]$ is sampled (i.e., no point of $\Psi_{[0,x]}$), whereas these points are usually critical, as is the case here. The set Ψ_0 is (part of) a circle, where the sampled points lie at an angular distance of $2\pi/p$. Condition* (2.35) *then guarantees that at least one sample is present with real part in the interval $[0,\ x]$.*

In general the shapes of the sets Ψ_0 and Ψ are much more complex than for the above example; however, it is clear that, to compensate for the effects of sampling, a lower bound for x should behave inversely proportional to p for large p, as is the case for (2.35). Formula (2.35) has been validated using a large number of benchmark problems.

2.2.3 ▪ Time-delay systems with commensurate delays

We show how the interdependence of the delays can be exploited for the case of commensurate delays [290], i.e.,

$$\tau_k = n_k \tau, \quad k = 1, \ldots, m, \tag{2.36}$$

where τ is a positive real number and $n_k \in \mathbb{N}$, $k = 1, \ldots, m$. In this way the characteristic matrix becomes

$$\Delta(\lambda) = \lambda I - A_0 - \sum_{k=1}^{m} A_k (e^{-\lambda\tau})^{n_k}. \tag{2.37}$$

First, we discuss the adaptation of the approach to determine N in the commensurate delays case. Second, we show how treating delays by commensurate delays may reduce the computational cost of locating characteristic roots for particular cases.

Adaptation of the method

The following results can be proven similarly to Proposition 2.3 and Corollary 2.4.

Proposition 2.7. *Assume that the delays satisfy* (2.36). *All characteristic roots satisfying $\Re(\lambda) \geq \xi$, $\xi \in \mathbb{R}$, belong to the set*

$$\Omega_{c_\xi} := \left\{ \bigcup \sigma\left(A_0 + \sum_{k=1}^{m} A_k z^{n_k} \right) : z \in \mathbb{C}, |z| \leq e^{-\xi\tau} \right\} \bigcap \{\lambda \in \mathbb{C} : \Re(\lambda) \geq \xi\}. \tag{2.38}$$

The boundary of Ω_{c_ξ} *is included in*

$$\Psi_{c_\xi} = \left(\bigcup_{\omega\in[0,2\pi)} \sigma\left(A_0 + \sum_{k=1}^{m} A_k e^{-\xi\tau n_k} e^{j\omega n_k} \right) \right) \bigcap \{\lambda | \Re(\lambda) \geq \xi\}. \tag{2.39}$$

Corollary 2.8. *Assume that the delays satisfy* (2.36). *Let* $x > 0$. *All characteristic roots in the right half plane* (2.21) *belong to a region whose boundary is included in*

$$\Psi_c = \Psi_{c_{[0,x]}} \bigcup \Psi_{c_x}, \tag{2.40}$$

where $\Psi_{c[0,x]}$ *and* Ψ_{c_x} *are given by*

$$\Psi_{c_{[0,x]}} = \left(\bigcup_{\omega\in[0,2\pi)} \sigma\left(A_0 + \sum_{k=1}^{m} A_k e^{j\omega n_k} \right) \right) \bigcap \{\lambda \in \mathbb{C} :\ 0 \leq \Re(\lambda) \leq x\} \tag{2.41}$$

and

$$\Psi_{c_x} = \left(\bigcup_{\omega\in[0,2\pi)} \sigma\left(A_0 + \sum_{k=1}^{m} A_k e^{-x\tau n_k} e^{j\omega n_k} \right) \right) \bigcap \{\lambda \in \mathbb{C} :\ \Re(\lambda) \geq x\}. \tag{2.42}$$

Using the above results, the procedure for determining N, discussed in Section 2.2.2, can be adapted, where the difference consists of replacing the estimate Ψ by Ψ_c with x determined as

$$x = \sin\left(\frac{2\pi}{p}\right) \times \left(\max\{\Re(\lambda) :\ \lambda \in \Psi_{c_0}\}\right).$$

The consequences are twofold:

1. In the discretization of (2.41) and (2.42) only a grid on the interval $[0, 2\pi)$ is needed. However, working with p grid points per period of the function to be approximated, as in Section 2.2.2, brings us to pn_m (where n_m is from $\tau_m = n_m\tau$) grid points over the interval $[0, 2\pi)$ because the function $\omega \mapsto e^{j\omega n_m}$ has period $2\pi/n_m$. Accordingly, the computational complexity of determining N changes from (2.34) to
$$\mathcal{O}(pn_m n^3). \tag{2.43}$$
2. The estimated region of the characteristic roots obtained from Ψ_c is in general more accurate than the estimate obtained from Ψ, because the interdependence of the delays is exploited. For comparison, note that $e^{j\omega_1}, \ldots, e^{j\omega_m}$ in (2.26)–(2.27) are treated as independent variables.

The latter property is illustrated with the following example:

Example 2.9. *Consider the equation*

$$\dot{x}(t) = -4x(t) - 5x(t-0.5) - 5x(t-1).$$

We take $p = 20$; *consequently,* $x = 0.309\left(\max\{\Re(\lambda) : \lambda \in \Psi_0 \text{ or } \Psi_{c_0}\}\right)$.

- *Treating the delays as commensurate delays,* $\tau_{1,2} = 0.5 n_{1,2}$*, for* $n_{1,2} = 1,2$*, yields*

$$\Psi_c = \Psi_{c_{[0,x]}} \bigcup \Psi_{c_x},$$

where $\Psi_{c[0,x]}$ *and* Ψ_{c_x} *are given by*

$$\Psi_{c_{[0,x]}} = \left(\bigcup_{\omega \in [0,2\pi)} \sigma\left(-4 - 5e^{j\omega} - 5\left(e^{j\omega}\right)^2\right)\right) \bigcap \{\lambda \in \mathbb{C} : 0 \leq \Re(\lambda) \leq x\}$$

and

$$\Psi_{c_x} = \left(\bigcup_{\omega \in [0,2\pi)} \sigma\left(-4 - 5e^{-0.5x} e^{j\omega} - 5e^{-x} e^{2j\omega}\right)\right) \bigcap \{\lambda \in \mathbb{C} : \Re(\lambda) \geq x\}.$$

- *Treating* $\tau_1 = 0.5$*,* $\tau_2 = 1$ *as independent delays yields*

$$\Psi := \Psi_{[0,x]} \bigcup \Psi_x,$$

where

$$\Psi_{[0,x]} := \left(\bigcup_{(\omega_1,\omega_2) \in [0,2\pi)^2} \sigma\left(-4 - 5e^{j\omega_1} - 5e^{j\omega_2}\right)\right) \bigcap \{\lambda \in \mathbb{C} : 0 \leq \Re(\lambda) \leq x\}$$

and

$$\Psi_x := \left(\bigcup_{(\omega_1,\omega_2) \in [0,2\pi)^2} \sigma\left(-4 - 5e^{-0.5x} e^{j\omega_1} - 5e^{-x} e^{j\omega_2}\right)\right) \bigcap \{\lambda \in \mathbb{C} : \Re(\lambda) \geq x\}.$$

In Figure 2.6 both of the sets Ψ *and* Ψ_c *are displayed.*

Approximation by commensurate delays for large values of m or n_m

The computational complexity of determining N for independent delays and commensurate delays is shown in (2.34) and (2.43), respectively. Hence, the method of determining N proposed in previous sections is particularly suitable if either m or n_m is small. If m or n_m is large, we can approximate the delays as multiples of a given number in the estimation of the roots' location: Given $q \in \mathbb{N} \setminus \{0\}$, we set a new basic delay $\tau_{new} = \tau_m / q$ and (re)calculate the associated integer $n_{k_{new}}$, i.e.,

$$n_{k_{new}} = [\tau_k / \tau_{new}] \quad \text{for } k = 1, \ldots, m.$$

Hence, the original delays are approximated as

$$\tau_{k_{new}} = n_{k_{new}} \tau_{new} \quad \text{for } k = 1, \ldots, m.$$

In the software this modification is done for independent delays if $m > 3$ and for commensurate delays if $n_m > 100$. The default value for q is $q = 100$. Note that, taking into account Assumption 1, we always get $\tau_{new} = 0.01$ and $n_{m_{new}} = 100$ in such a situation.

It is important to mention that in discretizing $\mathcal{A}$, in the computation of its eigenvalues and in the Newton corrections the exact delay values are used, whereas the delay approximations are only used in estimating the location of characteristic roots.

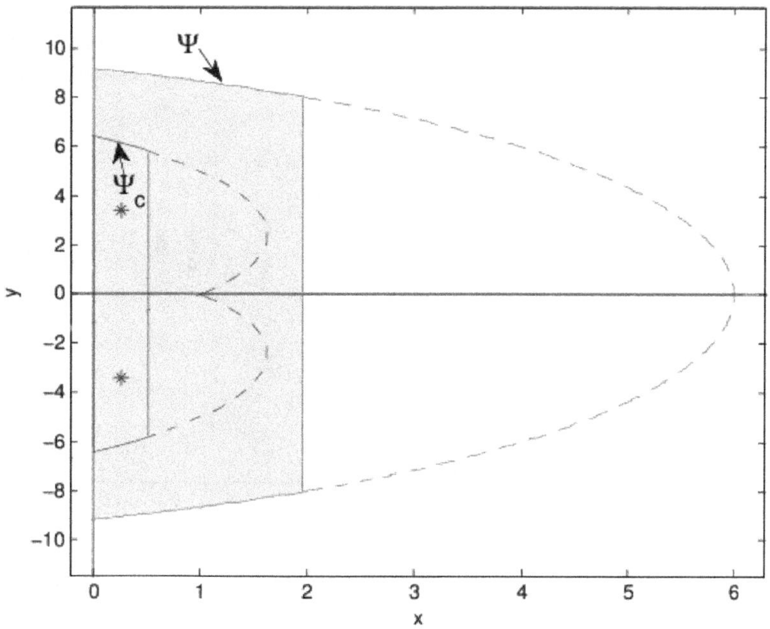

Figure 2.6. *The shadowed areas correspond to the sets Ψ and Ψ_c. The dashed curves correspond to Ψ_0 and Ψ_{c_0}, respectively* [409].

Example 2.10. *Consider a system with the following delays:* $\tau_1 = 0.5743$, $\tau_2 = 0.6753$, $\tau_3 = 0.8752$, $\tau_4 = 0.9390$, $\tau_5 = 0.9815$, $\tau_6 = 1.0000$.

- If we consider them as independent delays, the cost of selecting N has complexity $\mathcal{O}\left(p^6 n^3\right)$, according to (2.34).
- If we consider them as commensurate delays with basic delay $\tau = 0.0001$, consequently $n_6 = 10000$, the computational cost is $\mathcal{O}\left(10^4 p n^3\right)$, according to (2.43).
- If we set $\tau_{new} = 0.01$ and $n_{m_{new}} = 100$ to approximate the original delays, i.e., approximating them by $\tau_{1_{new}} = 0.57$, $\tau_{2_{new}} = 0.68$, $\tau_{3_{new}} = 0.88$, $\tau_{4_{new}} = 0.94$, $\tau_{5_{new}} = 0.98$, $\tau_{6_{new}} = 1$, the computational cost of selecting N is reduced to $\mathcal{O}\left(10^2 p n^3\right)$.

2.2.4 ▪ Numerical examples

We present examples to illustrate the effectiveness of the proposed procedure for computing all the characteristic roots in a given right half plane.

Example 2.11. *A system of DDEs with one delay which has been used in* [391]. *In this system,* $\tau_1 = 1$,

$$A_0 = \begin{bmatrix} -1 & 0 & 0 & 0 \\ 0 & 1 & 0 & 0 \\ 0 & 0 & -10 & -4 \\ 0 & 0 & 4 & -10 \end{bmatrix}, \quad \text{and} \quad A_1 = \begin{bmatrix} 3 & 3 & 3 & 3 \\ 0 & -1.5 & 0 & 0 \\ 0 & 0 & 3 & -5 \\ 0 & 5 & 5 & 5 \end{bmatrix}.$$

We aim at computing all characteristic roots of Example 2.11 with $\Re(\lambda) \geq -1.5$. Our procedure automatically chooses the number of discretization points as $N = 19$.

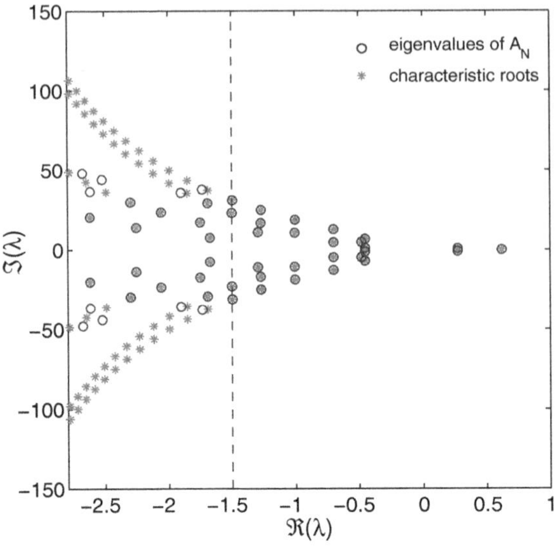

Figure 2.7. *Characteristic roots with* $\Re(\lambda) \geq -1.5$ [409].

Table 2.1. *Size of eigenvalue problem for* $\Re(\lambda) \geq r$.

r	DDE-BIFTOOL v.2.00	DDE-BIFTOOL v.2.03	Spectral method
0	184	48	16
−0.5	240	48	28
−1	332	60	36
−1.5	476	96	80
−2	716	140	136
−2.5	1104	220	204
−3	1740	360	340

Figure 2.7 shows that this number of N is sufficient. Also note that the selected N is very accurate: N is sufficient to compute all desired roots (i.e., with $\Re(\lambda) \geq -1.5$); however, it is not large enough to compute all roots with real part satisfying $\Re(\lambda) \geq -1.7$.

We compare the proposed method with the stability routine for equilibria of DDE-BIFTOOL v.2.00, and DDE-BIFTOOL v.2.03, in terms of the size of the discretized eigenvalue problem to be solved. Table 2.1 shows that the proposed procedure outperforms DDE-BIFTOOL 2.00 and leads to slightly better results than DDE-BIFTOOL 2.03.

Example 2.12. ***A system of DDEs with three delays,*** *where* $\tau_1 = 0.1$, $\tau_2 = 0.15$, $\tau_3 = 0.25$,

$$A_0 = \begin{bmatrix} -9.6713 & -9.7546 & -9.4913 \\ 1.8381 & 1.7961 & 9.5716 \\ 1.3647 & -2.7957 & -7.3561 \end{bmatrix}, \quad A_1 = \begin{bmatrix} 1.0115 & -9.3006 & 5.3222 \\ 7.2688 & -1.1960 & 9.9968 \\ 3.6508 & -1.2035 & -4.8507 \end{bmatrix},$$

$$A_2 = \begin{bmatrix} 7.7163 & 4.5911 & -5.5072 \\ -9.0056 & -0.0260 & -7.5404 \\ -3.3669 & 0.9332 & -0.2958 \end{bmatrix}, \quad A_3 = \begin{bmatrix} 7.4808 & -7.2571 & 9.4377 \\ 2.8285 & -7.1768 & -1.4221 \\ -1.0353 & 9.6519 & 5.1208 \end{bmatrix}.$$

To compute all characteristic roots of Example 2.12 with $\Re(\lambda) \geq -7$, the proposed procedure selects $N = 15$. Figure 2.8 shows that this value of N is sufficient; however, it

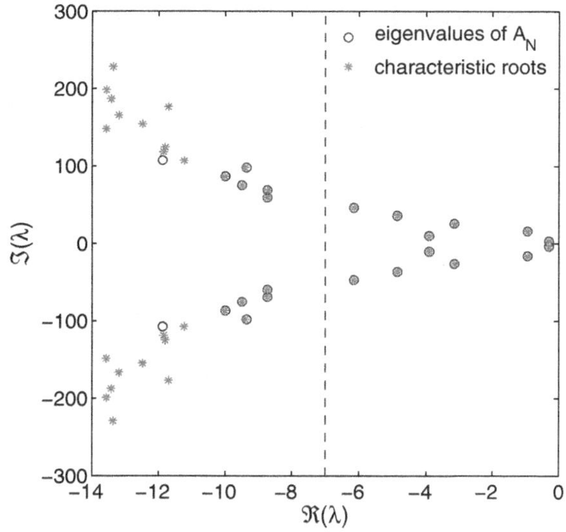

Figure 2.8. *Characteristic roots with* $\Re(\lambda) \geq -7$ [409].

Table 2.2. *Size of eigenvalue problem for* $\Re(\lambda) \geq r$.

r	DDE-BIFTOOL v.2.00	DDE-BIFTOOL v.2.03	Spectral method, independent delays	Spectral method, commensurate delays
−2	138	36	24	21
−3	156	36	27	24
−4	180	36	30	27
−5	207	36	33	30
−6	240	42	39	33
−7	282	48	48	39
−8	336	54	57	51
−9	399	66	66	57
−10	480	78	81	72

is also sufficient to compute all roots with real part larger than −10. The reason is that, when the number of delays increases, the estimation of the characteristic roots' location becomes more conservative.

Again we compared the proposed method with DDE-BIFTOOL v.2.00 and DDE-BIFTOOL v.2.03. In addition, we have also exploited the property that the delays in this example are commensurate delays:

$$\tau_{1,2,3} = 0.05 n_{1,2,3}, \quad \text{where } n_{1,2,3} = 2, 3, 5.$$

Table 2.2 shows the sizes of the discretized eigenvalue problems to be solved.

We use the proposed approach and DDE-BIFTOOL v.2.03 to compute all characteristic roots with $\Re(\lambda) \geq -1$. Our proposed procedure captures all the desired characteristic roots in the given right half plane, while DDE-BIFTOOL v.2.03 misses several desired roots and even the rightmost root. The reason lies in the large number of characteristic roots. The discretization of $\mathcal{T}(h)$ (where $h = 1/N$, N denotes the number of grid points in $[-\tau_m, 0]$) in DDE-BIFTOOL leads to approximations of $e^{\lambda h}$, where λ is the characteristic root. Hence, for small h the eigenvalues of $\mathcal{T}(h)$ are clustered around one, making the computation by discretizing $\mathcal{T}(h)$ numerically unstable.

2.3 ▪ Neutral systems and delay-differential algebraic equations

In Section 2.3.1 an approach for the computation of the characteristic roots is presented, which extends the spectral discretization approach of the previous section to the system of DDAEs (1.67). Connections with spectral discretization schemes for neutral equations are briefly discussed in Section 2.3.2. Finally, the computation of the robust spectral abscissa (1.58) is addressed in Section 2.3.3, motivated by the application to stability assessment.

2.3.1 ▪ A spectral discretization scheme

Consider the linear operator $\mathscr{A}$ on the space (1.76), defined by

$$\begin{aligned} &\mathscr{D}(\mathscr{A})=\left\{\phi\in X:\ \phi'\in\mathscr{C}([-\tau_m,\,0],\,\mathbb{R}^n),\ E\phi'(0)=A_0\phi(0)+\textstyle\sum_{k=1}^{m}A_k\phi(-\tau_k)\right\},\\ &\mathscr{A}\,\phi=\phi'. \end{aligned} \tag{2.44}$$

The eigenvalue problem for $\mathscr{A}$ is defined as

$$(\lambda I-\mathscr{A})z=0,\quad z\in\mathscr{D}(\mathscr{A}),\quad z\neq 0. \tag{2.45}$$

It can be verified that the characteristic roots of (1.67) are eigenvalues of the operator $\mathscr{A}$, as defined by (2.44).

The eigenvalue problem (2.45) can also be discretized using a spectral method (see, e.g., [241, 377, 43]). Given a positive integer N, we consider the grid (2.5) on the interval $[-\tau_m,\,0]$. In this way we can approximate the eigenvalue problem (2.45) by the generalized eigenvalue problem

$$(\lambda\mathbf{E}_N-\mathscr{A}_N)\mathbf{x}=0, \tag{2.46}$$

where

$$\mathbf{E}_N=\begin{bmatrix} I_{nN} & 0\\ 0 & E\end{bmatrix} \tag{2.47}$$

and matrix $\mathbf{A}_N$ is given by (2.6)–(2.7).

The following proposition provides an interpretation of the approximation of (2.45) by (2.46) in terms of an approximation of the characteristic matrix. It is the counterpart of Proposition 2.2.

Proposition 2.13. *For $\lambda\in\mathbb{C}$, let $p_N(\cdot;\,\lambda)$ be the polynomial of degree N satisfying*

$$\begin{cases} p_N(0;\,\lambda)=1,\\ p_N'(\theta_{N,i};\,\lambda)=\lambda p_N(\theta_{N,i};\,\lambda),\ \ 1\le i\le N.\end{cases} \tag{2.48}$$

Moreover, let

$$\Delta_N(\lambda):=\lambda E-A_0-\sum_{i=1}^{m}A_i p_N(-\tau_i;\,\lambda).$$

Then the following statements are equivalent:

1. $\exists\mathbf{x}\in\mathbb{C}^{(N+1)n},\ \mathbf{x}\neq 0:\ (\lambda\mathbf{E}_N-\mathscr{A}_N)\mathbf{x}=0;$ (2.49)
2. $\exists v\in\mathbb{C}^n,\ v\neq 0:\ \Delta_N(\lambda)v=0.$ (2.50)

Summing up the above results, the characteristic roots of (1.67) appear either as the solutions of the finite-dimensional nonlinear eigenvalue problem

$$\Delta(\lambda)v=0,\quad v\in\mathbb{C}^n,\quad v\neq 0, \tag{2.51}$$

or as the solutions of the infinite-dimensional linear eigenvalue problem (2.45), which can be discretized into (2.46). Both viewpoints can be combined in a computational scheme. The discretization of the linear infinite-dimensional problem allows us to obtain estimates for *all* characteristic roots in a region where the approximation is accurate. These estimates can subsequently be corrected by a local method acting on the nonlinear equations (2.51). This brings us to the following algorithm:

ALGORITHM 2.2. Computation of characteristic roots.

1. Fix N and compute the eigenvalues of the pencil $(\mathbf{E}_N, \mathbf{A}_N)$.

2. Correct these approximate characteristic roots by applying Newton's method to the equations (2.51).

In our implementation, the grid points in the discretization of (2.45) are chosen as (2.10); i.e., the nonzero grid points are the (scaled and shifted) zeros of the Chebyshev polynomial of the second kind and order N. With this choice of grid points the convergence of the individual eigenvalues of the pencil $(\mathbf{E}_N, \mathscr{A}_N)$ to corresponding characteristic roots of (1.67) is fast. More precisely, spectral accuracy (approximation error $\mathcal{O}(N^{-N})$) is obtained because asymptotic distribution is equal to the distribution of (scaled and shifted) zeros of Chebyshev polynomials. For the proof we refer the reader to the corresponding result for retarded time-delay systems in [43].

The interpretation of the discretized eigenvalue problem in Proposition 2.13, along with the fact that the functions (2.48) are independent on the system data (except for the dependence on τ_m, which can be removed by a preliminary scaling of the problem) allows us to choose the value of N necessary to capture all characteristic roots in a prescribed half plane $\{\lambda \in \mathbb{C} : \Re(\lambda) > \mu\}$, with $\mu > C_D$. In particular, the automatic selection of N in the software described in [241] is based on performing the following steps (which are similar to the procedure for retarded type systems in Section 2.2.2):

- Shifting the origin of the complex plane to $\lambda = \mu$. If necessary, an additional scaling is applied such that the maximum delay becomes one.

- Constructing a compact set Ω containing all characteristic roots in the closed right half plane. This set is derived from the property that all characteristic roots satisfying $\Re(\lambda) > \xi$, with $\xi \in \mathbb{R}$, belong to the set

$$\Omega_\xi = \Bigg\{ \lambda \in \mathbb{C} : \det\left(\lambda E - A_0 - \sum_{i=1}^{m} A_i z_i\right) = 0 \text{ for some } (z_1, \ldots, z_m) \in \mathbb{C}^m,$$
$$\text{satisfying } |z_i| \le e^{-\xi \tau_i}, i = 1, \ldots, m \Bigg\}.$$

- Selecting N as the smallest value for which $\Omega \subseteq S_N$, where

$$S_N := \left\{ \lambda \in \mathbb{C} : \max_{t \in [-1, 0]} \left| e^{\lambda t} - p_N(t; \lambda) \right| / |e^{\lambda t}| < \text{tol} \right\}$$

and tol > 0 is a tolerance. Note that the set S_N does not depend on the system matrices and delays and can be computed beforehand for a range of values of N.

By Proposition 2.13 the above procedure guarantees that the eigenvalues of the pencil $(\mathbf{E}_N, \mathbf{A}_N)$ in the set Ω have an interpretation as zeros of the characteristic equation, where the exponentials have been approximated with a relative accuracy better than tol. It is important to note that the tolerance should not be chosen too small, since this would blow up the size of the discretized eigenvalue problem. Moreover, the approximation error on the individual roots can still be corrected up to the desired precision in the second step of Algorithm 2.2. The only requirement on N is that all characteristic roots in Ω be sufficiently well approximated by eigenvalues of pencil $(\mathbf{E}_N, \mathbf{A}_N)$ such that Newton's method converges from these estimates. Taking into account this requirement, the form of the eigenfunctions, and the quality of the bounds that determine the set Ω, intensive experiments have led to the default value tol $= 0.05$ in the software.

2.3.2 ▪ Connection with the spectral discretization of neutral equations

We show that the discretization approach presented in Section 2.3.1 is advantageous compared to a (standard) spectral discretization of the neutral equations (1.77), whose spectrum consists of the spectrum of (1.67), possibly extended with a characteristic root at zero.

The approximate characteristic roots of the neutral equation (1.77), obtained by the spectral method described in [44], can be interpreted as the roots of the equation

$$\begin{aligned}\det\Bigg\{\lambda\begin{bmatrix} E^{(11)} & 0\\ 0 & -A_0^{(22)}\end{bmatrix} - \sum_{i=0}^{m}\begin{bmatrix} A_i^{(11)} & A_i^{(12)}\\ 0 & 0\end{bmatrix} p_N(-\tau_i;\,\lambda)\\ -\begin{bmatrix} 0 & 0\\ A_0^{(21)} & 0\end{bmatrix} p_N{}'(0;\,\lambda) - \sum_{i=1}^{m}\begin{bmatrix} 0 & 0\\ A_i^{(21)} & A_i^{(22)}\end{bmatrix} p_N{}'(-\tau_i;\,\lambda)\Bigg\},\end{aligned} \tag{2.52}$$

where the functions p_N are defined as in Proposition 2.13. A comparison of (2.52) with the characteristic equation of the neutral equation (1.77) shows that the effect of the discretization can be interpreted as the effect of the substitutions

$$e^{-\lambda\tau_i} \leftarrow p_N(-\tau_i;\,\lambda), \tag{2.53}$$

$$\left(\lambda e^{-\lambda\tau_i}\right) \leftarrow p_N{}'(-\tau_i;\,\lambda), \quad i = 1,\ldots,m, \tag{2.54}$$

in the characteristic matrix.

Similarly, the approximations obtained in the first step of Algorithm 2.2 correspond, by Proposition 2.13, to the roots of the equation

$$\det\left\{\lambda\begin{bmatrix} E^{(11)} & 0\\ 0 & 0\end{bmatrix} - \sum_{i=0}^{m}\begin{bmatrix} A_i^{(11)} & A_i^{(12)}\\ A_i^{(21)} & A_i^{(22)}\end{bmatrix} p_N(-\tau_i;\,\lambda)\right\} = 0, \tag{2.55}$$

which is obtained from the characteristic equation of (1.73) solely by the substitution

$$e^{-\lambda\tau_i} \quad\leftarrow\quad p_N(-\tau_i;\,\lambda),$$

implying

$$\left(\lambda e^{-\lambda\tau_i}\right) \quad\leftarrow\quad \lambda p_N(-\tau_i;\,\lambda), \quad i = 1,\ldots,m. \tag{2.56}$$

Clearly, the substitutions (2.56) are preferred on (2.54) because of the adverse effect of differentiating on the accuracy of a polynomial approximation. This is illustrated in

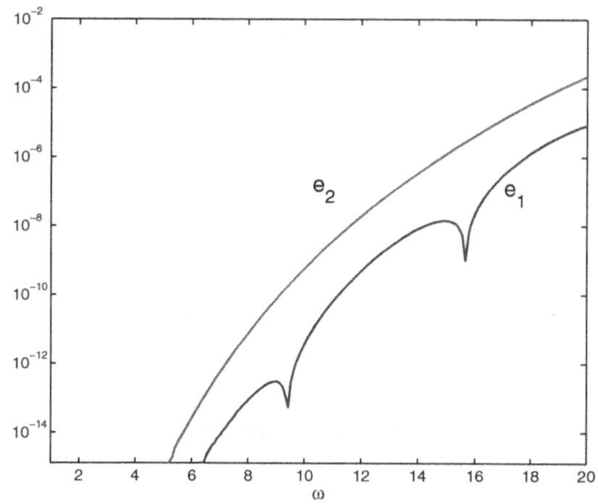

Figure 2.9. *Comparison of* (2.57) *and* (2.58) *as a function of* ω [241].

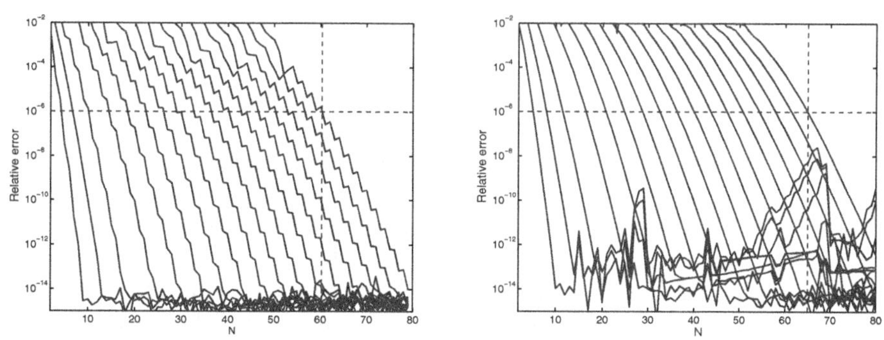

Figure 2.10. *Relative error on the* 32 *smallest characteristic roots obtained by a spectral discretization of* (2.59) *(left pane) and* (2.60) *(right pane)* [241].

Figure 2.9, which shows the results of the computation of the relative errors

$$e_1 := \frac{|j\omega e^{-j\omega} - j\omega p_N(-1;\, j\omega)|}{|j\omega e^{-j\omega}|} \tag{2.57}$$

and

$$e_2 := \frac{|j\omega e^{-j\omega} - p_N'(-1;\, j\omega)|}{|j\omega e^{-j\omega}|} \tag{2.58}$$

as a function of ω, for $N = 20$. Here, the functions p_N are defined on the grid (2.5), with $\tau_m = 1$.

The effect on the numerical stability of the type of equation discretized (DDAE or neutral equation) is illustrated with the following example:

Example 2.14. *Figure* 2.10 *(left) shows the relative error on the* 32 *smallest characteristic roots of the DDAE*

$$\begin{cases} \dot{x}_1(t) = x_2(t-1) \\ 0 = -x_2(t) + x_1(t) + \frac{1}{2}x_2(t-1), \end{cases} \tag{2.59}$$

approximated as in the first step of Algorithm 2.2, *in function of the number of discretization points, N. Differentiating the second equation in* (2.59) *leads to the neutral equation*

$$\begin{cases} \dot{x}_1(t) = x_2(t-1), \\ \dot{x}_2(t) = x_2(t-1) + \frac{1}{2}\dot{x}_2(t-1). \end{cases} \tag{2.60}$$

A spectral discretization of (2.60) *results in the approximation error displayed in Figure* 2.10 *(right). A comparison between the left and right panes of Figure* 2.10 *indicates that a method based on that discretizing* (2.60) *is much more prone to rounding errors. This can again be attributed to the approximation of derivatives in the right-hand side of* (2.60).

2.3.3 ▪ Computation of the robust spectral abscissa and stability assessment

Assessing the growth rate of solutions and stability (which both require checking the presence of characteristic roots in a right half plane) cannot always be reduced to computing the characteristic roots in a compact set, since system (1.67) may have an infinite series of characteristic roots whose imaginary parts tend to infinity, yet whose real parts have a finite limit. Moreover, the presence of such root chains may lead to a discontinuity of the spectral abscissa function with respect to the delays. The latter can be resolved by considering the robust spectral abscissa (1.58).

Based on the characterization (1.89), the robust spectral abscissa can be calculated by complementing the computation of characteristic roots, as presented in Section 2.3.1, with the evaluation of C_D. The latter is outlined in what follows.

We take a predictor-corrector approach to compute C_D, based on the second assertion of Proposition 1.51. In the prediction step we use the Dekker–Brent method [12] to find a zero of the function (1.83), where the function evaluations of f are approximated by restricting $\vec{\theta}$ in (1.84) to a grid. Note that for $m \geq 2$ we can write

$$f(\zeta;\, \vec{\tau}) = \max_{\vec{\theta}\in[0,\, 2\pi]^{m-1}} r_\sigma\left(H_1 e^{-\zeta\tau_1} + \sum_{k=2}^{m} H_k e^{-\zeta\tau_k} e^{j\theta_k}\right), \tag{2.61}$$

with

$$H_k = (U^T A_0 V)^{-1}(U^T A_k V), \quad k = 1, \ldots, m;$$

hence a grid on the space $[0,\, 2\pi]^{m-1}$ is sufficient. If a high accuracy of C_D is required, then one may want to use a local corrector, based on the equations

$$\begin{cases} \left(H_1 e^{-\zeta\tau_1} + \sum_{k=2}^{m} H_k e^{-\zeta\tau_k} e^{j\theta_k}\right) v = \lambda v, \\ u^*\left(H_1 e^{-\zeta\tau_1} + \sum_{k=2}^{m} H_k e^{-\zeta\tau_k} e^{j\theta_k}\right) = \lambda u^*, \\ n(u) = 1, \\ u^* v = 1, \\ \lambda^* \lambda = 1, \\ \Im\left(e^{-\zeta\tau_k} e^{j\theta_k}(u^* H_k v)\bar{\lambda}\right) = 0, \; k = 2, \ldots, m, \end{cases} \tag{2.62}$$

where $n(u) = 1$ is a normalization constraint. These equations express that for the desired value of ζ, the matrix

$$H_1 e^{-\zeta\tau_1} + \sum_{k=2}^{m} H_k e^{-\zeta\tau_k} e^{j\theta_k}$$

has an eigenvalue on the unit circle and that the derivatives of the modulus of this eigenvalue with respect to $\theta_2,\ldots,\theta_m$ are equal to zero. Since the (overdetermined) equations have an exact solution, the Gauss–Newton method exhibits quadratic convergence whenever the solution is isolated; see Section 10.2 of [305]. As an alternative, the systems of equations can be made square by the bordering technique.

Remark 2.15. *To assess strong stability of the null solution of* (1.67), *it is not necessary to compute* C_D *in addition to computing rightmost characteristic roots. By Theorem* 1.56 *it is only needed to check whether* $C_D < 0$, *which is equivalent to* $\gamma_0 < 1$. *Because*

$$\gamma_0 = f(0;\ \vec{\tau}),$$

this amounts to evaluating the function f *in one point, instead of finding the zero of* (1.83). *The value of* γ_0 *can be extracted from the solutions of* (2.61) *and* (2.62), *where* ζ *is set to zero and the equation* $\lambda^*\lambda = 1$ *in* (2.62) *is dropped.*

Remark 2.16. *The computational cost of* C_D *is dominated by the evaluation of the right-hand side of* (2.61), *where gridding leads to an exponential growth in the number of terms and restricts the approach to a small number of delays. However, in most practical problems, the number of delays to be considered in* (2.61) *is much smaller than the number of system delays,* m, *because most of the terms in* (2.61) *are zero. Note that in the context of feedback control a nonzero term corresponds to a high frequency feedthrough over the whole control loop.*

An alternative approach, based on a characterization in terms of positive polynomials matrices, is presented in [158].

2.4 ▪ Large-scale problems

We come back to the problem of computing characteristic roots of (2.1), where we assume that the matrices A_i, $i = 0,\ldots,m$, are large and sparse. The starting point is the representation (2.11)–(2.13) of the discretized eigenvalue problem.

2.4.1 ▪ Arnoldi's method

Arnoldi's method (first introduced in [7]) is one popular general purpose eigenvalue solver. In a traditional approach for computing characteristic roots, which we pursued in the previous sections, it is common to fix N and apply Arnoldi's method to an eigenvalue problem, which is here the generalized eigenvalue problem (2.11)–(2.13). This has the drawback that the matrices Σ_N and Π_N are large if N is large. Another drawback is that N has to be fixed before the iteration starts. The choice of N is a trade-off between computation time and accuracy, as the error decreases with growing N and the computation time grows with increasing N.

We will adapt a version of Arnoldi's method to the generalized eigenvalue problem $(\lambda\Pi_N - \Sigma_N)x = 0$ and exploit the structure in such a way that the method has the (somewhat remarkable) property of being, in a sense, independent of N. The constructed method is in this way a solution to the mentioned drawbacks and trade-offs of a traditional approach. It turns out that if we start Arnoldi's method corresponding to $\Pi_N^{-1}\Sigma_N$ in an appropriate way, the approximations after k iterations are identical to the approximations computed by k iterations of Arnoldi's method applied to the eigenvalue problem corresponding to *any* $N > k$. That is, it can be seen as carrying out an Arnoldi process associated with the limit $N \to \infty$. In light of this we will show in Section 2.4.3 that the method is also equivalent to Arnoldi's method applied to the infinite-dimensional operator $\mathscr{A}^{-1}$.

We will use the natural limit interpretation of Σ_N and Π_N. Let vec($\mathbb{C}^{n\times\infty}$) denote the set of all ordered infinite sequences of vectors of length n exponentially convergent to zero. The natural interpretation of the limits of the operators Σ_∞ and Π_∞ is with this notation $\Sigma_\infty : \mathrm{vec}(\mathbb{C}^{n\times\infty}) \to \mathrm{vec}(\mathbb{C}^{n\times\infty})$ and $\Pi_\infty : \mathrm{vec}(\mathbb{C}^{n\times\infty}) \to \mathrm{vec}(\mathbb{C}^{n\times\infty})$. We will call an element of $\mathrm{vec}(\mathbb{C}^{n\times\infty})$ an infinite vector and call Σ_∞ and Π_∞ infinite matrices. In many results and applications of Arnoldi's method, the method is implicitly equipped with the Euclidean scalar product. We will use a natural extension of the Euclidean scalar product to infinite vectors, $x^*y := \sum_{i=0}^{\infty} x_i^* y_i$, where x_i, y_i are the elements of x, y, respectively.

Arnoldi's method is a construction of an orthogonal basis of the set of linear combinations of a power sequence associated with matrix $A \in \mathbb{R}^{n\times n}$ and vector $b \in \mathbb{R}^n$,

$$\mathscr{K}_k(A,b) := \mathrm{span}\{b, Ab, \dots, A^{k-1}b\}.$$

This subspace is called a *Krylov subspace*. Arnoldi's method approximates eigenvalues of A by the eigenvalues of $H_k = V_k^* A V_k$ (which are called the Ritz values) where the columns of $V_k \in \mathbb{C}^{n\times k}$ form an orthonormal basis of $\mathscr{K}_k(A,b)$. The eigenvalues of H_k converge first to the extreme well-separated eigenvalues [338, Chapter VI]. In this book we are interested in eigenvalues close to the origin. In many applications those are not very well-separated, so convergence is expected to be slow. However, convergence can be drastically improved by applying Arnoldi's method to A^{-1}, as the eigenvalues of A near zero become typically well-separated extreme eigenvalues of A^{-1}, and so, fast convergence is expected. For this reason we will apply Arnoldi's method to form $\mathscr{K}_k(A^{-1},b)$ and call the inverse of the Ritz values the reciprocal Ritz values.

As in the definition of the Krylov subspace, the matrix vector product is an important component in Arnoldi's method. The underlying property which we will use next is the matrix vector product associated with the infinite matrix $\Sigma_\infty^{-1}\Pi_\infty$. It turns out to be structured in such a way that it has a closed form for a special type of vector. This is expressed in the following result of [177].

Theorem 2.17. *Suppose $\sum_{i=0}^{m} A_i$ is nonsingular. Let $\Sigma_\infty, \Pi_\infty$ be the limit of Σ_N and Π_N, defined by* (2.12)–(2.13), *and let $Y \in \mathbb{C}^{n\times k}$. Then*

$$\Sigma_\infty^{-1}\Pi_\infty \mathrm{vec}(Y, 0, \dots) = \mathrm{vec}(\hat{x}, Z, 0, \dots)$$

and

$$Z = Y L_k^T, \tag{2.63}$$

where $L_k \in \mathbb{R}^{k\times k}$ is given by

$$L_N = \frac{\tau_m}{4}\begin{bmatrix} 2 & 0 & -1 & & & \\ \frac{1}{2} & 0 & -\frac{1}{2} & & & \\ & \frac{1}{3} & 0 & \ddots & & \\ & & \frac{1}{4} & \ddots & -\frac{1}{N-2} & \\ & & & \ddots & 0 & \\ & & & & \frac{1}{N} & \end{bmatrix},$$

and

$$\hat{x} = \left(\sum_{j=0}^{m} A_j\right)^{-1} \left(\sum_{i=0}^{k-1} y_i - A_0 \sum_{i=0}^{k-1} z_i - \sum_{j=1}^{m} A_j \left(\sum_{i=0}^{k-1} T_{i+1}\left(1 - 2\frac{\tau_j}{\tau_m}\right) z_i\right)\right). \quad (2.64)$$

2.4.2 ▪ Algorithm

Now consider the power sequence for the operator $\Sigma_\infty^{-1}\Pi_\infty$ started with structured vector $\mathrm{vec}(w, 0, \ldots)$, $w \in \mathbb{R}^n$. From Theorem 2.17 we see that the nonzero part of the infinite vector grows by one vector (of length n) in each iteration such that at the jth step, the resulting infinite vector is $\mathrm{vec}(Y, 0, \ldots)$ where $Y \in \mathbb{R}^{n\times(j+1)}$.

Arnoldi's method builds the Krylov sequence vector by vector, where, in addition, the vectors are orthogonalized. In step k, the orthogonalization is a linear combination of the $k+1$st vector and the previously computed k vectors. Hence, the orthogonalization at the kth iteration does not change the general structure of the $k+1$st vector. This allows us to construct a scheme similar to Arnoldi's method where we dynamically increase the size of the basis vectors. In what follows we let V_k be the matrix consisting of the basis vectors and $v_{ij} \in \mathbb{C}^n$ be the vector corresponding to block element i, j.

The algorithm given in Algorithm 2.3 is (from the reasoning above) mathematically equivalent to Arnoldi's method applied to $\Sigma_\infty^{-1}\Pi_\infty$, as well as Arnoldi's method applied to the matrix $\Sigma_N^{-1}\Pi_N$, where N is larger than the total number of iteration steps taken. We use notation common for Arnoldi iterations: we let $\underline{H}_k \in \mathbb{C}^{(k+1)\times k}$ denote the dynamically constructed rectangular Hessenberg matrix, and let be $H_k \in \mathbb{C}^{k\times k}$ the corresponding $k \times k$ upper part.

ALGORITHM 2.3. A Krylov method for the delay eigenvalue problem.

Require $x_0 \in \mathbb{C}^n$ and time-delay system (2.1)
1. Let $v_1 = x_0/\|x_0\|_2$, $V_1 = v_1$, $k = 1$, $\underline{H}_0 =$ empty matrix,
2. Factorize $\sum_{i=0}^{m} A_i$
3. **for** $k = 1, 2, \ldots$ until converged
4. Let $\mathrm{vec}(Y) = v_k$
5. Compute Z according to (2.63) with sparse L_k
6. Compute $\hat{x}$ according to (2.64) using the factorization computed in step 2
7. Expand V_k with one block row (zeros)
8. Let $w_k := \mathrm{vec}(\hat{x}, Z)$, compute $h_k = V_k^* w_k$, and then $\hat{w}_k = w_k - V_k h_k$
9. Compute $\beta_k = \|\hat{w}_k\|_2$ and let $v_{k+1} = \hat{w}_k/\beta_k$
10. Let $\underline{H}_k = \begin{pmatrix} \underline{H}_{k-1} & h_k \\ 0 & \beta_k \end{pmatrix} \in \mathbb{C}^{(k+1)\times k}$
11. Expand V_k into $V_{k+1} = [V_k, v_{k+1}]$
12. **end for**
13. Compute the eigenvalues μ from the Hessenberg matrix H_k
14. Return approximations $1/\mu$

For the implementation aspects, including the extraction of eigenvectors and stopping criteria, we refer the reader to [177]

2.4.3 ▪ Equivalence with an infinite-dimensional operator setting

The original problem of finding λ is already a standard eigenvalue problem in the sense that λ is an eigenvalue of the infinite-dimensional operator $\mathcal{A}$. Since $\mathcal{A}$ is a linear operator, one can consider Arnoldi's method applied to $\mathscr{A}^{-1}$ in an abstract setting such that Arnoldi's method constructs a Krylov subspace of functions, i.e.,

$$\mathscr{K}_k(\mathscr{A}^{-1},\varphi) := \operatorname{span}\{\varphi,\mathscr{A}^{-1}\varphi,\ldots,\mathscr{A}^{-(k-1)}\varphi\}, \tag{2.65}$$

and projects on it. In this section we will see that Algorithm 2.3 has a complete interpretation in this setting if a scalar product is appropriately defined. The vector v_k in Algorithm 2.3 turns out to play the same role as the coefficients in the Chebyshev expansion. The Krylov subspace (2.65) is constructed for the inverse of $\mathscr{A}$. The inverse is explicitly given as follows:

Proposition 2.18. *The inverse of $\mathscr{A}$ exists if and only if $A_0+\sum_{i=1}^m A_i$ is nonsingular. Moreover, it is explicitly given as*

$$\begin{aligned}&\mathscr{D}(\mathscr{A}^{-1}) = \mathscr{C}([-\tau_m,0],\mathbb{R}^n),\\ &(\mathscr{A}^{-1}\phi)(\theta) = \int_0^\theta \phi(s)ds + C(\phi),\ \theta\in[-\tau_m,0],\quad \phi\in\mathscr{D}(\mathscr{A}^{-1}),\end{aligned} \tag{2.66}$$

where the constant $C(\phi)$ satisfies

$$C(\phi) = \left(A_0+\sum_{i=1}^m A_i\right)^{-1}\left[\phi(0)-\sum_{i=1}^m A_i\int_0^{-\tau_i}\phi(s)ds\right]. \tag{2.67}$$

Action and Krylov subspace equivalence

The key to the functional setting duality of this section is that we consider a scaled and shifted Chebyshev expansion of entire functions. Consider the expansion of two entire functions ψ and ϕ in a series of scaled Chebyshev polynomials,

$$\begin{aligned}\phi(t) &= \sum_{i=0}^\infty c_i T_i\left(2\frac{t}{\tau_m}+1\right)\\ \psi(t) &= \sum_{i=0}^\infty d_i T_i\left(2\frac{t}{\tau_m}+1\right),\quad t\in[-\tau_m,0].\end{aligned} \tag{2.68}$$

We will now see that the operation $\psi = \mathscr{A}^{-1}\phi$ can be expressed as a mapping of the coefficients, $c_0,c_1,\ldots$ and $d_0,d_1,\ldots$. This mapping turns out to reduce to the matrix vector product in Theorem 2.17. Suppose $\psi=\mathscr{A}^{-1}\phi$; then

$$\psi\in\mathscr{D}(\mathscr{A}), \tag{2.69}$$

$$\mathscr{A}\psi = \psi' = \phi. \tag{2.70}$$

From the fact that the derivative of a Chebyshev polynomial of the first kind can be expressed as a Chebyshev polynomial of the second kind, we note that

$$\psi'(t) = \sum_{i=1}^\infty \frac{2d_i i}{\tau_m} U_{i-1}\left(2\frac{t}{\tau_m}+1\right),$$

where U_i is the Chebyshev polynomial of the second kind and order i. Moreover, the relation between Chebyshev polynomials of the first kind and Chebyshev polynomials of the second kind, i.e., property

$$T_1(t) = \frac{1}{2}U_1(t),\quad T_i(t) = \frac{1}{2}U_i(t) - \frac{1}{2}U_{i-2}(t),\quad i\geq 2,$$

yields

$$\begin{aligned}\phi(t) &= c_0 U_0\left(2\frac{t}{\tau_m}+1\right) + \frac{1}{2}c_1 U_1\left(2\frac{t}{\tau_m}+1\right) + \sum_{i=2}^{\infty}\frac{c_i}{2}\left(U_i\left(2\frac{t}{\tau_m}+1\right)\right. \\ &\quad \left. - U_{i-2}\left(2\frac{t}{\tau_m}+1\right)\right) \\ &= c_0 U_0\left(2\frac{t}{\tau_m}+1\right) + \frac{1}{2}c_1 U_1\left(2\frac{t}{\tau_m}+1\right) \\ &\quad + \sum_{i=3}^{\infty}\frac{c_{i-1}}{2}U_{i-1}\left(2\frac{t}{\tau_m}+1\right) - \sum_{i=1}^{\infty}\frac{c_{i+1}}{2}U_{i-1}\left(2\frac{t}{\tau_m}+1\right).\end{aligned}$$

By matching coefficients in (2.70) we obtain the following recurrence relation for the coefficients:

$$d_i = \begin{cases} \frac{\tau_m}{4}(2c_0 - c_2), & i = 1, \\ \frac{\tau_m}{4}\frac{c_{i-1}-c_{i+1}}{i}, & i \geq 2. \end{cases} \tag{2.71}$$

From (2.69) and (2.70) we get

$$\phi(0) = A_0\psi(0) + \sum_{k=1}^{m} A_k \psi(-\tau_k).$$

Hence,

$$\sum_{i=0}^{\infty} c_i T_i(1) = \sum_{i=0}^{\infty}\sum_{k=0}^{m} A_k T_i\left(-2\frac{\tau_k}{\tau_m}+1\right) d_i = \sum_{i=0}^{\infty} R_i d_i. \tag{2.72}$$

By combining the results above and the fact that the Chebyshev coefficients of entire functions decay exponentially [377, Theorem 1] (as they are the Fourier coefficients of an entire function) we have proved the following relation:

Theorem 2.19. *Consider two entire functions ϕ and ψ and the associated Chebyshev expansion (2.68). Denote $c = (c_0^T, c_1^T, \ldots)^T$, $d = (d_0^T, d_1^T, \ldots)^T \in \operatorname{vec}(\mathbb{C}^{n\times\infty})$. Assume that $\sum_{i=0}^{m} A_i$ is nonsingular. Then the following two statements are equivalent:*

(i) $\psi = \mathscr{A}^{-1}\phi$.

(ii) $d = \Sigma_\infty^{-1}\Pi_\infty c$, *where c, d fulfill* (2.71)–(2.72).

Moreover, if $c = (c_0^T, \ldots, c_{k-1}^T, 0, \ldots, 0)^T = \operatorname{vec}(Y, 0, \ldots, 0)$, $Y \in \mathbb{R}^{n\times k}$, then $d = \operatorname{vec}(\hat{x}, Z, 0, \ldots, 0)$, where $\hat{x}$ and Z are the formulas in Theorem 2.17.

The equivalence between $\mathscr{A}^{-1}$ and $\Sigma_\infty^{-1}\Pi_\infty$ in Theorem 2.19 propagates to an equivalence between the corresponding Krylov subspaces. Let $\phi_0(t) = x_0$ be a constant function. Theorem 2.19 implies

$$\phi \in \mathscr{K}_k(\mathscr{A}^{-1}, \phi_0)$$

if and only if

$$\operatorname{vec}(c_0, c_1, \ldots, c_{k-1}) \in \mathscr{K}_k(\Sigma_k^{-1}\Pi_k, \operatorname{vec}(x_0, 0, \ldots, 0)),$$

where $c_0, \ldots, c_{k-1}$ are the Chebyshev coefficients of ϕ, i.e., (2.68).

Orthogonalization equivalence

We have seen that the matrix vector operation associated with $\Sigma_\infty^{-1}\Pi_\infty$ is equivalent to the operation $\mathscr{A}^{-1}$, in the sense that $\mathscr{A}$ applied to a function corresponds to a map between Chebyshev coefficients of two functions. The associated Krylov subspaces are also equivalent.

Arnoldi's method is a way to project on a Krylov subspace. In order to define the projection and compute the elements of H_k, we need to define a scalar product. In Algorithm 2.3 we use the natural way to define a scalar product, the Euclidean scalar product on the Chebyshev coefficients. In order to define a projection equivalent to Algorithm 2.3 in a consistent way, we define a scalar product in the function setting as

$$\langle \phi, \psi \rangle := c^* d = \sum_{i=0}^{\infty} c_i^* d_i, \tag{2.73}$$

where ϕ, ψ, c, and d are as in Theorem 2.19. We combine this with Theorem 2.19 to conclude that the Hessenberg matrix generated in Algorithm 2.3 and the Hessenberg matrix generated by the standard Arnoldi method applied to $\mathscr{A}^{-1}$ with the scalar product (2.73) are equal. A formal proof can, once again, be found in [177].

Theorem 2.20. *Let ϕ, ψ, c and d be as in Theorem 2.19. The Hessenberg matrix computed (with exact arithmetic) in Algorithm 2.3 is identical to the Hessenberg matrix of Arnoldi's method applied to $\mathscr{A}^{-1}$ with the scalar product (2.73) and the constant starting vector $\varphi(t) = x_0$.*

Using the above theorem, Algorithm 2.3 can be interpreted as Arnoldi's method applied to operator $\mathscr{A}^{-1}$ with the scalar product (2.73) for functions on $\mathscr{C}([-\tau_m, 0], \mathbb{R}^n)$ with a constant starting function, where the computation is carried out by mapping Chebyshev coefficients.

Summarizing, we have applied Arnoldi's method, which was originally designed for the standard linear eigenvalue problem, to nonlinear eigenvalue problems inferred from delay-differential equations. One of the key ingredients is the reformulation of the finite-dimensional nonlinear eigenvalue problem as an infinite-dimensional linear eigenvalue problem. This approach can be extended towards a broad class of nonlinear eigenvalue problem; see [179]. This includes also eigenvalue problems arising from neutral equations and delay-differential algebraic equations.

2.4.4 • Numerical examples

A scalar example

In order to illustrates some properties of Algorithm 2.3 we first consider the scalar DDE (also studied in [41])

$$\dot{x}(t) = (2 - e^{-2})x(t) + x(t-1). \tag{2.74}$$

When/if we apply Algorithm 2.3, the convergence of the reciprocal Ritz values is shown in Figure 2.11. First, we note that the application of the method does not involve a choice of N, which is consistent with the interpretation of the method as Arnoldi's method applied to the infinite-dimensional operator $\mathscr{A}^{-1}$. As a consequence, the number of converged roots can always be increased by doing more iterations, and the accuracy can be improved up to the desired precision. This is in contrast to a "discretize first approach," where the final accuracy is limited by the underlying approximation error. Second, in

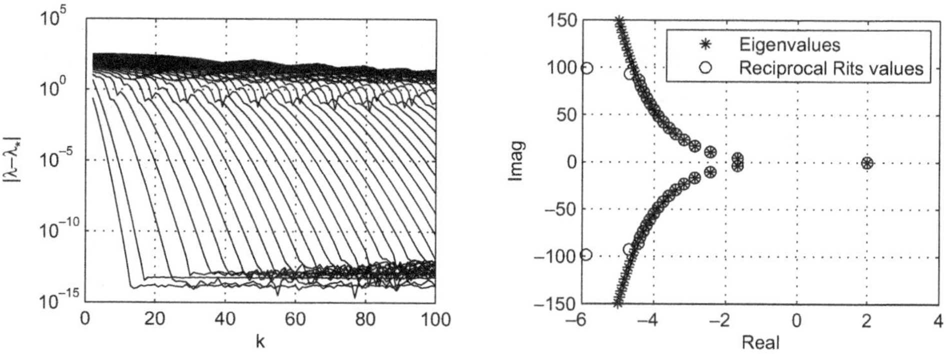

Figure 2.11. *Convergence and eigenvalue approximations of* $\mathcal{A}$ *for example* (2.74) *using Algorithm* 2.3. *(Left) Absolute error. (Right) Eigenvalues and reciprocal Ritz values* ($k = 50$) [177].

Figure 2.11 we observe geometric convergence. This observation is consistent with the convergence theory for the standard Arnoldi method [338] as well as the convergence theory for spectral discretization methods [377].

As the method corresponds to Arnoldi's method for $\mathcal{A}^{-1}$, it favors characteristic roots approximations close to the zero (or close to a given shift). Taking into account the typical location of the spectrum of retarded-type delay equations, described in Chapter 1, the method is suitable for computing rightmost eigenvalues and determining stability.

A large-scale problem

Numerical methods for sparse standard eigenvalue problems have turned out to be very useful because many applications are discretizations of partial differential equations (PDEs) which are (by construction) typically large and sparse. In order to illustrate that the presented method is particularly well suited for such problems, we consider a PDE with a delayed term. See [408] for many phenomena modeled as PDEs with delay, and see [45] for possibilities of combining a spatial discretization of the PDE with a discretization of the operator. Consider

$$\frac{\partial v(x,t)}{\partial t} = \frac{\partial^2 v(x,t)}{\partial x^2} + a_0(x)v(x,t) + a_1(x)v(\pi - x, t-1), \tag{2.75}$$

with $a_0(x) = -2\sin(x)$, $a_1(x) = 2\sin(x)$, and $v_x(0,t) = v_x(\pi,t) = 0$.

We discretize the PDE and construct a DDE by approximating the second derivative in space with central difference. This is a variant of a PDE considered in [45], where we have modified the delayed term. In the original formulation, the matrices are tridiagonal, which is not representative for our purposes. The convergence is visualized in Figure 2.12. For a system of size $n = 5000$, we carried out 100 iterations of Algorithm 2.3, in a total CPU time of 16.2s.

2.5 ▪ Notes and references

A spectral discretization of the infinitesimal generator of the solution operator is an established method for computing characteristic roots of linear time-delay systems. We proposed a procedure which can automatically select the number of discretization points, N, such that by solving the discretized eigenvalue problem, followed by Newton corrections on the resulting approximations, *all* roots in a given right half plane are computed with

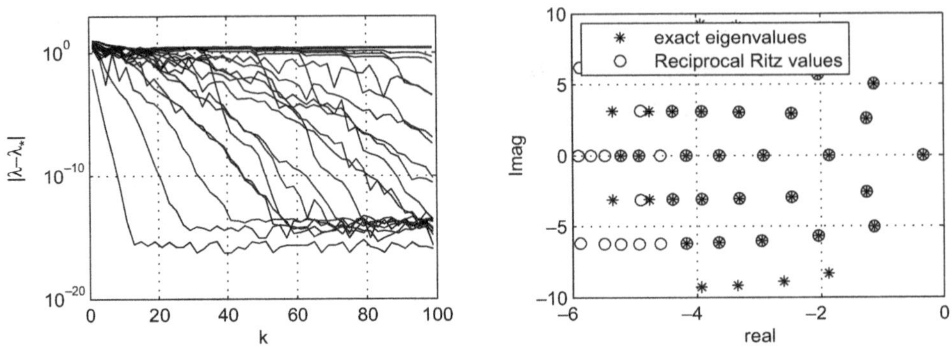

Figure 2.12. *Convergence and eigenvalue approximations of $\mathcal{A}$ for a spatial discretization of* (2.75) *with* $n = 5000$. *(Left) Absolute error. (Right) Eigenvalues and reciprocal of Ritz values* ($k = 50$) [177].

high accuracy. This makes the method very suitable in the context of stability assessment and bifurcation analysis, and it facilitates the use of the corresponding software.

Because from a computational point of view it is important not to overestimate the required number of discretization points, we have preferred developing and validating an accurate but reliable heuristic over constructing a (conservative) lower bound with a hard guarantee that all roots are found. However, the selection procedure has a solid mathematical foundation and builds on the relation between a spectral discretization and a rational approximation, where the heuristic aspect is reduced to the choice of one parameter, ε. Formulae (2.34) and (2.43) make it clear that the automatic selection of N is particularly suitable for systems with a small number of delays or with commensurate delays where the ratio between the largest delay and the basic delay (the value n_m in (2.36)) is small. However, problems with large m or n_m can be addressed by approximating the delays by commensurate delays in estimating the roots location, as we have seen.

The advantages of a discretization of the infinitesimal generator of the solution operator over a discretization of the solution operator are twofold. First, the proposed method yields directly characteristic roots approximations, while discretizing operator $\mathcal{T}(h)$ (with $h = 1/N$), which exhibits a clustering of the eigenvalues around one, may lead to a decrease of accuracy for large N, as we have illustrated. Second, the first-order form (1.7) is more suitable in the context of control problems (e.g., stabilization, $\mathcal{H}_2$ and $\mathcal{H}_\infty$ control), as control theory largely builds on a standard state-space representation of the system.

The principle used in Section 2.2.2 extends to compute all characteristic roots in any given compact set. The software corresponding to this book includes a routine for computing all characteristic roots in a rectangular region. Note that by subdivision, all roots in a very large region can be obtained; see, e.g., the problems addressed in [397].

We have also discussed the generalization to DDAEs. The approach follows along the same lines, yet special attention should be given to the potential sensitivity of stability with respect to small delay perturbations. In this context we outlined how to compute the robust spectral abscissa. Neutral equations can be addressed within the presented approach by transforming them to DDAEs, as explained in Section 1.3.

The method for determining all characteristic roots in a right half plane requires solving a large number of eigenvalue problems of dimension n-by-n (estimating the location of roots) and an eigenvalue problem of dimension $(N+1)n$-by-$(N+1)n$ (computing the characteristic roots). When the structure of the matrices is not exploited, the number of elementary operations amounts to $\mathcal{O}(n^3)$ and $\mathcal{O}((nN)^3)$. This precludes the applicability to large-scale problems, e.g., problems arising from the spatial discretization of PDEs with delay. Therefore, at the end of the chapter, we sketched an iterative, projection-based method, which only requires matrix-vector products and inner products of vectors and solving one linear system with the same, typically sparse, matrix of dimensions n-by-n in every iteration. This method has an interpretation as Arnoldi's method in a function setting. The method can be generalized to other types of eigenvalue problems exhibiting a nonlinearity in the eigenvalue parameter; see [179]. It is also closely related to the model reduction approach presented in [248].

This chapter is based on [409, 241, 177] and the references therein.

Chapter 3

Pseudospectra and robust stability analysis

3.1 ▪ Introduction

Closeness to instability is a key issue in understanding the behavior of physical systems subject to perturbations. The computation of pseudospectra has become an established tool in analyzing and gaining insight into this phenomenon (see, for instance, [378, 376, 164] and the references therein). More explicitly, pseudospectra of a system are sets in the complex plane to which eigenvalues or characteristic roots can be shifted under a perturbation of a given size. In the simplest case of a matrix (or linear operator) A, the ϵ-pseudospectrum $\Lambda_\epsilon(A)$ is defined as

$$\Lambda_\epsilon(A) := \{\lambda \in \mathbb{C} : \lambda \in \Lambda(A+P) \text{ for some } P \text{ with } \|P\| < \epsilon\}, \tag{3.1}$$

where Λ denotes the spectrum and $\|\cdot\|$ denotes an arbitrary matrix (or operator) norm. Equation (3.1) is known to be equivalent to the following:

$$\Lambda_\epsilon(A) = \{\lambda \in \mathbb{C} : \|R(\lambda, A)\| > 1/\epsilon\}, \tag{3.2}$$

where $R(\lambda, A) = (\lambda I - A)^{-1}$ denotes the corresponding resolvent operator.

Although most systems can be written in a first-order form, it is often advantageous to exploit the underlying structure in its analysis; for example, one may wish to compute pseudospectra of higher-order or delay-differential equations. In particular, this can be of importance in sensitivity investigations, where it is desirable to respect the structure of the governing system. For example, many physical problems involving vibration of structural systems and vibro-acoustics are modeled by second-order differential equations of the form

$$A_2\ddot{x}(t) + A_1\dot{x}(t) + A_0x(t) = 0,$$

where A_2, A_1, and A_0 represent mass, damping, and stiffness matrices, respectively. Stability is inferred from the characteristic roots, found as solutions of

$$\det(A_2\lambda^2 + A_1\lambda + A_0) = 0.$$

To understand the sensitivity of the characteristic roots with respect to complex perturbations with weights α_i applied to A_i, $i = 0, 1, 2$, the ϵ-pseudospectrum of the matrix polynomial $P(\lambda) = A_2\lambda^2 + A_1\lambda + A_0 \in \mathbb{C}^{n\times n}$ can be defined as

$$\begin{aligned}\Lambda_\epsilon(P) := \{\lambda \in \mathbb{C} \ : \ & (P(\lambda) + \Delta P(\lambda))x = 0 \text{ for some } x \neq 0 \\ & \text{and } \Delta P(\lambda) = \delta A_2\lambda^2 + \delta A_1\lambda + \delta A_0 \\ & \text{with } \delta A_i \in \mathbb{C}^{n\times n} \text{ and } \|\delta A_i\| < \epsilon\alpha_i,\ i = 0, 1, 2\}.\end{aligned} \tag{3.3}$$

We refer the reader to [374] for a survey on the quadratic eigenvalue problem, including numerical solutions and applications, and to [373] for pseudospectra of polynomial matrices. More recently, pseudospectra for nonlinear eigenvalue problems that correspond to linear time-delay systems have been defined and analyzed, starting with [125]. In its simplest form of one, fixed, discrete delay $\tau \in \mathbb{R}_+$, the characteristic matrix takes the form

$$Q(\lambda) := \lambda I - A_0 - A_1 e^{-\lambda\tau},$$

as we have seen in Chapter 1. Similar to (3.3) the associated pseudospectra is defined in [125] as

$$\begin{aligned}\Lambda_\epsilon(Q) := \{\lambda \in \mathbb{C} \ : \ & (Q(\lambda) + \Delta Q(\lambda))x = 0 \text{ for some } x \neq 0 \\ & \text{and } \Delta Q(\lambda) = \delta A_0 + \delta A_1 \epsilon^{-\lambda\tau} \\ & \text{with } \delta A_i \in \mathbb{C}^{n\times n} \text{ and } \|\delta A_i\| < \epsilon\alpha_i,\ i = 0,1\}.\end{aligned} \tag{3.4}$$

The aim of this chapter is twofold. The first goal is to present a unified theory for the definition and computation of pseudospectra of general nonlinear eigenvalue problems of the form

$$\det(F(\lambda)) = 0, \tag{3.5}$$

where the *characteristic matrix* F is given by

$$F(\lambda) := \sum_{i=0}^{m} A_i p_i(\lambda), \tag{3.6}$$

with $A_i \in \mathbb{C}^{n\times n}$ and p_i an entire function. The spectrum of F is defined as

$$\Lambda(F) := \{\lambda \in \mathbb{C} : \det(F(\lambda)) = 0\}. \tag{3.7}$$

It is easy to see that all the cases described above are included in this class of nonlinear eigenvalue problems. The second goal is to rephrase and interpret the results for the case of time-delay systems.

The chapter is organized as follows. In Section 3.2 we define and analyze pseudospectra for the nonlinear eigenvalue problem (3.5) under the assumption of complex perturbations on the coefficient matrices A_i. Thereby, the structure of the matrix function (3.6) is fully exploited. Various perturbation measures are discussed, and computable formulae are derived that are tractable from a numerical point of view. One of the practical applications concerns the associated complex *stability radius* of (3.6), that is, a measure of the distance to instability. See the book [164], and also [263, 324, 118], for the concept and various types of stability radii. Specifically, if we decompose $\mathbb{C}$ into two disjoint regions, a desired region $\mathbb{C}_d$ and an undesired region $\mathbb{C}_u$, the complex stability radius of (3.6) is defined as

$$\begin{aligned}r_{\mathbb{C}}(F;\ \mathbb{C}_d, \|\cdot\|_{\text{glob}}) := \inf_{\lambda\in\mathbb{C}_u} \inf_{\epsilon>0} \Bigg\{ \epsilon :\ & \det\left(\sum_{i=0}^{m}(A_i + \delta A_i)p_i(\lambda)\right) = 0 \\ & \text{for some } \Delta = (\delta A_0, \ldots, \delta A_m) \in \mathbb{C}^{n\times n\times(m+1)} \text{ with } \|\Delta\|_{\text{glob}} < \epsilon \Bigg\},\end{aligned} \tag{3.8}$$

where $\|\Delta\|_{\text{glob}}$ is a global measure of the perturbation Δ, which is a combination of the complex perturbations δA_i; this is discussed in detail in Section 3.2. In other words, $r_{\mathbb{C}}$

defines the norm of the smallest perturbation that destroys the $\mathbb{C}_d$-stability, that is, having all the roots confined to $\mathbb{C}_d$. Furthermore, it corresponds to the smallest ϵ value at which the ϵ-pseudospectrum has a nonempty intersection with $\mathbb{C}_u$. Note that for a system with continuous time, for example, the time-delay systems discussed in this book, $\mathbb{C}_d = \mathbb{C}_-$; for discrete time systems, $\mathbb{C}_d = \{\lambda \in \mathbb{C} : |\lambda| < 1\}$.

Section 3.3 in its turn is devoted to the definition and derivation of computable expressions for so-called *structured* pseudospectra and stability radii of (3.5). The main difference with the (unstructured) pseudospectra and stability radii discussed in Section 3.2 lies in the fact that, in *addition* to exploiting the structure of the matrix function (3.6), a particular structure can be imposed on the perturbations of the individual coefficient matrices A_i, at the price of a higher computational cost. The motivation stems from the observation that in many practical applications the coefficient matrices have a certain structure that should be respected in the sensitivity analysis [401]. In both Sections 3.2 and 3.3 the results are applied explicitly to time-delay systems. Two numerical examples are presented in Section 3.4.

In the analysis of Sections 3.2–3.4, we always assume that the uncertainty is concentrated in (elements of) the coefficient matrices. Therefore, in order to complete the presentation, we comment in Section 3.5 on stability radii in delay parameters. Some concluding remarks end the chapter.

3.2 ▪ Pseudospectra for nonlinear eigenvalue problems

Definitions and expressions for pseudospectra of (3.6) are presented. The connection with stability radii is clarified. Next, computational issues are discussed. Finally, the main results are rephrased for a class of time-delay systems.

3.2.1 ▪ Definition and expressions

We study the zeros of the equation given by (3.6), where $A_i \in \mathbb{C}^{n\times n}$, $i = 0,\dots,m$ and the functions $p_i : \mathbb{C} \to \mathbb{C}$, $i = 0,\dots,m$, are entire. In particular, we are interested in the effect of bounded perturbations of the matrices A_i on the position of the roots. For this, we analyze the perturbed equation,

$$\det\left\{\sum_{i=0}^{m}(A_i + \delta A_i)p_i(\lambda)\right\} = 0. \tag{3.9}$$

The first step in the robustness analysis is to define the class of perturbations under consideration, as well as a measure of the combined perturbation

$$\Delta := (\delta A_0,\dots,\delta A_m). \tag{3.10}$$

In this section we assume that the allowable perturbations δA_i, $i = 0,\dots,m$, are *complex* matrices, that is,

$$\Delta \in \mathbb{C}^{n\times n\times(m+1)}. \tag{3.11}$$

Introducing weights $w_i \in \overline{\mathbb{R}}_+, i = 0,\dots,m$, where $\overline{\mathbb{R}}_+ = \mathbb{R}_+ \cup \{\infty\}$, we define three global measures of the perturbations:

$$||\Delta||_{\text{glob}} := ||[w_0\ \delta A_0 \dots w_m\ \delta A_m]||_p \tag{3.12}$$

or

$$\|\Delta\|_{\text{glob}} := \left\| \begin{bmatrix} w_0\, \delta A_0 \\ \vdots \\ w_m\, \delta A_m \end{bmatrix} \right\|_p, \tag{3.13}$$

where $\|M\|_p$ is the induced matrix norm given by $\|M\|_p = \sup_{\|x\|_p=1} \|Mx\|_p$, $p \in \overline{\mathbb{N}} := \mathbb{N} \cup \{+\infty\}$. Notice that $w_j = \infty$ for some j means that no perturbation on A_j is allowed when the combined perturbation Δ is required to be bounded, that is, $w_j = \infty \Rightarrow \delta A_j = 0$ for some j. Finally, we also consider a measure of mixed type:

$$\|\Delta\|_{\text{glob}} := \left\| \begin{bmatrix} w_0 \|\delta A_0\|_{p_1} \\ \vdots \\ w_m \|\delta A_m\|_{p_1} \end{bmatrix} \right\|_{p_2}, \quad p_1, p_2 \in \overline{\mathbb{N}}_0. \tag{3.14}$$

For instance, when $p_2 = \infty$ and all weights are equal to one, the condition $\|\Delta\|_{\text{glob}} < \epsilon$ corresponds to the natural assumptions of taking perturbations satisfying $\|\delta A_i\|_{p_1} < \epsilon$, $i = 0, \ldots, m$. In this special case, (3.14) is also equal to the p_1 norm of the *block diagonal* perturbation matrix diag$(\delta A_0, \ldots, \delta A_m)$, considered in [324, 118] for polynomial matrices. Notice that, if all weights are finite, then the measures given by (3.12)–(3.14) are norms.

For any of the above definitions of $\|\Delta\|_{\text{glob}}$, we define the ϵ-pseudospectrum of (3.6) as follows:

Definition 3.1.

$$\begin{aligned} \Lambda_\epsilon(F; \|\cdot\|_{\text{glob}}) := \Bigg\{ \lambda \in \mathbb{C} :\ & \det\left(\sum_{i=0}^{m} (A_i + \delta A_i) p_i(\lambda) \right) = 0 \textit{ for some} \\ & \Delta = (\delta A_0, \ldots, \delta A_m) \in \mathbb{C}^{n \times n \times (m+1)} \textit{ with } \|\Delta\|_{\text{glob}} < \epsilon \Bigg\}. \end{aligned} \tag{3.15}$$

Defining the function $f : \mathbb{C} \to \overline{\mathbb{R}}^+$ as the inverse of the size of the smallest perturbation which shifts a root to λ, if such a perturbations exist, and zero otherwise, more precisely,

$$f(\lambda; \|\cdot\|_{\text{glob}}) = \begin{cases} 0 & \text{if } \det\left(\sum_{i=0}^m (A_i + \delta A_i) p_i(\lambda)\right) \neq 0 \ \forall \Delta \in \mathbb{C}^{n \times n \times (m+1)}, \\ +\infty & \text{if } \lambda \in \Lambda(F), \\ \left(\inf\left\{ \|\Delta\|_{\text{glob}} : \det\left(\sum_{i=0}^m (A_i + \delta A_i) p_i(\lambda)\right) = 0 \right\}\right)^{-1} & \text{otherwise}, \end{cases} \tag{3.16}$$

we can also define the ϵ-pseudospectra as

$$\Lambda_\epsilon(F; \|\cdot\|_{\text{glob}}) = \left\{ \lambda \in \mathbb{C} : f(\lambda; \|\cdot\|_{\text{glob}}) > \epsilon^{-1} \right\}. \tag{3.17}$$

The boundary of pseudospectra is thus formed by the level sets of the function f, which can be written in a computational form as follows [245, Theorem 1]:

Theorem 3.2. *For the perturbation measures* (3.12)–(3.14), *the function* (3.16) *satisfies*

$$f(\lambda;\, \|\cdot\|_{\text{glob}}) = \begin{cases} \left\|\left(\sum_{i=0}^{m} A_i p_i(\lambda)\right)^{-1}\right\|_{\alpha} \cdot \|w(\lambda)\|_{\beta}, & \lambda \notin \Lambda(F), \\ +\infty, & \lambda \in \Lambda(F), \end{cases}$$

where

$$w(\lambda) = \begin{bmatrix} \frac{p_0(\lambda)}{w_0} \\ \vdots \\ \frac{p_m(\lambda)}{w_m} \end{bmatrix} \tag{3.18}$$

and

$$\begin{array}{lll|l} \alpha = p, & \beta = p, & & \textit{perturbation measure } (3.12), \\ \alpha = p, & \beta = q, & \frac{1}{p}+\frac{1}{q}=1, & \textit{perturbation measure } (3.13), \\ \alpha = p_1, & \beta = q_2, & \frac{1}{p_2}+\frac{1}{q_2}=1 & \textit{perturbation measure } (3.14). \end{array}$$

The proof is frequency domain based. It relies on a feedback interpretation of the perturbed system and an explicit construction of worst case perturbations.

3.2.2 ▪ Connection with stability radii

As outlined in Section 3.1 the concept of stability radii given by (3.8) is closely related to pseudospectra. To clarify this relationship and to arrive at a computable formula, we need the following continuity property of the individual roots of (3.6) with respect to changes of matrices A_i, which can be once again shown using Rouché's theorem:

Proposition 3.3. *For all* $\mu > 0$ *and* $\lambda_0 \in \mathbb{C}$, *there exists a* $\nu > 0$ *such that for all* $\Delta = (\delta A_0, \ldots, \delta A_m) \in \mathbb{C}^{n\times n\times(m+1)}$ *with* $\|\Delta\|_{\text{glob}} < \nu$, (3.9) *has the same number of roots*[9] *as* (3.6) *in the disc* $\{\lambda \in \mathbb{C}:\, |\lambda - \lambda_0| < \mu\}$.

Assume that all the roots of (3.6) are in $\mathbb{C}_d$. Let Δ_c be an arbitrary perturbation with $\|\Delta_c\|_{\text{glob}}$ finite, for which there is at least one root in $\mathbb{C}_u$ (such perturbations always exist by Theorem 3.2). Next, apply the perturbation $\Delta := \epsilon\, \Delta_c$, where $\epsilon \geq 0$ is a parameter. Clearly, the function

$$\epsilon \in [0,\, 1] \rightarrow \epsilon\, \Delta_c$$

is continuous with respect to the measure $\|\cdot\|_{\text{glob}}$. Consequently, by Proposition 3.3 one of the following phenomena must happen to the roots of the perturbed system when ϵ is continuously varied from zero to one:

1. some roots move from $\mathbb{C}_d$ to $\mathbb{C}_u$;

2. roots coming from infinity appear in $\mathbb{C}_u$ (only for unbounded $\mathbb{C}_u$).

If the second case can be excluded, a loss of stability is always associated with roots on the boundary of $\mathbb{C}_d$, and it becomes sufficient to scan this boundary in the outer optimization of (3.8). In other words, the stability radius is the smallest value of ϵ for which an ϵ-pseudospectrum contour reaches the boundary of $\mathbb{C}_d$. Formally, using (3.16) one has:

[9] Multiplicity is taken into account.

Corollary 3.4. *Assume that all the roots of* (3.6) *are in* $\mathbb{C}_d$. *Then*

$$r_{\mathbb{C}}(F;\, \mathbb{C}_d, \|\cdot\|_{\text{glob}}) = \inf_{\lambda\in\Gamma_{\mathbb{C}_d}} \frac{1}{f(\lambda;\, \|\cdot\|_{\text{glob}})} = \frac{1}{\sup_{\lambda\in\Gamma_{\mathbb{C}_d}} f(\lambda;\, \|\cdot\|_{\text{glob}})}, \tag{3.19}$$

where $\Gamma_{\mathbb{C}_d}$ *is the boundary of the set* $\mathbb{C}_d$.

The following example from [245] demonstrates that Corollary 3.4 does not hold if perturbations create roots coming from infinity in $\mathbb{C}_u$.

Example 3.5. *The equation* $p(\lambda) = 0$, *with*

$$p(\lambda) = \lambda + 1 + \delta a\, e^{\lambda},$$

is $\mathbb{C}_-$*-stable for* $\delta a = 0$. *With* $\|\Delta\|_{\text{glob}} = |\delta a|$, *we have*

$$\inf_{\lambda\in\Gamma_{\mathbb{C}_-}} \frac{1}{f(\lambda;\, \|\cdot\|_{\text{glob}})} = \inf_{\omega\geq 0} \frac{|1+j\omega|}{|e^{j\omega}|} = 1;$$

that is, shifting roots to the imaginary axis requires $|\delta a| \geq 1$. *However, the stability radius is zero because for any real* $\delta a \neq 0$, *there are infinitely many roots in the open right half plane, whose real parts move off to plus infinity as* $|\delta a| \to 0+$. *To see this, note that* $p(-\lambda)$ *can be interpreted as the characteristic function of the delay-differential equation* $\dot{x}(t) = x(t) + \delta a\, x(t-1)$, *which has infinitely many characteristic roots located in a logarithmic sector of the left half plane, as we have seen in Chapter* 1.

3.2.3 ▪ Computational issues

Using (3.17) and Theorem 3.2, pseudospectra of (3.6) can be computed by evaluating

$$\left\|\left(\sum_{i=0}^{m} A_i\, p_i(\lambda)\right)^{-1}\right\|_{\alpha} \|w(\lambda)\|_{\beta}$$

for λ on a grid over a region of the complex plane. By using a contour plotter to view the results, the boundaries of ϵ-pseudospectra are then identified. Notice that for $\alpha = 2$, the left term can be computed as the inverse of the smallest singular value of $\sum_{i=0}^{m} A_i\, p_i(\lambda)$. Analogously, from Corollary 3.4 the complex stability radius can be computed using a grid, laid on the boundary of the stability region. Such an approach is taken for the numerical examples of Section 3.4.

It is important to mention that the computational complexity is similar to the complexity of computing pseudospectra of an n-by-n matrix (see formula (3.2)), since the characteristic matrix F has dimensions n-by-n. In particular, when the nonlinear eigenvalue problem stems from an infinite-dimensional linear system (such as the time-delay systems discussed in the next subsection), a "lifting" to a first-order representation leads to a formula of the form (3.2), with $\mathscr{A}$ an infinite-dimensional operator.

For computing the pseudospectral abscissa of time-delay systems, a level set algorithm has been proposed in [140], which is conceptually similar to the method for computing $\mathscr{H}_\infty$ norms described in Chapter 4. For large-scale problems the iterative method of [136] is applicable. However, this is beyond the scope of this book.

Table 3.1. *Special cases of Theorem 3.2/Corollary 3.4 treated in the literature* [245].

Reference	Problem	Perturbation Measure	Weights
[383]	matrix pencil	(3.12)	/
[118]	polynomial matrices	(3.12), (3.14) with $p_2 = \infty$	/
[324]	polynomial matrices	(3.12), (3.14) with $p_2 = \infty$	yes
[373]	polynomial matrices	(3.12), (3.14) with $p_2 = \infty$	yes
[125]	time-delay systems	(3.14) with $p_1 = 2$ and $p_2 = \infty$	yes
[245]	general case		

To conclude this section, Table 3.1 lists publications from which some of the results stated in Theorem 3.2 or Corollary 3.4 were obtained. Note that we have restricted ourselves to nonscalar problems (results for scalar polynomials are, for instance, described in [164, Section 5.4.2] and the references therein).

3.2.4 ▪ Application to time-delay systems

We apply the results of Section 3.2 to linear delay-differential equations of the form

$$\dot{x}(t) = A_0 x(t) + \sum_{i=1}^{m} A_i x(t - \tau_i), \tag{3.20}$$

where we assume that $0 < \tau_1 < \cdots < \tau_m$ and that the system matrices $A_i \in \mathbb{R}^{n\times n}$, $i = 0, \ldots, m$, are uncertain. In this particular case we have

$$F(\lambda) = \lambda I - A - \sum_{i=1}^{m} A_i e^{-\lambda \tau_i}. \tag{3.21}$$

Expressions

Pseudospectra and stability radii of (3.20), following the general definitions (3.8) and (3.15), can be computed as follows:

Theorem 3.6. *For perturbations $\delta A_i \in \mathbb{C}^{n\times n}$, $i = 0, \ldots, m$, measured by* (3.12)–(3.14), *the pseudospectrum Λ_ϵ of* (3.20) *satisfies*

$$\begin{aligned} \Lambda_\epsilon(F; \|\cdot\|_{\mathrm{glob}}) &= \Lambda(F) \\ &\cup \left\{ \lambda \in \mathbb{C} : \left\| \left(\lambda I - A_0 - \sum_{i=1}^{m} A_i e^{-\lambda \tau_i} \right)^{-1} \right\|_\alpha \cdot \|w(\lambda)\|_\beta > \epsilon^{-1} \right\} \end{aligned} \tag{3.22}$$

and, if the zero solution of (3.20) *is asymptotically stable, then the associated stability radius satisfies*

$$\begin{aligned} & r_{\mathbb{C}}(F; \mathbb{C}_-, \|\cdot\|_{\mathrm{glob}}) \\ &= \left(\left(\sum_{i=0}^{m} w_i^{-\beta} \right)^{\frac{1}{\beta}} \sup_{\omega \geq 0} \left\| \left(j\omega I - A_0 - \sum_{i=1}^{m} A_i e^{-j\omega \tau_i} \right)^{-1} \right\|_\alpha \right)^{-1}, \end{aligned} \tag{3.23}$$

where $w(\lambda) = \left[\frac{1}{w_0}\ \frac{e^{-\lambda\tau_1}}{w_1} \ldots \frac{e^{-\lambda\tau_m}}{w_m} \right]^T$ and α and β are defined as in Theorem 3.2.

Remark 3.7. *For the system*

$$\dot{x}(t) = (A + \delta A)x(t), \tag{3.24}$$

with $\|\Delta\|_{\text{glob}} = \|\delta A\|_2$, *expression* (3.22) *simplifies to*

$$\Lambda_\epsilon(F; \|\cdot\|_{\text{glob}}) = \Lambda(F) \cup \{\lambda \in \mathbb{C} : \|\mathcal{R}(\lambda, A)\|_2 > \epsilon^{-1}\}, \tag{3.25}$$

where $\mathcal{R}(\lambda, A) = (\lambda I - A)^{-1}$ *is the* resolvent *of* A. *As mentioned in the introduction, the right-hand side of* (3.25) *can also be considered as a definition for the* ϵ*-pseudospectrum of* (3.24). *In analogy to* (3.25), *one can alternatively define the* ϵ*-pseudospectrum of* (3.20) *as the set*

$$\Lambda(F) \cup \{\lambda \in \mathbb{C} : \|\mathcal{R}(\lambda, \mathcal{A})\| > \epsilon^{-1}\}, \tag{3.26}$$

where $\mathcal{A}$ *is the infinitesimal generator of the solution operator associated with the delay equation. Notice that definition* (3.26) *is related to the effect of* unstructured perturbations *of the* operator $\mathcal{A}$ *on stability. In this chapter we have chosen a more practical definition by directly relating pseudospectra to concrete perturbations on the system matrices. Notice that such a practical definition is typically used also for polynomial equations* [373, 324, 118].

Remark 3.8. *When* $\alpha = 2$, *computing the complex stability radius* (3.23) *requires the computation of the* $\mathcal{H}_\infty$*-norm of the transfer function*

$$H(j\omega) := \left(j\omega I - A_0 - \sum_{k=1}^{m} A_k e^{-\lambda \tau_k} \right)^{-1}, \tag{3.27}$$

for which we present a computational method in the next chapter.

Effect of weighting

Applying different weights to the system matrices A_i of (3.20), $i = 1, \ldots, m$, leads to changes in the pseudospectra. This can be understood by investigating the weighting function $w(\lambda) = w(\sigma + j\omega)$, where

$$\|w(\sigma + j\omega)\|_\beta = \left\| \left[\frac{1}{w_0}, \frac{e^{-\sigma\tau_1}}{w_1}, \ldots, \frac{e^{-\sigma\tau_m}}{w_m} \right]^T \right\|_\beta \quad \forall \sigma, \quad \omega \in \mathbb{R}. \tag{3.28}$$

Note that $w(\lambda)$ only depends on the real part σ, that is, $w(\lambda) \equiv w(\sigma)$. From (3.28) the following conclusions can be drawn:

1. Characteristic roots in the right half-plane are more sensitive to perturbations of the nondelayed term A_0.

2. Characteristic roots in the left half plane are more sensitive to perturbations of the delayed terms A_i, $i = 1, \ldots, m$.

3. Furthermore, the intersection of an ϵ-pseudospectrum contour with the imaginary axis is independent of the weights, provided that the β norm of $w(\lambda) = w(0)$ is constant. As a consequence, under this condition, the stability radius is also independent of the weights.

Asymptotic properties

In order to characterize boundedness properties of pseudospectra, we investigate the behavior of

$$f(\lambda; \|\cdot\|_{\text{glob}}) := \begin{cases} \left\| \left(\lambda I - A_0 - \sum_{i=1}^m A_i e^{-\lambda \tau_i}\right)^{-1} \right\|_\alpha \|w(\lambda)\|_\beta, & \lambda \notin \Lambda(F), \\ +\infty & \lambda \in \Lambda(F) \end{cases}$$

as $|\lambda| \to \infty$. This leads to the following two results from [245, Section 3.3]:

Proposition 3.9. *For all* $\mu \in \mathbb{R}$,

$$\lim_{R\to\infty} \inf \left\{ f(\lambda, \|\cdot\|_{\text{glob}})^{-1} : \Re(\lambda) > \mu, \ |\lambda| > R \right\} = \infty. \tag{3.29}$$

As a consequence, the cross-section between any pseudospectrum and any right half-plane is bounded.

Proposition 3.10. *Let*

$$\Psi_\gamma := \left\{ \lambda \in \mathbb{C} : \Re(\lambda) < -\gamma, \ |\lambda| < e^{-(\Re(\lambda)+\gamma)\tau_m} \right\}. \tag{3.30}$$

If A_m *is regular, then*

$$\begin{aligned} &\forall \epsilon \in \left(0, \frac{w_m}{\|A_m^{-1}\|_\alpha}\right), \ \exists \gamma > 0 \textit{ such that } \Psi_\gamma \cap \Lambda_\epsilon(F; \|\cdot\|_{\text{glob}}) = \phi, \\ &\forall \epsilon > \frac{w_m}{\|A_m^{-1}\|_\alpha}, \ \exists \gamma > 0 \textit{ such that } \Psi_\gamma \subset \Lambda_\epsilon(F; \|\cdot\|_{\text{glob}}). \end{aligned} \tag{3.31}$$

If A_m *is singular, then*

$$\forall \epsilon > 0, \ \exists \gamma > 0 \textit{ such that } \Psi_\gamma \subset \Lambda_\epsilon(F; \|\cdot\|_{\text{glob}}). \tag{3.32}$$

In the case of a singular A_m, the pseudospectrum Λ_ϵ thus stretches out along the negative real axis for *any* value of $\epsilon > 0$. Conversely, for the case of a regular A_m, this only happens for $\epsilon > w_m/\|A_m^{-1}\|_\alpha$. As a consequence, *infinitesimal* perturbations may result in the introduction of characteristic roots with *small* imaginary parts (but large negative real parts).

The two cases detailed above are connected as follows: when the matrix A_m is regular, we have

$$\inf_{\delta A_m \in \mathbb{C}^{n\times n}} \{\|\delta A_m\|_\alpha : \det(A_m + \delta A_m) = 0\} = 1/\|A_m^{-1}\|_\alpha,$$

that is, the smallest rank reducing perturbation has size $1/\|A_m^{-1}\|_\alpha$. Furthermore, the smallest perturbation $\Delta = (\delta A_0, \ldots, \delta A_m)$ on the delay equation (3.20), which introduces a characteristic root with a predetermined very large negative real part but small imaginary part, can be decomposed into a minimal size perturbation $\Delta_c = (0, \ldots, 0, \delta A_m)$ which makes A_m singular (due to the weights, we have $\|\Delta_c\|_{\text{glob}} = w_m/\|A_m^{-1}\|_\alpha$), together with a very small perturbation to place the characteristic root, according to (3.32).

3.3 ▪ Structured pseudospectra for nonlinear eigenvalue problems

Structured pseudospectra of (3.6) are defined and motivated. Next, various computational expressions are presented and discussed. Finally, the results are applied to the time-delay system (3.20).

3.3.1 ▪ Motivation

A number of stability and robustness problems for linear systems leads to the study of the eigenvalues of a matrix A with a certain structure, for example, a block structure, which should be respected in the sensitivity analysis [378]. For this, perturbations of the form $A+DPE$ have been considered in [163], where the fixed matrices D and E describe the perturbation structure and P is a complex perturbation matrix. This approach has been further developed in [400] for perturbations of the form $A+\sum D_i P_i E_i$, which, in particular, allow one to deal with elementwise perturbations. On the other hand, specific classes of systems, such as higher-order systems or systems with time-delays, lead to the study of the zeros of nonlinear eigenvalue problems of the form (3.5), as we have seen in the previous section. In this section we *combine* these two approaches for exploiting a system's structure. In light of this, we define pseudospectra for the nonlinear eigenvalue problem (3.5) and derive computable formulae, where, in addition to exploiting the form of the nonlinear eigenvalue problem, a particular structure can be imposed on the perturbations of the individual coefficient matrices A_i. The motivation stems from the fact that in a lot of applications the different coefficient matrices have a certain structure that should naturally be respected in a sensitivity analysis, since unstructured perturbations may lead to irrelevant or nonphysical effects. One example is discussed in [125], where the emergence of unbounded pseudospectra of a delay system in certain directions is explained by non-physical perturbations that destroy an intrinsic property, namely the singular nature, of one of the coefficient matrices. Another example from laser physics will be discussed in Section 3.4.2.

3.3.2 ▪ Definition and expressions

The definition (3.15) for the ϵ-pseudospectrum of the nonlinear eigenvalue problem (3.5), for complex perturbations measured with (3.14), where $p_1=2$ and $p_2=\infty$, is equivalent to

$$\Lambda_\epsilon(F) := \left\{ \lambda\in\mathbb{C} :\ \det\left(\sum_{i=0}^{m}(A_i+\delta A_i)p_i(\lambda)\right)=0,\ \text{for some } \delta A_i\in\mathbb{C}^{n\times n} \right. \\ \left. \text{with } w_i\|\delta A_i\|_2<\epsilon,\ 0\le i\le m \right\}. \tag{3.33}$$

Observe that the perturbations δA_i considered in (3.33) lead to an additive uncertainty on the characteristic matrix (3.6) given by

$$\delta F(\lambda) := \sum_{i=0}^{m} \delta A_i\ p_i(\lambda). \tag{3.34}$$

Although the structure of the expression (3.6) is explicitly taken into account in the definition (3.33), the perturbations δA_i applied to the different matrices A_i are unstructured. In other words, the elementwise structure of A_i is not taken into account when using the corresponding perturbation δA_i. We now present a framework for the definition and computation of pseudospectra in which various types of structures on the perturbation matrices can be imposed, too. For this we assume a more general additive uncertainty on (3.6) than what (3.34) allows. This uncertainty takes the form

$$\delta F(\lambda) := \sum_{j=0}^{f} D_j(\lambda)\Delta_j E_j(\lambda) + \sum_{j=0}^{s} d_j G_j(\lambda) H_j(\lambda). \tag{3.35}$$

In this expression, $\Delta_j \in \mathbb{C}^{k_j \times l_j}$ and $d_j \in \mathbb{C}$ denote the underlying unstructured perturbations, and $D_j \in \mathbb{C}^{n\times k_j}$, $E_j \in \mathbb{C}^{l_j \times n}$, $G_j \in \mathbb{C}^{n\times m_j}$, and $H_j \in \mathbb{C}^{m_j\times n}$ are appropriate shape matrices, whose elements are entire functions. We further assume that $m_j \geq 2$, G_j has full column rank, and H_j has full row rank, for all $j = 0,\dots,s$. We define the *uncertainty set* $\mathbf{\Delta}$ as

$$\begin{aligned}\mathbf{\Delta} := \Big\{\operatorname{diag}(\Delta_0,\dots,\Delta_f, d_0 I_{m_0},\dots,d_s I_{m_s}):\ \Delta_i \in \mathbb{C}^{k_i\times l_i},\ d_j \in \mathbb{C},\\ 0\leq i\leq f,\ 0\leq j\leq s\}\end{aligned} \tag{3.36}$$

endowed with the following norm:

$$\begin{aligned}\|\Delta\|_{\text{glob}} = \max\Big\{\|\Delta_0\|_2,\dots,\|\Delta_f\|, |d_0|,\dots,|d_s|\Big\},\\ \Delta = \operatorname{diag}(\Delta_0,\dots,\Delta_f, d_0 I_{m_0},\dots,d_s I_{m_s}) \in \mathbf{\Delta}.\end{aligned}$$

The structured ϵ-pseudospectrum $\Lambda_\epsilon^{\text{str}}(F)$ of F with respect to the uncertainty (3.35) can then be defined as follows:

Definition 3.11.

$$\begin{aligned}\Lambda_\epsilon^{\text{str}}(F;\ \mathbf{\Delta}) := \{\lambda\in\mathbb{C} : \det(F(\lambda)+\delta F(\lambda)) = 0 \textit{ for some } \delta F \textit{ as in}\\ (3.35) \textit{ with } \|\Delta_j\|_2 < \epsilon,\ 0\leq j\leq f, \textit{ and } |d_j| < \epsilon,\ 0\leq j\leq s\}.\end{aligned} \tag{3.37}$$

Let us make a comparison with Definition 3.1. In addition to the fact that the structure of (3.35) is exploited in (3.37), we notice that the following:

- The underlying norm used in Definition 3.11 is of mixed type and similar to (3.14) with $p_1 = 2$ and $p_2 = \infty$. Although other types of norms can be used without any problem, we restrict ourselves to this type because it is most relevant from an application point of view, and, as we shall see, it gives rise to directly computable expressions in terms of structured singular values.
- Scalar weights of the perturbations are not used in Definition 3.11, as they can always be absorbed in the shape matrices in (3.35).

In a similar way the complex-structured stability radius of (3.6) w.r.t. the perturbations (3.35)–(3.36) and w.r.t. the desired "stability" region $\mathbb{C}_d$ is given by:

Definition 3.12.

$$\begin{aligned}r_{\mathbb{C}}^{\text{str}}(F;\ \mathbb{C}_d, \mathbf{\Delta}) := \inf_{\lambda\in\mathbb{C}_u}\inf_{\epsilon>0}\{\epsilon :\ \det(F(\lambda)+\delta F(\lambda)) = 0 \textit{ for some } \delta F \textit{ as in}\\ (3.35) \textit{ with } \|\Delta_j\|_2 < \epsilon,\ 0\leq j\leq f \textit{ and } |d_j| < \epsilon,\ 0\leq j\leq s\Big\}.\end{aligned}$$

Several problems encountered in the literature fit into the general framework presented above such as, for instance, the following:

- The reference [324] deals with perturbations of matrix polynomials (3.6) of the form
$$\sum_{i=0}^{m}\left(A_i+\sum_{j=0}^{f} D_j\Delta_j E_{ij}\right) p_i(\lambda) \tag{3.38}$$
in the context of stability radii for polynomial matrices. The shape matrices D_j and E_{ij} in (3.38) can be used to perturb only a submatrix of A_i, to assign weights to perturbations of rows, columns, or elements of each A_i, and to weight the perturbations applied to the matrices $A_0,\dots,A_m$ with respect to each other. The uncertainty in (3.38) leads to the additive perturbation
$$\delta F(\lambda)=\sum_{j=0}^{f} D_j\Delta_j\left(\sum_{i=0}^{m} E_{ij}p_i(\lambda)\right).$$

- In many applications the characteristic matrix of an uncertain system is given by $\sum_{i=0}^{m}\tilde{A}_i p_i(\lambda)$, where the matrices $\tilde{A}_i$ depend linearly on a number of uncertain scalar parameters, say
$$\tilde{A}_i=A_i+\sum_j \theta_j P_{ij},$$
with $\theta_j\in\mathbb{C}$ describing the uncertainties on these parameters—assuming that one wishes to investigate the possible positions of the characteristic roots when $|\theta_j|\le\epsilon$ for all j. It follows that we are in the framework of (3.6), (3.35), and (3.37), as we can express
$$\begin{aligned}\sum_{i=0}^{m}\tilde{A}_i p_i(\lambda)&=F(\lambda)+\sum_j\theta_j\left(\sum_{i=1,P_{ij}\neq 0}^{m} U_{ij}V_{ij}^*\,p_i(\lambda)\right)\\ &=F(\lambda)+\sum_j\theta_j\,[\cdots U_{ij}\cdots][\cdots V_{ij}\,\bar{p}_i(\lambda)\cdots]^*,\end{aligned} \tag{3.39}$$
where each U_{ij} has full column rank and U_{ij} and V_{ij} can be computed, for instance, from a singular value decomposition of P_{ij}. Notice that (3.39) leads to $s>0$ in the general expression (3.35) if and only if one of the matrices P_{ij} has rank larger than one or if one of the parameters explicitly appears in different matrices $\tilde{A}_i$.

- Structured perturbations of the form (3.35) can sometimes be used for classes of systems with a nonlinear dependence on the parameters. As an illustration, the uncertain system
$$\dot{x}(t)=(A+\delta A)x(t)+(B+\delta B)(C+\delta C)x(t-\tau)$$
can be rewritten in a descriptor form as
$$\begin{aligned}\dot{x}(t)&=(A+\delta A)x(t)+(B+\delta B)y(t),\\ 0&=(C+\delta C)x(t-\tau)-y(t),\end{aligned}$$
see also [112] and the references therein. This system has a nominal characteristic matrix
$$F(\lambda)=\begin{bmatrix}\lambda I-A & -B\\ Ce^{-\lambda\tau} & -I\end{bmatrix},$$

to which we may apply the structured perturbations

$$\delta F(\lambda) = -\begin{bmatrix} I \\ 0 \end{bmatrix} \delta A [I \ 0] - \begin{bmatrix} I \\ 0 \end{bmatrix} \delta B [0 \ I] - \begin{bmatrix} 0 \\ I \end{bmatrix} \delta C \left[e^{-\lambda\tau} I \ 0 \right].$$

A reformulation of (3.37) in terms of *structured singular values* gives rise to a computable form for $\Lambda_\epsilon^{\mathrm{str}}(F;\, \boldsymbol{\Delta})$ (the formula for $r_{\mathbb{C}}(F;\, \mathbb{C}_d, \boldsymbol{\Delta})$ follows straightforwardly). See Section A.2 of the appendix for a short introduction to the concept and computation of structured singular value.

Theorem 3.13. *Consider the characteristic matrix* (3.6) *with additive uncertainty* (3.35)–(3.36). *Defining*

$$\begin{aligned} M(\lambda) &:= \left[E_1(\lambda)^T \ldots E_f(\lambda)^T \ H_1(\lambda)^T \ldots H_s(\lambda) \right]^T, \\ N(\lambda) &:= [D_1(\lambda) \cdots D_f(\lambda) \ G_1(\lambda) \cdots G_s(\lambda)], \end{aligned} \tag{3.40}$$

and

$$T(\lambda) := M(\lambda) F(\lambda)^{-1} N(\lambda),$$

we have

$$\Lambda_\epsilon^{\mathrm{str}}(F;\, \boldsymbol{\Delta}) = \Lambda \cup \left\{ \lambda \in \mathbb{C} :\ \mu_{\boldsymbol{\Delta}}(T(\lambda)) > \epsilon^{-1} \right\}, \tag{3.41}$$

where $\mu_{\boldsymbol{\Delta}}(\cdot)$ is the structured singular value with respect to the uncertainty set $\boldsymbol{\Delta}$.

The proof is based on an appropriate feedback interpretation of the perturbed system and can be found in [401].

3.3.3 ▪ Computational issues and special cases

From (3.41) the boundaries of structured ϵ-pseudospectra can be defined as level sets of the function

$$\mu_{\boldsymbol{\Delta}}(T(\lambda)), \quad \lambda \in \mathbb{C}. \tag{3.42}$$

In general the computation of the structured singular value of a matrix with respect to the uncertainty set (3.36) is a difficult problem. Indeed, it is known to belong to the class of NP hard problems [375], which makes it inefficient for large problems. In many cases, however, the structured singular value can be computed efficiently by exploiting the structure of $T(\lambda)$. Some results in this sense will be presented below. Furthermore, for the examples presented in [401, 399], a good performance was found of numerical algorithms that approximate the structured singular value. There, lower and upper bounds on the structured singular value were computed by solving eigenvalue optimization problems, based on the ideas outlined in Appendix A.2. Such bounds are sharp in many cases. Note, in particular, that, if the additional restriction $f + 2s \leq 3$ holds for the uncertainty set (3.36) and the full blocks are square, then an exact computation of $\mu_{\boldsymbol{\Delta}}(\cdot)$ is always possible by solving a convex optimization problem. Hence, we propose the use of the structured singular value for computing structured pseudospectra. Note also that the computation of upper bounds for the structured singular value in (3.42) gives lower bounds for the ϵ-values, for which a point lies in the structured ϵ-pseudospectrum. This is in agreement with the common use of pseudospectra (or directly related values such as stability radii) in a worst case analysis. In this context the above approach can be used to give rigorous bounds for the behavior of eigenvalues under perturbations.

In some cases the particular structure of $T(\lambda)$ can be exploited for an efficient computation of the structured singular value in (3.42). This is illustrated by the following result of [401], which slightly generalizes one of the assertions of Theorem3.2 and is also related to Proposition 3.4 of [324]:

Theorem 3.14. *We consider the characteristic matrix* (3.6) *with uncertainty* (3.35). *Furthermore, we assume that* $s=0$ *and that there exist matrices* D *and* E *and functions* $q_j:\mathbb{C}\to\mathbb{C}$ *such that*

$$\begin{aligned} D_j(\lambda) &= D(\lambda),\\ E_j(\lambda) &= E(\lambda)\,q_j(\lambda),\ \ 1\leq j\leq f. \end{aligned}$$

By defining $T(\lambda)$ *and* $\boldsymbol{\Delta}$ *as in Theorem* 3.2, *the following holds:*

$$\mu_{\boldsymbol{\Delta}}(T(\lambda)) = \left\|E(\lambda)F^{-1}(\lambda)D(\lambda)\right\|_2\left(\sum_{j=0}^{f}|q_j(\lambda)|\right),\ \ \lambda\notin\Lambda. \tag{3.43}$$

Note that, in addition to the availability of a directly computable formula, the dimensions of $E(\lambda)F^{-1}(\lambda)D(\lambda)$ are f *times smaller* than the dimensions of $T(\lambda)$.

To conclude this section we give an overview of special types of problems encountered in the pseudospectra literature for which the combination of Theorems 3.13 and 3.14 ensures an efficient computation of pseudospectra:

- Unstructured pseudospectra, in the sense of Definition 3.1, for the measure (3.14) with $p_1=2$, $p_2=\infty$ and unit weights.

  With $F(\lambda)+\delta F(\lambda)=\sum_{i=0}^{m}(A_i+\delta A_i)p_i(\lambda)$ we have

  $$\delta F(\lambda)=\sum_{i=0}^{m}\underbrace{I}_{D_i(\lambda)}\underbrace{\delta A_i}_{\Delta_i}\underbrace{p_i(\lambda)I}_{E_i(\lambda)}.$$

  An application of Theorem 3.14 yields

  $$\mu_{\boldsymbol{\Delta}}(T(\lambda))=\left\|F(\lambda)^{-1}\right\|_2\sum_{i=0}^{m}|p_i(\lambda)|. \tag{3.44}$$

  In combination with Theorem 3.13 one of the results of Theorem 3.2 is recovered.

- Pseudospectra in the sense of [373].

  With $F(\lambda)+\delta F(\lambda)=\sum_{i=0}^{m}(A_i+D\delta A_iE_i)p_i(\lambda)$ we get

  $$\delta F(\lambda)=\underbrace{D}_{D(\lambda)}\underbrace{[\delta A_1\ldots\delta A_m]}_{\Delta}\underbrace{\left[E_1^Tp_1(\lambda)\cdots E_m^Tp_m(\lambda)\right]^T}_{E(\lambda)},$$

  and by Theorem 3.14,

  $$\mu_{\boldsymbol{\Delta}}(T(\lambda))=\left\|\begin{bmatrix}p_1(\lambda)E_1\\ \vdots\\ p_m(\lambda)E_m\end{bmatrix}F(\lambda)^{-1}D\right\|_2.$$

For the special case $E_i = E$, $i = 1,\dots,m$, this expression can also be simplified to

$$\mu_\Delta(T(\lambda)) = \|EF^{-1}(\lambda)D\|_2\sqrt{\sum_{i=0}^{m}|p_i(\lambda)|^2}. \tag{3.45}$$

Furthermore, if we consider the dual problem, $F(\lambda)+\delta F(\lambda) = \sum_{i=0}^{m}(A_i + D_i\delta A_i E)p_i(\lambda)$, instead, we have

$$\delta F(\lambda) = \underbrace{[D_1 p_1(\lambda)\cdots D_m p_m(\lambda)]}_{D(\lambda)}\underbrace{[\delta A_1^T \cdots \delta A_m^T]^T}_{\Delta}\underbrace{E}_{E(\lambda)}$$

and

$$\mu_\Delta(T(\lambda)) = \left\|EF(\lambda)^{-1}[p_1(\lambda)D_1\cdots p_m(\lambda)D_m]\right\|_2,$$

which can for the special case $D_i = D$, $i = 1,\dots,m$, be simplified to (3.45), as expected.

Notice that expression (3.45) does not reduce to expression (3.44) if $D = I$ and $E = I$, although in that case they concern the same additive perturbation on F. This is explained by the fact that both formulae are based on a different way of measuring the perturbations: $\max_{1\le i\le m}\|\delta A_i\|_2$ for (3.44) and $\|[\delta A_1 \dots \delta A_m]\|_2$ for (3.45).

3.3.4 ▪ Application to time-delay systems

Rephrasing Theorems 3.13–3.14 for the time-delay system (3.20) yields:

Theorem 3.15. *Consider the characteristic matrix* (3.6) *with additive uncertainty* (3.35). *Let F be defined as in* (3.21) *and let the uncertainty set $\mathbf{\Delta}$ be as* (3.36). *Then*

$$\Lambda_\epsilon^{\mathrm{str}}(F;\,\mathbf{\Delta}) = \Lambda(F) \cup \left\{\lambda\in\mathbb{C}:\ \mu_\Delta\left(M(\lambda)\left(\lambda I - A_0 - \sum_{i=1}^{m}A_i e^{-\lambda\tau_i}\right)^{-1}N(\lambda)\right) > \epsilon^{-1}\right\},$$

where M and N are given by (3.40) *and $\mu_\Delta(\cdot)$ is the structured singular value with respect to the uncertainty set $\mathbf{\Delta}$. If the zero solution of* (3.20) *is asymptotically stable, then the associated complex stability radius satisfies*

$$r_{\mathbb{C}}^{\mathrm{str}}(F;\,\mathbb{C}_-,\mathbf{\Delta}) = \left\{\sup_{\omega\ge 0}\mu_\Delta\left(M(j\omega)\left(j\omega I - A_0 - \sum_{i=1}^{m}A_i e^{-j\omega\tau_i}\right)^{-1}N(j\omega)\right)\right\}^{-1}.$$

If, furthermore, $s = 0$, and there exist matrices D and E and functions $q_j : \mathbb{C}\to\mathbb{C}$ such that

$$D_j(\lambda) = D(\lambda),$$
$$E_j(\lambda) = E(\lambda)\,q_j(\lambda),\quad 1\le j\le f,$$

then

$$\Lambda_\epsilon^{\mathrm{str}}(F;\,\mathbf{\Delta}) = \Lambda(F)\cup\left\{\lambda\in\mathbb{C}:\ \left\|E(\lambda)\left(\lambda I - A - \sum_{i=1}^{m}A_i e^{-\lambda\tau_i}\right)^{-1}D(\lambda)\right\|_2 \left(\sum_{j=0}^{f}|q_j(\lambda)|\right) > \epsilon^{-1}\right\}$$

and

$$r_{\mathbb{C}}^{\text{str}}(F;\ \mathbb{C}_-, \Delta) = \left\{ \sup_{\omega \geq 0} \left\| E(j\omega)\left(j\omega I - A - \sum_{i=1}^{m} A_i e^{-j\omega\tau_i} \right)^{-1} D(j\omega) \right\|_2 \left(\sum_{j=0}^{f} |q_j(j\omega)| \right) \right\}^{-1}.$$

In Chapter 9 we will describe an algorithm for the related synthesis problem of optimizing complex stability radii as a function of system or controller parameters.

3.4 ▪ Illustrative examples

We illustrate the results obtained in this chapter by means of the time-delay system

$$\dot{x}(t) = A_0 x(t) + A_1 x(t-1), \tag{3.46}$$

for which we have

$$F(\lambda) = \lambda I - A_0 - A_1 e^{-\lambda}.$$

We take two different sets of parameter values. With the first set we illustrate the results of Section 3.2. With the second set, which comes from a laser physics application, we motivate and illustrate the results of Section 3.3.

3.4.1 ▪ Second-order system

Consider the system (3.46), where

$$A_0 = \begin{bmatrix} -5 & 1 \\ 2 & -6 \end{bmatrix} \quad \text{and} \quad A_1 = \begin{bmatrix} -2 & 1 \\ 4 & -1 \end{bmatrix}. \tag{3.47}$$

Figure 3.1(a) shows the rightmost characteristic roots of (3.46)–(3.47). The system is shown to be stable with all characteristic roots confined to the open left half plane. To investigate how stability may change under perturbations of the matrices A_0 and A_1, we need to compute the corresponding pseudospectra. Towards this end, we consider unstructured perturbations of A_0 and A_1 using the global measure (3.14) with $p_1 = 2$ and $p_2 = \infty$. Pseudospectra can then be computed using Theorems 3.6 and 3.2 with $\alpha = 2$ and $\beta = 1$. Specifically (for $\lambda \notin \Lambda$),

$$f(\lambda;\ \|\cdot\|_{\text{glob}}) = \left\| \left(\lambda I - A_0 - A_1 e^{-\lambda} \right)^{-1} \right\|_2 \left(\frac{1}{w_0} + \frac{|e^{-\lambda}|}{w_1} \right). \tag{3.48}$$

By evaluating f on a grid over a region of the complex plane, and by using a contour plotter, we have identified the boundaries of ϵ-pseudospectra $\Lambda_\epsilon(F, \|\cdot\|_{\text{glob}})$.

Figures 3.1(b)–(d) show the ϵ-pseudospectra of (3.47) where different weights have been applied to A_0 and A_1. Specifically, $(w_0, w_1) = (\infty, 1)$ (b), $(w_0, w_1) = (2, 2)$ (c), and $(w_0, w_1) = (1, \infty)$ (d). In each panel, from outermost to innermost (or rightmost to leftmost if the curve is not closed), the curves correspond to boundaries of ϵ-pseudospectra with $\epsilon = 10^{1.25}, 10^{1.0}, 10^{0.75}, 10^{0.5}, 10^{0.25}, 10^{0}$, and $10^{-0.5}$. It can be seen that the conclusions drawn in Section 3.2.4 on the effect of weighting hold; that is, perturbations of A_0 stretch pseudospectra lying in the right half plane (d), while perturbations applied to A_1 stretch

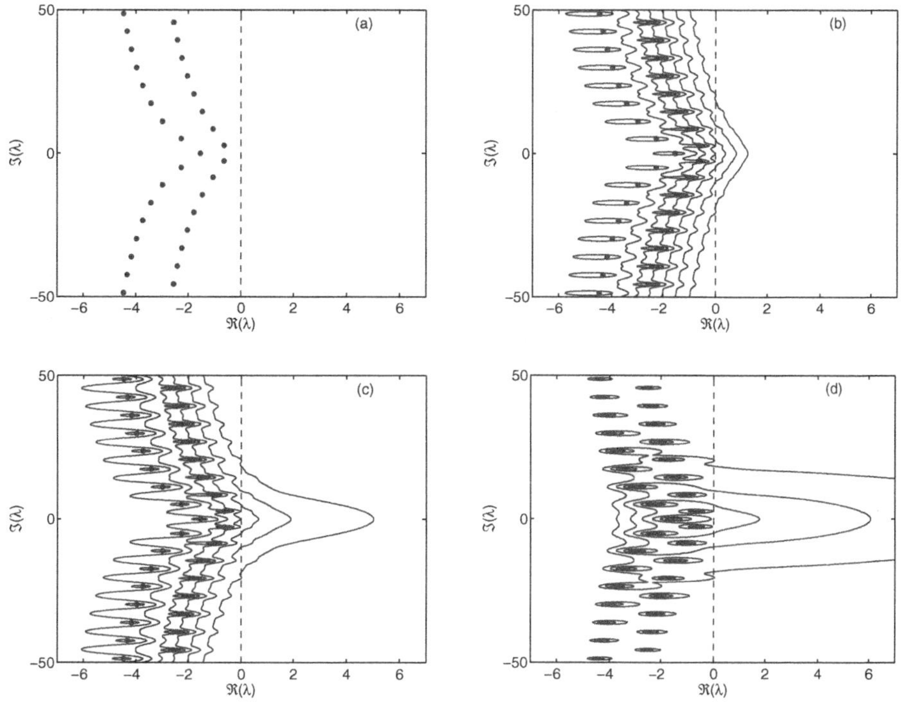

Figure 3.1. *Weighted pseudospectra of* (3.46)–(3.47). *Panel* (a) *shows the spectrum of the unperturbed problem. In all other panels, from rightmost to leftmost, the contours correspond to* $\epsilon = 10^{1.25}, 10^{1.0}, 10^{0.75}, 10^{0.5}, 10^{0.25}, 10^{0},$ *and* $10^{-0.5}$. *From* (b) *to* (d), *the weights* w_0 *and* w_1 *applied to the* A_0 *and* A_1 *matrices were* $(w_0, w_1) = (\infty, 1)$, $(w_0, w_1) = (2, 2)$, *and* $(w_0, w_1) = (1, \infty)$, *respectively* [245].

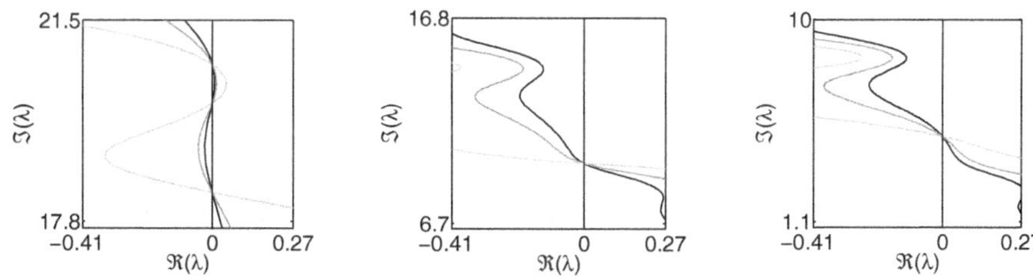

Figure 3.2. *Crossings of* ϵ*-pseudospectrum curves with the imaginary axis for* $\epsilon = 10^{1.25}$ *(left),* $\epsilon = 10$ *(middle), and* $\epsilon = 10^{0.75}$ *(right). In all three cases the darkest contour corresponds to the weights* $(w_0, w_1) = (\infty, 1)$, *the next-to-darkest curve to* $(2, 2)$, *and the lightest curve to* $(1, \infty)$ [245].

the pseudospectra lying in the left half plane (b). Furthermore, Figure 3.2 shows the intersection of ϵ-pseudospectrum, shown in panels (b)–(d) of Figure 3.1, with the imaginary axis. The darkest curve corresponds to an ϵ-pseudospectrum curve of Figure 3.1(a), the next-to-darkest curve corresponds to that of Figure 3.1(b), and the lightest curve corresponds to an ϵ-pseudospectrum curve of Figure 3.1(c). Specifically, Figure 3.2(a) shows the intersection of the three curves for $\epsilon = 10^{1.25}$, Figure 3.2(b) for $\epsilon = 10^{1.0}$, and Figure 3.2(c) for $\epsilon = 10^{0.75}$. For a given ϵ, these curves are seen to intersect the imaginary axis at the same point, independent of the weighting applied to the system matrices, and thus demonstrating the third conclusion.

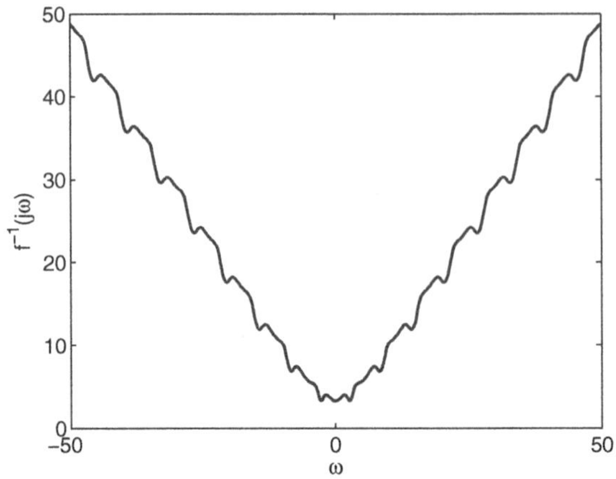

Figure 3.3. *The function* $\omega \to f^{-1}(j\omega;\ \|\cdot\|_{\text{glob}})$. *The minimum is the complex stability radius* $r_{\mathbb{C}}(F;\ \mathbb{C}_-, \|\cdot\|_{\text{glob}})$ [245].

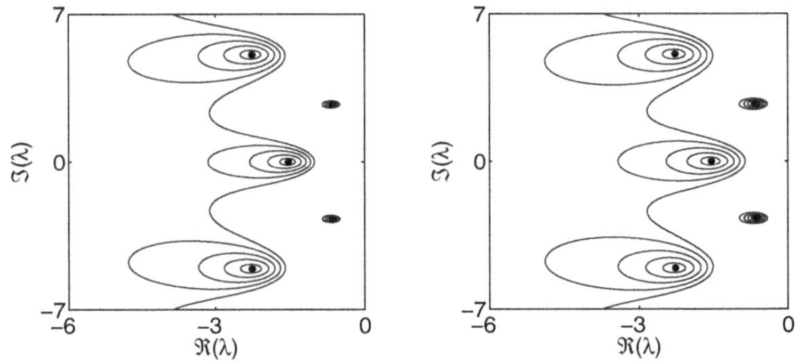

Figure 3.4. *(Left)* ϵ*-pseudospectrum curves for* $(w_0, w_1) = (\infty, 1)$ *and* $\epsilon = 0.1, 0.2, 0.3, 0.4, 0.5$. *(Right)* ϵ*-pseudospectrum curves for* $(w_0, w_1) = (2, 2)$ *and* $\epsilon = 0.2, 0.4, 0.6, 0.8, 1$ [245].

Figure 3.3 shows which ϵ-pseudospectrum curve intersects the imaginary axis at $\lambda = j\omega$, that is, $f^{-1}(j\omega,\ \|\cdot\|_{\text{glob}})$, for each $\omega \in [-50,\ 50]$. The minimum of this curve represents the complex stability radius of the system,

$$r_{\mathbb{C}}(F;\ \mathbb{C}_-, \|\cdot\|_{\text{glob}}) \approx 3.28011.$$

Since the minimum is reached for $\omega = 0$, the smallest destabilizing perturbations shift a characteristic root to the origin.

In Figure 3.4(a) we show ϵ-pseudospectra for the weights $(w_0, w_1) = (\infty, 1)$ and $\epsilon = 0.1, 0.2, 0.3, 0.4, 0.5$. In Figure 3.4(b) we take $(w_0, w_1) = (2, 2)$ and $\epsilon = 0.2, 0.4, 0.6, 0.8, 1$. In both cases, only the ϵ-pseudospectrum for the largest value of ϵ stretches out infinitely far along the negative real axis. This is in accordance with Proposition 3.10 since $w_1/\|A_1^{-1}\|_2 \approx 0.4282\ w_1$.

3.4.2 ▪ Feedback controlled semiconductor laser

In [125] pseudospectra have been applied to the analysis of the robustness of stability of a model for a semiconductor laser subject to optical feedback. For certain fixed model

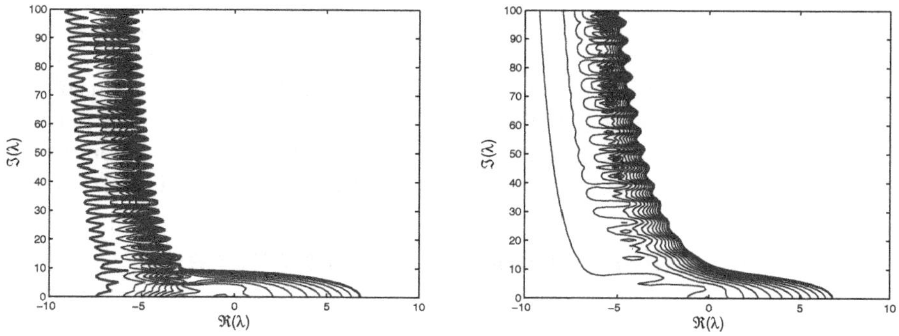

Figure 3.5. *Structured (left) and unstructured (right) pseudospectra for the systems* (3.46) *and* (3.49). *The contours corresponding to* $\epsilon = 0.001$ *to* $\epsilon = 0.27$, *in intervals of* 0.01, *are depicted. The contours corresponding to* $\epsilon = 0.27$ *are shown in bold.*

parameters, the problem leads to the study of the delay-differential equation (3.46), where

$$A_0 = \begin{bmatrix} -0.8498 & 0.1479 & 44.37 \\ 0.003756 & -0.2805 & -229.2 \\ -0.1754 & 0.02296 & -0.3608 \end{bmatrix}, \quad A_1 = \begin{bmatrix} 0.28 & 0 & 0 \\ 0 & -0.28 & 0 \\ 0 & 0 & 0 \end{bmatrix}. \qquad (3.49)$$

We investigate the effect of an uncertainty in specific elements of A_0 and A_1 on the characteristic roots by computing structured pseudospectra. From physical considerations, an important requirement on the uncertainty is that in A_1 only the elements on positions (1,1) and (2,2) are nonzero and remain opposite to each other. Physically, these elements describe the feedback process of the laser; see [384] for the details. We can take this structure into account by considering perturbations on A_1 of the form $\mathrm{diag}(\delta a, -\delta a, 0)$, with $\delta a \in \mathbb{C}$, in addition to unstructured perturbations on A_0. The resulting additive uncertainty on F has the general form (3.35), namely

$$\delta F(\lambda) = \underbrace{-I_3}_{D_1(\lambda)} \; \delta A_0 \; \underbrace{I_3}_{E_1(\lambda)} + \delta a \underbrace{\begin{bmatrix} -1 & 0 \\ 0 & 1 \\ 0 & 0 \end{bmatrix}}_{G_1(\lambda)} \underbrace{\begin{bmatrix} 1 & 0 & 0 \\ 0 & 1 & 0 \end{bmatrix} e^{-\lambda}}_{H_1(\lambda)}, \qquad (3.50)$$

and the uncertainty set $\mathbf{\Delta}$ is the set of complex block diagonal 5×5 matrices with one full 3×3 block and one repeated scalar 2×2 block. An application of Theorem 3.15 yields

$$\Lambda_\epsilon^{\mathrm{str}}(F; \mathbf{\Delta}) = \left\{ \lambda \in \mathbb{C} : \mu_{\mathbf{\Delta}} \left(\begin{bmatrix} & I_3 & \\ \hline e^{-\lambda} & 0 & 0 \\ 0 & e^{-\lambda} & 0 \end{bmatrix} F(\lambda)^{-1} \left[-I_3 \left| \begin{matrix} -1 & 0 \\ 0 & 1 \\ 0 & 0 \end{matrix} \right. \right] \right) > \frac{1}{\epsilon} \right\}.$$

In this case ($f = s = 1$), the structured singular value can be computed exactly as the solution of a convex optimization problem; see Appendix A.2. We have combined the (22)mussv routine of MATLAB with a contour plotter to visualize the structured pseudospectra, and the results are shown in Figure 3.5 (left).

For comparison, unstructured pseudospectra of (3.46) and (3.49) are shown in Figure 3.5 (right), corresponding to the measure (3.14) with $p_1 = 2$, $p_2 = \infty$, and unity weights. An application of Theorem 3.6 then yields

$$\Lambda_\epsilon(F; \|\cdot\|_{\mathrm{glob}}) = \left\{ \lambda \in \mathbb{C} : \|F(\lambda)^{-1}\|_2 \left(1 + |e^{-\lambda}|\right) > \frac{1}{\epsilon} \right\}.$$

As a significant qualitative difference, the (unstructured) ϵ-pseudospectra stretch out infinitely far along the negative real axis, even for arbitrarily small values of ϵ, following from the singularity of A_1 and Corollary 3.10. In Section 3.4, this phenomenon is related to the behavior of characteristic roots, which are introduced by perturbations that make the matrix A_1 nonsingular. Such perturbations are, however, nonphysical and, as we have demonstrated, can be excluded by applying the structured uncertainty (3.50).

3.5 ▪ Stability radii in delay parameter spaces

3.5.1 ▪ Concept

We address the uncertain system

$$\dot{x}(t)=A_0x(t)+\sum_{i=1}^{m}A_ix(t-\tau_i-w_i\delta\tau_i), \tag{3.51}$$

where $x(t)\in\mathbb{R}^n$, $A_i\in\mathbb{R}^{n\times n}$, and $\tau_i\geq 0$. The weights $\vec{w}=(w_1,\ldots,w_m)$ are used in what follows to determine the relative size of the allowable perturbations on the different delays. We assume that the zero solution of the corresponding unperturbed system

$$\dot{x}(t)=A_0x(t)+\sum_{i=1}^{m}A_ix(t-\tau_i) \tag{3.52}$$

is asymptotically stable. Letting F be the characteristic matrix of this system, we define the *stability radius* w.r.t. the delays as

$$\begin{aligned} r_{\vec{\tau}}(F):=\sup\{\gamma\geq 0:\ &\text{the zero solution of (3.51) is asymptotically stable}\\ &\text{for all } \delta\vec{\tau} \text{ satisfying } |\delta\tau_i|\leq\gamma \text{ and } \tau_i+w_i\delta\tau_i\geq 0,\ i=1,\ldots,m\}. \end{aligned}$$

Due to the nonlinear dependence of F on the delays τ_i, a computational expression for $r_{\vec{\tau}}(F)$ is difficult to obtain. Instead, we focus on the derivation of lower bounds on the stability radius. Such lower bounds correspond to *robust stability conditions.*

3.5.2 ▪ Feedback interconnection point of view

We factorize

$$A_i=B_iC_i,\ B_i\in\mathbb{R}^{n\times n_i},\ C_i\in\mathbb{R}^{n_i\times n},\quad i=1,\ldots,m, \tag{3.53}$$

where all B_i have full column rank, all C_i have full row rank, and we let $\hat{n}=\sum_{i=1}^{m}n_i$.

For $u\in\mathscr{L}_2([0,\ \infty],\mathbb{R}^{\hat{n}})$, let $y=\mathscr{G}u$ be defined by

$$\begin{cases} \dot{x}(t)=A_0x(t)+\sum_{i=1}^{m}A_ix(t-\tau_i)+\sum_{i=1}^{m}[B_1\cdots B_m]u(t),\ x(\theta)=0,\ \theta\leq 0,\\ y(t)=[w_1C_1^T\cdots w_mC_m^T]^T\dot{x}(t). \end{cases}$$

Clearly $y\in\mathscr{L}_2([0,\infty],\mathbb{R}^{\hat{n}})$. By the asymptotic stability of the unperturbed system and Parseval's theorem, the $\mathscr{L}_2$-induced norm of $\mathscr{G}$ satisfies

$$\|\mathscr{G}\|_{\mathscr{L}_2}=\|G(j\omega)\|_{\mathscr{H}_\infty},$$

where

$$G(j\omega)=j\omega\begin{bmatrix} w_1C_1\\ \vdots\\ w_mC_m \end{bmatrix}\left(j\omega I-A_0-\sum_{i=1}^{m}A_ie^{-j\omega\tau_i}\right)^{-1}[B_1\ldots B_m]. \tag{3.54}$$

Next, we let

$$\mathcal{S}_i^{\nu} : \mathcal{L}_2([0,\infty),\mathbb{R}^{\nu}) \mapsto \mathcal{L}_2([0,\infty),\mathbb{R}^{\nu}),$$
$$(\mathcal{S}_i^{\nu}\xi)(t) = \frac{1}{w_i}\int_{t-\tau_i-w_i\delta\tau_i}^{t-\tau_i} \tilde{\xi}(s)ds,$$

where $\nu \in \mathbb{N}$, $i \in \{1,\ldots,m\}$, and $\tilde{\xi} \in \mathcal{L}_2(\mathbb{R},\mathbb{R}^{\nu})$ satisfies

$$\tilde{\xi}(t) = \begin{cases} \xi(t), & t \geq 0, \\ 0, & t < 0. \end{cases} \tag{3.55}$$

By defining

$$\begin{aligned} \mathcal{D} &: \mathcal{L}_2([0,\ \infty),\mathbb{R}^{\hat{n}}) \mapsto \mathcal{L}_2([0,\ \infty),\mathbb{R}^{\hat{n}}), \\ &(\mathcal{D}\xi)(t) = \mathrm{diag}((\mathcal{S}_1^{n_1}\xi_1)(t),\ldots,(S_m^{n_m}\xi_m)(t)), \end{aligned} \tag{3.56}$$

where $\xi(t) = [\xi_1^T(t)\cdots\xi_m^T(t)]^T$, with $\xi_i(t) \in \mathbb{R}^{n_i}$, $i = 1,\ldots,m$, we can interpret the system (3.51) as a *feedback interconnection* of $\mathcal{G}$ and $\mathcal{D}$.

3.5.3 ▪ Lower bounds on the stability radius

Let the entire functions s_i be defined as

$$s_i(\lambda) = \begin{cases} e^{-\lambda\tau_i}\frac{1-e^{-\lambda(w_i\delta\tau_i)}}{w_i\ \lambda}, & \lambda \neq 0, \\ 1, & \lambda = 0, \end{cases} \qquad i = 1,\ldots,m.$$

Then the transfer function associated with the operator $\mathcal{D}$ defined in (3.56) is given by

$$D(\lambda) := \mathrm{diag}(s_1(\lambda)I_{n_1},\ldots,s_m(\lambda)I_{n_m}).$$

Note that we have

$$|s_i(j\omega)| \leq |\delta\tau_i|, \quad i = 1,\ldots,m, \tag{3.57}$$

implying in its turn that

$$\|D(j\omega)\|_2 \leq \|\delta\vec{\tau}\|_{\infty}. \tag{3.58}$$

The following result holds:

Proposition 3.16. *Define the uncertainty set*

$$\mathbf{\Delta} := \left\{\mathrm{diag}(d_1 I_{n_1},\ldots,d_m I_{n_m}) : d_i \in \mathbb{C}, 1 \leq i \leq m\right\}. \tag{3.59}$$

Then

$$r_{\vec{\tau}}(F) \geq \left(\sup_{\omega\geq 0} \mu_{\mathbf{\Delta}} G(j\omega)\right)^{-1}, \tag{3.60}$$

where $\mu_{\mathbf{\Delta}}(\cdot)$ is the structured singular value w.r.t. (3.59).

Proof. The characteristic equation of (3.51) can be written on the imaginary axis as

$$\det\left(j\omega I - A_0 - \sum_{i=1}^{m} A_i e^{-j\omega\tau_i}\right)\det(I - G(j\omega)D(j\omega)) = 0, \tag{3.61}$$

where the first factor is nonzero for all $\omega \geq 0$ because the unperturbed system is assumed to be asymptotically stable. From (3.61) and the fact that $D(j\omega) \in \mathbf{\Delta}$ for all $\omega \geq 0$, a sufficient stability condition is given by

$$||D(j\omega)||_2 < \frac{1}{\mu_{\mathbf{\Delta}}(G(j\omega))} \quad \forall \omega \geq 0.$$

The assertion follows taking into account (3.58). □

Further refinements can be found in [244] which are based on a tighter bound on $|s_i|$ than (3.57).

3.5.4 ▪ Combined uncertainty in coefficient matrices and delays

We consider a system with both uncertain matrices and uncertain delays,

$$\dot{x}(t) = (A_0 + D_0\, \delta A_0\, E_0)\, x(t) + \sum_{i=1}^{m} (A_i + D_i\, \delta A_i\, E_i)\, x(t - \tau_i - w_i \delta\tau_i), \tag{3.62}$$

under appropriate initial conditions. The uncertainty is expressed by real-valued perturbations[10] δA_i and $\delta \tau_i$. The matrices D_i and E_i, $i = 0, \dots, m$, allow us to impose a certain structure on the perturbations of the system matrices A_i. Once again, the weights on the delay perturbation bounds are denoted by $\vec{w}$. The combined stability radius of the unperturbed system (3.52) w.r.t. the *combined* uncertainty in (3.62) is defined as

$$\begin{aligned} r_{\text{comb}}(F) := \sup\{\gamma \geq 0 :\ & \text{the zero solution of (3.62) is asymptotically stable} \\ & \forall \text{ real-valued } \delta A_i \text{ satisfying } ||\delta A_i||_2 \leq \gamma,\ i = 0, \dots, m, \text{ and} \\ & \delta\tau_i \text{ satisfying } \delta\tau_i \leq \gamma \text{ and } \tau_i + w_i \delta\tau_i \geq 0,\ i = 1, \dots, m\}. \end{aligned} \tag{3.63}$$

From an analysis point of view the main difference w.r.t. the case discussed so far is the *nonlinear* but polynomial dependence of the right-hand side of (3.62) on the uncertainty, in particular, on the products of δA_i and $x(t - \tau_i - w_i \delta \tau_i)$. This problem can be overcome by introducing additional inputs and outputs. First, let $B_i, C_i, \tilde{A}_i, \tilde{D}_i, \tilde{E}_i$ be such that

$$A_i + D_i\, \delta A_i(t)\, E_i = B_i \left(\tilde{A}_i + \tilde{D}_i\, \delta A_i(t)\, \tilde{E}_i\right) C_i, \quad i = 1, \dots, m,$$

where each $C_i \in \mathbb{R}^{\tilde{n}_i \times n}$ has full row rank. A trivial choice is given by

$$B_i = C_i = I, \quad \tilde{A}_i = A_i, \quad \tilde{D}_i = D_i,\ \tilde{E}_i = E_i, \qquad i = 1, \dots, m,$$

yet it is beneficial if a decomposition can be chosen where $\text{rank}(C_i) = \tilde{n}_i < n$ (as this leads to *smaller* block sizes in the uncertainty structure). Next, we write (3.62) as the feedback interconnection of the system

$$\left\{ \begin{array}{l} \dot{x}(t) = A_0 x(t) + \sum_{i=1}^{m} A_i x(t - \tau_i) + D_0 \tilde{u}_0(t) \\ \qquad + \sum_{i=1}^{m} B_i \tilde{D}_i \tilde{u}_i(t) + \sum_{i=1}^{m} B_i \tilde{A}_i u_i(t), \\ \tilde{y}_0(t) = E_0 x(t), \\ \tilde{y}_i(t) = \tilde{E}_i C_i x(t - \tau_i) + \tilde{E}_i u_i,\ i = 1, \dots, m, \\ y_i(t) = -\zeta w_i C_i \dot{x}(t),\ i = 1, \dots, m, \end{array} \right. \tag{3.64}$$

[10] The obtained lower bounds on the stability radii will remain valid when complex-valued perturbations are allowed.

where $\zeta > 0$ and the control loop is closed with

$$\begin{cases} \tilde{u}_i(t) = \delta A_i(t)\, \tilde{y}_i(t), & i = 0,\dots,m, \\ u_i(t) = \frac{1}{\zeta}(\mathscr{S}_I^{\tilde{n}_i}\, y_i)(t), & i = 1,\dots,m. \end{cases} \tag{3.65}$$

Using this feedback interconnection point of view, lower bounds on the stability radii can be derived analogously as in the case where only the delays are uncertain. Therefore, we restrict ourselves to formulating the main result:

Proposition 3.17 (see [244]). *Define the uncertainty set*

$$\begin{aligned} \mathbf{\Delta} := \Big\{ \operatorname{diag}(\Delta_0,\dots,\Delta_m, d_1 I_{\tilde{n}_1},\dots,d_m I_{\tilde{n}_m}) :\ & \Delta_i \in \mathbb{C}^{n_i \times n_i}, d_j \in \mathbb{C}, \\ & i = 0,\dots,m,\ 1 \le j \le m \Big\}. \end{aligned} \tag{3.66}$$

Then

$$r_{\text{comb}}(F) \ge \left(\sup_{\omega \ge 0} \mu_{\mathbf{\Delta}}(G(j\omega;\, 1)) \right)^{-1},$$

where G is defined as

$$\begin{aligned} G(\lambda;\, \zeta) = & \begin{bmatrix} E_0 \\ e^{-\lambda\tau_1}\tilde{E}_1 C_1 \\ \vdots \\ e^{-\lambda\tau_m}\tilde{E}_m C_m \\ -\zeta w_1 \lambda C_1 \\ \vdots \\ -\zeta w_m \lambda C_m \end{bmatrix} \left[\lambda I - A_0 - \sum_{i=1}^{m} A_i e^{-\lambda\tau_i} \right]^{-1} \\ & \times \left[D_0\ B_1\tilde{D}_1 \cdots B_m\tilde{D}_m\ B_1\tilde{A}_1 \cdots B_m\tilde{A}_m \right] \\ & + \begin{bmatrix} 0 & \cdots & 0 & & & 0 \\ & & 0 & \tilde{E}_1 & & \\ & & & & \ddots & \\ & & & & & \tilde{E}_m \\ 0 & & & & & 0 \\ \vdots & & & & & \vdots \\ 0 & & & & \cdots & 0 \end{bmatrix}. \end{aligned}$$

3.6 ▪ Notes and references

Motivated by the observation that the eigenvalue problems encountered in the study of linear systems described by higher-order differential equations, differential algebraic equations, and delay-differential (algebraic) equations have a similar structure, we first presented a unifying treatment of pseudospectra and stability radii of nonlinear eigenvalue problems, where the structure of the problem is fully exploited. Various perturbation measures were considered and efficient formulae for the computation of both pseudospectra and stability radii were derived, whose computational complexity is similar to the complexity of computing the pseudospectra of an n-by-n matrix, with n-by-n the dimension of the characteristic matrix. The results were applied to classes of time-delay systems, which are inherently infinite-dimensional. Qualitative properties of the pseudospectra of such systems were investigated, with the emphasis on boundedness properties and the effect of the weight factors in the definition.

Next, we presented an approach for computing structured pseudospectra and stability radii of nonlinear eigenvalue problems and applied the results to time-delay systems. The proposed method allows one towards direct perturbations to specific elements (or groups of elements) of the individual coefficient matrices of the eigenvalue problem under consideration, thus to exploit the structure of these matrices, in addition to exploiting the structure of the eigenvalue problem itself. A general formula was presented which is based on the computation of appropriately defined structured singular values. It was outlined how, for special cases, the computational efficiency can be improved by exploiting additional information on the system or the uncertainty set. In fact, this reveals an inherent trade-off between the extent to which one wants to impose structure on the perturbations on the one hand, and the computational complexity of the resulting pseudospectra computations on the other hand.

Two illustrative examples were presented, including one particular example from laser physics, where structured pseudospectra were necessary to exclude physically nonrealistic perturbations, which would otherwise have a considerable impact on the pseudospectra.

Concerning the uncertainty on coefficient matrices, we have restricted ourselves to complex-valued perturbations. Some extensions to the case where only real-valued perturbations are allowed can be found in [118, 259, 330, 355, 164]. Although the extension of Theorem 3.13 to real-valued perturbations, based on a feedback interconnection point of view and appropriately defined structured singular values, is trivial from a conceptional point of view, the complexity of computing or approximating the involved structural singular values typically increases considerably. On the other hand, we recall that considering a larger class of perturbations leads to upper bounds on the sensitivity of the characteristic roots and lower bounds on the stability radii (which correspond to sufficient stability conditions). This is in agreement with the common use of pseudospectra and stability radii in a worst case analysis.

Finally, in Section 3.5 we briefly addressed stability radii in delay parameters. As the characteristic matrix depends on the delays in a nonlinear way, exact computational expressions, similar to those for uncertain coefficient matrices, are hard to obtain. Instead, we took a robust control point of view, where the delay related terms are encapsulated in bounded "uncertainty." This led us to lower bounds on stability radii—also known as sufficient robust stability conditions. It should be noted that the interconnection setting outlined in Section 3.5.2 is still applicable in a time-domain analysis, where, using techniques from $\mathscr{L}_2$ gain analysis [382], expressions for stability radii can be obtained for the case where the perturbations are allowed to vary in time [244]. Finally, if time-independent uncertainty affects one or two delays, the exact stability regions can be computed in the delay space (see also [133]). This is explained in Chapter 6 (geometric methods).

The presented results are based on [245, 401, 399, 244] and the references therein.

Chapter 4

Computation of $\mathcal{H}_2$ and $\mathcal{H}_\infty$ norms

4.1 ▪ Introduction

In the field of robust control of linear systems, stability and performance criteria are often expressed by means of $\mathcal{H}_2$ or $\mathcal{H}_\infty$ norms of appropriately defined transfer functions [414]. Therefore, the availability of robust computational methods is essential in a computer aided control system design.

$\mathcal{H}_\infty$ norms appear, for instance, in criteria for robust stability and robust performance, where a deterministic, worst case setting is taken. $\mathcal{H}_2$ criteria are related to linear quadratic optimal control (see, e.g., [5, 414]) to smooth relaxations of stability criteria (see, e.g., [386]), and they can be used as similarity measures in model reduction [95, 52]. When designing controllers, optimizing criteria expressed in terms of $\mathcal{H}_2$ norms are particularly useful if the system is affected by additive disturbances that can be accurately modeled by (filtered) white noise [414, Chapter 4].

In the first part of this chapter, Section 4.2, we present an approach to compute the $\mathcal{H}_\infty$ norm of the transfer function

$$G(j\omega) = C\left(j\omega I - A_0 - \sum_{i=1}^{m} A_i e^{-j\omega\tau_i}\right)^{-1} B + De^{-j\omega\tau_0}, \tag{4.1}$$

where $A_i \in \mathbb{R}^{n\times n}$, $0 \le i \le m$, $B \in \mathbb{R}^{n\times n_u}$, $C \in \mathbb{R}^{n_y\times n}$, $D \in \mathbb{R}^{n_y\times n_u}$ are the system matrices, and the nonnegative numbers $(\tau_1,\ldots,\tau_m)$ correspond to time-delays. The $\mathcal{H}_\infty$ norm of the transfer function (4.1) is finite if and only if the zeros of the equation

$$\det\left(\lambda I - A_0 - \sum_{i=1}^{m} A_i e^{-\lambda\tau_i}\right) = 0$$

are confined to the open left half complex plane. Under this condition it can be expressed as

$$\|G(j\omega)\|_{\mathcal{H}_\infty} = \sup_{\omega\ge 0} \sigma_1(G(j\omega)),$$

where $\sigma_1(\cdot)$ denotes the largest singular value [414, 244].

The commonly used methods for computing $\mathcal{H}_\infty$ norms and related robustness measures for systems without delay belong to the class of level set methods. They are based on the duality between the singular value plot of the transfer function and the position of the spectrum of an appropriately defined Hamiltonian matrix with respect to the imaginary axis, as expressed in the following result from [36] (see also [60]):

Proposition 4.1. *Let $G_f(j\omega) = C(j\omega I - A)^{-1}B + D$. Let $\xi > 0$ be such that the matrix $D^T D - \xi^2 I$ is nonsingular. For $\omega \geq 0$, the matrix $G_f(j\omega)$ has a singular value equal to ξ if and only if $j\omega$ is an eigenvalue of the matrix*

$$L_\xi = \begin{bmatrix} A & -B(D^T D - \xi^2 I)^{-1} B^T \\ -C^T C + C^T D(D^T D - \xi^2 I)^{-1} D^T C & -A^T \end{bmatrix}.$$

From Proposition 4.1 we get

$$\|G_f(j\omega)\|_{\mathcal{H}_\infty} = \inf\{\xi > \sigma_1(D) : L_\xi \text{ has no imaginary axis eigenvalues}\}.$$

This result directly leads to a bisection algorithm on the parameter ξ for computing the $\mathcal{H}_\infty$ norm of $G_f(j\omega)$, as outlined in [36]. Quadratically convergent algorithms based on a search in a two-parameter space (ω, ξ) are presented in [35, 49]. A similar algorithm for computing pseudospectral abscissa for systems without delays is proposed in [57].

Our approach for computing the $\mathcal{H}_\infty$ norm of (4.1) builds on a generalization of Proposition 4.1 to time-delay systems. Due to the fact that a time-delay system is inherently infinite-dimensional, the singular value curves of (4.1) can no longer be related to the imaginary axis eigenvalues of a matrix but to the imaginary axis eigenvalues of an infinite-dimensional linear operator $\mathcal{L}_\xi$, as we shall see. This leads to a two-step approach for the computation of the $\mathcal{H}_\infty$ norm of (4.1). In the first step (the prediction step), an approximation of the $\mathcal{H}_\infty$ norm of (4.1) is computed based on a finite-dimensional approximation of the system, induced by a discretization of the operator $\mathcal{L}_\xi$. Because this operator is a derivative operator on a function space with nonlocal boundary condition, the discretization is done using a spectral method [377], well established for this type of operator; see [44] and the references therein. In the next step (the correction step) the approximation of the $\mathcal{H}_\infty$ norm is improved up to the desired accuracy with a local method by solving a set of nonlinear equations. These are obtained from the reformulation of the eigenvalue problem for the linear infinite-dimensional operator $\mathcal{L}_\xi$ as a nonlinear eigenvalue problem of finite dimension.

The proposed method for computing $\mathcal{H}_\infty$ norms has several similarities with some existing methods for computing characteristic roots of time-delay systems, although the underlying problems are totally different. First, the characteristic roots solve an infinite-dimensional linear eigenvalue problem as well as a finite-dimensional nonlinear eigenvalue problem (induced by the characteristic equation). As we have seen in Chapter 2, this may also lead to a two-step approach. Approximations of the characteristic roots are obtained by discretizing the infinite-dimensional linear eigenvalue problem and solving the resulting matrix eigenvalue problem. The approximate characteristic roots are corrected subsequently by Newton iterations on the nonlinear characteristic equation.

In the second part of the chapter, Section 4.3, we consider the computation of the $\mathcal{H}_2$ norm of (4.1). The $\mathcal{H}_2$ norm of the exponentially stable time-delay system can be defined analogous to the delay-free case, i.e.,

$$\|G(j\omega)\|_{\mathcal{H}_2}^2 := \frac{1}{2\pi}\int_{-\infty}^{\infty} \operatorname{Tr}(G(i\omega)^* G(i\omega))\, d\omega.$$

The corresponding time-domain representation is the energy functional

$$\|G(j\omega)\|_{\mathcal{H}_2}^2 = \int_{t=0}^{\infty} \operatorname{Tr}(h(t)^T h(t))\, dt, \tag{4.2}$$

where h denotes the impulse response. In order to have a finite $\mathcal{H}_2$ norm we will assume that $D \neq 0$ throughout Section 4.3.

A fundamental and elegant property of the $\mathcal{H}_2$ norm for systems without delay is that it can be computed from the solutions of Lyapunov equations (see, e.g., [414, Lemma 4.6]). We will see that this is also the case for systems with delay. Instead of solving the Lyapunov equations we need to solve the *delay Lyapunov equations*, which fortunately have already been studied in the literature (see, e.g., [327, 130, 192, 388, 309, 308, 193, 189]), but mostly in the context of stability assessment with Lyapunov's second method. We will present a formula for the $\mathcal{H}_2$ norm in terms of the solution to the delay Lyapunov equations. We will outline how these equations can be solved by a spectral discretization. In the case of commensurate delays, an explicit expression for the $\mathcal{H}_2$ norm can be obtained, containing only fundamental linear algebra operations on matrices of finite dimension.

Throughout the chapter the following assumption is made:

Assumption 4.2. *The delays are ordered such that* $0 < \tau_1 < \cdots < \tau_m = 1$.

Note that Assumption 4.2 can be taken without any loss of generality because the variable ω and the system matrices in (4.1) can always be rescaled. It will allow us to significantly simplify the notations.

4.2 ▪ Computing $\mathcal{H}_\infty$ norms

4.2.1 ▪ Theoretical basis

We briefly present the theoretical results on which the algorithm for the $\mathcal{H}_\infty$ norm computation is based. We refer the reader to [247] for the proofs.

Relations with a Hamiltonian eigenvalue problem

The following lemma extends Proposition 4.1 to time-delay systems:

Lemma 4.3. *Let* $\xi > 0$ *be such that the matrix*

$$D_\xi := D^T D - \xi^2 I$$

is nonsingular. For $\omega \geq 0$*, the matrix* $G(j\omega)$ *has a singular value equal to* ξ *if and only if* $\lambda = j\omega$ *is a solution of the equation*

$$\det H(\lambda, \xi) = 0, \tag{4.3}$$

where

$$H(\lambda,\xi) := \lambda I - M_0 - \sum_{i=1}^{m}\left(M_i e^{-\lambda\tau_i} + M_{-i}e^{\lambda\tau_i}\right) - \left(N_1^{-\lambda\tau_0} + N_{-1}e^{\lambda\tau_0}\right), \tag{4.4}$$

with

$$M_0 = \begin{bmatrix} A_0 & -BD_\xi^{-1}B^T \\ -C^TC + C^TDD_\xi^{-1}D^TC & -A_0^T \end{bmatrix},$$

$$M_i = \begin{bmatrix} A_i & 0 \\ 0 & 0 \end{bmatrix},\ M_{-i} = \begin{bmatrix} 0 & 0 \\ 0 & -A_i^T \end{bmatrix},\ 1 \leq i \leq N,$$

$$N_1 = \begin{bmatrix} 0 & 0 \\ 0 & C^TDD_\xi^{-1}B^T \end{bmatrix},\ N_{-1} = \begin{bmatrix} -BD_\xi^{-1}D^TC & 0 \\ 0 & 0 \end{bmatrix}.$$

For a fixed value of ξ, the solutions of (4.3) can be found by solving the nonlinear eigenvalue problem

$$H(\lambda,\xi)\,v=0,\quad \lambda\in\mathbb{C},\ v\in\mathbb{C}^{2n},\ v\neq 0. \tag{4.5}$$

This nonlinear eigenvalue problem can be "linearized" to an infinite-dimensional linear eigenvalue problem. For this, we consider the space

$$X:=\mathscr{C}\left(\left[-\max_{0\le i\le m}\tau_i,\ \max_{0\le i\le m}\tau_i\right],\mathbb{C}^{2n}\right)=\mathscr{C}([-1,\ 1],\mathbb{C}^{2n}),$$

where we have taken into account Assumption 4.2. We let the operator $\mathscr{L}_\xi$ on X be defined by

$$\mathscr{D}(\mathscr{L}_\xi)=\left\{\phi\in X:\ \phi'\in X,\ \phi'(0)=M_0\phi(0)+\sum_{i=1}^{m}(M_i\phi(-\tau_i)+M_{-i}\phi(\tau_i))\right. \tag{4.6}$$

$$\left.+N_1\phi(-\tau_0)+N_{-1}\phi(\tau_0)\right\},$$

$$\mathscr{L}_\xi\,\phi=\phi',\quad \phi\in\mathscr{D}(\mathscr{L}_\xi). \tag{4.7}$$

The eigenvalue problem for this linear operator is defined as

$$(\lambda I-\mathscr{L}_\xi)u=0:\ \lambda\in\mathbb{C},\ u\in X,\ u\neq 0. \tag{4.8}$$

The eigenvalues of $\mathscr{L}_\xi$ have a one-to-one correspondence to the eigenvalues of the nonlinear eigenvalue problem (4.5):

Proposition 4.4. *Let H be defined by (4.4). Let $\xi>0$ be such that $D^TD-\xi^2I$ is nonsingular. Then we have*

$$\exists v\in\mathbb{C}^{2n},\ v\neq 0:\ H(\lambda,\xi)v=0\Leftrightarrow\exists u\in X,\ u\neq 0:\ (\lambda I-\mathscr{L}_\xi)u=0.$$

Furthermore, if (λ,u) satisfies (4.8), then u has the form

$$u(\theta)=e^{\lambda\theta},\ \theta\in[-1,\ 1], \tag{4.9}$$

where $v\in\mathbb{C}^{2n}$ and (λ,v) satisfies (4.5). Conversely, if (λ,v) satisfies (4.5), then (λ,u) satisfies (4.8) with u given by (4.9).

By combining Lemma 4.3 and Proposition 4.4 we arrive at:

Theorem 4.5. *Let $\xi>0$ be such that the matrix $D^TD-\xi^2I$ is nonsingular. For $\omega\ge 0$, the matrix $G(j\omega)$ has a singular value equal to ξ if and only if $\lambda=j\omega$ is an eigenvalue of the operator $\mathscr{L}_\xi$, defined by (4.6) and (4.7).*

Corollary 4.6.

$$\|G(j\omega)\|_{\mathscr{H}_\infty}=\inf\{\xi>\sigma_1(D^TD):\ \textit{operator}\ \mathscr{L}_\xi\ \textit{has no imaginary axis eigenvalues}\}.$$

Properties of the eigenvalue problem

Although the operator $\mathscr{L}_\xi$ generally has an infinite number of eigenvalues, the number of eigenvalues on the imaginary axis is always finite; see [247].

The set of eigenvalues of $\mathscr{L}_\xi$ is symmetric w.r.t. the imaginary axis, as expressed in the following proposition (for comparison, in the delay-free case the operator $\mathscr{L}_\xi$ reduces to a Hamiltonian matrix):

Proposition 4.7. *A complex number λ is an eigenvalue of $\mathscr{L}_\xi$ if and only if $-\bar{\lambda}$ is an eigenvalue of $\mathscr{L}_\xi$.*

Interpretation in a two parameter space

We consider the equation

$$\det H(j\omega,\xi)=0 \tag{4.10}$$

as the central equation, and look at its solutions in the two-parameter space $(\omega,\xi)\in\mathbb{R}_+\times\mathbb{R}_+$. From Lemma 4.3 and Proposition 4.4 we have the following theorem:

Theorem 4.8. *Let $\omega\geq 0$ and $\xi\geq 0$ such that the matrix D_ξ is nonsingular. The following statements are equivalent:*

1. $\det H(j\omega,\xi)=0$.
2. *$j\omega$ is an eigenvalue of the operator $\mathscr{L}_\xi$.*
3. *ξ is a singular value of the matrix $G(j\omega)$.*

Graphically, equation (4.10) defines a set of curves in the (ω,ξ) parameter space. The intersections with horizontal lines (ξ fixed) can be found by computing the imaginary axis eigenvalues of $\mathscr{L}_\xi$. Similarly, the intersections with vertical lines (ω fixed) can be found by computing the singular values of $G(j\omega)$. This is shown in Figure 4.1.

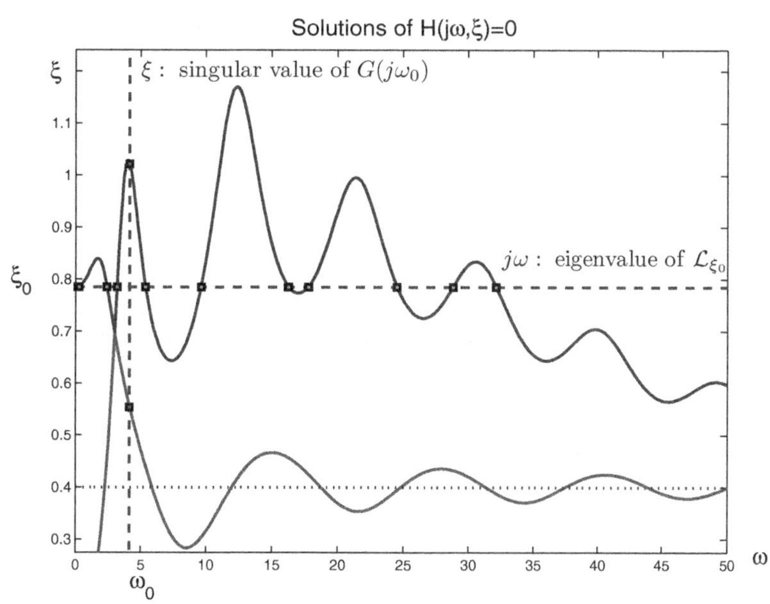

Figure 4.1. *Solutions of equation* (4.10) *for the problem data* (4.35) [247].

4.2.2 ▪ Finite-dimensional approximation

The numerical methods for computing $\mathcal{H}_\infty$ norms presented in Section 4.2.3 are strongly based on Corollary 4.6. Because the operator $\mathcal{L}_\xi$, defined by (4.6)–(4.7), is infinite-dimensional, these algorithms will involve a discretization of this operator. In this section we outline a discretization approach and discuss its properties.

Discretization

Following the approach of [43, 44], we discretize the operator $\mathcal{L}_\xi$ using a *spectral method* (see Chapter 2, where a spectral discretization is used for computing characteristic roots).

Given a positive integer N, we consider a mesh Ω_N of $2N+1$ distinct points in the interval $[-1, 1]$:

$$\Omega_N = \{\theta_{N,i},\ i=-N,\dots,N\}, \tag{4.11}$$

where

$$-1 \le \theta_{N,-N} < \cdots < \theta_{N,-1} < \theta_{N,0} = 0 < \theta_{N,1} < \cdots < \theta_{N,N} \le 1$$

and

$$\theta_{N,-i} = -\theta_{N,i}, \quad i=1,\dots,N. \tag{4.12}$$

The mesh Ω_N allows us to replace the continuous space X with a space X_N of discrete functions. More precisely, a function $\phi \in X$ is discretized into a block vector $x = [x_{-N}^T \cdots x_N^T]^T \in X_N$ with components

$$x_i = \phi(\theta_{N,i}) \in \mathbb{C}^{2n}, \quad i=-N,\dots,N.$$

When defining $\mathcal{P}_N x$, $x \in X_N$, as the unique $\mathbb{C}^{2n}$-valued interpolating polynomial of degree less than or equal to $2N$ satisfying

$$\mathcal{P}_N x(\theta_{N,i}) = x_i, \quad i=-N,\dots,N,$$

we can approximate the operator $\mathcal{L}_\xi$ over X with the matrix $\mathcal{L}_\xi^N : X_N \to X_N$, defined as

$$\begin{aligned}
\left(\mathcal{L}_\xi^N x\right)_i &= (\mathcal{P}_N x)'(\theta_{N,i}), && i=-N,\dots,-1,\\
\left(\mathcal{L}_\xi^N x\right)_0 &= M_0 \mathcal{P}_N x(0) + \sum_{i=1}^m \left(M_i \mathcal{P}_N x(-\tau_i) + M_{-i}\mathcal{P}_N x(\tau_i)\right)\\
&\quad + N_1 \mathcal{P}_N x(-\tau_0) + N_{-1}\mathcal{P}_N x(\tau_0),\\
\left(\mathcal{L}_\xi^N x\right)_i &= (\mathcal{P}_N x)'(\theta_{N,i}), && i=1,\dots,N.
\end{aligned} \tag{4.13}$$

An *explicit expression* for the elements of the matrix $\mathcal{L}_\xi^N$ can be obtained by using the Lagrange representation of $\mathcal{P}_N x$,

$$\mathcal{P}_N x = \sum_{k=-N}^{N} l_{N,k}\, x_k, \tag{4.14}$$

where the Lagrange polynomials $l_{N,k}$ are real-valued polynomials of degree $2N$ satisfying

$$l_{N,k}(\theta_{N,i}) = \begin{cases} 1, & i=k,\\ 0, & i\neq k. \end{cases}$$

By substituting (4.14) in (4.13) we obtain the expression

$$\mathcal{L}_\xi^N = \begin{bmatrix} d_{-N,-N} & \cdots & d_{-N,N} \\ \vdots & & \vdots \\ d_{-1,-N} & \cdots & d_{-1,N} \\ a_{-N} & \cdots & a_N \\ d_{1,-N} & \cdots & d_{1,N} \\ \vdots & & \vdots \\ d_{N,-N} & \cdots & d_{N,N} \end{bmatrix} \in \mathbb{R}^{(2N+1)(2n)\times(2N+1)2n},$$

where

$$\begin{aligned} d_{i,k} &= l'_{N,k}(\theta_{N,i})I, \quad i\in\{-N,\dots,-1,1,\dots,N\},\ k\in\{-N,\dots,N\}, \\ a_0 &= M_0 + \sum_{k=1}^{m}\left(M_k l_{N,0}(-\tau_k) + M_{-k} l_{N,0}(\tau_k)\right) + N_1 l_{N,0}(-\tau_0) + N_{-1} l_{N,0}(\tau_0), \\ a_i &= \sum_{k=1}^{m}\left(M_k l_{N,i}(-\tau_k) + M_{-k} l_{N,i}(\tau_k)\right) + N_1 l_{N,i}(-\tau_0) + N_{-1} l_{N,i}(\tau_0), \\ & i\in\{-N,\dots,-1,1,\dots,N\}. \end{aligned}$$

It is important to note that all the problem specific information and the parameter ξ are concentrated in the middle row of $\mathcal{L}_\xi^N$, i.e., the elements $(a_{-N},\dots,a_N)$, while all other elements of $\mathcal{L}_\xi^N$ can be computed beforehand.

We outline some properties of the matrix $\mathcal{L}_\xi^N$. First, analogously to the continuous case, the (linear) eigenvalue problem for $\mathcal{L}_\xi^N$,

$$\mathcal{L}_\xi^N x = \lambda x, \quad \lambda\in\mathbb{C},\ x\in\mathbb{C}^{(2N+1)2n},\ x\neq 0, \tag{4.15}$$

has a nonlinear eigenvalue problem of dimension $2n$ as counterpart. To clarify this we need the following definition:

Definition 4.9. *For $\lambda\in\mathbb{C}$, let $p_N(\cdot;\ \lambda)$ be the polynomial of degree $2N$ satisfying*

$$\begin{aligned} &p_N(0;\ \lambda) = 1, \\ &p'_N(\theta_{N,i};\ \lambda) = \lambda p_N(\theta_{N,i};\ \lambda),\ \ i\in\{-N,\dots,-1\}\cup\{1,\dots,N\}. \end{aligned} \tag{4.16}$$

Note that the polynomial $p_N(t;\ \lambda)$ is an approximation of $\exp(\lambda t)$ on the interval $[-1;\ 1]$. Indeed, the first equation of (4.16) is an interpolation requirement at zero; the other equations are collocation conditions for the differential equation $\dot z = \lambda z$, of which $\exp(\lambda t)$ is a solution. The following main result, proven in [247], relates a spectral discretization with a direct approximation of exponentials:

Proposition 4.10. *The following statements are equivalent:*

$$\exists x\in\mathbb{C}^{(2N+1)2n},\ x\neq 0:\ \left(\lambda I - \mathcal{L}_\xi^N\right)x = 0 \Leftrightarrow \exists v\in\mathbb{C}^{2n},\ v\neq 0:\ H_N(\lambda,\xi)v = 0,$$

where

$$\begin{aligned} H_N(\lambda,\xi) := \lambda I - M_0 &- \sum_{i=1}^{m}\left(M_i p_N(-\tau_i;\ \lambda) + M_{-i} p_N(\tau_i,\ \lambda)\right) \\ &- \left(N_1 p_N(-\tau_0;\ \lambda) + N_{-1} p_N(\tau_0;\ \lambda)\right). \end{aligned} \tag{4.17}$$

Notice that $H_N(\lambda,\xi)$ can be obtained from $H(\lambda,\xi)$, by making the substitution

$$e^{-\lambda\tau_i} \leftarrow p_N(-\tau_i;\,\lambda), \quad i=-N,\dots,N.$$

As we shall see in Section 4.2.3, the functions $p_N(-\tau_i;\,\lambda)$ are proper rational functions of the parameter λ. Thus, the effect of a spectral discretization of the operator $\mathcal{L}_\xi$ can be interpreted as the effect of a *rational approximation* of $H(\lambda,\xi)$.

Second, the spectral property described in Proposition 4.7 is preserved, due to the symmetry of the grid:

Proposition 4.11. *A complex number λ is an eigenvalue of $\mathcal{L}_\xi^N$ if and only if $-\bar{\lambda}$ is an eigenvalue of $\mathcal{L}_\xi^N$.*

Interpretation in a two-parameter space

Similarly as in Section 4.2.1 we characterize the solutions of the two-parameter problem

$$\det H_N(j\omega,\xi)=0, \tag{4.18}$$

where $\omega\geq 0$ and $\xi\geq 0$. The counterpart of Theorem 4.8 reads as follows (for the proof, we once more refer the reader to [247]):

Theorem 4.12. *Let $\omega\geq 0$ and let $\xi\geq 0$ such that the matrix D_ξ is nonsingular. The following statements are equivalent:*

1. $\det H_N(j\omega,\xi)=0$.
2. *$j\omega$ is an eigenvalue of $\mathcal{L}_\xi^N$.*
3. *ξ^2 is an eigenvalue of the matrix*

$$M_N(j\omega):=\begin{bmatrix} X_N(j\omega)+X_N(j\omega)^*+Y_N(j\omega)^*Y_N(j\omega)+D^TD & \sqrt{r(\omega)}\,X_N(j\omega)^* \\ \sqrt{r(\omega)}\,X_N(j\omega) & D^TD \end{bmatrix} \tag{4.19}$$

where

$$X_N=D^TC\left(j\omega I-A_0-\sum_{i=1}^{m}A_i p_N(-\tau_i;\,j\omega)\right)^{-1}Bp_N(\tau_0;\,j\omega),$$

$$Y_N=C\left(j\omega I-A_0-\sum_{i=1}^{m}A_i p_N(-\tau_i;\,j\omega)\right)^{-1}B,$$

$$r_N(\omega)=\frac{1}{|p_N(-\tau_0;\,j\omega)|^2}-1.$$

The graphical interpretation of Theorem 4.12 is as follows. The intersections between the curves in the (ω,ξ) plane, defined by (4.18), with horizontal lines (ξ fixed), can be found by computing the imaginary axis eigenvalues of the matrix $\mathcal{L}_\xi^N$. Similarly, the intersections with vertical lines (ω fixed) can be found by computing the positive real eigenvalues of the matrix $M_N(j\omega)$. This is illustrated in Figure 4.2.

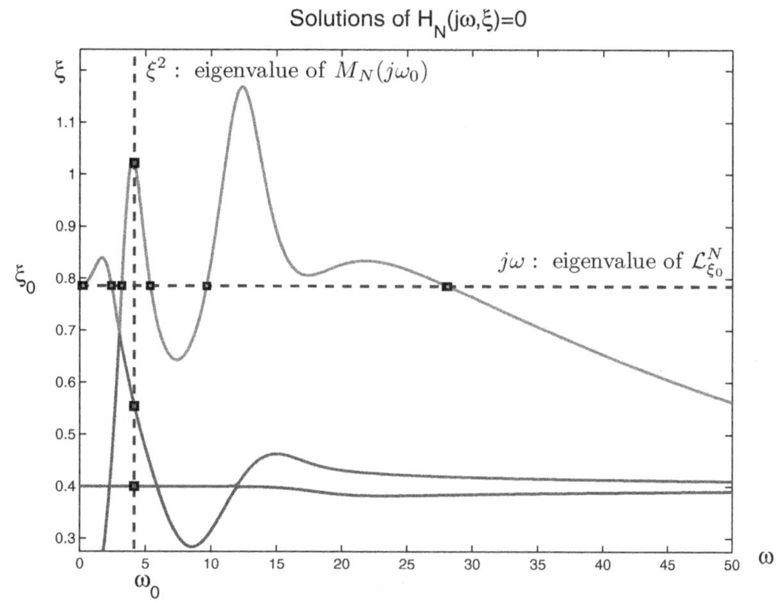

Figure 4.2. *Solutions of equation* (4.18) *with* $N = 10$ *for the problem data* (4.35) [247].

In the remainder of the section, we discuss some spectral properties of the matrix $M_N(j\omega)$, defined in (4.19).

Proposition 4.13. *The following statements hold:*

1. *The eigenvalues of* $M_N(j\omega)$ *are real or appear in complex conjugate pairs.*

2. *The matrix* $M_N(j\omega)$ *has* $2n_u$ *eigenvalues* $\lambda_k(M_N(j\omega))$, $1 \le k \le 2n_u$, *satisfying*

$$\lim_{\omega\to\infty} \lambda_k(M_N(j\omega)) = (\sigma_l(D))^2 \text{ for some } l \in \{1,\dots,n_u\}.$$

3. *If* $r(\omega) \ge 0$ *or, equivalently,* $|p_N(-\tau_0, j\omega)| \le 1$, *then the matrix* $M_N(j\omega)$ *is Hermitian.*

4. *If* $\tau_0 = 0$, *then*

$$M_N(j\omega) = \begin{bmatrix} G_N(j\omega)^* G_N(j\omega) & 0 \\ 0 & D^T D \end{bmatrix},$$

where

$$G_N(j\omega) := C\left(j\omega I - A_0 - \sum_{i=1}^{m} A_i p_N(-\tau_i;\, j\omega)\right)^{-1} B + D, \tag{4.20}$$

with p_N *defined by* (4.16).

The fourth assertion of Proposition 4.13 is of particular interest, because it shows that under the condition $\tau_0 = 0$ the curves defined by (4.18) can be interpreted as the singular value curves of a rational approximation $G_N(j\omega)$ of $G(j\omega)$.

To conclude this section, we summarize the relations between the eigenvalue problems defined in Sections 4.2.1–4.2.2 in Table 4.1.

Table 4.1. *Relations between $\mathcal{L}_\xi$, $H(j\omega,\xi)$, and $\|G(j\omega)\|_{\mathcal{H}_\infty}$, as well as their discrete counterparts. The latter are all induced by a spectral discretization of the operator $\mathcal{L}_\xi$ into the matrix $\mathcal{L}_\xi^N$* [247].

	Implicit two-parameter problem	Horizontal search (explicit in ω)	Vertical search (explicit in ξ)	Corresponding $\mathcal{H}_\infty$ problem
continuous (Section 4.2.1)	$\det H(j\omega,\xi)=0$	$(j\omega I-\mathcal{L}_\xi)u=0$	$(\xi^2 I-G(j\omega)^*G(j\omega))\,v=0$	$\|G(j\omega)\|_{\mathcal{H}_\infty}$
discretized (Section 4.2.2)	$\det H_N(j\omega,\xi)=0$	$(j\omega I-\mathcal{L}_\xi^N)x=0$	$(\xi^2 I-M_N(j\omega))v=0$	$\|G_N(j\omega)\|_{\mathcal{H}_\infty}$ if $\tau_0=0$

4.2.3 ▪ Algorithm

Similarly to the algorithm for the computation of characteristic roots proposed in Chapter 2, we propose an algorithm for computing $\mathcal{H}_\infty$ norms that relies on a two-step approach. In the first step, we approximate (predict) the $\mathcal{H}_\infty$ norm based on the discretization of the operator $\mathcal{L}_\xi$ into the matrix $\mathcal{L}_\xi^N$. In the second step, we correct the results based on the reformulation of the eigenvalue problem for $\mathcal{L}_\xi$ as a nonlinear eigenvalue problem of finite dimension. Finally, we discuss the choice of the discretization stepsize in the prediction step.

Throughout this section we assume that the grid Ω_N, employed in the discretization of $\mathcal{L}_\xi$, consists of Chebyshev extremal points in the discretization, that is,

$$\theta_{N,i}=\cos\left(\frac{(N-i)\pi}{2N}\right),\quad i=-N,\dots,N. \tag{4.21}$$

Prediction of the $\mathcal{H}_\infty$ norm

Inspired by Corollary 4.6, a natural way to approximate $\|G(j\omega)\|_{\mathcal{H}_\infty}$ consists of computing

$$g_{\max}(N):=\inf\left\{\xi>\sigma_1(D^TD):\ \text{matrix } \mathcal{L}_\xi^N \text{ has no imaginary axis eigenvalues}\right\}, \tag{4.22}$$

where N is fixed. The following properties lay the groundwork for the corresponding algorithms.

Proposition 4.14. *If the condition*

$$|p_N(-\tau_0;\, j\omega)|\le 1\quad \forall\omega\ge 0 \tag{4.23}$$

is satisfied, then the following statements hold:

1 *The quantity $g_{\max}(N)$, defined in (4.22), is finite.*

2 *The matrix $\mathcal{L}_\xi^N$ has eigenvalues on the imaginary axis for all*

$$\xi\in(\sigma_1(D),\ g_{\max}(N)]$$

and no eigenvalues on the imaginary axis for $\xi>g_{\max}(N)$.

3 *The matrix $M_N(j\omega)$ has $2n_u$ real eigenvalues $\lambda_k(M_N(j\omega)$, $1\le k\le 2n_u$ for all $\omega\ge 0$, satisfying*

$$\lim_{\omega\to\infty}\lambda_k(M_N(j\omega))=(\sigma_l(D))^2 \textit{ for some } l\in\{1,\dots,n_u\}.$$

Moreover, if $\tau_0 = 0$*, then*

$$g_{\max}(N) = \|G_N(j\omega)\|_{\mathscr{H}_\infty},$$

where $G_N(j\omega)$ *is defined as in* (4.20).

Notice that the condition (4.23) is always satisfied if $\tau_0 = 0$, which implies $p_N(-\tau_0;\ j\omega) \equiv 1$. If $\tau_0 \neq 0$, then it is also satisfied by the choice (4.21) of the grid Ω_N.

Definition (4.22) and properties 1–2 described in Proposition 4.14 naturally lead to a bisection algorithm on the parameter ξ to compute $g_{\max}(N)$, similarly to the algorithm presented in [60]. However, based on the interpretations described in Section 4.2.2 and property 3 of Proposition 4.14, the efficiency can be improved by performing a criss-cross search in the two-parameter space (ω, ξ), instead of a search in the parameter ξ only. More precisely, an adaptation of the algorithm presented in [49] results in the following (see also [57]):

ALGORITHM 4.1.

Input: system data, N*, symmetric grid* Ω_N *defined by* (4.21)*, candidate critical frequency* ω_t *if available, tolerance tol*
Output: $g_{\max}(N)$

1. compute a lower bound ξ_l on $g_{\max}(N)$,
   e.g., $\xi_l = \max\left\{\sigma_1(G(0)), \sigma_1(D), \text{tol}, \sqrt{\max(\lambda_1(M_N(j\omega_t)), 0)}\right\}$

2. repeat until break

   2.1 set $\xi := \xi_l(1 + 2\,\text{tol})$

   2.2 compute the set of eigenvalues $\mathscr{E}_\xi$ of the matrix $\mathscr{L}_\xi^N$ on the positive imaginary axis, $\mathscr{E}_\xi := \left\{j\omega^{(1)}, j\omega^{(2)}, \ldots\right\}$, with $0 \le \omega^{(1)} < \omega^{(2)} < \cdots$

   2.3 if $\mathscr{E}_\xi = \phi$, break
   else
   $\mu^{(i)} := \sqrt{\omega^{(i)}\omega^{(i+1)}}$, $i = 1, 2, \cdots$

   set $\xi_l := \max_i \sqrt{\max\left(\lambda_1(M_N(j\mu^{(i)})), \sigma_1(D)^2\right)}$,

   {result: estimate $(\xi + \xi_l)/2$ for $g_{\max}(N)$}

The underlying idea is illustrated in Figure 4.3, where some steps of the algorithm are visualized.

Algorithm 4.1 relies on the evaluation of the matrix $M_N(\lambda)$ and, hence, on the evaluation of the functions

$$\lambda \mapsto p_N(\pm\tau_i;\ \lambda), \quad i = 0, \ldots, m,$$

for specific values of λ. For this, we represent $p_N(\cdot;\ \lambda)$ in a polynomial basis (with T_i denoting the Chebyshev polynomial of the first kind and order i):

$$p_N(t;\ \lambda) = \sum_{i=0}^{2N} \alpha_i T_i(t), \quad t \in [-1,\ 1],$$

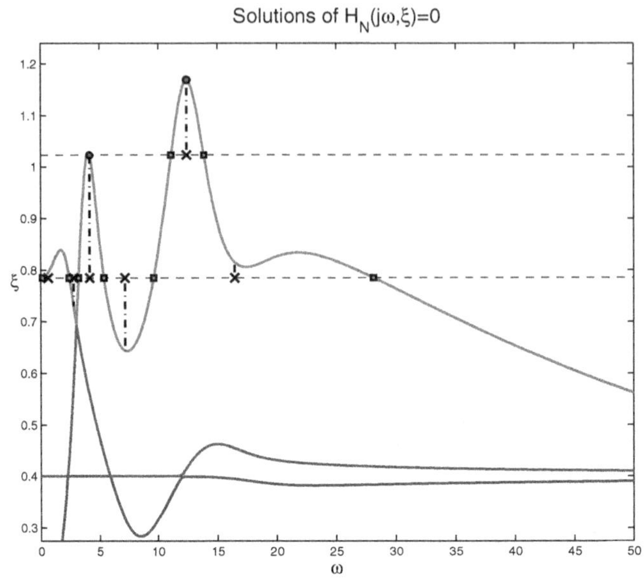

Figure 4.3. *Some steps of Algorithm* 4.1, *for the problem data* (4.35) [247].

where we suppress the dependence of the coefficients α_i on λ in the notation. For $\lambda \neq 0$, conditions (4.16) can be written as

$$(\lambda T - U)\begin{bmatrix} \alpha_0 \\ \vdots \\ \alpha_{2N} \end{bmatrix} = R, \tag{4.24}$$

where

$$T = \begin{bmatrix} T_0(\theta_{N,-N}) & \cdots & T_{2N}(\theta_{N,-N}) \\ \vdots & & \vdots \\ T_0(\theta_{N,-1}) & \cdots & T_{2N}(\theta_{N,-1}) \\ T_0(0) & \cdots & T_{2N}(0) \\ T_0(\theta_{N,1}) & \cdots & T_{2N}(\theta_{N,1}) \\ \vdots & & \vdots \\ T_0(\theta_{N,N}) & \cdots & T_{2N}(\theta_{N,N}) \end{bmatrix}, \quad U = \begin{bmatrix} T_0'(\theta_{N,-N}) & \cdots & T_{2N}'(\theta_{N,-N}) \\ \vdots & & \vdots \\ T_0'(\theta_{N,-1}) & \cdots & T_{2N}'(\theta_{N,-1}) \\ 0 & \cdots & 0 \\ T_0'(\theta_{N,1}) & \cdots & T_{2N}'(\theta_{N,1}) \\ \vdots & & \vdots \\ T_0'(\theta_{N,N}) & \cdots & T_{2N}'(\theta_{N,N}) \end{bmatrix}, \quad R = \begin{bmatrix} 0 \\ \vdots \\ 0 \\ \lambda \\ 0 \\ \vdots \\ 0 \end{bmatrix}.$$

The matrix T is always invertible, which can easily be deduced from a representation in a Lagrange basis. After solving (4.24) for a given value of λ we can evaluate

$$p_N(\pm\tau_i;\, \lambda) = \sum_{i=0}^{2N} \alpha_i T_i(\pm\tau_i), \quad i = 0, \ldots, m.$$

Our implementation is based on a representation in an orthogonal basis of Chebyshev polynomials.

Remark 4.15. *We can formally write*

$$p_N(t;\, \lambda) = S(t)(\lambda T - U)^{-1}R,$$

where $S(t) = [T_0(t) \cdots T_{2N}(t)]$. This shows that the functions $p_N(\pm\tau_i;\ \lambda)$ are a proper rational function of λ. Note that the coefficients of this rational function never need to be explicitly computed.

Correction of the $\mathcal{H}_\infty$ norm

We describe how an approximation of $\|G(j\omega)\|_{\mathcal{H}_\infty}$ can be corrected to the actual value. This is done by solving a set of nonlinear equations, which we derive first.

Let $\hat{\xi} \geq 0$ and $\hat{\omega} \geq 0$ be such that

$$\|G(j\omega)\|_{\mathcal{H}_\infty} = \hat{\xi} = \sigma_1(G(j\hat{\omega})), \tag{4.25}$$

and assume for the moment that the singular value $\sigma_1(G(j\hat{\omega}))$ has multiplicity one. Then the nonlinear eigenvalue problem (4.5), with $\xi = \hat{\xi}$, has a double non-semisimple eigenvalue $\lambda = j\hat{\omega}$ (see [170] for the definition of multiple eigenvalues of a nonlinear eigenvalue problem). This property is clarified in Figure 4.4. Therefore, setting

$$h(\lambda, \xi) = \det H(\lambda, \xi),$$

the pair $(\hat{\omega}, \hat{\xi})$ satisfies

$$h(j\omega, \xi) = 0, \quad \frac{\partial h}{\partial \lambda}(j\omega, \xi) = 0. \tag{4.26}$$

These complex-valued equations seem overdetermined, but this is not the case due to the spectral properties of the operator $\mathcal{L}_\xi$. As a corollary of Proposition 4.7 we namely get:

Corollary 4.16. *For $\omega \geq 0$, we have*

$$\Im\, h(j\omega, \xi) = 0 \tag{4.27}$$

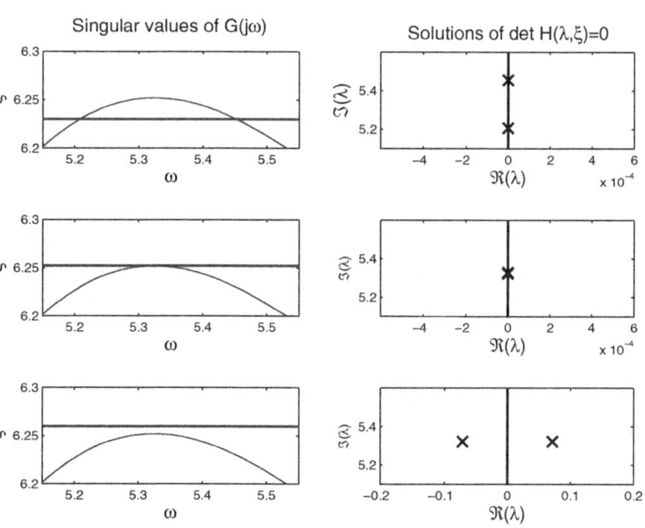

Figure 4.4. *(Left) Intersections of the singular value plot of G with the horizontal line $\xi = c$, for $c < \hat{\xi}$ (top), $c = \hat{\xi}$ (middle), and $c > \hat{\xi}$ (bottom). (Right) Corresponding eigenvalues of the problem* (4.5) [247].

and

$$\Re \frac{\partial h}{\partial \lambda}(j\omega,\xi)=0. \tag{4.28}$$

Using Corollary 4.16 we can simplify conditions (4.26) to

$$\begin{cases} \Re\, h(j\omega,\xi)=0, \\ \Im \frac{\partial h}{\partial \lambda}(j\omega,\xi)=0. \end{cases} \tag{4.29}$$

In this way the pair $(\hat{\omega},\hat{\xi})$ satisfying (4.25) can be directly computed from the two equations (4.29), for example, using Newton's method, provided that good starting values are available.

The drawback of working directly with (4.29) is that an explicit expression for the determinant of $H(\lambda,\xi)$ is required. To avoid this, let $v_1,v_2 \in \mathbb{C}^n$ be such that

$$H(j\omega,\xi)\begin{bmatrix} v_1 \\ v_2 \end{bmatrix}=0, \quad n(v_1,v_2)=0,$$

where $n(v_1,v_2)=0$ is a normalizing condition. Given the structure of $H(j\omega,\xi)$ it can be verified that a corresponding left null vector is given by $[-v_2^*\ v_1^*]$. According to [170] a necessary condition for the eigenvalue $\lambda=j\omega$ to be double but non-semisimple is given by[11]

$$[-v_2^*\ v_1^*]\,\frac{\partial H}{\partial \lambda}(j\omega,\xi)\begin{bmatrix} v_1 \\ v_2 \end{bmatrix}=0. \tag{4.30}$$

A simple computation results in

$$[-v_2^*\ v_1^*]\,\frac{\partial H}{\partial \lambda}(j\omega,\xi)\begin{bmatrix} v_1 \\ v_2 \end{bmatrix}=2\Im\left\{v_2^*\left(I+\sum_{i=1}^{p}A_i\tau_i e^{-j\omega\tau_i}+BD_\xi^{-1}D^T C\tau_0 e^{j\omega\tau_0}\right)v_1\right\}.$$

This expression is always real, which is a property inferred from (4.28). In this way we end up with $4n+3$ real equations

$$\begin{cases} H(j\omega,\ \xi)\left[\begin{smallmatrix} v_1 \\ v_2 \end{smallmatrix}\right]=0, \\ n(v_1,v_2)=0, \\ \Im\left\{v_2^*\left(I+\sum_{i=1}^{p}A_i\tau_i e^{-j\omega\tau_i}+BD_\xi^{-1}D^T C\tau_0 e^{j\omega\tau_0}\right)v_1\right\}=0 \end{cases} \tag{4.31}$$

in the $4n+2$ unknowns $\Re(v_1)$, $\Im(v_1)$, $\Re(v_2)$, $\Im(v_2)$, ω, and ξ. These equations are still overdetermined because property (4.27) is not explicitly exploited in the formulation, unlike property (4.28). However, this property makes equations (4.31) exactly solvable.

In our implementation we solve equation (4.31) in a least squares sense using the Gauss–Newton algorithm, which can be shown to be quadratically converging in the case under consideration, where the residual in the desired solution is zero. An alternative would consists of *bordering* the system, i.e., defining a slack variable in order to equate the number of unknown and the number of equations [342]. The program gives a warning when the correction involves a relative change larger than 10%, because this indicates that the approximation in the prediction step might not be accurate enough.

The above results can be used to compute the $\mathcal{H}_\infty$ norm of $G(j\omega)$ in the following way. Given an approximation of a pair $(\hat{\xi},\hat{\omega})$ satisfying (4.25), and given corresponding

[11] Condition (4.30) guarantees the existence of a Jordan chain of length larger than one.

estimates of v_1 and v_2, we can find the exact values by solving the nonlinear equations (4.31) in a least squares sense. The fact that the residual must be zero in the desired solution can be used as an additional optimality certificate. The approximation $(\hat{\xi}, \hat{\omega})$ can be obtained by the prediction step outlined in this section. When using Algorithm 4.1 for the prediction step the total algorithm becomes:

ALGORITHM 4.2.

Input: system data, N, symmetric grid Ω_N defined by (4.21), *candidate critical frequency ω_t if available, tolerance tol for prediction step*
Output: $||G(j\omega)||_{\mathcal{H}_\infty}$

Prediction step:

Apply Algorithm 4.1.

Correction step:

1. determine all eigenvalues $\{j\omega^{(1)}, \ldots, j\omega^{(p)}\}$ of $\mathcal{L}_{\xi_l}^N$ on the positive imaginary axis;
2. for all $i \in \{1, \ldots, p\}$, solve (4.31) with starting values

$$\omega = \omega^{(i)}, \quad \xi = \xi_l, \left[\begin{array}{c} v_1 \\ v_2 \end{array} \right] = \arg\min ||H(j\omega^{(i)}, \xi_l)v||/||v||;$$

   denote the solution with $(\tilde{u}^{(i)}, \tilde{v}^{(i)}, \tilde{\omega}^{(i)}, \tilde{\xi}^{(i)})$
3. set $||G(j\omega)||_{\mathcal{H}_\infty} := \max_{1 \le i \le p} \tilde{\xi}^{(i)}$

Number of discretization points

An important aspect in the application of Algorithm 4.2 consists of choosing N or, equivalently, the number of grid points for the discretization of $\mathcal{L}_\xi$ into $\mathcal{L}_\xi^N$. On the one hand, from a computational point of view, N should be as small as possible, given that the computational cost of Algorithm 4.2 is dominated by the computation of the eigenvalues of the matrix $\mathcal{L}_\xi^N$ with dimensions $(2N+1)2n$. On the other hand, N should be sufficiently large to generate starting values for which the corrector converges to the desired values. The relations established in Sections 4.2.1–4.2.2 and summarized in Table 4.1 turn out to be very useful in the determination of N. This is explained in what follows.

From Definition 4.16 the function

$$t \in [-1, 1] \mapsto p_N(t; \lambda) \tag{4.32}$$

is an approximation of the function

$$t \in [-1, 1] \mapsto e^{\lambda t}. \tag{4.33}$$

Moreover, a comparison between $H(\xi, \lambda)$, defined by (4.4), and $H_N(\xi, \lambda)$, defined by (4.17), shows that the effect of approximating $\mathcal{L}_\xi$ with $\mathcal{L}_\xi^N$ can be interpreted as the effect of approximating the exponential functions $e^{\pm\lambda\tau_i}$ with the rational functions $p_N(\pm\tau_i; \lambda)$

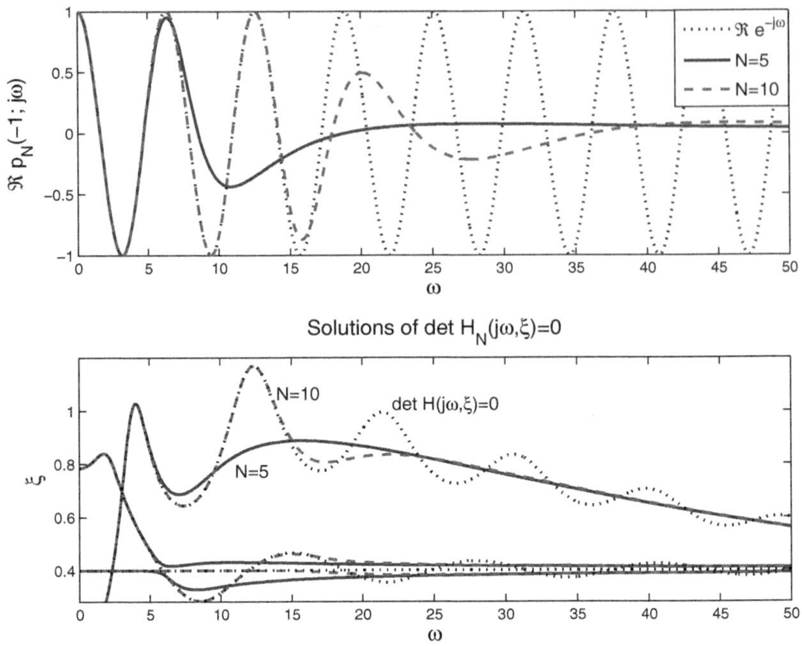

Figure 4.5. *(top) Comparison between the functions* $\omega \mapsto \exp(j\omega)$ *and* $\omega \mapsto p_N(-1;\, j\omega)$ *for* $N=5$ *and* $N=10$*. (bottom) Corresponding comparison between the solutions of* $\det H(j\omega,\xi)=0$ *(the singular value plot of G) and the solutions of* $\det H_N(j\omega;\, \xi)=0$*, with problem data (4.35). An approximation with* $N=10$ *is sufficient to cover the highest peak, unlike an approximation with* $N=5$ [247].

for $i=0,\ldots,m$. Hence, the number N in the prediction step of Algorithm 4.2 should be chosen in such a way that

$$p_N(\pm\tau_i;\, j\omega) \approx e^{\pm j\omega\tau_i}, \quad i=0,\ldots,m,$$

in the relevant frequency range, that is, where the highest peak values in the singular value plot of $G(j\omega)$ occur. This is illustrated in Figure 4.5.

In Figure 4.6 we depict the "cut-off frequency"

$$\omega_c^{\delta}(N) := \min\left\{\omega \geq 0 :\ \max_{t\in[-1,\,1]} |e^{j\omega t} - p_N(t;\, j\omega)| \geq \delta\right\} \tag{4.34}$$

as a function of N for different values of δ. The importance is as follows: if $\omega \in [0, \omega_c^{\delta}(N)]$, then it is guaranteed that $p_N(\pm\tau_i; j\omega)$ approximates $e^{\pm j\omega\tau_i}$ with relative error smaller than δ for $i=0,\ldots,m$. Because ω_c^{δ} is *independent* of the problem, Figure 4.6 can assist in determining a suitable value of N when information about the relevant frequency range in the singular value plot of G is available. It is not necessary to choose δ very small, because the prediction step in Algorithm 4.2 is followed by a correction step; hence, it is sufficient to capture qualitatively the highest peaks in the singular value plot. In our implementation of Algorithm 4.2 the value $\delta = 0.1$ is taken. The user has the option to specify the corresponding cut-off frequency $\omega_c^{0.1}$, from which the value of N is automatically determined. Otherwise the default value $N=15$ is used. If in the prediction step a peak value is computed at a frequency ω_p satisfying $\omega_p > \omega_c^{0.1}$, then a warning is generated to increase the cut-off frequency.

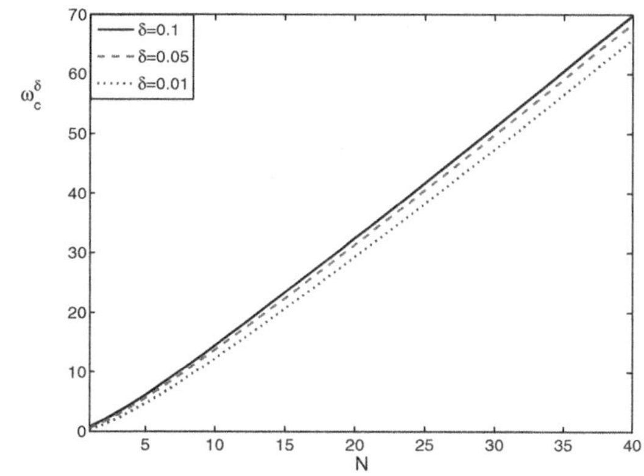

Figure 4.6. *The cut-off frequency ω_c^δ as a function of N for $\delta = 0.01$, $\delta = 0.05$, and $\delta = 0.10$.*

Our experience from extensive benchmarking shows that in most practical problems a very small value of N can be taken (the default $N = 15$ is largely sufficient). This can be motivated as follows. First, via Corollary 4.6 the $\mathscr{H}_\infty$ problem is transformed into a problem of checking the eigenvalues of $\mathscr{L}_\xi$ on the imaginary axis. Because these eigenvalues are typically among the *smallest* eigenvalues of $\mathscr{L}_\xi$, and the individual eigenvalues of $\mathscr{L}_\xi^N$ exhibit *spectral convergence* to the corresponding eigenvalues of $\mathscr{L}_\xi$ (this can be shown following along the lines of [44]), a small value of N already yields good approximations of the imaginary axis eigenvalues. Second, in most applications the delay parameters are critical from a stability point of view, in the sense that increasing the delay parameters eventually destabilizes the system, but stability is maintained by decreasing the delay parameters (often referred to in the literature by the term delay-dependent stability). In [247] the connection between delay-dependent stability and the possibility of working with a relatively small value of N in the $\mathscr{H}_\infty$ computation is demonstrated.

4.2.4 ▪ Numerical examples

So far, the obtained results and the ideas behind Algorithm 4.2 have been illustrated with the problem data

$$A_0 = \begin{bmatrix} 108 & 110 & 18 \\ -107 & -109 & -17 \\ -217 & -217 & -37 \end{bmatrix}, \quad A_1 = \begin{bmatrix} 46.5 & 46.5 & 1.5 \\ -46.5 & -46.5 & -1.5 \\ -93 & -93 & -3 \end{bmatrix},$$

$$A_2 = \begin{bmatrix} -0.3 & 0.3 & -0.3 \\ 0.3 & -0.3 & 0.3 \\ 0 & 0 & 0 \end{bmatrix}, \quad B = \begin{bmatrix} 0.5 & -90 \\ -0.5 & 90 \\ 0 & 180 \end{bmatrix},$$

$$C = \begin{bmatrix} 1 & -1 & 1 \\ 0.1833 & 0.1833 & 0.1833 \end{bmatrix}, \quad D = \begin{bmatrix} 0.4 & 0 \\ 0 & 0 \end{bmatrix},$$

$$\tau_0 = 0.5, \quad \tau_1 = 0.667, \quad \tau_2 = 1.$$

In particular, the singular value plot of the corresponding transfer function G is shown in Figure 4.1, and the solutions of $\det H_N(j\omega, \xi) = 0$ in Figure 4.2, for $N = 10$. The two-parameter search in the prediction step is illustrated in Figure 4.3. The effect of the

Table 4.2. *Benchmarks for the $\mathcal{H}_\infty$ norm computation* [247].

Plants	(n,m)	N	ξ_{pred}	ξ_{corr} ($\|G(j\omega)\|_{\mathcal{H}_\infty}$)
$G1$	(3,1)	1	10.0235	10.0235
$G2$	(3,1)	3	3.3693	3.3709
$G3$	(1,1)	2	0.7158	0.7196
$G4$	(1,1)	3	1.9230	1.9883
$G5$	(3,3)	2	0.8852	0.8848
$G6$	(3,3)	5	0.8974	0.9356
$G7$	(4,3)	10	1.6259	1.6283
$G8$	(10,7)	4	22.2979	22.3195
$G9$	(20,9)	13	1.2827	1.2903
$G10$	(40,3)	3	811.0898	814.6221
$G11$	(3,2)	9	1.1579	1.1696
$G12$	(3,2)	18	1.1626	1.1696

approximation in the prediction step is shown in Figure 4.5. With $N = 10$, the predicted $\mathcal{H}_\infty$ norm is $\xi_{pred} = 1.1626$ and the corrected $\mathcal{H}_\infty$ norm is given by $\xi_{corr} = 1.1696$.

In Table 4.2 we present the results of benchmarking of our code with 12 problems. The second column shows the size of matrices A_i, n, and the number of state delays, m. The third column gives the minimum value of N such that in the correction step the desired solution is computed. The fourth and fifth columns contain the predicted and corrected $\mathcal{H}_\infty$ norms of the corresponding time-delay system. The plant $G12$ correspond to the problem data (4.35).

For the plant $G12$ a warning is generated when using the default value $N = 15$, indicating that the corresponding cut-off frequency might be too small. With $N = 18$ no warning is given and the results are correct.

The problem data for the above benchmark examples and a MATLAB implementation of our code for the $\mathcal{H}_\infty$ norm computation are available at the website
http://twr.cs.kuleuven.be/research/software/delay-control/hinf/

4.3 ▪ Computing $\mathcal{H}_2$ norms

4.3.1 ▪ Characterization of $\mathcal{H}_2$ norm in terms of Lyapunov matrices

When $D = 0$, expression (4.1) corresponds to the transfer function of the system

$$\dot{x}(t) = A_0 x(t) + \sum_{k=1}^{m} A_k x(t-\tau_k) + Bu(t), \tag{4.35}$$

$$y(t) = Cx(t). \tag{4.36}$$

The fundamental solution, denoted $K : \mathbb{R}_+ \to \mathbb{R}^{n\times n}$, is defined as the solution of the matrix delay-differential equation started with identity at time zero and zero elsewhere, i.e.,

$$\dot{K}(t) = A_0 K(t) + \sum_{k=1}^{m} A_k K(t-\tau_k), \tag{4.37}$$

$$K(0) = I, \ K(\theta) = 0 \text{ when } \theta < 0. \tag{4.38}$$

There are several definitions of the *delay Lyapunov matrix*. We will use the energy functional definition[12]

$$U(t) := \int_{s=0}^{\infty} K^T(s) C^T C K(s+t)\, ds. \tag{4.39}$$

It now turns out that the formula for the $\mathcal{H}_2$ norm of systems without delays extends to systems with delay, where $U(0)$ plays the same role as the solution to the standard Lyapunov equations for systems without delay. The following result combines the relation between $\mathcal{H}_2$ norms and delay Lyapunov matrices from [180] with the uniqueness of solutions of delay Lyapunov equations proven in [327, Theorem 6.31], [192, Theorem 4].

Theorem 4.17. *Assume that the time-delay system* (4.35)–(4.36) *is exponentially stable. Then the $\mathcal{H}_2$ norm satisfies*

$$||G(j\omega)||^2_{\mathcal{H}_2} = \mathrm{Tr}(B^T U(0) B) \tag{4.40}$$

$$= \mathrm{Tr}(C V(0) C^T), \tag{4.41}$$

where $U(t)$, $V(t)$ are the unique solutions of the delay Lyapunov equation

$$\dot{U}(t) = U(t)A_0 + \sum_{k=1}^{m} U(t-\tau_k)A_k, \quad t \geq 0, \tag{4.42a}$$

$$U(-t) = U^T(t), \tag{4.42b}$$

$$-C^T C = U(0)A_0 + A_0^T U(0) + \sum_{k=1}^{m} U(-\tau_k)A_k + A_k^T U(\tau_k) \tag{4.42c}$$

and the dual delay Lyapunov equation

$$\dot{V}(t) = V(t)A_0^T + \sum_{k=1}^{m} V(t-\tau_k)A_k^T, \quad t \geq 0, \tag{4.43a}$$

$$V(-t) = V^T(t), \tag{4.43b}$$

$$-BB^T = V(0)A_0^T + A_0 V(0) + \sum_{k=1}^{m} V(-\tau_k)A_k^T + A_k V(\tau_k). \tag{4.43c}$$

We note that the formula for the $\mathcal{H}_2$ norm in Theorem 4.17 is a generalization of a formula for the $\mathcal{H}_2$ norm of a dynamical system without delay. The formula for the $\mathcal{H}_2$ norm of a stable dynamical system described by the triple (C, A, B) is namely given by (see, e.g., [414, Lemma 4.6])

$$||G(j\omega)||^2_{\mathcal{H}_2} = \mathrm{Tr}(B^T U B) = \mathrm{Tr}(C V C^T),$$

where

$$-C^T C = UA + A^T U \text{ and } -BB^T = AV + VA^T. \tag{4.44}$$

If the delays tend to zero, (4.42)–(4.43) reduce to the algebraic constraints (4.42c) and (4.43c).

Theorem 4.17 shows a major complication with respect to the delay-free case: instead of solving an algebraic Lyapunov equations, we need to solve a matrix-valued boundary value problem.

[12] In [192] and related works, where the delay Lyapunov matrix is used to construct complete Lyapunov–Krasovskii functionals, U is typically defined as $U(t) := \int_{s=0}^{\infty} K^T(s) W K(s+t)\, ds$, with arbitrary symmetric W. In the context of the $\mathcal{H}_2$ norm it is natural to fix W as in (4.39).

4.3.2 ▪ Computational schemes based on solving delay Lyapunov equations

In order to apply Theorem 4.17 we only need to know the function value of U at the point $t=0$. We outline two approaches for computing $U(0)$, which are both based on solving (4.42).

Vectorization and explicit solution for commensurate delays

A well-known explicit approach to analyze and sometimes solve the standard Lyapunov equation (4.44) is based on vectorizing the equation, i.e., reformulating the linear matrix equation into a standard linear equation of squared dimension. In [180] it is shown that the analogy of this *vectorization approach* extends to time-delay systems if we assume that the system has commensurate delays. Recall that a time-delay system is said to have commensurate delays if all delays are integer multiples of some basic delay. In what follows we work out the vectorization approach for the single delay case and comment on the generalization.

For $m=1$ we denote τ_1 simply by τ. Let us define

$$\begin{aligned} U_0(t) &= U(t), \\ U_{-1}(t) &= U(t-\tau), \quad t\in[0,\ \tau]. \end{aligned} \tag{4.45}$$

Hence, the function $U(t)$, $t\in[-\tau,\ \tau]$, is decomposed into two segments. From (4.42a), (4.42b), and (4.45) we get

$$\begin{aligned} \dot U_0(t) &= U_0(t)A_0 + U_{-1}(t)A_1 \\ \dot U_{-1}(t) &= \dot U(t-\tau) \\ &= -\dot U(\tau - t)^T \\ &= -[U(\tau-t)A_0 + U(-t)A_1]^T \\ &= -A_0^T U_{-1}(t) - A_1^T U_0(t). \end{aligned}$$

These matrix-valued ODEs can be written in vectorized form as

$$\dot z(t) = Mz(t), \tag{4.46}$$

where

$$z(t)=\begin{bmatrix} \operatorname{vec} U_0(t) \\ \operatorname{vec} U_{-1}(t) \end{bmatrix}, \quad M=\begin{bmatrix} A_0^T\otimes I & A_1^T\otimes I \\ -I\otimes A_1^T & -I\otimes A_0^T \end{bmatrix}.$$

Similarly, a vectorization of the algebraic boundary condition (4.42c), combined with the relation $U_0(0)=U_{-1}(\tau)$, yields

$$-\begin{bmatrix} \operatorname{vec} C^TC \\ 0 \end{bmatrix} = N_1 z(0) + N_2 z(\tau),$$

where

$$N_1=\begin{bmatrix} A_0^T\otimes I & A_1^T\otimes I \\ I & 0 \end{bmatrix}, \quad N_2=\begin{bmatrix} I\otimes A_1^T & I\otimes A_0^T \\ 0 & -I \end{bmatrix}. \tag{4.47}$$

From (4.46) and (4.47) we obtain

$$z(0) = -(N_1 + N_2 e^{M\tau})^{-1}\begin{bmatrix} \operatorname{vec} C^TC \\ 0 \end{bmatrix}$$

(see [180] for a proof of the invertibility of $N_1 + N_2 e^{M\tau}$) and, finally,

$$\begin{aligned} \|G(j\omega)\|_{\mathcal{H}_2}^2 &= \mathrm{Tr}\left(B^T \mathrm{mat}([I\ 0]\ z(0))\ B\right) \\ &= \left(\mathrm{vec}\ B^T B\right)^T [I\ 0] z(0). \end{aligned} \tag{4.48}$$

A similar expression can be obtained by vectorization of the dual Lyapunov equation.

In [180] the above approach is generalized to the case $m > 1$. The main difference is that, in order to transform the delay Lyapunov equation into ODEs, the function $U(t)$, $t \in [-\tau_m, \tau_m]$, needs to be decomposed into $2m$ function segments over time intervals of length $h = \tau/m$, assuming that

$$\tau_i = i\tau_1, \quad i = 1, \ldots, m.$$

After vectorization, the system of ODEs has dimension $2mn^2$. For further reading on vectorization delay Lyapunov equations, mostly in the context of constructing complete-type Lyapunov–Krasovskii functionals, we refer the reader to [327, Chapter 6] and [192, 388, 309].

When using a general purpose approach to compute the matrix exponential as, e.g., in [269], we can assume that the computational complexity of the matrix exponential is roughly cubic. Then the computational complexity of evaluating the $\mathcal{H}_2$ norm using the explicit expression is $\mathcal{O}((mn^2)^3) = \mathcal{O}(m^3 n^6)$, which significantly restricts the problem size from a computational point of view. For comparison, for systems without delay the computational complexity using the standard, Bartels–Stewart algorithm for solving the Lyapunov equation is $\mathcal{O}(n^3)$. Even though the latter algorithm also involves squaring of dimensions by vectorization, the structure of the resulting linear system can be exploited, which seems difficult in the delay case.

As we have stressed in Chapter 1, time-delay systems form a class of infinite-dimensional systems. Therefore, it is somewhat remarkable that formula (4.48) is finite-dimensional, in the sense that it is a construction allowing us to compute the $\mathcal{H}_2$ norm using only elementary linear algebra operations applied to vectors and matrices (of finite dimension). This finite-dimensional characterization of the $\mathcal{H}_2$ norm is conceptually similar to the characterization of critical delay values, i.e., values for which there are characteristic roots on the imaginary axis, as we shall discover in Chapter 6.

Numerical solution for the general case

Another approach, also applicable to the case where the delays are not commensurate, is based on discretizing the delay Lyapunov equation. This idea was first adopted by Ochoa and coworkers; see, e.g., [309].

In [180] a spectral discretization approach is proposed, where the function $U(t)$, $t = [0, \tau_m]$ is approximated by a matrix polynomial $\tilde{U}$ of degree N, expressed in a Chebyshev basis, scaled and shifted to the interval $[0, \tau_m]$. The corresponding approximation on the interval $[-\tau_0, 0]$ is induced by the symmetry property (4.42b). The coefficients of the polynomial are determined by imposing the algebraic condition (4.42c), supplemented with collocation requirements, which express that the polynomial $\tilde{U}$ satisfies (4.42a) on a grid of Chebyshev extremal points,

$$\theta_i = \frac{\tau_m}{2}(\chi_i + 1), \quad \chi_i = \cos\left(\pi - \frac{i\pi}{N-1}\right), \quad i = 0, \ldots, N-1. \tag{4.49}$$

The advantages of the choice of polynomial basis and grid are twofold. First, it is widely known that the basis used in a representation of a polynomial has a big impact on the numerical stability. A polynomial may be very sensitive to rounding errors in the coefficients if it is represented in an unsuitable basis, e.g., the nominal basis. (See, e.g., [162, Chapter 5].) A nice property of Chebyshev polynomials is that if they are used as basis functions, then the polynomial is typically not sensitive to rounding errors in the coefficients. Second, the choice of the discretization grid influences the convergence speed. The rough motivation for the faster convergence with a Chebyshev distribution is that it is denser at the boundaries—preventing large oscillations around the boundaries often present for an equidistant grid. The Chebyshev distribution is in a sense optimal in a more formal setting, as it has an associated potential with minimal energy [377, Chapter 5].

The convergence of the $\mathcal{H}_2$ norm approximation, induced by a spectral discretization of U, as a function of the number of discretization points N is also investigated in [180]. Inherently to a spectral discretization, the convergence rate turns out to be closely related to the degree of regularity of the approximated solution. In the single delay case, the function $U(t)$ is smooth on the interval $[0, \tau_m]$, which can be seen from (4.46). In this case, a spectral discretization is optimal, and so-called *spectral convergence*, i.e., a convergence rate $\mathcal{O}\left(N^{-N}\right)$, is observed. In the multiple delay case, $m > 1$, it can be shown that the function $U(t)$, $t \in [0,\ \tau_m]$, is in general not smooth,[13] exhibiting a second derivative which is only piecewise continuous, but of bounded variation. This nonsmoothness affects the convergence speed of the spectral discretization and results in a convergence rate of the $\mathcal{H}_2$ norm approximation of $\mathcal{O}\left(N^{-2}\right)$.

The computationally dominating part of the approach consists of solving a system of $N+1$ linear matrix equations. With a standard method for linear systems, the computational complexity amounts to $\mathcal{O}(N^3 n^6)$. Hence, the number of operations does not depend on the number of delays, m, unlike the exact vectorization approach, which applies as long as the delays are commensurate. The dependence on the dimension n is still proportional to n^6.

4.3.3 ▪ Computing the $\mathcal{H}_2$ norm using a spectral discretization

The approach which we now summarize is based on approximating the time-delay system (4.35) by a standard linear time-invariant system. For the details we refer the reader to [385].

Reformulation into state-space form

The reformulation of a time-delay system as an ODE over the Banach space $\mathscr{C}([-\tau_m,\ 0],\ \mathbb{R}^n)$, which we discussed in Section 1.1.1, is not suitable in the presence of inputs. The reason is that the system dynamics appear in the definition of the domain of the operator $\mathscr{A}$, which leads to a domain depending on the input. We start by outlining the alternative formulation of [84], in which the presence of inputs and outputs can be addressed in a natural way. Denote the state space

$$\mathscr{X} := \mathscr{L}_2([-\tau_m,\ 0), \mathbb{R}^n) \times \mathbb{R}^n$$

and consider the operator $\mathscr{A}$ on $\mathscr{X}$ whose domain is defined by

$$D(\mathscr{A}) := \left\{ \left(\begin{array}{c} \phi \\ r \end{array} \right) \in \mathscr{X} \ \middle|\ \begin{array}{l} \phi \text{ absol. continuous on } [-\tau_m, 0],\ \phi' \in \mathscr{L}_2([-\tau_m, 0], \mathbb{R}^n) \\ r = \phi(0) \end{array} \right\}, \tag{4.50}$$

[13] In the case when the delays are commensurate, the function is still piecewise smooth. This property is exploited by the vectorization approach.

and whose action is given by

$$\mathscr{A}\xi = \mathscr{A}\begin{pmatrix} \phi \\ r \end{pmatrix} := \begin{pmatrix} \phi' \\ A_0 r + \sum_{k=1}^{m} A_k \phi(-\tau_k) \end{pmatrix}. \tag{4.51}$$

Similarly, we define $\mathscr{B} : \mathbb{R}^{n_u} \to \mathscr{X}$ as the operator spanning a vector $z \in \mathbb{R}^{n_u}$ onto $\binom{0}{Bz}$, and define $\mathscr{C} : \mathscr{X} \to \mathbb{R}^{n_y}$ as the operator mapping $\binom{\phi}{r} \in \mathscr{X}$ onto Cr. With these three operators, we can bring the time-delay system (4.35) into the following format:

$$\begin{cases} \frac{d}{dt}\xi(t) = \mathscr{A}\xi(t) + \mathscr{B}u(t), \\ y(t) = \mathscr{C}\xi(t), \end{cases} \tag{4.52}$$

where $\xi(t) \in \mathscr{D}(\mathscr{A}) \subset \mathscr{X}$. More precisely, the relation between corresponding solutions of (4.52) and (4.35) is described by

$$\xi_1(t)(\theta) \equiv x(t+\theta), \quad \theta \in [-\tau_m, 0),$$
$$\xi_2(t) = x(t),$$

and hence the state at time t can be seen as an arrow, where the head corresponds to $x(t)$ and the tail to the past information. Compared to the approach of Section 1.1.1, the advantage of splitting the state into two components is that the information about the dynamics, including the connection with inputs, is moved from the domain of the operator to the action of the operator.

A spectral discretization

The spectral discretization of (4.52) follows along the same line as the approach in Section 2.2.1. Let us consider the grid (2.5) and define matrix $\mathbf{A}_N$ by (2.6). In the same fashion we can approximate the operators $\mathscr{B}$ and $\mathscr{C}$ with the matrices $\mathbf{B}_N \in \mathbb{R}^{(N+1)n \times n_u}$ and $\mathbf{C}_N \in \mathbb{R}^{n_y \times (N+1)n}$, namely by constructing them as

$$\mathbf{B}_N := \begin{bmatrix} 0 & \cdots & 0 & I_n \end{bmatrix}^{\top} B \quad \text{and} \quad \mathbf{C}_N := C \begin{bmatrix} 0 & \cdots & 0 & I_n \end{bmatrix}. \tag{4.53}$$

We now have all the ingredients needed to approximate the system (4.52) by the following linear system of dimension $(N+1)n$:

$$\begin{cases} \xi_N'(t) = \mathbf{A}_N \xi)_N(t) + \mathbf{B}_N u(t), \\ y(t) = \mathbf{C}_N \xi_N(t). \end{cases} \tag{4.54}$$

The $\mathscr{H}_2$ norm of G can be approximated as

$$\|G(j\omega)\|_{\mathscr{H}_2} \approx \gamma_n, \tag{4.55}$$

where γ_N is the $\mathscr{H}_2$ norm of (4.54) and satisfies

$$\gamma_n^2 = \operatorname{Tr}(\mathbf{B}_N^T U_N \mathbf{B}_N) = \operatorname{Tr}(\mathbf{C}_N V_N \mathbf{C}_N^T), \tag{4.56}$$

where

$$-\mathbf{C}_N^T \mathbf{C}_N = U_N \mathbf{A}_N + \mathbf{A}_N^T U_N,$$
$$-\mathbf{B}_N \mathbf{B}_N^T = \mathbf{A}_N V_N + V_N \mathbf{A}_N^T.$$

In [385] a detailed convergence analysis of the approximation (4.55) is performed for $N \to \infty$, based on the relation between a spectral discretization and a rational approximation described by Proposition 2.2. As a result, if the grid points have the same asymptotic distribution as Chebyshev points, as is the case with the grid (2.10), the order of convergence is $\mathcal{O}\left(N^{-3}\right)$. The number of operations needed to compute γ_N, using a stand Lyapunov equation solver, is $\mathcal{O}(n^3 N^3)$.

4.3.4 ▪ Numerical examples

As a first example we consider a scalar system with one delay,

$$\begin{aligned} \dot{x}(t) &= -ax(t-\tau) + bu(t), \\ y(t) &= cx(t), \end{aligned} \tag{4.57}$$

where $a > 0$. In this case formula (4.48) can be worked out analytically, resulting in

$$\|G(j\omega)\|_{\mathcal{H}_2} = \left(\frac{c^2 b^2}{2a} \frac{\cos(a\tau)}{1 - \sin(a\tau)} \right)^{\frac{1}{2}}.$$

For $a = b = c = \tau = 1$ we get $\gamma := \|G(j\omega)\|_{\mathcal{H}_2} = 1.3054163$. For these parameter values we show in Figure 4.7 the errors

$$|\mu_N - \gamma|, \quad |\gamma_N - \gamma|,$$

as a function of N, where μ_N is the approximation, obtained by discretizing the delay Lyapunov equation using a spectral method on the grid (4.49), and γ_N is the approximation (4.56), obtained by discretizing the delay system itself. Since there is only one delay, spectral convergence, that is, a rate $\mathcal{O}(N^{-N})$, is obtained in the former case. In the latter case the convergence order is $\mathcal{O}(N^{-3})$.

As a second example we consider the system

$$\begin{aligned} \dot{x}(t) &= A_0 x(t) + A_1 x(t-\tau_1) + A_2 x(t-\tau_2) + Bu(t), \\ y(t) &= Cx(t), \end{aligned} \tag{4.58}$$

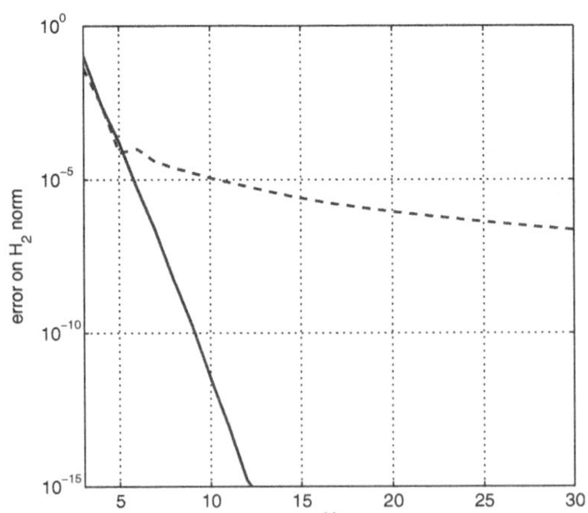

Figure 4.7. *Approximation error on the $\mathcal{H}_2$ norm as a function of the number of discretization points for system (4.57). The full curve corresponds to μ_N-spectral discretization of the delay Lyapunov equation. The dashed curve corresponds to γ_N-spectral discretization of the delay equation.*

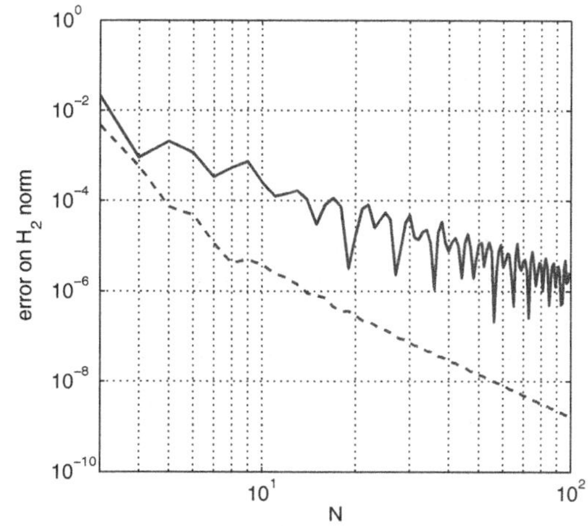

Figure 4.8. *Approximation error on the $\mathcal{H}_2$ norm as a function of the number of discretization points for system (4.58)–(4.59). The full curve corresponds to μ_N-spectral discretization of the delay Lyapunov equation. The dashed curve corresponds to γ_N-spectral discretization of the delay equation.*

where

$$A_0 = \begin{bmatrix} -1 & 1 & 2 \\ 1 & -3 & 2 \\ 0 & 0 & -1 \end{bmatrix}, \; A_1 = \frac{1}{5}\begin{bmatrix} -3 & 0 & 1 \\ 1 & -2 & 0 \\ 0 & 2 & -2 \end{bmatrix}, \; A_2 = \frac{1}{5}\begin{bmatrix} -4 & 1 & 0 \\ 0 & -2 & 1 \\ 2 & 1 & -3 \end{bmatrix},$$
$$B^T = C = [1\ 1\ 1], \; \tau_1 = \pi/10, \; \tau_2 = 1. \tag{4.59}$$

The $\mathcal{H}_2$ norm is equal to $\gamma := 3.8299451$. Since the delays are not commensurate, the vectorization approach is not applicable. In Figure 4.8 we plot the errors on μ_N and γ_N, which are defined in the same way as for the previous example. Because there are two delays, the Lyapunov matrix $U(t)$ is not smooth on the interval $[0, 1]$, exhibiting a discontinuity in the second derivative. As a result, a spectral discretization of the delay Lyapunov equation yields the convergence order $\mathcal{O}(N^{-2})$. When discretizing the delay system itself, the convergence rate is still $\mathcal{O}(N^{-3})$.

4.4 ▪ Notes and references

In this chapter we described algorithms for the computation of $\mathcal{H}_\infty$ and $\mathcal{H}_2$ norms for time-delay systems.

The method for computing $\mathcal{H}_\infty$ norms adopts a two-step approach, in analogy with methods for computing characteristic roots. In the prediction step an estimate is computed based on a finite-dimensional approximation, which is subsequently corrected by a locally converging algorithm. The choice of predictor is motivated by the following arguments:

- An approach based on a spectral discretization of an appropriately defined derivative operator with nonlocal boundary conditions is known to be not only an accurate but also a numerically stable way to solve infinite-dimensional eigenvalue problems in the context of time-delay system (see Chapter 2 for a discussion on various methods for the "dual" problem of computing characteristic roots). On the contrary, working with an explicit rational approximation of G may lead to an ill-conditioned Hamiltonian matrix when applying Proposition 4.1 to the resulting finite-dimensional system. This is due to potentially large differences in magnitudes of the coefficients in rational approximations of high order (a high order is necessary for "globally" capturing the transfer function in the relevant frequency range).

- The dimensions of matrix $\mathcal{L}_\xi^N$, $(2N+1)2n \times (2N+1)2n$, are *independent* of the number of delays in the problem, m. This can also be seen from Remark 4.15, which shows that the poles of the rational functions $\lambda \mapsto p_N(-\tau_i; \lambda)$ are independent of τ_i; that is, one can interpret the effect of discretizing $\mathcal{L}_\xi$ as the effect of an approximation of all exponential functions by rational functions with common poles. With Padé and many other types of rational approximations these poles are not the same, and the dimension of the discretized systems will become proportional to the number of delays.

- An important advantage of a direct rational approximation of the exponential functions in (4.1) is that a high accuracy in a relevant frequency range can easily be guaranteed by the choice of the order of the approximation. We have demonstrated that this is also possible when working with a spectral discretization of the operator $\mathcal{L}_\xi$, once again via the interpretation of the effect of its discretization as the effect of a rational approximation (although the coefficients of the rational functions $p_N(-\tau_i; \lambda)$ never needed to be explicitly computed). This property was used in Section 4.2.3 for the determination of the number of discretization points.

A similar algorithm for computing *pseudospectral abscissa* is proposed in [140]. See also [136] for another algorithm suitable for large-scale problems.

In the second part of the chapter (Section 4.3) we discussed methods for computing $\mathcal{H}_2$ norms. First, we established an important relation between $\mathcal{H}_2$ norms and Lyapunov matrices. Next we presented two classes of computational schemes. The first is based on computing Lyapunov matrices by solving the corresponding matrix-valued boundary value problem. For the case of one delay we presented an explicit expression in terms of a matrix exponential and outlined its generalization to the multiple, commensurate delay case. We also described a general scheme based on spectral collocation. It should be noted that these methods for computing Lyapunov equations are also applicable to neutral-type systems (see [180] for the generalization) and, as a consequence, to systems described by delay-differential algebraic equations. Furthermore, these methods are also very valuable for the construction of complete-type Lyapunov matrices (see, e.g., [190]). The second class of computational schemes is based on a "discretize first" approach. The main advantage of this approach is that, with a general purpose solver, the complexity is cubic in the dimension n, instead of proportional to n^6. In contrast to the method for computing $\mathcal{H}_\infty$ norms, it seems difficult to remove the effect of the approximation on the error on the $\mathcal{H}_2$ norm.

All methods presented in this chapter have been benchmarked extensively. They turn out to be robust and reliable, yet the price to pay is that they are computationally demanding. Indeed, both the predictor in the computation of the $\mathcal{H}_\infty$ norm and the discretize first approach for computing $\mathcal{H}_2$ norms are based on a system approximation of dimension proportional to Nn. The other methods for computing $\mathcal{H}_2$ norms are implicitly or explicitly based on the vectorization of matrices, which squares the dimensions. In all cases the development of fast and reliable iterative methods, suitable for large-scale problems, is a largely open research topic.

This chapter is based on [247, 180, 385] and the references therein.

Chapter 5

Computation of stability regions in parameter spaces

5.1 ▪ Introduction

As in the finite-dimensional case, the stability of a linear time-delay system is given by its spectrum location. The next important step is represented by the characterization of *spectrum behavior* as a function of the parameters variation. To perform such an analysis, three ingredients are essential:

(a) the continuity property of the spectrum with respect to the parameters as discussed in the first chapter;

(b) the detection, and the explicit computation (if any!), of the characteristic roots located on the imaginary axis;

(c) the behavior of the characteristic roots located on the imaginary axis in a very small neighborhood if one or several parameters vary.

The first problem was largely discussed in Chapter 1, and the main ideas are clear (for instance, in the retarded case, the continuity properties of the rightmost characteristic roots with respect to all the parameters of the system including also the delay parameters). The next two problems represent the main topics in what follows. More precisely, we start by presenting several approaches and methodologies to describe the characteristic roots behavior when they are located on the imaginary axis. Some basic ideas concerning the detection and the explicit computation of the characteristic roots located on the imaginary axis also will be presented and discussed in some particular cases. A deeper analysis will be proposed in the next chapter (stability in the delay-parameter space including also the characterization of delay-interference phenomena).

We start by analyzing the properties of some generic characteristic root located on the imaginary axis as a function of its parameters. Such a method is certainly not new, and it proved its interest in the case of finite-dimensional systems. Next, the crossing direction characterization w.r.t. the parameters is largely treated, and several intuitive approaches are presented (Jacobi's formulae for computing the derivative of a determinant; perturbation analysis). Particular attention will be paid to the simple and double characteristic roots cases. In the case of double roots, we will discuss both cases: semisimple and non–semisimple function of the geometric multiplicity of the corresponding characteristic root. Finally, explicit crossing direction computation in some particular cases that will be useful in the next chapters are also proposed. Although the whole presentation concerns time-delay systems of retarded type, the underlying ideas can be easily adapted to time-delay systems described by neutral functional differential equations or by delay-differential algebraic equations under appropriate assumptions concerning the stability of the corresponding delay-difference operator. However, for the sake of brevity such cases are not treated in this chapter.

The chapter is organized as follows. Section 5.2 includes some basic notions and definitions. Next, Section 5.3 is devoted to various existing methods and approaches for deriving the stability crossing boundaries. The presentation is intuitive and starts with some historical perspective concerning the D-decomposition method with an exemplification in the scalar case. Next, we present the "dual" τ-decomposition method, and still use the scalar system as an illustrative example. The difference between these approaches mainly lies in the way in which the delays are treated. Finally, we discuss numerical continuation methods. Roughly speaking, the idea behind the numerical continuation method in our case is to find a one-dimensional curve in the parameter space and to "follow" (or "continue") it in the parameter space starting from one or several "first-points" found on the curve. The next section is devoted to the crossing direction characterization. In the next-to-last section of the chapter, Section 5.5, we review an extension of the root-locus method to the time-delay systems.

Section 5.6 ends the chapter with some notes and comments, including also a list of references related to the topics treated in the chapter.

5.2 ▪ Basic notions and definitions

Consider the following class of delay systems:

$$\dot{x}(t)=A_0(p)x(t)+\sum_{i=1}^{m}A_i(p)x(t-\tau_i(p)), \tag{5.1}$$

under appropriate initial conditions, with

$$p:=(p_1,\dots,p_{n_p})\in\mathcal{D}\subset\mathbb{R}^{n_p}$$

the parameters, under the constraint that

$$\tau_i(p)\geq 0 \quad \forall p\in\mathcal{D},$$

and such that there exists at least one point $\tilde{p}$ in the parameter set $\mathcal{D}$, and at least one positive integer i_0, $1\leq i_0\leq n_p$, such that

$$\tau_{i_0}(\tilde{p})>0, \quad A_{i_0}\neq 0.$$

In other words, we assume that (5.1) is always a time-delay system in some neighborhood of $\tilde{p}$ in the corresponding parameter space. We also assume that the matrices A_i and delays τ_l smoothly depend on p.

Next, define the characteristic matrix associated to (5.1) as the mapping $M:\mathbb{C}\times\mathcal{D}\mapsto\mathbb{C}^{n\times n}$ given by

$$M(\lambda;\ p):=\lambda I-A_0(p)-\sum_{i=1}^{m}A_i(p)e^{-\lambda\tau_i(p)}, \tag{5.2}$$

and define the characteristic function as the mapping $H:\mathbb{C}\times\mathcal{D}\mapsto\mathbb{C}$ described by

$$H(\lambda;\ p):=\det\left(\lambda I-A_0(p)-\sum_{i=1}^{m}A_i(p)e^{-\lambda\tau_i(p)}\right). \tag{5.3}$$

As defined in the previous chapters, the roots of the characteristic equation $H(\lambda;\ p) = 0$ will be called *characteristic roots* of the delay system under consideration. With these notations, we introduce the following notions for stability domains (or regions) and stability crossing boundaries:

Definition 5.1.

(1) *The set of values $p \in \mathscr{D}$ of the parameter space $\mathbb{R}^{n_p}$ such that the delay system* (5.1) *is exponentially stable is called the* stability domain *or* stability region.

(2) *The set of values $p \in \mathscr{D}$ of the parameter space $\mathbb{R}^{n_p}$ is called the* stability crossing boundary *of the delay system* (5.1) *if for any point p_0 of the boundary, the characteristic function $H(\lambda;\ p_0)$ has at least one root on the imaginary axis.*

The continuity properties discussed in the previous chapter allow us to divide the parameter space into domains characterized by the same number of characteristic roots in the open right half plane for all the parameters points of the domain. The "separation" between domains with a distinct number of unstable characteristic roots is given by the so-called *stability crossing boundaries*. Such an idea is exploited by the so-called *D-decomposition* and *τ-decomposition* methods that are discussed in Section 5.3.

5.3 ▪ From D-decomposition to numerical continuation

With the notations, definitions, and comments above, we first concentrate on the way in which stability crossing boundaries are defined in the parameter space and on the corresponding crossing direction from one domain to another. We start with a brief discussion on the D-decomposition method. Next, we will present the so-called τ-decomposition method, and we discuss the differences between these two methods. Finally, we focus on numerical continuation.

5.3.1 ▪ D-decomposition and stability crossing boundaries

To the best of the authors' knowledge, the first method for characterizing the stability regions of time-delay systems in a parameter space was proposed by Neimark [288] for quasi-polynomials at the end of the 1940s. It consists in computing a particular decomposition of the parameter space in regions such that the number of unstable characteristic roots is invariant with respect to all the points of the parameter space inside the region, and such that for each point of the boundaries the corresponding characteristic equation has at least one root on the imaginary axis. This method is also known as the *D-decomposition* (or *D-subdivision*) *method.* In order to describe the algorithm, some supplementary assumptions are needed. More precisely, we assume that the delays τ_i, $i = 1, \ldots, m$, do *not depend* on the parameters $p \in \mathbb{R}^{n_p}$, and, furthermore, they are considered as *fixed.* In other words, our system is rewritten as

$$\dot{x}(t) = A_0(p)x(t) + \sum_{i=1}^{m} A_i(p)x(t-\tau_i),$$

under appropriate initial conditions. The characteristic function becomes

$$H(\lambda;\ p) = \det\left(\lambda I - A_0(p) - \sum_{i=1}^{m} A_i(p)e^{-\lambda\tau_i}\right).$$

With the notations and remarks above, the algorithm of the D-decomposition method can be resumed as follows:

(i) first, the equation
$$H(j\omega;\ p)=0$$
is solved for p as a function of $j\omega$ (including the origin of the complex plane) in order to find (stability crossing) surfaces in the parameter space $\mathbb{R}^{n_p}$ such that for each p on such a surface, there exists at least one characteristic root on the imaginary axis;

(ii) second, these surfaces divide the parameter space into several regions, and sometimes it is possible to conclude, by using appropriate additional arguments, for which region the stability is guaranteed. As additional arguments, we can find, for example, a particular point (for instance, on the axis of the parameter space) for which the stability analysis becomes easier to perform (e.g., a finite-dimensional system). Each region derived in this way is characterized by the same number of strictly unstable characteristic roots for all the points of the corresponding domain. Such a number is also called the *instability degree* of the domain or region (see also [196]).

As briefly discussed in Chapter 1, it is worth mentioning that the instability degree is always finite in the case of time-delay systems of retarded type. Under appropriate assumptions concerning the stability of the associated delay-difference operator, such a property (i.e., finite instability degree) holds also for linear dynamical systems described by neutral delay-differential equations or delay-differential algebraic equations. However, such a method is quite computationally involved, and it allows us to derive a complete description of the stability regions in the parameter space for relatively simple systems. Unfortunately, for high-order systems, including a large number of parameters, the method loses its efficiency.

D-decomposition applied to a scalar system including one delay

Consider now the simplest example of a time-delay system, that is, the first-order system including a single delay,
$$\dot{x}(t)=-ax(t)-bx(t-\tau), \tag{5.4}$$
under appropriate initial conditions with $(a,b)\in\mathbb{R}^2$ and $\tau>0$. The characteristic equation of our scalar system simply is
$$H(\lambda;\ a,b)=\lambda+a+b^{-\lambda\tau}=0. \tag{5.5}$$
In the parameter space Oab, the zero-characteristic root defines the line $a+b=0$. Such a line is the *only* crossing curve separating the stable domain from the unstable one of the system free of delay, that is, of the system $\dot{x}(t)=-(a+b)x(t)$.

Next, for $\omega\neq 0$, $H(j\omega;\ a,b)=0$ if and only if the following conditions are satisfied simultaneously:
$$\begin{cases} a+b\cos(\omega\tau)=0,\\ b\sin(\omega\tau)=\omega, \end{cases} \tag{5.6}$$
which simply leads to the following parametrization of the (a,b)-curves:
$$\begin{cases} a=-\dfrac{\omega\cos(\omega\tau)}{\sin(\omega\tau)},\\ b=\dfrac{\omega}{\sin(\omega\tau)}. \end{cases} \tag{5.7}$$

Without any loss of generality, we can restrict our analysis to only $\omega \in \mathbb{R}_+$. Next, the parameters a and b as functions of ω, that is, $a = a(\omega)$ and $b = b(\omega)$, are correctly defined if and only if $\omega \neq k\pi$, with $k \in \mathbb{N}$. Thus, we can define the set of crossing frequencies,

$$\Omega = \bigcup_{k \in \mathbb{N}} \Omega_k = \bigcup_{k \in \mathbb{N}} (k\pi, (k+1)\pi),$$

and the mapping $p : \Omega \mapsto \mathbb{R}^2$ given by the relations (5.7) defines all the (stability crossing) curves in the parameter space Oab. Denote by C_k the corresponding curve defined on the frequency interval Ω_k, where k is a positive integer.

Based on the construction above, we need to detect whether the curves C_k intersect the line $a + b = 0$. Since $b(\omega)$ can be prolonged by continuity at $\omega = 0$, and the corresponding limit value is $b(0) = 1/\tau$, it follows that the point $(-1/\tau, 1/\tau)$ corresponds to the intersection of the line $a+b = 0$ with C_0. For all $k \geq 1$, the curves C_k do not intersect this line. It is important to point out that the point $(-1/\tau, 1/\tau)$ corresponds to a double characteristic root at zero. Along the curve C_0 we have a pair of complex conjugate characteristic roots located on the imaginary axis, and along the line $a + b = 0$ we have a characteristic root at zero.

To resume, the only curve intersecting $a + b = 0$ is C_0. Next, it is not difficult to see that the curves C_k never intersect Oa. However, the curves C_k always intersect the axis Ob for all positive integer k. Next, based on the sinus sign on each interval, it is easy to see that the curves C_{2k} intersect Ob for positive b, and C_{2k+1} intersect Ob for negative b. Furthermore, two independent crossing curves C_{k_1} and C_{k_2} ($k_1, k_2 \in \mathbb{N}$, $k_1 \neq k_2$) defined on the disjoint frequency sets Ω_{k_1} and Ω_{k_2}, respectively, *do not intersect* each-other. Finally, the distance between this intersection point ($C_{2k} \cap Ob$ and, respectively, $C_{2k+1} \cap Ob$) and the origin increases when k increases.

In this way, we have been able to define a *first domain* $\mathscr{D}_0$, which is bounded by the line $a+b = 0$ and by the curve C_0 given by (5.7) evaluated on Ω_0. The remaining domains are defined by the curves C_k corresponding to the frequency intervals Ω_k, $k \in \mathbb{N}$ ($k \neq 0$), and the line $a + b = 0$. Thus, we performed steps (i) of the procedure above.

Consider for instance $\mathscr{D}_0$. It is easy to see that such a domain includes the (positive) axis, Oa. Now, since for $b = 0$ the system simply becomes a finite-dimensional system of the form $\dot{x}(t) = -ax(t)$, the situation where $a > 0$ corresponds to a stable system, it follows that the domain $\mathscr{D}_0$ constructed above is nothing else than a *stability domain*. Finally, it is important to point out that this region is the *only stability domain* for the corresponding delay system above. Starting from this point, one needs to analyze first the domains having a common boundary with $\mathscr{D}_0$. One of the approaches that can be used in such a case consists in analyzing the way in which the characteristic roots cross the imaginary axis if some small variations are applied to the parameters. All these aspects will be discussed in the next paragraphs. Note also that we make a distinction between the cases with and without a characteristic root at zero. The explanation for such a distinction is easy to understand, since a characteristic root at zero is invariant under delay changes. Such an *invariance property* will play an important role in stability/instability characterization with respect to delay parameters, as we shall discuss in depth in the next chapter.

In Figure 5.1 we show the stability regions in the (a, b)-parameter space for the system (5.4) with $\tau = 1$. Further examples on different applications of the method for low-order systems can be found in [359, 197, 196]. In Chapter 13 we will discuss the stability regions in some appropriate parameter space of a second-order system encountered in modeling congestion in high-performance networks by using fluid approximations.

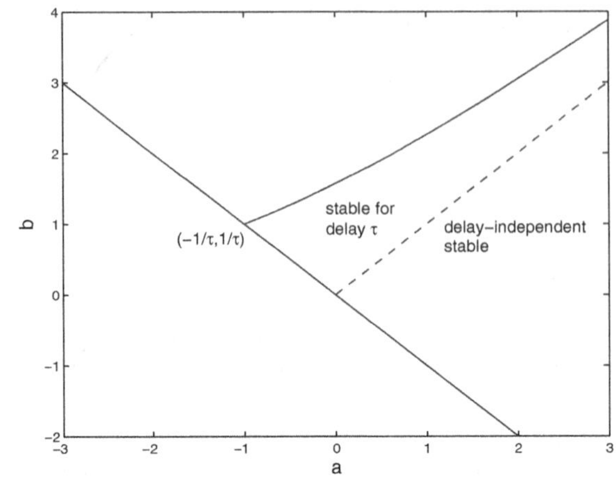

Figure 5.1. *The stability regions in the parameter space* Oab *of the scalar system* $\dot{x}(t) = -ax(t) - bx(t-\tau)$ *for the fixed delay value* $\tau = 1$.

(Algebraic) multiplicity of the characteristic roots

From the discussion above, we found that, for some particular choices of the parameters (a, b), the (algebraic) multiplicity of the characteristic roots may be larger than 1. More precisely, for example, if $a + b = 0$ and $a = -1/\tau$, we will have a *double root* at the *origin.* In this context, it is natural to see whether or not there exist other characteristic roots having the multiplicity larger than 1 and what is the largest (algebraic) multiplicity we may expect for such a scalar delay-differential equation. Some simple computations prove that the characteristic roots cannot have algebraic multiplicity larger than 2 and, except for the origin, there exists another particular point (still on the real axis) for which such a property holds.

Indeed, if one considers $a = 0$, we rewrite the characteristic function $H(\lambda;\ 0, b)$ as $H(\lambda;\ 0, b) := \lambda + be^{-\lambda\tau}$. It is easy to see that $\lambda = -1/\tau$ represents a characteristic root if $b\tau = e^{-1}$. Furthermore, $H'(\lambda;\ 0, b) = 1 - b\tau e^{-\lambda\tau}$, and it becomes 0 for the choices mentioned above, and $H''(\lambda;\ 0, b) = b\tau^2 e^{-\lambda\tau} \neq 0$. A simple analysis proves that if $b\tau < e^{-1}$, then we have two real roots x_1 and x_2, both negative, and if we assume $x_1 < x_2$, then when $b \to 0_+$, we have $x_1 \to -\infty$ and $x_2 \to 0_-$. Finally, for $\beta\tau > e^{-1}$, we do not have real roots. This last result simply says that the solution $x(t)$ of the scalar delay-differential equation is *oscillatory* (that is, it may have arbitrarily large zeros). For further characterizations and discussions of such a property and their applications in biology, we refer the reader to [351] and [123]. It is worth mentioning that such a result is still valid when $a > 0$. Indeed, in this case, the conditions above is rewritten as

$$b\tau = e^{-(a\tau+1)},$$

and the corresponding double root on the real axis becomes $\lambda = -a - 1/\tau$.

Rightmost characteristic roots and stability/instability issues in the scalar case

Let us focus on a particular region of the parameter space defined by $a \geq |\, b\,|$, $a + b > 0$. As discussed above, by using the D-decomposition method this particular region corresponds to a stable system. Consider now the *rightmost characteristic root* of the equation $H(\lambda;\ a, b) = 0$, and denote it as λ_r. Such a root is characterized by the fact that its real part equals the spectral abscissa, that is, $\lambda_r = \alpha + j\omega_r$, with

$$\alpha = \max\{\Re(\lambda): \quad \lambda \in \mathbb{C}, \quad H(\lambda; a, b) = 0\}.$$

In what follows, we shall develop a simplified argument for proving by contradiction that $\lambda_r \in \mathbb{C}_-$ in the case under consideration. In this sense, assume $\lambda_r \in \mathbb{C}_+ \cup j\mathbb{R}$. Then $r \geq 0$. Since λ_r is a characteristic root, it should satisfy the following (real, imaginary part) conditions:

$$\begin{cases} r + a + be^{-r\tau}\cos(\omega_r \tau) = 0, \\ \omega - be^{-r\tau}\sin(\omega_r \tau) = 0. \end{cases} \tag{5.8}$$

Simple computations lead to the following relation:

$$(r+a)^2 + \omega_r^2 = b^2 e^{-2r\tau}. \tag{5.9}$$

Let us consider first the case $r > 0$. Then

$$a^2 < (r+a)^2 + \omega_r^2 = b^2 e^{-2r\tau} < b^2,$$

since a is positive by hypothesis, and $e^{-2r\tau} < 1$ due to the assumption that $r > 0$ ($\tau > 0$). In conclusion, we arrive at the condition $a^2 < b^2$, which cannot be satisfied for the region we considered. In conclusion, we need $r \leq 0$. Let us analyze now the case $r = 0$, that is, at least one root of the characteristic equation on the imaginary axis. Note that $\omega_r \neq 0$, since $\lambda = 0$ is not a characteristic root. Using the same argument as in the previous case, we arrive at

$$a^2 < a^2 + \omega_r^2 = b^2, \tag{5.10}$$

since $|\omega_r| > 0$, and $e^{-2r\tau} = 1$ ($r = 0$) in (5.9). Once again, we arrive at $a^2 < b^2$, which contradicts the initial hypothesis.

In conclusion, all the roots of the characteristic equation $H(\lambda; a, b) = 0$ should be located in $\mathbb{C}_-$, if (a, b) satisfies the condition $a \geq |b|$, $a + b > 0$. Furthermore, this stability condition does *not* include any information on the *delay* size, which is equivalent to saying that the stability is of *delay-independent* type.

Next, consider the case where $|a| < |b|$. Then the scalar system is *unstable* for sufficiently large delays. The corresponding bounds as a function of the parameters (a, b) will be detailed in the next section. Here we are interested in computing some (first) estimation of the domain of the complex plane in $\mathbb{C}_+$ where the rightmost root is located. Assume that for some $\tau = \tau_0$ the scalar system is unstable. Then, without any loss of generality, we can consider $r > 0$ and $\omega_r > 0$. Conditions (5.8) lead to the estimates

$$\begin{aligned} 0 &\leq \Re(\lambda_r) = r \leq -a + |b|, \\ 0 &\leq \Im(\lambda_r) = \omega_r \leq |b| \end{aligned}$$

by using the properties of sinus, cosinus, and exponential functions for real arguments. In conclusion, the rightmost root is located in some *appropriate rectangle* in $\mathbb{C}_+$, with the imaginary axis as a left boundary. As expected from the theory presented in Chapter 1, the upper limit for the real part of the rightmost root ($\Re(\lambda_r)$) is *finite*. In a more general setting (state-space representation), the construction above still works by using appropriate matrix measures (see Section 5.6 for further comments).

Finally, consider the particular situation when $a+b<0$, and let us focus on the existence of some particular unstable roots: *positive* characteristic roots. In such a case, λ_r is rewritten as $\lambda_r = r$, and condition (5.8) reduces to

$$r+a+b^{-r\tau}=0. \tag{5.11}$$

Since $a+b<0$, it is easy to see that $f_a(0)<0$, and $f_a(+\infty)=+\infty$, where the (auxiliary) function f_a is defined by $f_a:\mathbb{R}_+\mapsto\mathbb{R}$, $f_a(x)=x+a+be^{-x\tau}$. Since f_a is a continuous mapping, the conditions above simply say that its graphics should intersect the real axis for some *finite* r-value. In conclusion, the parameters choice above always leads to some instability conditions, and this property holds *independently* of the *delay* value. In other words, we will have a particular delay-independent instability property. Such an idea holds for more general quasi-polynomials and will be discussed later in a different framework (controlling linear SISO systems by using appropriate delay blocks).

5.3.2 ▪ τ-decomposition and delay stability intervals

To the best of the authors' knowledge, such a method has its origin in the works of Sokolov and Miasnikov from the 1950s and is discussed in Popov's book [328] (see also [198] for further discussions). The method is devoted to the stability analysis only with respect to the *delay parameter* by assuming that all the other parameters of the system are fixed.

Without any loss of generality, the method can be resumed as follows:

(i) first, "decompose" the delay axis $O\tau$ into (delay) intervals

$$\mathbb{R}_+^* = \bigcup_{k\in\mathbb{N}}(\tau_k,\tau_{k+1}),$$

with $\tau_0=0$, such that within each interval the number of unstable characteristic roots is invariant for all the points of the interval, that is, the same instability degree for all the delay values inside the corresponding interval;

(ii) second, investigate the change of the number of roots at the end points of the delay intervals computed at the previous step.

This method can be seen as "dual" to the D-decomposition method discussed in the previous section. Indeed, here we concentrate our attention only on *one parameter* (delay), and the domains (or regions) are reduced to intervals such that each end point of the interval corresponds to a characteristic root crossing with respect to the imaginary axis. The only point for which such a crossing does not necessarily exist is the origin of the delay axis $\tau_0=0$. Indeed, for $\tau_0=0$, two situations may occur, depending on the presence or absence of characteristic roots on the imaginary axis. The first situation fits well within the description above. In the second case, the first delay interval is of the form $[0,\tau_1)$, and its stability (instability) is given by the stability (instability) of the system free of delay. As mentioned in the previous sections, for any positive delay value, the number of unstable characteristic roots of the corresponding characteristic equation (i.e., the instability degree) is always finite. Furthermore, the property above holds also for time-delay systems described by neutral delay-differential equations or by delay-differential algebraic equations under appropriate stability assumptions for the associate delay-difference operator.

τ-decomposition applied to the scalar case

In order to illustrate the method, we reconsider once again the scalar system including a single (constant) delay,

$$\dot{x}(t) = -ax(t) - bx(t-\tau), \tag{5.12}$$

under appropriate initial conditions with $a, b \in \mathbb{R}$ and $\tau > 0$. As mentioned above, a and b should be fixed, and the only "free" parameter is the delay τ. The characteristic equation $\lambda + a + be^{-\lambda\tau} = 0$ admits a solution $\lambda = j\omega$ on the imaginary axis if and only if the following conditions are satisfied simultaneously:

$$\begin{cases} \omega^2 + a^2 = b^2, \\ a + b\cos(\omega\tau) = 0. \end{cases}$$

The first condition is nothing else than the modulus condition (for complex numbers), and it has an interesting property in the sense that we can derive the crossing frequency $j\omega$ *independently* on the delay value. Such an idea can be applied to high-order systems also, and it can be generalized to delay systems in a state-space representation (using matrix pencil techniques).

Depending on the values of a and b, three situations may occur:

(a) $|a| > |b|$. In such a case, there does not exist any frequency ω satisfying the modulus condition mentioned above. In conclusion, there is no crossing with respect to the imaginary axis; that is, the scalar system is stable or unstable depending on the stability (or instability) of the corresponding system free of delay. In other words, the (stability, instability) property is of *delay-independent* type. If there exists at least one unstable characteristic root, that is, the instability degree is strictly positive, then the system is called *hyperbolic* (see, for instance, [150, 290, 289] for further discussions).

(b) $|a| = |b| \neq 0$, that is, $a = \pm b$. If $a = -b$, it easy to see that $\lambda = 0$ is a solution of the characteristic equation for $\tau = 0$ and, furthermore, such a solution is *invariant* with respect to the delay τ (since $e^{0\cdot\tau} = 1$ for all $\tau \geq 0$). In conclusion, the system cannot be asymptotically stable due to the characteristic root at zero. Now, if $a = b$, then it follows that the pair (ω, τ) should satisfy simultaneously $\omega\tau = (2k+1)\pi$, for some positive integer k, and $\omega = 0$, conditions which are impossible for finite delays. Thus, the stability (instability) of the scalar system is still defined in this case by the stability (or instability) property of the system free of delay. In other words, the (stability, instability) property is still of *delay-independent* type. The main difference with respect to the previous situation is that the line $a = b$ defines the *boundary* of the *delay-independent stability (instability) domain* with respect to the dual one, the so-called *delay-dependent stability (instability) domain*. Indeed, it is easy to see that there exists some perturbation on the pair (a, a) such that the delay-independent stability (instability) is lost at $(a, a+\varepsilon)$ for ε sufficiently small, where $\text{sign}(a) = \text{sign}(\varepsilon)$.

(c) $|a| < |b|$. In this case, crossing frequency exists; that is, characteristic roots will cross the imaginary axis if the delay is increased starting from 0. It is worth mentioning that, in this case, there exists only one "critical" characteristic root located on the imaginary axis and it is *simple*.

The situations described above have a very simple graphical representation by using the modulus condition above. Define the transfer

$$H_{yu}(\lambda) := -\frac{b}{\lambda + a},$$

which corresponds to a first-order system, and next represent the intersections between the graph of $H_{yu}(j\omega)$ with the unit circle $\mathscr{C}(0,1)$ in the complex plane for $\omega \geq 0$. Since $\lim_{\omega\to\infty} | H_{yu}(j\omega) | = 0$, and using the monotonicity of $| H_{yu}(j\omega) |$ as a function of ω for $\omega \in \mathbb{R}_+$, it follows that explicit crossing exists if and only if $| a | < | b |$. Case (b) above corresponds to the situation when the graph starts tangentially from the point $(-1,0)$ of the unit circle, and, except for this (particular) point, the whole graph is *inside* the unit circle. The corresponding stability is also called *weak delay-independent stability* (see, for instance, [290] for further discussions on the topics).

Crossing direction characterization

Let us detail the τ-decomposition method in this last case since crossing with respect to the imaginary axis exists. More precisely, there exists only *one (crossing) frequency* (in $\mathbb{R}_+$) that is given by

$$\omega_c = \sqrt{b^2 - a^2}.$$

As a first observation, we remark that the crossing frequency is *independent* of the delay, and it is nothing else than the (positive) root of some appropriate real polynomial. This is also confirmed by the graphical representation in Figure 5.2 (only one intersection point). The corresponding delays for which characteristic roots cross the imaginary axis are given by the relation

$$\tau_c = \min_{\ell \in \mathbb{N}} \left\{ \frac{1}{\sqrt{b^2 - a^2}} \left[\cos^{-1}\left(-\frac{a}{b}\right) + 2\ell\pi \right] > 0 \right\},$$

$$\tau_{k+1} = \tau_c + \frac{2k\pi}{\sqrt{b^2 - a^2}}, \qquad k \in \mathbb{N}.$$

In this way, we have directly the construction of all the intervals (τ_k, τ_{k+1}), where k is a positive integer and with the convention $\tau_0 = 0$, whose union defines the whole $\mathbb{R}_+$. Since $\omega_c \neq 0$, it follows that the first interval is of the form $[0, \tau_1)$, with $\tau_1 = \tau_c$.

We need to analyze the behavior of the crossing root $j\omega_c$ when τ is increased from τ_c to $\tau_c + \varepsilon$, where $\varepsilon > 0$, but very small. Based on the developments in Chapter 1, it follows that the mapping $\lambda : \mathbb{R}_+ \mapsto \mathbb{C}$, defined by $\lambda = \lambda(\tau)$ is a differentiable function. Furthermore, since the root $\lambda = j\omega_c$ is *simple*, the derivative $d\lambda/d\tau$ evaluated at $\lambda = j\omega_c$, $\tau = \tau_1$ exists, and it is different from 0. Thus, we can use the implicit function theorem (see, for instance, [135]) for computing $d\lambda/d\tau$ in some appropriate neighborhood of τ_c. Let us differentiate the corresponding characteristic function:

$$\begin{aligned} \frac{d}{d\tau} H(\lambda(\tau);\, \tau) &= \frac{d}{d\tau}\lambda + b e^{-\lambda\tau}\left(-\tau \frac{d}{d\tau}\lambda - \lambda\right) \\ &= \left(1 - b\tau e^{-\lambda\tau}\right)\frac{d}{d\tau}\lambda - b\lambda e^{-\lambda\tau}. \end{aligned} \tag{5.13}$$

Since at $\tau = \tau_1$, the critical root $\lambda = j\omega_c$ is *simple*, it follows that evaluating (5.13) at $\tau = \tau_1$ will lead to

$$\left(1 - b e^{-j\omega_c \tau_1}\right) \frac{d}{d\tau} \lambda(\tau_1) = j\omega_c b e^{-j\omega_c \tau_1},$$

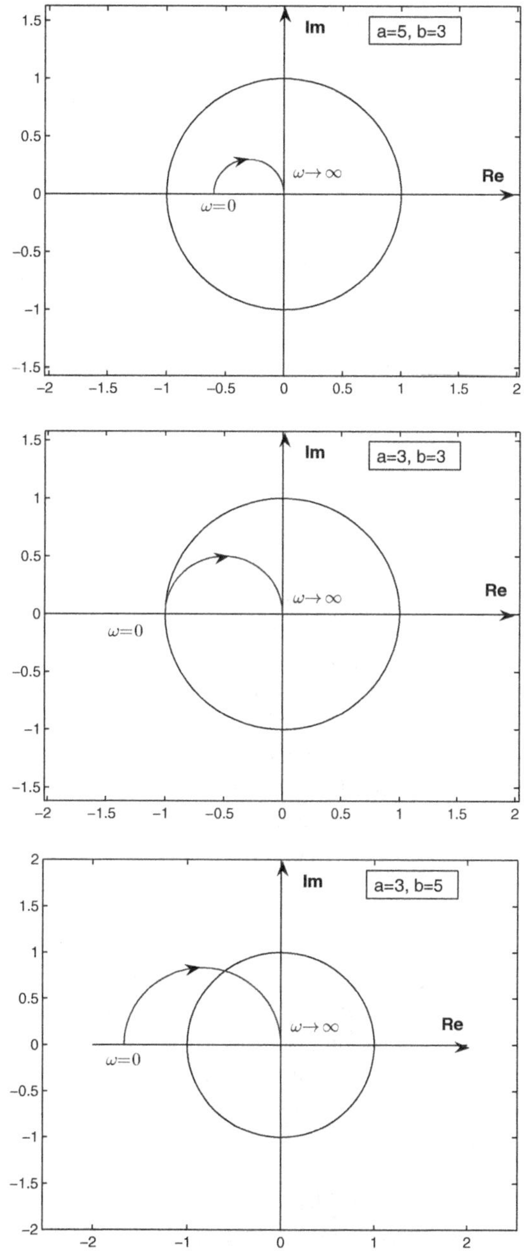

Figure 5.2. *The intersection in the complex plane of the transfer $H_{yu}(\lambda) = -b/(\lambda + a)$ evaluated on the imaginary axis $j\mathbb{R}_+$ with the unit circle $\mathscr{C}(0,1)$ for different values of the parameters a and b covering all the possible situations.*

which is simply rewritten as

$$\left[\frac{d}{d\tau}\lambda(\tau_1)\right]^{-1} = -\frac{\tau_1}{j\omega_c} - \frac{1}{j\omega_c b e^{-j\omega_c \tau_1}},$$

because $d\lambda/d\tau$ exists at $\tau = \tau_1$ and it is different from 0 since the corresponding critical root $j\omega_c$ on the imaginary axis is simple. It is important to note that $\omega_c \neq 0$ since $a+b \neq 0$. Next,

$$e^{-j\omega_c\tau_1} = -\frac{j\omega_c + a}{b},$$

since $j\omega_c$ is a characteristic root, and thus we get the following evaluation of the derivative at $\tau = \tau_1$:

$$\left[\frac{d}{d\tau}\lambda(\tau_1)\right]^{-1} = -\frac{\tau_1}{j\omega_c} + \frac{\omega_c + ja}{\omega_c(a^2+\omega_c^2)}. \tag{5.14}$$

It is easy to see that the crossing direction given by an increment $\varepsilon > 0$ on $\tau = \tau_1$ is given by the sign of $\Re\left[\frac{d}{d\tau}\lambda(\tau)\right]_{\tau=\tau_1}$. In our case, one needs to evaluate the real part of the quantities in (5.14), that is,

$$\Re\left[\frac{d}{d\tau}\lambda(\tau_1)\right]^{-1} = \frac{1}{a^2+\omega_c^2}, \tag{5.15}$$

a quantity which always *exists*, and it is always *positive*. Furthermore, this evaluation does *not depend* on the *critical delay* value τ_1, but only on the critical characteristic root value.

In other words, for all the critical delays τ_k, $k \geq 1$, the *crossing direction is always towards instability*. In conclusion, the only delay interval guaranteeing (asymptotic) stability is the *first delay interval* $[0, \tau_1)$ if the system free of delays is asymptotically stable. The bound τ_1 is also known as the *delay margin*. In such a situation, increasing the delay value does not improve the stability of the system, but it induces more roots to cross the imaginary axis towards instability. Such an *effect* is usually called *delay-induced instability* or *destabilizing effect* of the delay.

From the graphical representation above, it follows that the crossing directions at the end delay points of all intervals should be in the *same* direction, a fact that is confirmed by the relation above, which proves that the crossing direction is *invariant* with respect to the delay crossing value. Indeed, a change of crossing direction should be associated to the existence of at least *one more* critical characteristic root on the imaginary axis, a fact which is impossible in the scalar case (first-order polynomial in ω^2). However, the existence of at least two critical roots on the imaginary axis does not mean that the increase of the delay parameter will necessarily induce stability. These aspects will be discussed later in the next chapter.

Remark 5.2. *Brauer* [39] *has also analyzed the stability of the linear scalar single delay case with respect to the delay term τ by defining a "special" quantity called* characteristic return time τ_c *to the equilibrium $x \equiv 0$:*

$$\tau_c = \frac{-\tau}{\alpha},$$

where α is the spectral abscissa. Next, by analyzing the monotonicity properties of this "quantity" (characteristic return time) with respect to τ, we may decide whether there exist other stability regions by increasing the delay parameter τ for the case $b >| a |$. This leads to the following simple properties:

(i) *If $b < 0$ and $a + b > 0$, then the characteristic return time is a monotone increasing function of τ.*

(ii) *If $b > 0$, then the characteristic return time is a decreasing function of τ for all $\tau \in [0, \tau_c)$, where τ_c is defined by*

$$b\tau^* e^{a\tau_c} = e^{-1}.$$

Note that it is an increasing function of τ for $\tau > \tau_c$ and remains finite for all τ for which the zero solution is asymptotically stable.

Although the idea of analyzing the behavior of such stability regions is quite attractive, it seems difficult to generalize for more complicated (high-order) systems.

5.3.3 ▪ Numerical continuation

We present the main ideas behind the numerical continuation approach for the automatic computation of branches in a two-parameter space that correspond to characteristic roots on the imaginary axis.

Note that (5.1) has a characteristic root at $j\omega$, $\omega > 0$, if and only if

$$\begin{cases} \Re(M(j\omega;\, p)v) = 0, \\ \Im(M(j\omega;\, p)v) = 0, \\ \Re(n(v)) = 0, \\ \Im(n(v)) = 0, \end{cases} \tag{5.16}$$

where $v \in \mathbb{C}^{n\times 1}$ is the right eigenvector of M (more precisely, $ve^{\lambda\theta}$, $\theta \in [-\tau_m, 0]$, is the right eigenfunction of the infinitesimal generator of the solution operator corresponding to (5.1); see Chapter 1), and $n(v) = 0$ is a *normalizing* condition for the eigenvector. For instance, one can take $n(v) = a^T v = 0$ with $a \in \mathbb{C}^n$. Assume that there are two parameters, that is, $p = (p_1, p_2)$. Then (5.16) consists of $2n+2$ equations in the $2n+3$ unknowns (p_1, p_2, ω, v) (note that both $\Re(v)$ and $\Im(v)$ have n components). So, if p_1 and p_2 are free, then (5.16) locally defines a branch under the assumptions of the implicit function theorem. For the problem under consideration, the purpose of numerical continuation is to compute a (discrete approximation of a) branch of solutions by starting from some point(s) on the branch and following the branch. The starting point(s) are typically obtained by freezing one parameter or adding an extra condition, such that the number of unknowns becomes equal to the number of parameters, and by solving the resulting system using Newton's method.

The so-called *predictor-corrector methods* essentially boil down to the following main steps:

- the predictor step, consisting of estimating a new point on the branch;
- the correction step, consisting of computing a new point on the branch by correcting the predicted point.

We illustrate this with the commonly used *secant predictor*, combined with a correction based on an *arclength* or *pseudo-arclength* parametrization of the branch.

Assume that we have computed two nearby points on the branch, namely $P^{(j-1)}$ and $P^{(j)}$,

$$P^{(j-1)} : (p_1^{(j-1)}, p_2^{(j-1)}, \omega^{(j-1)}, v^{(j-1)}),$$
$$P^{(j)} : (p_1^{(j)}, p_2^{(j)}, \omega^{(j)}, v^{(j)}),$$

where the superscript refers to point number. Then a prediction $\hat{P}^{(j+1)}$ of a new point

$P^{(j+1)}$ can be obtained from the linear approximation of the branch through $P^{(j-1)}$ and $P^{(j)}$ as follows:

$$\hat{P}^{(j+1)} : (\hat{p}_1^{(j+1)}, \hat{p}_2^{(j+1)}, \hat{\omega}^{(j+1)}, \hat{v}^{(j+1)}),$$

where

$$\begin{aligned}
\hat{p}_1^{(j+1)} &= p_1^{(j)} + \frac{\epsilon^{(j)}}{\|p^{(j)} - p^{(j-1)}\|_2} (p_1^{(j)} - p_1^{(j-1)}), \\
\hat{p}_2^{(j+1)} &= p_2^{(j)} + \frac{\epsilon^{(j)}}{\|p^{(j)} - p^{(j-1)}\|_2} (p_2^{(j)} - p_2^{(j-1)}), \\
\hat{\omega}^{(j+1)} &= \omega^{(j)} + \frac{\epsilon^{(j)}}{\|p^{(j)} - p^{(j-1)}\|_2} (\omega^{(j)} - \omega^{(j-1)}), \\
\hat{v}^{(j+1)} &= v^{(j)} + \frac{\epsilon^{(j)}}{\|p^{(j)} - p^{(j-1)}\|_2} (v^{(j)} - v^{(j-1)}),
\end{aligned} \tag{5.17}$$

and $\epsilon^{(j)}$ is the step length taken in the jth step. Note that, as it appears in (5.17), the latter corresponds to the distance between (the projections of) $\hat{P}^{(j+1)}$ and $P^{(j)}$ in the (p_1, p_2)-plane.

The next step consists of correcting $\hat{P}^{(j+1)}$ to a new point $P^{(j+1)}$ on the branch. This can be done by applying Newton's method to (5.16), with (5.17) as starting values. However, since p_1 and p_2 are freed, the solutions of (5.16) define a branch, as we have seen. Therefore, an extra equation has to be added first to (5.17) in order to uniquely specify the solution. One possibility consists of adding

$$\left(p_1 - p_1^{(j)}\right)^2 + \left(p_2 - p_2^{(j)}\right)^2 = \left(\varepsilon^{(j)}\right)^2. \tag{5.18}$$

This is called an arclength condition, as it expresses that the new point $P^{(j+1)}$ must lie at a distance $\varepsilon^{(j)}$ from $P^{(j)}$ in the (p_1, p_2)-plane and $\sum_j \varepsilon^{(j)}$ can be seen as an approximation of the parameter corresponding to an arclength parametrization of the branch in the (p_1, p_2)-plane. In combination with (5.17), it is natural to take

$$\varepsilon^{(j)} = \epsilon^{(j)}. \tag{5.19}$$

Because the condition (5.18) is nonlinear in the parameters and it contains no information about the direction in which the new point must be found along the branch, usually a linearization of (5.18) is employed, leading to a so-called pseudo-arclength condition. For instance, one can replace (5.18)–(5.19) with

$$\left\langle p^{(j)} - p^{(j-1)}, p - \hat{p}^{(j+1)} \right\rangle = 0,$$

which expresses that the new point lies on a line through the prediction $\hat{p}^{(j+1)}$ and perpendicular to the approximation of the tangent vector at $p^{(j)}$, given by $p^{(j)} - p^{(j-1)}$.

For more details on numerical computation we refer the reader to [342, 203] and the references therein.

5.4 ▪ Computing the crossing direction of characteristic roots

Assume that $\lambda_0 = j\omega_0 \in \sigma(M)$ is a characteristic root on the imaginary axis with $\omega_0 \neq 0$ for some parameter vector $p = p^0 \in \mathbb{R}^{n_p}$, that is, $\lambda_0 = \lambda(p^0)$. Assume also that the functions $\tau_i : \mathbb{R}^{n_p} \mapsto \mathbb{R}_+$, $i = 1, \ldots, m$, and the matrix functions $A_i : \mathbb{R}^{n_p} \mapsto \mathbb{R}^{n \times n}$ are continuously differentiable functions of the parameters $p := (p_1\ p_2, \cdots, p_{n_p}) \in \mathbb{R}^{n_p}$.

5.4.1 ▪ Simple crossing characteristic roots

We start with the assumption that the characteristic roots on the imaginary axis are *simple*. Due to the differentiability property of $A_0(\cdot), A_1(\cdot), \ldots, A_m(\cdot), \tau_1(\cdot), \tau_2(\cdot), \ldots, \tau_m(\cdot)$ it follows that $\lambda_0 = j\omega_0 = \lambda(p)$ ($p \in \mathbb{R}^{n_p}$) is also a differentiable function, and its derivative exists in some neighborhood of $p^0 \in \mathbb{R}^{n_p}$.

There are several ways to compute the derivative of the function of $p \mapsto \lambda(p)$ evaluated at $p = p^0$ with respect to any of the directions p_i, $i = 1, \ldots, n_p$, in the parameter space. Formally, simple computations lead to

$$0 = \frac{\partial}{\partial p_i} H(\lambda(p);\, p) = \frac{\partial}{\partial \lambda} H(\lambda;\, p) \cdot \frac{\partial \lambda}{\partial p_i} + \frac{\partial}{\partial p_i} H(\lambda;\, p)$$

evaluated at (λ_0, p^0), and we get (using the implicit function theorem; see, for instance, [135]) that, *locally*, in some neighborhood of the point $p = p^0$ such that $\lambda_0 = \lambda(p^0)$, the following condition holds:

$$\frac{\partial \lambda}{\partial p_i} := -\frac{\dfrac{\partial}{\partial p_i} H(\lambda;\, p)}{\dfrac{\partial}{\partial \lambda} H(\lambda;\, p)} \tag{5.20}$$

evaluated at the same point. It is important to note that $\frac{\partial}{\partial \lambda} H(\lambda;\, p)\,|_{\lambda=\lambda_0, p=p^0} \neq 0$ since the corresponding characteristic root is simple.

In conclusion, in order to explicitly evaluate this derivative, one needs to evaluate the derivative of the determinant $H(\lambda;\, \tau) = 0$ at $\lambda = \lambda_0$ and $\tau = \tau(p^0)$. In what follows, we shall propose two independent approaches for estimating such a crossing characteristic root derivative: one based on the Jacobi's formula, and the other based on the perturbation theory for matrices.

Jacobi's formula-based approach

Consider also the corresponding right (left) eigenvectors u_0 (v_0^*) associated with the characteristic root $\lambda_0 = j\omega_0$. If $\lambda_0 = \lambda(p^0)$ is a *simple characteristic root*, then $\lambda = \lambda(p)$ is a differentiable function around $p = p^0$, and furthermore the eigenvectors are also differentiable.[14] We can apply this last argument in order to derive the corresponding derivative. However, in what follows, we shall present a different method based on Jacobi's formula for the determinant's derivative.

We have the following result:

Proposition 5.3. *If the characteristic root $\lambda_0 = j\omega_0 = \lambda(p^0) \neq 0$ is simple, then for any positive integer i, $1 \leq i \leq m$, the following holds:*

$$\frac{\partial \lambda}{\partial p_i} = -\frac{v_0^* \dfrac{\partial M}{\partial p_i} u_0}{v_0^* \dfrac{\partial M}{\partial \lambda} u_0}, \tag{5.21}$$

[14] Please note that such a property does not necessarily hold if the corresponding characteristic root is not simple.

where the partial derivatives of the characteristic matrix $M(\lambda;\ p)$ are evaluated at $\lambda = \lambda_0$, and $p = p^0$. The vectors v_0^* and u_0 are the left and right eigenvectors[15] *corresponding to the characteristic root λ_0.*

Proof. From the definition of the characteristic function, it follows that $\lambda = \lambda_0$ is a simple root if and only if $\operatorname{rank}(M(\lambda_0, p^0)) = n-1$. The use of Jacobi's formula

$$d\det(M) = Tr\,(\operatorname{Adj}(M)dM) \tag{5.22}$$

leads to the relation

$$\begin{aligned}\frac{\partial}{\partial p_i}\det(M(\lambda;\ p)) &= Tr\left(\operatorname{Adj}(M(\lambda;\ p))\frac{\partial}{\partial p_i}M(\lambda;\ p)\right)\\ &= \operatorname{Tr}\left[u_0 v_0^*\left(\frac{\partial}{\partial\lambda}M(\lambda;\ p)\frac{\partial\lambda}{\partial p_i}+\frac{\partial}{\partial p_i}M(\lambda;\ p)\right)\right]\\ &= v_0^*\left(\frac{\partial}{\partial\lambda}M(\lambda;\ p)\frac{\partial\lambda}{\partial p_i}+\frac{\partial}{\partial p_i}M(\lambda;\ p)\right)u_0,\end{aligned} \tag{5.23}$$

where we used the standard property of the adjoint of a matrix $M \in \mathbb{C}^{n\times n}$ of rank $(n-1)$ to be a matrix of rank one [168], expressed by its left and right null vectors as $u_0 v_0^*$.

Next, using the same argument as in deriving (5.20), the condition (5.21) evaluated at $\lambda = \lambda_0$ and $p = p^0$ follows straightforwardly from (5.23), and $v_0^*\frac{\partial M}{\partial\lambda}u_0 \neq 0$ at $\lambda = \lambda_0$ and $p = p^0$ follow since the characteristic root is simple. □

An explicit computation of the derivative leads to the formula

$$\frac{\partial\lambda}{\partial p_i} = \frac{v_0^*\left(\dfrac{\partial A_0(p)}{\partial p_i}+\displaystyle\sum_{k=1}^{m}\frac{\partial A_k(p)}{\partial p_i}e^{-\lambda\tau_k(p)}-\lambda\sum_{k=1}^{m}A_k(p)\frac{\partial\tau_k(p)}{\partial p_i}e^{-\lambda\tau_k(p)}\right)u_0}{v_0^*\left(I_n+\displaystyle\sum_{k=1}^{m}A_k\tau_k(p)e^{-\lambda\tau_k(p)}\right)u_0}, \tag{5.24}$$

evaluated at $\lambda = \lambda_0$ and $p = p^0$.

Perturbation theory–based approach

As mentioned at the beginning of the section, there exist several ways to compute the derivative of the function $\lambda(p)$ evaluated at $p = p^0$. Here, we focus on a different way to derive such a derivative by using the *perturbation theory* for (analytic) operators (see, for instance, [187]).

Assume now that the function $\varepsilon : \mathbb{R} \mapsto \mathbb{R}^{n_p}$, $\varepsilon \mapsto p(\varepsilon)$ is sufficiently smooth in some neighborhood of $p = p^0$, and assume that $p(0) = p^0$. Let $d \in \mathbb{R}^{n_p}$ be the *direction vector* in the parameter space defined as follows:

$$d := \frac{d}{d\varepsilon}p(0) = \left[\begin{array}{cccc}\frac{d}{d\varepsilon}p_1(0) & \frac{d}{d\varepsilon}p_2(0) & \cdots & \frac{d}{d\varepsilon}p_{n_p}(0)\end{array}\right]. \tag{5.25}$$

[15] More precisely, the function segments $u_0 e^{\lambda_0\theta}$, $\theta \in [-\tau,\ 0]$, and $v_0^* e^{\bar{\lambda}\theta}$, $\theta \in [0,\ \tau]$, are right and left eigenfunctions of the corresponding infinitesimal generator.

Without any loss of generality, we can use the following normalization condition:

$$||d||_2 := \sqrt{\sum_{k=1}^{n_p} d_k^2} = 1.$$

For instance, for the sake of simplicity, we may consider the simplest case, i.e.,

$$p(\varepsilon) = p^0 + \varepsilon d.$$

Using now the smoothness of $p(\varepsilon)$, it follows that the characteristic matrix $M(\lambda;\ p)$ is rewritten as

$$M(\lambda;\ p(\varepsilon)) = M_{1,0}(\lambda) + \sum_{i=1}^{r} \varepsilon^i M_{i,1}(\lambda) + \cdots \tag{5.26}$$

for some positive integer $r > 1$. The construction of the matrices $M_{1,0}$ and $M_{i,1}$, $1 \leq i \leq r$, follows straightforwardly. For instance,

$$\begin{aligned}
M_{1,0}(\lambda) &= M(\lambda;\ p^0),\\
M_{1,1}(\lambda) &= \sum_{k=1}^{n_p} \frac{\partial}{\partial p_k} M(\lambda;\ p^0) d_k\\
&= -\sum_{k=1}^{n_p} \frac{\partial}{\partial p_k} A_0(p^0) d_k - \sum_{l=1}^{m} \sum_{k=1}^{n_p} \frac{\partial}{\partial p_k} A_l(p^0) d_k e^{-\lambda \tau_l(p^0)}\\
&\quad + \sum_{l=1}^{m} \sum_{k=1}^{n_p} \lambda d_k \frac{\partial}{\partial p_k} \tau_l(p^0) A_l(p^0) e^{-\lambda \tau_l(p^0)}.
\end{aligned}$$

In the case of simple characteristic roots, the corresponding *Puiseux formulae*[16] (see, e.g., [187]) lead to the following expansion of the characteristic root λ and of the corresponding right-eigenvector u:

$$\lambda = \lambda_0 + \sum_{i=1}^{r} \varepsilon^i \lambda_i + \cdots, \tag{5.27}$$

$$u = u_0 + \sum_{i=1}^{r} \varepsilon^i u_i + \cdots. \tag{5.28}$$

Assume now that λ_0 is located on the imaginary axis ($\lambda_0 = j\omega_0 \neq 0$). The way in which the characteristic root is moving with respect to the imaginary axis (that is, towards stability/instability regions) is given by the complex λ_1 that needs an explicit evaluation.

More precisely:

(i) If $\Re(\lambda_1) > 0$, then the crossing direction at $\varepsilon > 0$ is *towards instability*.

(ii) If $\Re(\lambda_1) < 0$, then the crossing direction at $\varepsilon > 0$ is *towards stability*.

(iii) Finally, if $\Re(\lambda_1) = 0$, then we need a *second-order analysis* in order to see whether the corresponding characteristic root will cross the imaginary axis or will stay in the same half plane. In other words, we need the explicit computation of λ_2 in the corresponding λ-expansion.

[16]Throughout the book, we also use the term *Puiseux formulae* for describing the corresponding terms of Taylor series.

In order to compute λ_1, we need a further expansion of (5.26) in some neighborhood of $\lambda_0 = j\omega_0 = \lambda(p^0) = \lambda(p(0))$, by using the mapping $\lambda \circ p : \mathbb{R} \mapsto \mathbb{C}$, and defined by $(\lambda \circ p)(\varepsilon) = \lambda(p(\varepsilon))$. For the sake of brevity, we shall "identify" $\lambda(p(\varepsilon))$ as $\lambda(\varepsilon)$. It is easy to see that (5.26) is rewritten as

$$M(\lambda(\varepsilon);\ p(\varepsilon)) = M_0 + \sum_{i=1}^{r} \varepsilon^i M_i + \cdots, \tag{5.29}$$

with

$$M_0 = M(\lambda_0, p^0).$$

Simple algebraic manipulations lead to

$$\begin{aligned} M_1 = {} & \lambda_1 \left(I_n + \sum_{l=1}^{m} \tau_l(p^0) A_l e^{-\lambda_0 \tau_l(p^0)} \right) \\ & - \sum_{k=1}^{n_p} \left(\frac{\partial}{\partial p_k} A_0(p^0) d_k + \sum_{l=1}^{m} \frac{\partial}{\partial p_k} A_l(p^0) d_k e^{-\lambda_0 \tau_l(p^0)} \right) \\ & + \sum_{k=1}^{n_p} \sum_{l=1}^{m} \lambda_0 d_k A_l(p^0) \frac{\partial}{\partial p_k} \tau_l(p^0) e^{-\lambda_0 \tau_l(p^0)}. \end{aligned} \tag{5.30}$$

Since the characteristic root $\lambda_0 = j\omega_0$ is *simple*, it follows that

$$M_0 u_0 = 0, \quad v_0^* M_0 = 0, \quad v_0^* u_0 \neq 0.$$

The last condition is a consequence of the fact that the zero eigenvalue of the complex matrix M_0 is (algebraically) simple. Without any loss of generality, we can use the following normalization condition:

$$v_0^* u_0 = 1.$$

Next, by identifying the coefficients, we arrive to the following:

Proposition 5.4. *Assume that the characteristic root $\lambda_0 = \lambda(p^0)$ located on the imaginary axis is simple. Then, for a given direction vector $d \in \mathbb{R}^{n_p}$ with $||d||_2 = 1$, the characteristic root $\lambda = \lambda(p)$ crosses the imaginary axis at $p = p^0$ towards instability (stability) with respect to the direction d if $\Re(\lambda_1) > 0$ ($\Re(\lambda_1) < 0$), where*

$$\lambda_1 = \sum_{k=1}^{n_p} \frac{\partial}{\partial p_k} \lambda(p^0) d_k, \tag{5.31}$$

where $\partial\lambda/\partial p_i$ for $i = 1, \ldots, n_p$ are given by (5.24).

There is still one case to be discussed, that is, the case where $\Re(\lambda_1) = 0$. Then we have two distinct situations depending on whether $\lambda_1 = 0$:

(i) The first case ($\lambda_1 = 0$) corresponds to the situation where the direction vector d is tangent to the "surface" in the parameter space that corresponds to the presence of characteristic roots on the imaginary axis.

(ii) The second case ($\lambda_1 \neq 0$: $\Re(\lambda_1) = 0$ and $\Im(\lambda_1) \neq 0$) corresponds to the situation where the characteristic root arrives "*tangentially*" to the imaginary axis (1-contact point in the sense mentioned by [48]), and as mentioned above, a second-order analysis is needed, that is, an explicit computation of $\Re(\lambda_2)$. As expected, two situations are possible:

(ii.1) the characteristic root will stay in the same domain, that is, the corresponding half plane of $\mathbb{C}$ (the so-called "mirror" effect), or

(ii.2) it will cross the imaginary axis depending on the sign of $\Re(\lambda_2)$.

Finally, if $\Re(\lambda_2) = 0$, the procedure should follow the same steps as above, and a new iteration will be needed.

For the brevity of the chapter, we do not discuss the second-order analysis in the general case. However, we mention the existence of such an analysis in the commensurate delay case as reported in [69] (for further discussions, see also some of the examples presented in the next chapter).

5.4.2 ▪ Double characteristic roots

Consider now the case where the characteristic root $\lambda_0 = j\omega_0 \neq 0$ is *not simple*, and assume for the sake of simplicity that its algebraic multiplicity μ_0 is equal to 2. In such a situation, the characteristic function is rewritten as

$$H(\lambda;\, p^0) = (\lambda - \lambda_0)^2 H_1(\lambda;\, p^0),$$

with $H_1(\lambda_0;\, p^0) \neq 0$.

Two situations may appear depending on the geometric multiplicity of the characteristic root as an eigenvector of the corresponding infinitesimal generator (number of linearly independent eigenfunctions):

(i) The characteristic root λ_0 is *double semisimple*; that is, there exists two linearly independent eigenvectors.

(ii) The characteristic root λ_0 is *double, non-semisimple*; that is, there exists only one linearly independent eigenvector.

The two cases will be presented separately below. Consider the following notations:

$$\begin{aligned}
M_0 &= M(\lambda_0, p^0) = \left(\lambda_0 I_n - A_0(p^0) - \sum_{i=1}^{m} A_i(p^0) e^{-\lambda_0 \tau_i(p^0)}\right), \\
M_1^0 &= \frac{\partial}{\partial \lambda} M(\lambda_0, p^0) = \left(I_n + \sum_{i=1}^{m} A_i(p^0)\tau(p^0) \mathrm{e}^{-\lambda_0 \tau(p^0)}\right), \\
M_1^1 &= \sum_{k=1}^{n_p} d_k \frac{\partial}{\partial p_k} M(\lambda_0, p^0) = \sum_{k=1}^{n_p} d_k \frac{\partial}{\partial p_k} A_0(p^0) \\
&\quad + \sum_{i=1}^{m}\sum_{k=1}^{n_p} d_k \left(\frac{\partial}{\partial p_k} A_i(p^0)\right) \mathrm{e}^{-\lambda_0 \tau(p^0)} \\
&\quad - \sum_{i=1}^{m}\sum_{k=1}^{n_p} \lambda_0 d_k \frac{\partial}{\partial p_k} (\tau(p^0)) A_i(p^0) \mathrm{e}^{-\lambda_0 \tau(p^0)}, \\
M_2^0 &= \frac{\partial^2}{\partial \lambda^2} M(\lambda_0, p^0) = -\sum_{i=1}^{m} A_i(p^0) \tau^2(p^0) \mathrm{e}^{-\lambda_0 \tau(p^0)}.
\end{aligned} \tag{5.32}$$

Double semisimple characteristic root

If the root is double and semisimple, then the normalized left $(v_{0,1}^*, v_{0,2}^*)$ and right $(u_{0,1}, u_{0,2})$ eigenvectors should satisfy the following equations (see, for instance, [395] for general result and [277] for some discussions in the matrix case):

$$\begin{cases} M_0\, u_{0,l} = 0, \\ v_{0,k}^*\, M_0 = 0, \\ v_{0,k}^*\, u_{0,l} = \delta_{l,k} \quad \forall\, k, l \in \{1, 2\}, \end{cases} \tag{5.33}$$

where $\delta_{l,k}$ is the Kronecker symbol (1 if $k = l$, and 0 otherwise). It is worth noting that linear combination of the eigenvectors $u_{0,1}$ and $u_{0,2}$ is also an eigenvector for $M(\lambda_0, p^0)$. In this case, the perturbations of the root λ_0, as well as the corresponding eigenvectors in terms of the power series, are

$$\lambda = \lambda_0 + \sum_{i=1}^{r} \varepsilon^i\, \lambda_i + \cdots, \tag{5.34}$$

$$u = u_0 + \sum_{i=1}^{r} \varepsilon^i\, u_i + \cdots. \tag{5.35}$$

Since the characteristic root λ and the associated right eigenvector u satisfy

$$M(\lambda(\varepsilon), p(\varepsilon))\, u(p(\varepsilon)) = 0, \tag{5.36}$$

then, by substituting (5.33)–(5.35) as well as the Taylor expansion of the characteristic matrix M in the eigenvalue problem (5.36), one gets the following expressions (for the first two orders ε^0, ε^1):

$$\begin{aligned} &M_0\, u_0 = 0, \\ &M_0\, u_1 + \left(\lambda_1 M_1^0 + M_1^1\right) u_0 = 0. \end{aligned} \tag{5.37}$$

The first equation in (5.37) simply says that u_0 is an eigenvector for $M(\lambda_0, p^0)$. Thus, for some appropriate scalars γ_1 and γ_2, we have

$$u_0 = \gamma_1\, u_{0,1} + \gamma_2\, u_{0,2}. \tag{5.38}$$

Consider now the second equation in (5.37) (written in shorter form by using the notations in (5.32)). Then, by multiplying it by $v_{0,1}^*$ and $v_{0,2}^*$, we have

$$\begin{aligned} v_{0,1}^* M_0\, u_1 + v_{0,1}^* \left(\lambda_1 M_1^0 + M_1^1\right) u_0 = \lambda_1\, v_{0,1}^* M_1^0\, u_0 + v_{0,1}^* M_1^1\, u_0 = 0, \\ v_{0,2}^* M_0\, u_1 + v_{0,2}^* \left(\lambda_1 M_1^0 + M_1^1\right) u_0 = \lambda_1\, v_{0,2}^* M_1^0\, u_0 + v_{0,2}^* M_1^1\, u_0 = 0. \end{aligned} \tag{5.39}$$

Taking into account the expression of u_0 given by (5.38), we obtain the linear system

$$\left(\lambda_1 \mathscr{M} + \mathscr{N}\right) \begin{bmatrix} \gamma_1 \\ \gamma_2 \end{bmatrix} = 0, \tag{5.40}$$

where

$$\mathscr{M} = \begin{bmatrix} v_{0,1}^* M_1^0\, u_{0,1} & v_{0,1}^* M_1^0\, u_{0,2} \\ v_{0,1}^* M_1^0\, u_{0,2} & v_{0,2}^* M_1^0\, u_{0,2} \end{bmatrix} \quad \text{and} \quad \mathscr{N} = \begin{bmatrix} v_{0,1}^* M_1^1\, u_{0,1} & v_{0,1}^* M_1^1\, u_{0,2} \\ v_{0,2}^* M_1^1\, u_{0,1} & v_{0,2}^* M_1^1\, u_{0,2} \end{bmatrix}. \tag{5.41}$$

In other words, the double semisimple case is reduced to solving the generalized eigenvalue problem (5.40):

$$\det(\lambda \mathcal{M} + \mathcal{N}) = 0. \tag{5.42}$$

If in addition, the matrix $\mathcal{M}$ is nonsingular, then the double semisimple case is reduced to solving the following eigenvalue problem:

$$-\mathcal{M}^{-1}\mathcal{N}\begin{bmatrix} \gamma_1 \\ \gamma_2 \end{bmatrix} = \lambda \begin{bmatrix} \gamma_1 \\ \gamma_2 \end{bmatrix}. \tag{5.43}$$

In conclusion, with the notations above, we have the following result:

Proposition 5.5. *Assume that the characteristic root* $\lambda_0 = \lambda(p^0)$ *located on the imaginary axis is double and semisimple.*

Furthermore, assume that $\det(\mathcal{M})\det(\mathcal{N}) \neq 0$, *where* $\mathcal{M}$ *and* $\mathcal{N}$ *are given by* (5.41). *Then, for a given direction vector* $d \in \mathbb{R}^{n_p}$ *with* $\|d\|_2 = 1$, *the double characteristic root* $\lambda = \lambda(p)$ *at* $p = p^0$ *splits up into two simple characteristic roots with respect to the direction* d, *and the corresponding crossing directions towards instability (stability) are given by* $\Re(\lambda_{1,i}) > 0$ *(*$\Re(\lambda_{1,i}) < 0$, $i = 1, 2$*), where* $\lambda_{1,i}$ *are the eigenvalues of the matrix* $\mathcal{M}^{-1}\mathcal{N}$.

Remark 5.6. *It is worth mentioning that Proposition* 5.5 *simply describes the way a double semisimple characteristic root located on the imaginary axis will split by a variation of the parameters defined by the direction* d. *The corresponding behavior is nothing else than the behavior of two* almost independent *simple crossing roots and it is also known as the* weak interaction *of the characteristic roots (see also* [343]*).*

Double non–semisimple characteristic root

The fact that the eigenvalue λ_0 is double and non–semisimple implies that there exist vectors u_0, v_0, u_1, and v_1 such that the following conditions hold simultaneously:

$$\begin{cases} M_0 u_0 = 0, \\ M_0 u_1 + u_0 = 0 \end{cases} \tag{5.44}$$

and

$$\begin{cases} v_0^* M_0 = 0, \\ v_1^* M_0 + v_0^* = 0. \end{cases} \tag{5.45}$$

Furthermore, u_0, v_0, u_1, and v_1 should satisfy the conditions

$$v_0^* u_0 = 0, \qquad v_1^* u_0 = v_0^* u_1 \neq 0.$$

As in the double semisimple characteristic root case, and without any loss of generality, we can impose the following regularity condition:

$$v_1^* u_1 = 0, \qquad v_0^* u_1 = v_1^* u_0 = 1. \tag{5.46}$$

In this case, the Puiseux formulae (see, for instance, [187]) lead to the following expansion of λ and of the corresponding right eigenvalue u:

$$\lambda = \lambda_0 + \sum_{i=1}^{r} \varepsilon^{\frac{i}{2}} \lambda_i + \cdots, \tag{5.47}$$

$$u = u_0 + \sum_{i=1}^{r} \varepsilon^{\frac{i}{2}} w_i + \cdots. \tag{5.48}$$

Since the characteristic root λ and the associated right eigenvector u satisfy

$$M(\lambda(\varepsilon), p(\varepsilon))\, u(\varepsilon) = 0, \tag{5.49}$$

similarly to the previous case study, then, substituting (5.47) and (5.48) into the eigenvalue problem (5.49) and collecting terms of equal powers of ε lead to (only the first three orders $\varepsilon^0, \varepsilon^{\frac{1}{2}}, \varepsilon^1$)

$$\begin{cases} M_0\, u_0 = 0, \\ M_0\, w_1 + \lambda_1 M_1^0\, u_0 = 0, \\ M_0\, w_2 + \lambda_1 M_1^0\, w_1 + \left(\lambda_2 M_1^0 + \frac{\lambda_1^2}{2} M_2^0 - M_1^1\right) u_0 = 0. \end{cases} \tag{5.50}$$

Moreover, for the sake of uniqueness of u, we use the normalization $v_1^* u = 1$ which can be written in terms of the expansion of u as follows:

$$v_1^* \left(u_0 + \varepsilon^{\frac{1}{2}}\, w_1 + w_2\, \varepsilon + \cdots\right) = 1.$$

Now, from the second equation from (5.46), that is, $v_1^*\, u_0 = 1$, we get the following normalization condition:

$$v_1^*\, w_1 = v_1^*\, w_2 = 0. \tag{5.51}$$

With this last normalization, one can compute the explicit expressions of λ_1, λ_2, w_1, and w_2. For instance, in what follows, we focus on deriving appropriate formulae for λ_1 and w_1. To determine w_1, one uses the second equation from (5.44) with the second equation from (5.50). After some tedious algebraic computations, one gets

$$w_1 = \lambda_1 M_1^0\, u_1. \tag{5.52}$$

Finally, using the third equation from (5.50) and substituting the expression of w_1 given by (5.52) and multiplying by v_0^* lead to

$$\lambda_1^2\, v_0^* \left(M_1^{0^2}\, u_1 + \frac{1}{2} M_2^0\, u_0\right) + v_0^* \left(\lambda_2 M_1^0 - M_1^1\right) u_0 = 0. \tag{5.53}$$

But it is well known that $M_0\, u_1 + M_1^0\, u_0 = 0$, (see, for instance, [151] or [63]). Thus, by using (5.44), one simply concludes that $M_1^0\, u_0 = -M_0\, u_1 = u_0$, leading to the following expression of λ_1:

$$\lambda_1 = \pm \sqrt{-\frac{v_0^*\, M_1^1\, u_0}{v_0^* \left[\left(M_1^0\right)^2 u_1 + \frac{1}{2} M_2^0\, u_0\right]}}. \tag{5.54}$$

In conclusion, we have the following result:

Proposition 5.7. *Assume that the characteristic root $\lambda_0 = \lambda(p^0)$ located on the imaginary axis is double and non-semisimple. Then, for a given direction vector $d \in \mathbb{R}^{n_p}$ with $\|d\|_2 = 1$, the double characteristic root $\lambda = \lambda(p)$ at $p = p^0$ splits up into two simple characteristic roots with respect to the direction d, and the corresponding crossing directions towards instability (stability) are given by $\Re(\lambda_{1,\pm}) > 0$ ($\Re(\lambda_{1,\pm}) < 0$), where $\lambda_{1,\pm}$ is given by*

$$\lambda_{1,\pm} = \pm \sqrt{-\frac{v_0^* \left(\sum_{k=1}^{n_p} d_k \frac{\partial}{\partial p_k} M(\lambda_0, p^0)\right) u_0}{v_0^* \left[\left(\frac{\partial}{\partial \lambda} M(\lambda_0, p^0)\right)^2 u_1 + \frac{1}{2} \frac{\partial^2}{\partial \lambda^2} M(\lambda_0, p^0)\, u_0\right]}},$$

where u_0, u_1, and v_0 are appropriate normalized eigenvectors of M_0 satisfying (5.44)–(5.45).

5.4.3 ▪ Delay interdependence and crossing direction evaluation

Consider now the delay dependence on the parameters. Recall that the mapping $\tau : \mathbb{R}^{n_p} \mapsto \mathbb{R}^m$ was assumed continuous and differentiable, and was subjected to the constraint $\tau_i(p) \geq 0$ for all $1 \leq i \leq m$ and for all $p \in \mathscr{D} \subset \mathbb{R}^{n_p}$. As discussed at the beginning of this chapter, there exist some simple cases of *delay interdependence* for which the conditions above are automatically satisfied. Since the interdependence of the delays plays an important role in computing stability crossing boundaries, and related stability regions, we need to have the explicit expressions of the corresponding crossing direction:

Corollary 5.8 (delay interdependence: crossing direction evaluation). *Under the assumptions of Proposition* 5.3 *the following expressions hold with respect to the delays:*

(i) *[Independent delays: $n_p = m$, $\tau_i = p_i$, $i = 1,\dots,m$] The crossing direction with respect to the delay τ_i is given by*

$$\frac{\partial \lambda}{\partial \tau_i} = \frac{v_0^* \left(\dfrac{\partial A_0(\tau)}{\partial \tau_i} + \displaystyle\sum_{k=1}^{m} \dfrac{\partial A_k(\tau)}{\partial \tau_i} e^{-\lambda \tau_k} - \lambda A_i(\tau) e^{-\lambda \tau_i} \right) u_0}{v_0^* \left(I_n + \displaystyle\sum_{k=1}^{m} A_k \tau_k e^{-\lambda \tau_k} \right) u_0} \tag{5.55}$$

evaluated at $\lambda = \lambda_0$ and $\tau_i = \tau_i^0$ for all $i = 1,\dots,m$.

(ii) *[Dependent delay: $n_p < m$, $\tau_i = \sum_{k=1}^{n_p} \gamma_{ik} p_k$, $i = 1,\dots,m$] The crossing direction with respect to the delay p_i is given by*

$$\frac{\partial \lambda}{\partial p_i} = \frac{v_0^* \left(\dfrac{\partial A_0(p)}{\partial p_i} + \displaystyle\sum_{k=1}^{m} \dfrac{\partial A_k(p)}{\partial p_i} e^{-\lambda \tau_k(p)} - \lambda \sum_{k=1}^{m} \gamma_{ki} A_k(p) e^{-\lambda \tau_k(p)} \right) u_0}{v_0^* \left(I_n + \displaystyle\sum_{k=1}^{m} A_k \tau_k(p) e^{-\lambda \tau_k(p)} \right) u_0} \tag{5.56}$$

evaluated at $\lambda = \lambda_0$ and $p = p^0$.

(iii) *[Commensurate delays: $\tau_k = k\tau$, $k = 1,\dots,m$] The crossing direction with respect to the delay τ is given by*

$$\frac{d\lambda}{d\tau} = \frac{v_0^* \left(\dfrac{dA_0(\tau)}{d\tau} + \displaystyle\sum_{k=1}^{m} \dfrac{dA_k(\tau)}{d\tau} e^{-k\lambda\tau} - \lambda \sum_{k=1}^{m} k A_k(\tau) e^{-k\lambda\tau} \right) u_0}{v_0^* \left(I_n + \tau \displaystyle\sum_{k=1}^{m} k A_k e^{-k\lambda\tau} \right) u_0} \tag{5.57}$$

evaluated at $\lambda = \lambda_0$ and $\tau = \tau^0$.

The main difference between the formulas above and the general formula (5.24) lies in the different forms of the corresponding numerators. As we shall see in the next chapter, in the cases when the matrices A_0 and A_k, $k = 1,\dots,m$, do not explicitly depend on the parameters defined by the vector p, the crossing direction evaluations will have some interesting *invariance properties* (with respect to the delays) that will be exploited in defining the stability regions and the corresponding stability crossing boundaries.

5.5 ▪ Computing the root-locus

In what follows, we outline the approach of [142] and the references therein. It is worth mentioning that this method can be used in characterizing all the stabilizing pairs (controller gain, delay) for an SISO linear system free of delays. Such a problem will be discussed in Chapter 10, where an appropriate geometric approach is proposed.

5.5.1 ▪ Problem formulation

We consider a rational, proper SISO plant with an input or output time-delay, which has the transfer function representation

$$G(\lambda)e^{-h\lambda}, \tag{5.58}$$

where

$$G(\lambda)=\alpha\frac{(\lambda-z_1)(\lambda-z_2)\dots(\lambda-z_m)}{(\lambda-p_1)(\lambda-p_2)\dots(\lambda-p_n)}. \tag{5.59}$$

The gain, zeros, and poles of $G(\lambda)$ are $\alpha\in\mathbb{R}$, $z_r\in\mathbb{C}$, $r=1,\dots,m$, and $p_i\in\mathbb{C}$, $i=1,\dots,n$, respectively.

The *root-locus equation* corresponding to (5.58) is given by

$$f(\lambda;\ k,h)=0, \tag{5.60}$$

where

$$f(\lambda;\ k,h):=1+kG(\lambda)e^{-\lambda h}.$$

The *root-locus region* is defined as a complex right half plane

$$\mathbb{C}_{\sigma_0}=\{\lambda\in\mathbb{C}:\Re(\lambda)>\sigma_0\}, \tag{5.61}$$

with the value of σ_0 depending on the analysis requirements. We consider the following two root-locus problems.

Root-locus w.r.t. the controller gain: Compute the trajectories of the characteristic roots of the root-locus equation (5.60) inside the root-locus region $\mathbb{C}_{\sigma_0}$ for a constant positive time-delay and a gain interval $k\in[0,k_{\max}]$, where $k_{\max}$ is a given positive real number.

Root-locus w.r.t. the time-delay: Compute the trajectories of the characteristic roots of the root-locus equation (5.60) inside the root-locus region $\mathbb{C}_{\sigma_0}$ for a delay range $h\in[0,h_{\max}]$, where $h_{\max}$ is a given positive real number.

5.5.2 ▪ Approach

The computation of the root-locus involves two main tasks:

- computing the *critical points* of the root-locus;
- following the root-locus trajectories.

Critical points

Critical values of the parameter are values for which there is a root entering or leaving the root-locus region, or values corresponding to branch points, where multiple trajectories intersect inside $\mathbb{C}_{\sigma_0}$. The behavior of the root-locus around branch points is described by the following result.

Proposition 5.9. *Assume that $\tilde{\lambda}$ is a solution of the root-locus equation for parameter $p = \tilde{p}$ (with $p = k$ or $p = h$) and has multiplicity N. Then the root-locus has N intersecting trajectories at $\lambda = \tilde{\lambda}$, and the angle of direction change of a root incoming to and outgoing from point branch point $\tilde{\lambda}$ is 0 or $-\frac{\pi}{N}$ if N is odd or even, respectively.*

In [142] frequency-sweeping-based algorithms are presented for determining *directly* roots entering or leaving the root-locus region, as well as the corresponding parameter values and the direction of roots (entering or leaving) when the parameter is changed. If the delay is the free parameter, then the method corresponds to [137] and extends the approach for computing imaginary axis characteristic roots, presented in the next chapter.

Branch points can be detected while following a root-locus trajectory using a so-called test function (see, for instance, [342]). In the case when the free parameter is the gain k, they can also be computed directly. Indeed, a branch point satisfies the following two conditions:

$$f(\lambda;\, k,h)=0, \qquad f'(\lambda;\, k,h)=0.$$

The second equation reduces to $G'(\lambda)-hG(\lambda)=0$, which, for a given h, has only finitely many solutions that can be computed using a standard polynomial root finder.

5.5.3 ▪ Computing a root-locus trajectory

An intuitive idea for computing a root-locus trajectory is to parametrize it directly in the root-locus parameter, k or h. However, this may create numerical problems, especially in the vicinity of a branch point. To illustrate this, we consider the root-locus of an oscillator system, where

$$G(\lambda)=\frac{\lambda^2}{(\lambda^2+\omega_1^2)(\lambda^2+\omega_2^2)}, \tag{5.62}$$

with respect to the controller gain k, for $w_1 = w_2 = 1$ and $h = \frac{4\pi}{3}$. In the left pane of Figure 5.3, we show characteristic roots as a function of k near $k = 0$, where a branch point occurs. The plot reveals the high sensitivity in the neighborhood of the branch point. As a consequence, if one tries to follow trajectory 1 or 2, by continuation or by changing the parameter in steps, problems are expected around $k = 0$, due to the branch point which behaves as a turning point.

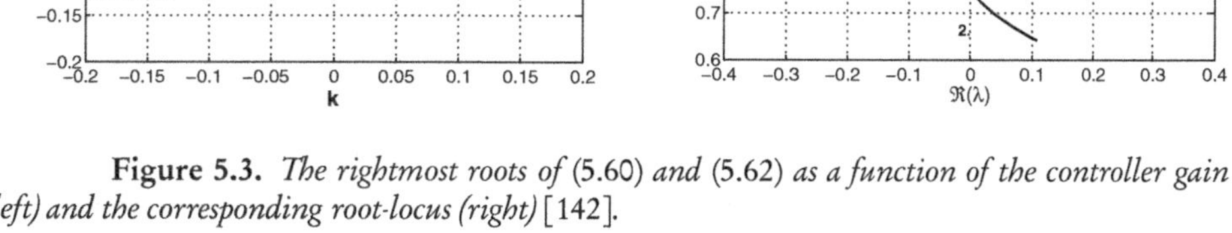

Figure 5.3. *The rightmost roots of* (5.60) *and* (5.62) *as a function of the controller gain k (left) and the corresponding root-locus (right)* [142].

However, the two root trajectories in Figure 5.3 are smooth curves in the complex plane, as shown in the right pane. This property is exploited in [142], where, in the computational approach, the root-locus trajectories are parametrized in terms of the arc-length γ in the combined root (λ) and locus parameter (k or h) space, i.e., $\lambda = \lambda(\gamma)$ and $k = k(\gamma)$ ($\tau = \tau(\gamma)$) for $\gamma > 0$. This parametrization, commonly used in the context of bifurcation analysis, eliminates the numerical problems sketched above and is numerically robust. In [142], the branches are followed using a predictor-corrector approach with a variable step-length [342].

5.5.4 ▪ Illustrative numerical example

We consider the SISO dead-time system (5.58), where

$$G(\lambda) = \frac{\lambda^2 - 10\lambda + 50}{\lambda^3 + 4\lambda^2 + 4.25\lambda + 1.25} \tag{5.63}$$

and plot its root-locus trajectories for the controller gain interval $k \in [0,\ 5]$ inside the root-locus region $\{\lambda \in \mathbb{C} : \Re(\lambda) \geq -3.5\}$. Figure 5.4 illustrates the general behavior of the root-locus trajectories. As the controller gain increases, more characteristic roots cross the boundary $\Re(\lambda) = -3.5$, and asymptotic root trajectories get closer to the imaginary axis. Figure 5.5 shows the local behavior around the starting points, $\lambda = -0.5, -1, -2.5$. On the left, we see that the characteristic roots cross the imaginary axis at $k = 0.07$ and the closed-loop system becomes unstable. On the right, the root-locus has a branching point, $\lambda = -0.6976$ at $k = 0.0009$, and the trajectory starting with $\lambda = -2.5$ leaves the region at $k = 0.0023$.

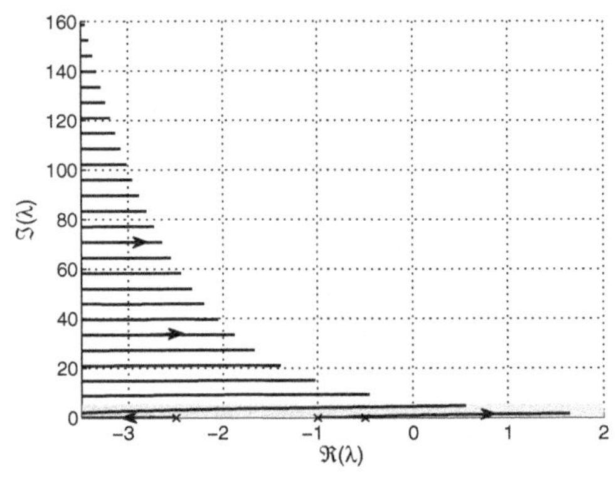

Figure 5.4. *The root-locus of* (5.63) *for* $k \in [0,5]$ *inside* $\Re(\lambda) \geq -3.5$ [142].

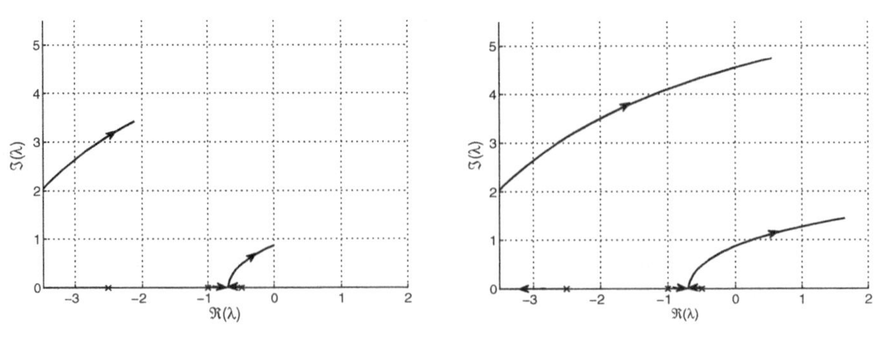

Figure 5.5. *The zoomed root-locus of* (5.63) *for* $k \in [0, 0.07]$ *(left) and* $k \in [0,5]$ *(right)* [142].

5.6 ▪ Notes and references

The characterization of stability domains by using the D-decomposition (sometimes called also D-subdivision or D-partition) method has been largely presented in the literature starting in the 1950s. However, to the best of the authors' knowledge, the fundamental ideas of such a method go back to the 19th century and can be found in the works of Vyshegradsky, where he described the structure of the stability domains for some low-order polynomial in the space of the coefficients parameters (see, for instance, [128] and the references therein). An excellent overview of the method for finite-dimensional systems can be found in [129]. The presentation here follows along the lines considered in [290, 93]. For the brevity of this chapter, we decided to present only the essential facts, results, and ideas. An interesting discussion can be found in the almost forgotten monograph of Pinney [326].

In the case of scalar systems including one delay, the complete characterization (for fixed delay) of all crossing curves in the parameter space can be found, for instance, in [93]. For a different approach based on Rouché's theorem, we refer the reader to the overview proposed by Nussbaum [307]. Although the scalar neutral delay-differential equation is not discussed, the proposed methods can be applied to such a case. Indeed, for example, for the neutral system

$$\dot{x}(t) + \beta \dot{x}(t-\tau) = -ax(t) - bx(t-\tau)$$

(under appropriate initial conditions), the characteristic function is written as

$$H(\lambda; a, b, \beta, \tau) := \lambda(1 + \beta e^{-\lambda\tau}) + a + be^{-\lambda\tau}.$$

Under the assumption that $|\beta| < 1$ (stable difference-operator) and $a + b > 0$, the results proposed in Chapter 1 allow us to conclude that the system is still stable for small delay values. Now, if $a \geq |b|$, the system is stable independently of the delay value. Finally, if $|a| < b$, some straightforward algebraic calculations prove that, similarly to the retarded case, there exists only one crossing frequency $\omega_c = \sqrt{b^2 - a^2}/\sqrt{1-\beta^2}$, and thus the stability will be guaranteed for the first delay interval $[0, \tau_c)$, where the delay margin can be easily computed (see also [116]). In this context, an interesting remark is the following: if β is seen as *free-parameter*, the delay margin may become larger than that derived in the retarded case. As depicted in Figure 5.6, it is easy to see that for $\beta \in (0, 1)$, the delay margin is larger than that corresponding to $\beta = 0$ (retarded case). Next, it is worth mentioning

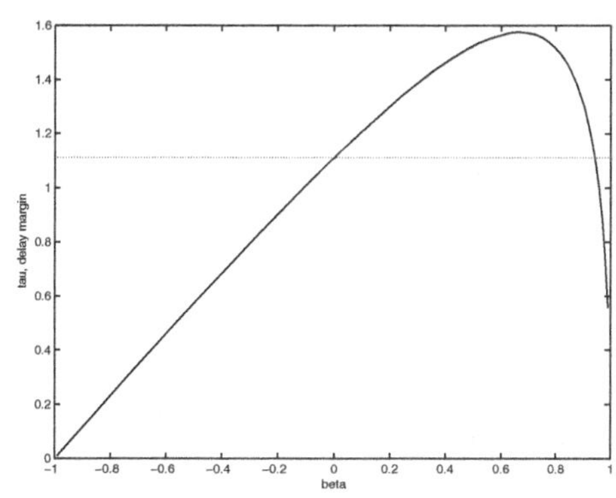

Figure 5.6. *Delay margin τ_c versus parameter β for $a = -1/\sqrt{2}$, $b = 1$* [116].

that the multiplicity of the roots on the imaginary axis (except for the origin) is not larger than 1. However, we will have a different situation on the real axis and we may have a root with multiplicity equal to 3 (recall that in the retarded case, the maximal algebraic multiplicity is 2). For further discussions on the estimation of the algebraic multiplicity of the root at the origin of the characteristic function of more general delay systems, the reader is referred to [33]. Finally, the simplest delay-differential algebraic system is represented by the scalar lossless propagation system including five independent parameters. The corresponding stability analysis (delay-independent/delay-dependent cases) will be presented in the next chapter as an appropriate illustrative example.

The ideas for developing simple estimates for the stability domain of delay systems in state-space representation can be found in Mori and Kokame [276] (and the references therein). Their idea simply makes use of the maximum principle of a harmonic (or subharmonic) complex function. Without entering into details, their stability results can be resumed as follows: consider the system

$$\dot{x}(t) = A_0 x(t) + A_1 x(t-\tau),$$

under appropriate initial conditions, and let $\mu(A_0)$ be the corresponding matrix measure with respect to some p-(matrix) norm ($p = 1, 2, \ldots, \infty$). If $\mu(A_0) + \|A_1\| < 0$, the system above is stable independently of the delay size. Assume now that $\mu(A_0) + \|A_1\| > 0$. Then, as suggested by [276], the system is still asymptotically stable if there does not exist any root located in the Γ rectangle bounded by the imaginary axis, by the line $\Re(\lambda) = \mu(A_0) + \|A_1\|$, and the lines $\Im(\lambda) = \pm(\mu(-jA_0) + \|A_1\|)$. However, the construction of Mori and Kokame is more general in the sense that if the system is unstable, the rightmost root together with all other unstable characteristic roots should be located in the *domain* Γ defined above. It is worth mentioning that Proposition 1.12 (Chapter 1) allows a better estimation of such a domain. Furthermore, there exists an abundant literature concerning the computation of the number of unstable roots for a given time-delay system. Among the existing results and for the sake of brevity, we mention only Stépán's monograph [359] that includes also a long list of examples (see also [156] for a recent generalization).

Next, the property concerning the existence of strictly unstable positive roots for scalar systems for some particular choice of the systems' parameters also holds for more general quasi-polynomials. To the best of the authors' knowledge, such a property was first mentioned in the literature in the 1960s and can be found in [123, 359] (and the references therein). Some remarks in this sense can be found also in the applications part of this book (Part III), in particular Chapter 10 devoted to the delayed output feedback control problem for single-input single-output (SISO) linear systems. Without any loss of generality, let us consider the following analytic function ($\tau > 0$):

$$p(\lambda;\ \tau) := Q(\lambda) + P(\lambda)e^{-\lambda\tau}.$$

with P and Q analytic, and assume that

$$\operatorname{sign}\left[(P(0) + Q(0))Q(+\infty)\right] = -1,$$

that is, $P(0)+Q(0)$ and $Q(+\infty)$ have opposite signs. Then, using the same argument as in the scalar case, it simply follows that the characteristic equation $p(\lambda;\ \tau) = 0$ always has a characteristic root located on the *positive real axis*, and this property holds independently of the delay size; that is, we have a *delay-independent unstable* system. In the case where P and Q are polynomials with $\deg(P) < \deg(Q)$, the condition above simply says that the coefficient of the dominant term of the polynomial Q should have an opposite sign to $P(0) + Q(0)$, as discussed for instance in [359].

The "dual" method of D-decomposition, the τ-decomposition method, is largely discussed by Lee and Hsu [205] in the context of quasi-polynomials including one delay parameter. As in the presentation of the previous method, we focused more on ideas and essential facts by adding, in particular, some simple geometric interpretations that are useful for the other developments of the chapter. The characterization of the only (stability) delay interval follows along the lines of [290], where the reader can find further discussions on the boundary of the delay-independent stability domain together with the relations existing with strong/weak delay-independent stability.

The τ-decomposition in the most general case for analytical functions with respect to only one delay-parameter can be found in the paper of Cooke and van den Driessche [81], which is at the origin of various research results in the control literature published within the last 15 years. Cooke and van den Driessche [81] considered the following analytic function ($\tau > 0$):

$$p(\lambda;\ \tau) := Q(\lambda) + P(\lambda)e^{-\lambda\tau},$$

with P and Q analytic, and focused on the first-order analysis of the crossing roots with respect to the delay-parameter, that is, the definition of all delay intervals for which stability/instability is guaranteed. Independently, a similar analysis in the case of polynomials P and Q was proposed in the control literature by Walton and Marshall [402] one year later. Corrections of [81] for the case of neutral systems (see, for instance, Chapter 1 for definitions and related stability results) have been reported by Boese [29] (and, to the best of the authors' knowledge, mentioned first by Kuang [202] in his monograph a few years before).

Although it seems that almost everything has been said for the scalar system including one constant delay, we would like to mention two particular approaches that are quite simple to apply in the stability analysis: the *pseudodelay technique* and the *Lambert functions*-based approach.

Consider now the following quasi-polynomial:

$$p(\lambda;\ \tau) := Q(\lambda) + P(\lambda)e^{-\lambda\tau}.$$

The idea of the first approach (pseudodelay technique) consists in *finding* two parameter-dependent polynomials $A_0(\lambda T)$ and $A_1(\lambda T)$ such that the intersection in the complex plane of the ratio curve $-Q(j\omega)/P(j\omega)$ with the unit circle (described by $e^{-j\omega\tau}$) is *reduced* to the roots location of the following parameter-dependent polynomial:

$$p(\lambda;\ T) := Q(\lambda)A_1(\lambda T) + P(\lambda)A_0(\lambda T)$$

for $\lambda = j\omega$, when T increases from 0 to ∞. The last polynomial was derived by "replacing" the delay element $e^{-\lambda\tau}$ by $A_0(\lambda T)/A_1(\lambda T)$ in the quasi-polynomial $p(\lambda;\ \tau)$. The "quantity" or the "parameter" T is called *pseudodelay*. In conclusion, the analysis of the roots distribution with respect to the imaginary axis of the corresponding characteristic equation is reduced to the analysis of the roots of some polynomials of higher order.

Such a transformation allows us to reformulate the *crossing roots* characterization of quasi-polynomials as a *crossing roots* analysis for some parameter-dependent polynomials. Hence, the stability analysis of a delay systems is reduced to the analysis of the roots distribution of some appropriate parameter-dependent polynomials. To the best of the authors' knowledge, such an idea was firstly used in control by Rekasius at the beginning of 1980s [331] (see also the comments in [370]). The result of Rekasius [331] makes use of the transformation $(1 - j\omega T)/(1 + j\omega T)$, which maps $[0, \infty)$ into a semicircle. Various related results (as well as some appropriate corrections) can be found in [160, 220, 402, 371], etc. Note that similar (bilinear) transformations are used in signal processing (see, for instance, [314] and the references therein). Further comments on the method can be found in MacDonald's monograph devoted to the stability analysis of some biological models [223]. Finally, a more deep analysis of this approach and various extensions are proposed by Sipahi [348] (see also [313]). Finally, it is worth mentioning that such a method detects the existence of some crossing frequencies but not necessarily the multiplicity of the corresponding root located on the imaginary axis.

The second approach mentioned above can be found in [82, 176] (see also [72, 224] for some applications), where the authors used an *analytical* method based on Lambert functions for solving the corresponding transcendental characteristic equation. Recall that a Lambert function is a complex function $L(\lambda)$ satisfying

$$L(\lambda)e^{L(\lambda)} = \lambda,$$

and the corresponding characteristic equation in the single delay case can be rewritten in this form. The next step is to compute the (infinite) branches of the complex function $L(\lambda)$. Note that the (first) "critical" roots will correspond to the principal branch (see, e.g., [224] and the references therein).

The continuation method presented in this chapter is inspired by [342, 100, 94]. We refer the reader to [342, 203] for the general theory of numerical continuation and to the work of Engelborghs and coworkers for an application to time-delay systems [101]. See also [100] for a software package for continuation and bifurcation analysis of time-delay systems (a brief description can be found in Section A.5 of our appendix).

Concerning the computation of the crossing direction of the characteristic roots, the proof idea in Proposition 5.3 based on the use of Jacobi's formula for the derivative of a determinant was first used in the context of commensurate delays in [296]. Next, the perturbation theory approach follows along the classical lines of the approach (see, for instance, [187] and also [347] for a comprehensive introduction). The case of simple characteristic roots was considered in the literature by various authors [313, 296], mainly in the commensurate delays case or in the delay-parameter space [348]. The approach presented here generalizes the ideas in [296]. For the analysis of the sensitivity and crossing directions of multiple characteristic roots we refer the reader to [170, 256] and the references therein.

We note that for the case of PID controlled plants, specific methods for computing stability regions in the controller parameter space, based on Pontryagin-type criteria [197], can be found in [346]. The approach proposed here can be seen as an alternative to these methods.

Finally, we sketched the predictor-corrector approach of [142] for computing the root-locus. It relies on a combination of numerical continuation of characteristic root trajectories with the computation of critical points, corresponding to either roots on a vertical line in the parameter space or to characteristic roots with multiplicity larger than one.

Chapter 6

Stability regions in delay-parameter spaces

6.1 ▪ Introduction

In the previous chapters, we focused on the behavior of the system

$$\dot{x}(t)=A_0(p)x(t)+\sum_{i=1}^{m}A_i(p)x(t-\tau_i(p)) \tag{6.1}$$

in some neighborhood of some (generic) stability crossing boundary (that is, the set of parameters such that the corresponding characteristic equation has at least one root on the imaginary axis) in the parameter space $\mathscr{D}\subset\mathbb{R}^{n_p}$. Some properties of the crossing direction of the characteristic roots on the imaginary axis were presented and discussed.

In this chapter, we concentrate on the particular case where the parameter space is defined *only* by the delays. In the first part of the chapter, we start by presenting some important *invariance* properties concerning the characteristic roots on the imaginary axis and/or concerning the crossing direction. Next, we discuss in detail the stability in the delay-parameter space for a particular ray defined by *commensurate* delays. We shall see that the particular structure of the delay system allows us to use some elimination principle that leads to an explicit computation of the delay intervals guaranteeing stability. The emphasis will be put on the use of matrix pencil techniques. The presentation will be completed with the analysis of some particular problems such as hyperbolicity (no characteristic roots on the imaginary axis for all positive delays), and the stability characterization of a particular class of quasi-polynomials. Although most of the presentation concerns time-delay systems of retarded type, the analysis of systems described by delay-differential algebraic equations also will be addressed by using the matrix pencil technique.

In the part starting with Section 6.4, a full characterization of the stability crossing curves will be made for a class of quasi-polynomials including two distinct delays. Such an approach will be adopted in later chapters to the stability analysis of immune dynamics models in leukemia or to the delay sensitivity analysis of Smith predictors. The computation of the stability radius in the delay-parameter space also will be discussed. It is worth mentioning that the presentation will be based on the duality between geometric and algebraic approaches. Such a choice is motivated by some facility to understand the methodologies and techniques that we are describing. The chapter ends with some insight into the characterization of the delay ratio sensitivity, that is, the analysis of perturbed rays in the delay-parameter space.

The chapter is organized as follows: some important invariance properties of characteristic roots are presented and discussed in Section 6.2. The stability analysis in the commensurate delay case is presented in Section 6.3 (both retarded and delay-differential algebraic cases). Section 6.4 is devoted to the geometric approach applied to quasi-polynomials with two distinct delays. Delay ratio sensitivity and the characterization of the so-called *interference phenomenon* proposed in Section 6.5 complete the presentation. Some notes and references end the chapter in Section 6.6.

6.2 ▪ Invariance properties

We focus on the interactions between characteristic roots on the imaginary axis and the delays. More precisely, we discuss two problems: the invariance of characteristic roots on the imaginary axis w.r.t. particular delay shifts, and the crossing direction invariance in the commensurate case or in the case where we have independent delays with one varying and the other ones fixed. It is worth mentioning that such properties also hold for time-delay systems described by neutral differential equations or by delay-differential algebraic equations. For the sake of brevity, in this section we focus only on the retarded case.

Consider the system (6.1) and assume that the delay vector $\vec{\tau}$ defines the parameter space $\mathscr{D} \subset \mathbb{R}^{n_p}$, that is, $n_p = m$ and $\mathscr{D} = \mathbb{R}_+^m$. Then the system is rewritten as follows:

$$\dot{x}(t) = A_0 x(t) + \sum_{i=1}^{m} A_i x(t - \tau_i). \tag{6.2}$$

Define the following *shift function*: $f_s : \mathbb{R}^m \times \mathbb{Z}^m \times \mathbb{R} \to \mathbb{R}^m$,

$$(\vec{\alpha}, \vec{\ell}, r) \mapsto f_s(\vec{\alpha}, \vec{\ell}, r) := (\alpha_1 + \ell_1 r, \alpha_2 + \ell_2 r, \ldots, \alpha_m + \ell_m r). \tag{6.3}$$

Such a shift function allows us to associate with a given vector $\vec{\alpha} \in \mathbb{R}^m$ another vector in $\mathbb{R}^m$ situated on the ray defined by the direction $\vec{\ell} \in \mathbb{Z}^m$ and the point $\vec{\alpha}$.

6.2.1 ▪ Delay shifts and characteristic roots

Let $\lambda = \lambda_0 := j\omega_0 \neq 0$ be a zero of the characteristic function

$$p(\lambda;\, \vec{\tau}) := \det\left(\lambda I_n - A_0 - \sum_{i=1}^{m} A_i e^{-\lambda \tau_i}\right)$$

for some delays $\vec{\tau} = \vec{\tau}_0$, and let u_0 (v_0) be the corresponding right (left) eigenvector. Then we have the following result:

Proposition 6.1. *For any integer vector $\vec{\ell} \in \mathbb{Z}^m$, a zero $\lambda = j\omega_0 \neq 0$ of the characteristic function $p(\lambda;\, \vec{\tau})$ corresponding to some delays $\vec{\tau}_0 \in \mathbb{R}_+^m$ is invariant under the delay shift:*

$$\vec{\tau} := f_s\left(\vec{\tau}_0, \vec{\ell}, \frac{2\pi}{\omega_0}\right). \tag{6.4}$$

Proof. The result straightforwardly follows from the invariance of the exponential function $\alpha \in \mathbb{R} \mapsto e^{-j\alpha} \in \mathbb{C}$ under the shift $\alpha \leftarrow \alpha + 2\pi\ell_k$, with $\ell_k \in \mathbb{Z}$. □

6.2.2 ▪ Crossing direction invariance

Using the developments of the previous chapter, it follows that $\partial\lambda/\partial\tau_k$, evaluated at a simple characteristic root $\lambda=\lambda_0$ and $\vec{\tau}=\vec{\tau}_0$ satisfies

$$\frac{\partial\lambda}{\partial\tau_k}=-\lambda_0\frac{v_0^*A_k u_0 e^{-\lambda_0\tau_{0,k}}}{v_0^*\left(I_n+\sum_{i=1}^m A_i\tau_{0,i}e^{-\lambda_0\tau_{0,i}}\right)u_0}, \tag{6.5}$$

where u_0 (v_0) is the right (left) eigenvector corresponding to λ_0. Under the assumption of a root on the imaginary axis, $\lambda_0=j\omega_0$, the condition (6.5) leads to

$$\left[\frac{\partial\lambda}{\partial\tau_k}\right]^{-1}=-\frac{v_0^*u_0}{j\omega_0 v_0^*A_k u_0 e^{-j\omega_0\tau_{0,k}}}-\sum_{i=1,i\neq k}^{m}\frac{\tau_{0,i}}{j\omega_0}\cdot\frac{v_0^*A_i u_0}{v_0^*A_k u_0}e^{-j\omega_0(\tau_{0,i}-\tau_{0,k})}-\frac{\tau_{0,k}}{j\omega_0}.$$

Since the crossing direction is given by the sign of $\Re(\partial\lambda/\partial\tau_k)$, evaluated at $\lambda=j\omega_0$ and $\vec{\tau}=\vec{\tau}_0$, we arrive at:

Proposition 6.2. *Under the assumption of a simple crossing, the crossing direction at* $\lambda=\lambda_0:=j\omega_0\neq 0$ *and* $\vec{\tau}=\vec{\tau}_0$ *w.r.t. the delay parameter* τ_k, $k\in\{1,2,\ldots,m\}$, *is invariant under the delay shift*

$$\vec{\tau}:=f_s\left(\vec{\tau}_0,he_k,\frac{2\pi}{\omega_0}\right), \tag{6.6}$$

where e_k *is the* k*th row of* I_n *and* $h\in\mathbb{Z}$.

Proof. The result follows from the observation that the delay shift f_s described by (6.6) does not affect the real part of the quantity $[\partial\lambda/\partial\tau]^{-1}$. Indeed, the quantities involving $\tau_{0,k}$ are on the imaginary axis (for instance, $-\tau_{0,k}/(j\omega_0)$) or appear in some exponential terms $e^{-j\omega_0(\tau_{0,i}-\tau_{0,k})}$, for which the considered delay shift does not change the value. □

Consider next a particular case that will be largely treated in what follows: *commensurate* delays. If the delays satisfy $\tau_i:=i\tau$, $i=1,\ldots,m$, with $\tau\geq 0$, then the crossing direction $\partial\lambda/\partial\tau$, evaluated at a simple characteristic root $\lambda=\lambda_0$ and $\tau=\tau_0$, becomes

$$\frac{\partial\lambda}{\partial\tau}:=-\lambda_0\frac{\sum_{i=1}^m i v_0^*A_i u_0 e^{-i\lambda_0\tau_0}}{v_0^*\left(I_n+\sum_{i=1}^m A_i i\tau_0 e^{-i\lambda_0\tau_0}\right)u_0}, \tag{6.7}$$

where u_0 (v_0) is the right (left) eigenvector corresponding to the characteristic root λ_0. If $\lambda_0=j\omega_0$, $\omega_0>0$, then the condition (6.7) leads to

$$\left[\frac{\partial\lambda}{\partial\tau}\right]^{-1}=-\frac{v_0^*u_0}{j\omega_0 v_0^*\sum_{i=1}^m iA_i u_0 e^{-j\omega_0 i\tau_{0,k}}}-\frac{\tau_0}{j\omega_0}. \tag{6.8}$$

In conclusion, we have the following result:

Corollary 6.3. *Assume that the delays satisfy* $\tau_i=i\tau$, $i=1,2,\ldots,m$. *Assume that* $\lambda=\lambda_0:=j\omega_0\neq 0$ *is a simple characteristic root for* $\tau=\tau_0$. *The crossing direction w.r.t. the delay parameter* τ *is invariant under the delay shift*

$$\tau:=\tau_0+h\frac{2\pi}{\omega_0}, \tag{6.9}$$

where $h\in\mathbb{Z}$, *such that* $\tau_0+h2\pi/\omega_0>0$.

6.2.3 ▪ Multiple imaginary characteristic roots and invariance properties

Consider now the case of some characteristic root located on the imaginary axis having its algebraic multiplicity larger than 1. We are interested in analyzing the way the invariance properties mentioned above are still valid. For the sake of brevity, we consider the following time-delay system including a single delay:

$$\dot{x}(t) = A_0 x(t) + A_1 x(t-\tau), \tag{6.10}$$

under appropriate initial conditions, where A_0 and A_1 are constant matrices, and we focus on the case of a double characteristic root non-semisimple $\lambda_0 = j\omega_0 \in j\mathbb{R}$ at $\tau = \tau_0 > 0$. The characteristic function of the system (6.10) is a quasi-polynomial of the form

$$f(\lambda, \tau) = a_0(\lambda) + a_1(\lambda)e^{-\tau\lambda} + \cdots + a_k(\lambda)e^{-k\tau\lambda}, \tag{6.11}$$

where $k \in \mathbb{N}_+$ and $a_0(\lambda), \dots, a_k(\lambda)$ are appropriate polynomials in λ.

Based on the results proposed in the previous paragraphs, if $\lambda_0 = j\omega_0 \in j\mathbb{R}$ is a characteristic root for the delay τ_0, then λ_0 is also a root for the delays $\tau_k = \tau_0 + 2k\pi/\omega_0$ for any $k \in \mathbb{Z}$ such that $\tau_k > 0$ (i.e., invariance of the characteristic imaginary roots w.r.t. particular delay shifts). Next, to analyze the crossing direction invariance, we need to understand the explicit behavior of the characteristic root λ_0 in some neighborhood of τ_k. We have the following result:

Proposition 6.4. *For the system* (6.10), *assume that $\lambda_0 = j\omega_0 \neq 0$ is a double imaginary root non-semisimple at $\tau = \tau_0$ and that $\frac{\partial}{\partial\tau} f(\lambda_0, \tau_0) \neq 0$.*

Then, near (λ_0, τ_0), the evolution of λ with respect to τ can be expressed as

$$\lambda = \lambda_0 \pm \sqrt{-2\frac{\frac{\partial}{\partial\tau} f(\lambda_0, \tau_0)}{\frac{\partial^2}{\partial\lambda\partial\lambda} f(\lambda_0, \tau_0)}}\,\varepsilon^{\frac{1}{2}} + o(\varepsilon^{\frac{1}{2}}). \tag{6.12}$$

Furthermore, near (λ_0, τ_k), for all $k \in \mathbb{Z}^*$, such that $\tau_k > 0$, the evolution of λ with respect to τ can be expressed as

$$\lambda = \lambda_0 - \frac{\frac{\partial}{\partial\tau} f(\lambda_0, \tau_k)}{\frac{\partial}{\partial\lambda} f(\lambda_0, \tau_k)}\varepsilon + o(\varepsilon). \tag{6.13}$$

If $\Re\left[\sqrt{-2\frac{\partial}{\partial\tau} f(\lambda_0, \tau_0)/\frac{\partial^2}{\partial\lambda\partial\lambda} f(\lambda_0, \tau_0)}\right] \neq 0$, then, for all $k \in \mathbb{Z}^*$ such that $\tau_k > 0$, the number change of unstable roots caused by the splitting of the imaginary root $\lambda = \lambda(\tau_k)$ as τ increases from $\tau_k - \epsilon$ to $\tau_k + \epsilon$ for some small $\epsilon > 0$ is the same as for $\lambda = \lambda(\tau_0)$.

Proof. Since the characteristic function $f(\lambda, \tau)$ is analytical w.r.t. both variables, then in some neighborhood of (λ_0, τ_0), it can be expanded as an appropriate convergent power series. Then, by using the fact that the characteristic root λ_0 is double and non-semisimple, i.e.,

$$\lambda = \lambda_0 + \lambda_1 \varepsilon^{\frac{1}{2}} + \cdots,$$

and by some simple but tedious algebraic manipulation (see also the basic ideas of the perturbation-based approach developed in Chapter 5), we obtain the value

$$\lambda_1 = \sqrt{-2\frac{\partial}{\partial\tau} f(\lambda_0, \tau_0)/\frac{\partial^2}{\partial\lambda\partial\lambda} f(\lambda_0, \tau_0)}$$

(see also [178, 207, 69]). Consider now the case when $\tau = \tau_k + \varepsilon$ for $k \in \mathbb{Z}^*$ such that $\tau_k > 0$. Then it is easy to see that in such a case the corresponding mapping $\varepsilon \in \mathbb{R} \mapsto$

$\lambda(\varepsilon) \in \mathbb{C}$ is differentiable at $\varepsilon = 0$, and furthermore the characteristic root $\lambda(0)$ located on the imaginary axis is simple. Thus, the first coefficient of the expansion can be easily obtained (see also Chapter 5) and the formula (6.13) follows straightforwardly.

Now, for the invariance property characterization, the following cases occur: (1) If $\Im\left[\frac{\partial^2}{\partial\lambda\partial\lambda}f(\lambda_0,\tau_0)/\frac{\partial}{\partial\tau}f(\lambda_0,\tau_0)\right] > 0$, then for critical delays $\tau_k > \tau_0$ (i.e., $k > 0$) the root path touches the imaginary axis from above and in the left half plane. For delays $\tau_k < \tau_0$ ($k < 0$, but such that $\tau_k > 0$) the imaginary axis is touched from the left and upward; (2) analogously, if $\Im\left[\frac{\partial^2}{\partial\lambda\partial\lambda}f(\lambda_0,\tau_0)/\frac{\partial}{\partial\tau}f(\lambda_0,\tau_0)\right] < 0$, the root path for critical delays $\tau_k < \tau_0$ touches the imaginary axis in the left half plane and for delays $\tau_k > \tau_0$ touches the imaginary axis from above and in the right half plane. Thus, by inspecting all the possible cases (double but not triple characteristic root), it follows that, for all $k \in \mathbb{Z}^*$ such that $\tau_k > 0$, the number change of unstable roots caused by the splitting of the imaginary root λ at τ_k as τ increases from $\tau_k - \epsilon$ to $\tau_k + \epsilon$ for some small $\epsilon > 0$ is the same as for λ at τ_0. The proof is complete. □

In conclusion, under some appropriate assumptions on the characteristic function (6.11) ($\frac{\partial}{\partial\tau}f(\lambda_0,\tau_0) \neq 0$), the presence of a double characteristic root on the imaginary axis $\lambda_0 = j\omega_0$ for some delay value τ_0 implies that this root (λ_0) is simple for other delay values τ_k ($k \in \mathbb{Z}^*$ such that $\tau_k > 0$) and the number change of unstable roots caused by the splitting of the imaginary root $\lambda = \lambda(\tau_k)$ as τ increases from $\tau_k - \epsilon$ to $\tau_k + \epsilon$ for some small $\epsilon > 0$ is the same as for $\lambda = \lambda(\tau_0)$. Finally, it is worth mentioning that such an invariance property holds also for more general cases when the multiplicity of the characteristic on the imaginary axis is larger than 2 (see, for instance, [207]).

6.3 ▪ Algebraic methods

Throughout this section we assume that the delays τ_i, $i = 1,\dots,m$, are commensurate. Without loss of generality we may assume that $\tau_i = i\tau$, $i = 1,\dots,m$, with $\tau \geq 0$. Then the system (6.1) it rewritten as

$$\dot{x}(t) = A_0 x(t) + \sum_{i=1}^{m} A_i x(t - i\tau), \tag{6.14}$$

and the corresponding characteristic function becomes

$$p(\lambda;\ \tau) := \det\left(\lambda I_n - A_0 - \sum_{i=1}^{m} A_i e^{-\lambda i \tau}\right). \tag{6.15}$$

6.3.1 ▪ Elimination principle: Basic ideas

We briefly discuss two particular approaches for the computation and/or detection of characteristic roots on the imaginary axis: *two variable-based* approaches and approaches based on *bilinear transformations* (including 2-D polynomials and the *pseudodelay* technique).

Two variable-based approaches

We start by rewriting the characteristic function (6.15) as

$$p_1(\lambda;\ z) := \det\left(\lambda I_n - A_0 - \sum_{i=1}^{m} A_i z^i\right), \tag{6.16}$$

where we have formally replaced $e^{-\lambda i\tau}$ by z^i, since λ and $e^{-\lambda\tau}$ can be seen as algebraically independent variables if $\lambda \neq 0$. The function p_1 is a *bivariate* polynomial. The relation between the "original" characteristic function p and the "associated" function p_1 becomes clear when we consider the case of characteristic roots on the imaginary axis. Indeed, let $j\omega_s \in j\mathbb{R}^*$ be a zero of p_1 for some $z_s \in \mathscr{C}(0,1)$, that is, $p_1(j\omega_s;\ z_s)=0$. Then it is easy to see that $j\omega_s$ is a zero of the characteristic function p of the original delay system (6.14) if the delay τ belongs to the set $\mathscr{T}_{\omega_s}$ given by

$$\mathscr{T}_{\omega_s} := \left\{ \frac{1}{\omega_s}[\angle(\bar{z})+2\ell\pi] > 0,\ \ell \in \mathbb{Z} \right\}.$$

By reciprocity, a zero $j\omega_s \in j\mathbb{R}^*$ of the characteristic function p is a zero of the characteristic function p_1 for $z_s = e^{-j\omega_s}$.

The particular relationship between the zeros of p and p_1 suggests the idea of computing the characteristic roots on the imaginary axis by *exploiting* the particular *form, structure*, and *dependence* of the characteristic function p_1 with respect to the two *variables*: $j\omega$ on the imaginary axis, and z on the unit circle. More precisely, the idea is to *eliminate* one of the variables, leading to two types of solutions: μ-analysis and matrix pencil-based solutions. Notes and comments on the first solution can be found in [290] (see also [130]). The matrix pencil approach will be presented to some extent since it leads to explicit algorithms for the computation of stability regions in the delay-parameter space.

Bilinear transformations and related methods

The computation of the characteristic function leads to an expression of the form

$$p(\lambda;\ \tau) = \sum_{i=0}^{n_d} p_i(\lambda) e^{-i\lambda\tau}, \tag{6.17}$$

where p_i are polynomials of degree at most n, and $n_d \geq m$ represents the number of commensurate delays in the quasi-polynomial representation. As in the previous section, we can interpret (6.17) as a bivariate polynomial p_1:

$$p_1(\lambda;\ z) := \sum_{i=0}^{n_d} p_i(\lambda) z^i, \tag{6.18}$$

where z formally replaces $e^{-\lambda\tau}$.

2-D polynomials

The representation of the characteristic function p as a bivariate polynomial leads to the relatively simple approach of analyzing the stability of p in terms of a 2-D polynomial. We consider the bilinear transformation

$$\lambda := \frac{1+w}{1-w},$$

which maps the open unit disk onto the open left half plane. Next, we construct the 2-D polynomial

$$p_2(w;\ z) := (1-w)^n p_1\left(\frac{1+w}{1-w};\ z\right).$$

It is evident that p_1 has zeros on $j\mathbb{R} \times \mathscr{C}(0,1)$ if and only if p_2 has zeros on $\mathscr{C}(0,1) \times \mathscr{C}(0,1)$. Further comments and discussions can be found in [130].

Pseudodelay technique

This approach was already commented on in the notes and references of the previous chapter. Although it is not considered in detail in this monograph, the main ideas are presented since there exist natural connections with the generalized eigenvalue distribution of some appropriate matrix pencils, as we shall see later. Based on the bilinear transformation,

$$z := \frac{1-\lambda T}{1+\lambda T}, \quad T \geq 0,$$

we construct the parameter-dependent polynomial

$$p_3(\lambda;\, T) := (1+\lambda T)^{n_d} p_1\left(\lambda;\, \frac{1-\lambda T}{1+\lambda T}\right) = \sum_{i=0}^{n_d} p_i(\lambda)(1-T\lambda)^i(1+T\lambda)^{n_d-i}. \tag{6.19}$$

We have the following result (see, e.g., [370]):

Proposition 6.5. *The quasi-polynomial $p(\lambda;\, \tau)$ has $\lambda = j\omega$, $\omega > 0$, as a characteristic root for some delay value $\tau \geq 0$ if and only if the parameter-dependent polynomial $p_3(\lambda;\, T)$ has $j\omega$ as a root for some $T \geq 0$.*

Further comments and discussions can be found in [348]. It is worth mentioning that such a result allows us to detect crossing with respect to the imaginary axis (if any), but it does not give any explicit information concerning the (algebraic) multiplicity of the corresponding root.

6.3.2 ▪ Matrix pencil approach and crossing characterization

Recall the system (6.14) and its characteristic function (6.15). Introduce the following matrix pencil [70, 289]:

$$\Lambda(z) := zM + N, \tag{6.20}$$

where $M,\, N \in \mathbb{R}^{(2mn^2)\times(2mn^2)}$ are given by

$$M = \begin{bmatrix} I_{n^2} & 0 & \dots & 0 & 0 \\ 0 & I_{n^2} & \dots & 0 & 0 \\ & & \ddots & & \\ 0 & 0 & \dots & I_{n^2} & 0 \\ 0 & 0 & \dots & 0 & B_m \end{bmatrix}, \quad N = \begin{bmatrix} 0 & -I_{n^2} & 0 & \dots & 0 \\ 0 & 0 & -I_{n^2} & \dots & 0 \\ & & & \ddots & \\ 0 & 0 & 0 & \dots & -I_{n^2} \\ B_{-m} & B_{-m+1} & B_{-m+2} & \dots & B_{m-1} \end{bmatrix}$$

and B_{-k} $(k = 1,\dots,m)$, B_i $(i = 1,\dots,m)$ are defined as $B_{-k} = I_n \otimes A_k^T$, $B_i = A_i \otimes I_n$, $B_0 = A \oplus A^T$, where the operators $\otimes$ and $\oplus$ denote the Kronecker product and sum (see, e.g., [124]).

Frequency crossing set characterization

The following result gives a complete description of characteristic imaginary roots in terms of the generalized eigenvalue distribution of the matrix pencil Λ. The proof follows closely the ideas mentioned in [70, 289] and was presented in [296].

Proposition 6.6. *Assume that the matrix pencil* Λ *is regular. The characteristic equation* $p(\lambda;\ \tau)=0$ *has a root* $j\omega_0$, $\omega_0>0$, *for some positive delay value* τ *if and only if there exists a complex number*

$$z_0\in\sigma(\Lambda)\cap\mathscr{C}(0,1) \tag{6.21}$$

such that

$$j\omega_0\in\sigma\left(A+\sum_{i=1}^{m}A_i z_0^i\right). \tag{6.22}$$

Furthermore, the corresponding delay values are given by

$$\mathscr{T}_{\omega_0}=\left\{\frac{\angle\overline{z}_0}{\omega_0}+\frac{2\pi\ell}{\omega_0}>0:\ z_0\in\sigma(\Lambda)\cap\mathscr{C}(0,1),\right. \\ \left. j\omega_0\in\sigma\left(A_0+\sum_{i=1}^{m}A_i z_0^i\right),\ \ell\in\mathbb{Z}\right\}. \tag{6.23}$$

Proof. We start with the first assertion.

$\Leftarrow$ Let (ω_0,z_0) satisfy (6.22). It remains to prove that there exists at least one positive delay value τ_0 such that $z_0=e^{-j\omega_0\tau_0}$. This is true since the general solution of this equation is given by

$$\tau=\frac{\mathrm{Log}(\overline{z_0})}{j\omega_0}+\frac{2\pi\ell}{\omega_0},$$

with $\ell\in\mathbb{Z}$, and we can always take ℓ such that τ is positive.

$\Rightarrow$ Assume that the characteristic function $p(\lambda;\ \tau)$ has at least one (nonzero) root on the imaginary axis, $j\omega_0$, for some delay $\tau_0>0$, that is, $p(j\omega_0;\ \tau_0)=0$. Let z_0 be defined as $z_0=e^{-j\omega_0\tau_0}$. It is clear that

$$\det\left(j\omega_0 I_n-A-\sum_{i=1}^{m}A_i z_0^i\right)=0 \tag{6.24}$$

and

$$\det\left(-j\omega_0 I_n-A^T-\sum_{i=1}^{m}A_i^T\overline{z}_0^i\right)=0. \tag{6.25}$$

Expressions (6.24)–(6.25) imply that

$$D(z_0):=\det\left[\left(A+\sum_{i=1}^{m}A_i z_0^i\right)\oplus\left(A+\sum_{i=1}^{m}A_i z_0^i\right)^*\right]=0. \tag{6.26}$$

Some simple computations lead us to $D(z_0)=z_0^m\det(z_0M+N)$. Since $z_0\neq 0$, Λ has at least one generalized eigenvalue z_0 on the unit circle, for which (6.22) is satisfied. Next, to prove the second assertion we have to find all values of τ such that $p(j\omega_0;\ \tau)=0$. From the arguments spelled out in the proof of the first assumption (more precisely, the implication $\Rightarrow$) it follows that $e^{-j\omega_0\tau}\in\sigma(\Lambda)\cap\mathscr{C}(0,1)$. Expression (6.23) then follows straightforwardly. $\square$

The following definition is related to the elimination technique used in the proof of Proposition 6.6:

Definition 6.7. *A complex number* z_0 *satisfying* (6.21) *and for which* (6.22) *is satisfied for some* $\omega_0 > 0$ *is called a crossing generator. The set of all crossing generators is denoted with* σ_g. *The frequency crossing set* Ω *is defined as*

$$\Omega := \left\{ \omega \in \mathbb{R}_+^* : j\omega \in \sigma\left(A_0 + \sum_{i=1}^{m} A_i z_0^i\right), \quad z_0 \in \sigma_g \right\} \tag{6.27}$$

and the set

$$\mathcal{T} := \bigcup_{\omega \in \Omega} \mathcal{T}_\omega \tag{6.28}$$

is called the delay crossing set.

In terms of Definition 6.7, Proposition 6.6 simple says that the existence of characteristic roots on the imaginary axis for some delay values is equivalent to the property that the frequency crossing set is not empty. By construction, Ω is not empty if and only if the set σ_g of all crossing generators is not empty. For the sake of brevity, it is worth mentioning that we will focus only on positive frequencies $\omega \in \Omega \cap \mathbb{R}_+^*$ since if $j\omega$ represents a characteristic root of the corresponding system, the same property holds for $-j\omega$. As a final observation, it is easy to see that under the assumption of asymptotic stability of the delay-free system, two situations occur:

(i) $\mathcal{T} =$. In such a case, the system is *delay-independent* asymptotically stable;

(ii) $\mathcal{T} \neq$. In this case, the quantity "min $\mathcal{T}$" represents the corresponding *delay margin.*

Remark 6.8. Not all *the generalized eigenvalues of the matrix pencil* Λ *on the unit circle correspond to* crossing generators. *Indeed, by the properties of the Kronecker sum, the solutions* z *located on the unit circle of the equation*

$$\det\left[\left(A_0 + \sum_{i=1}^{m} A_i z^i\right) \oplus \left(A_0 + \sum_{i=1}^{m} A_i z^i\right)^*\right] = 0$$

include not only the crossing generators, but all the values of z *on the unit circle for which the matrix* $A_0 + \sum_{i=1}^{m} A_i z^i$ *has some eigenvalues symmetric w.r.t. the origin of the complex plane (which is more general than having eigenvalues* on *the positive imaginary axis)* [70, 289].

Proposition 6.6 directly leads to the following algorithm for the computation of the delay crossing set:

ALGORITHM 6.1. Computation of delay crossing set.

A. Compute the generalized eigenvalues of the matrix pencil Λ on the unit circle of the complex plane (under the assumption of regularity).

B. For each generalized eigenvalue $z \in \mathscr{C}(0,1) \cap \sigma(\Lambda)$, compute the eigenvalues on the imaginary axis of the complex matrix $A_0 + \sum_{i=1}^{m} A_i z^i$.

C. Compute the frequency crossing set Ω from (6.27) and the delay crossing set $\mathcal{T}$ from (6.28) and (6.23).

Crossing direction characterization: Simple characteristic roots

We first consider the characterization of a *simple* root crossing. The multiple semisimple crossing case will be considered separately. We have the following result:

Proposition 6.9. *Let $j\omega_0$, $\omega_0 > 0$, be a characteristic root of the system* (6.14) *for some delay value τ_0. Let $z_0 \in \sigma_g$ be the corresponding crossing generator. If the delay is increased, then the characteristic root crosses the imaginary axis towards instability (stability) if and only if*

$$\Re\left\{\frac{\omega_0}{j v_0^* u_0}\sum_{i=1}^{m} i z_0^i (v_0^* A_i u_0)\right\} > 0 (< 0), \tag{6.29}$$

where v_0 and u_0 are the left and right eigenvectors of the complex matrix $A_0 + \sum_{i=1}^m A_i z_0^i$, corresponding to the eigenvalue $j\omega_0$.

The proof is omitted since it follows straightforwardly from Corollary 5.8.

Crossing direction characterization: Semisimple characteristic roots

The result below is taken from [68], where the authors handled the problem by using a slightly different argument from the one proposed in the previous chapter:

Proposition 6.10. *Let $j\omega_0$, $\omega_0 > 0$, be a semisimple characteristic root of the system* (6.14) *for some delay value τ_0, with multiplicity equal to q. Let $z_0 \in \sigma_g$ be a corresponding crossing generator. For τ sufficiently close to τ_0, the characteristic roots corresponding to $j\omega_0$ can be expanded as*

$$j\omega_0 - \lambda_k\left(R^*\left(j\omega_0\sum_{i=1}^{m} iA_i z_0^i\right)Q\right)(\tau-\tau_0) + O\left((\tau-\tau_0)^2\right), \qquad k = 1, 2, \ldots, q, \tag{6.30}$$

where $Q = [u_1\ u_2\ \cdots\ u_q]$, $R = [v_1\ v_2\ \cdots\ v_q]$, and the notation $\lambda_k(\cdot)$ stands for the kth eigenvalue. The vectors u_i and v_i, $i = 1, 2, \ldots, q$, constitute a normalized set of right and left eigenvectors of the matrix $A_0 + \sum_{i=1}^m A_i z_0^i$, corresponding to the eigenvalue $j\omega_0$.

Thus, for τ sufficiently close to τ_0 but $\tau > \tau_0$, there are at least ℓ ($\ell \le q$) characteristic roots in the open right half plane if l of the following inequalities are satisfied:

$$\Re\left\{\lambda_k\left[R^*\left(j\omega_0\sum_{i=1}^{m} iA_i z_0^i\right)Q\right]\right\} < 0, \quad k = 1, \ldots, q. \tag{6.31}$$

6.3.3 ▪ Particular cases and other elimination techniques

In what follows, we consider some special cases for which the analysis presented in the previous section simplifies significantly.

Hyperbolicity and delay-independent stability

If there are no characteristic roots on the imaginary axis, for any delay value, then the system is called *hyperbolic*, in the sense mentioned by Hale, Infante, and Tsen in [150]. In other words, the number of strictly unstable characteristic roots is *constant* for all delay values, including also the case that is free of delay. We have the following result [289]:

Proposition 6.11. *Assume that the system* (6.14) *that is free of delays has no characteristic roots on the imaginary axis. Then it is hyperbolic if and only if the frequency crossing set* Ω *is empty.*

The characterization of *delay-independent* asymptotic *stability* is similar. The only difference is that asymptotic stability of the delay-free system needs to be imposed instead of having no characteristic roots on the imaginary axis.

Rank one matrices and frequency-sweeping tests

In Chapter 16 on applications in biosciences, we will encounter models of the form (6.14), where the matrices $A_1,\dots,A_m$ have rank one. In what follows, we present the main ideas for analyzing such problems by means of the single delay case. Therefore, we study the system

$$\dot{x}(t) = A_0 x(t) + BC^T x(t-\tau), \tag{6.32}$$

where B and C are column matrices. For the sake of conciseness, we assume that the system that is free of delay is asymptotically stable and that the matrix A_0 has no eigenvalues on the imaginary axis. Define the analytic function

$$p_a(\lambda;\ \tau) := 1 - a(\lambda)e^{-\lambda\tau}, \tag{6.33}$$

where a is given by $a(\lambda) := C^T(\lambda I - A_0)^{-1}B$. The particular structure of the rank one matrix BC^T allows us to derive the following result:

Proposition 6.12. *The characteristic function of* (6.32) *and the function* $p_a(\lambda;\ \tau)$ *have the same zeros in a neighborhood* $\mathcal{V}_\delta$ *of the imaginary axis, where*

$$\mathcal{V}_\delta := \{\lambda \in \mathbb{C}: \quad \delta \geq \Re(\lambda) > -\delta\}$$

for some $\delta > 0$.

Proof. Since A_0 has no eigenvalues on the imaginary axis, the continuity property of the roots of the corresponding characteristic equation leads to the existence of some $\delta > 0$ such that $\lambda I - A_0$ is invertible in $\mathcal{V}_\delta$. Next, for all $\lambda \in \mathcal{V}_\delta$, we have

$$\begin{aligned}\det(\lambda I_n - A_0 - BC^T e^{-\lambda\tau}) &= \det(\lambda I_n - A_0)\det(I_n - (\lambda I_n - A_0)^{-1}BC^T e^{-\lambda\tau}) \\ &= p_a(\lambda;\ \tau)\det(\lambda I_n - A_0),\end{aligned}$$

where we used the properties of the Schur complement. □

If p_a has a zero $j\omega_0$, $\omega_0 > 0$, for some delay value τ_0, then $a(j\omega_0) = e^{j\omega_0\tau}$, which implies that the frequency crossing set Ω consists of the strictly positive zeros of the function

$$f: \mathbb{R}_+ \to \mathbb{R}, \quad \omega \mapsto f(\omega) = 1 - |a(j\omega)|^2, \tag{6.34}$$

while the delay crossing set can be computed from the phase information of $a(j\omega_0) = e^{j\omega_0\tau}$. Furthermore, if we differentiate $p_a(\lambda(\tau);\ \tau) = 0$ w.r.t. τ (under the assumption of a simple root) we get

$$\left[\frac{d\lambda}{d\tau}\right]^{-1} = -\frac{\tau}{\lambda} + \frac{a'(\lambda)}{\lambda a(\lambda)},$$

which in the case $\lambda = j\omega_0$, $\omega_0 \geq 0$, yields

$$\Re\left[\frac{d\lambda}{d\tau}\right]^{-1} = \Re\frac{a'(j\omega_0)}{j\omega_0 a(j\omega_0)} = \frac{1}{2\omega_0|a(j\omega_0)|^2}f'(\omega_0).$$

In this way, we arrive at the following result:

Proposition 6.13. *Consider the system (6.32). Assume that A_0 has no eigenvalues on the imaginary axis and that $1-a(0)\neq 0$. Assume further that the zeros of f are simple. The system (6.32) has a characteristic root $j\omega_0$, $\omega_0 > 0$, for some delay value if and only if $f(\omega_0) = 0$. Furthermore, the set of corresponding delay values is given by*

$$\mathcal{T}_{\omega_0} = \left\{\frac{1}{\omega_0}[-j\operatorname{Log}(a(j\omega_0)) + 2\pi l] \geq 0,\ l \in \mathbb{Z}\right\}. \tag{6.35}$$

When increasing the delay, the corresponding crossing direction of the characteristic root is towards instability (stability) if $f'(\omega) > 0$ (< 0).

Consider now the case of multiple roots on the imaginary axis. Introduce the two-variable expression $p_a(\lambda, z)$ by letting $z = e^{-\tau\lambda}$, that is,

$$p_a(\lambda, z) := 1 - C^T(\lambda I_n - A)^{-1}Bz. \tag{6.36}$$

In light of (6.36), for each $\lambda = j\omega$ ($\omega \in \mathbb{R}$), we have a solution of z, denoted by $z(j\omega)$ such that $p_a(\lambda, z) = 0$. Define now a *frequency-sweeping curve* as the mapping $\omega \in \mathbb{R} \mapsto |z(j\omega)| \in \mathbb{R}_+$. Now, if $\lambda_0 = j\omega_0$ is a characteristic root of $p_a(\lambda, \tau)$ and if we denote by τ_k $k \in \mathbb{N}$ the corresponding critical delays, it follows that there exists a pair ($\lambda_0 = j\omega_0, z_0 = e^{-j\omega_0\tau_0}$) such that $p_a(\lambda_0, z_0) = 0$. Note that $e^{-j\omega_0\tau_k}$ has only one value for all $k \in \mathbb{N}$ and is equal to z_0.

We have the following result (see also [207] for further discussions on the frequency-sweeping approaches):

Proposition 6.14. *Assume that $\lambda_0 = j\omega_0$ is a critical imaginary root with any multiplicity for $\tau = \tau_k$ ($k \in \mathbb{N}$) of the quasi-polynomial $p_a(\lambda, \tau)$.*

Then, at all critical delays τ_k, the number change of unstable roots caused by the splitting of the imaginary root λ_0 at τ_k as τ increases from $\tau_k - \epsilon$ to $\tau_k + \epsilon$ for small $\epsilon > 0$ is characterized by the frequency-sweeping curve $|z(j\omega)|$ as follows:

(i) The characteristic root crosses the imaginary axis towards instability (stability) if and only if as ω increases, $|z(j\omega_0)| - 1$ changes its sign near ω_0 from negative to positive (from positive to negative).

(ii) The characteristic root does not cross the imaginary axis if and only if as ω increases, $|z(j\omega)| - 1$ does not change its sign near ω_0.

Proof. By a simple transformation, p_a can be rewritten as

$$f(\lambda;\ \tau) = e^{-\tau\lambda} + \varphi(\lambda) = 0$$

with $\varphi(\lambda) = (-C^T(\lambda I_n - A)^{-1}B)^{-1}$. Since the matrix functions $(\lambda I_n - A)$ and $(-C^T(\lambda I_n - A)^{-1}B)$ are invertible near a critical pair (λ_0, τ_k), with $k \in \mathbb{N}$, it is easy to see that $\varphi(\lambda)$ is analytic at $\lambda = \lambda_0$. The remaining part of the proof can be done by applying Lemma 1 of [207, Appendix, p. 894] (see also [205]). $\square$

Example 6.15. *In order to illustrate the last result, consider the following system having the realization (A,B,C) (see also [207]):*

$$A = \begin{pmatrix} \frac{1}{\pi} & \frac{1-\pi}{\pi} \\ \frac{1+\pi}{\pi} & \frac{1}{\pi} \end{pmatrix}, \qquad B = \begin{pmatrix} \frac{1}{\pi} \\ \frac{\pi+1}{\pi} - \frac{\pi+1}{\sqrt{(\pi^2-1)}} \end{pmatrix}, \qquad C = \begin{pmatrix} \frac{\pi}{\sqrt{(\pi^2-1)}} + 1 \\ \frac{1}{\pi+1} \end{pmatrix}.$$

In this case, some simple computations show that $\lambda = j$ is a characteristic root for $\tau = (2k+1)\pi$, with $k \in \mathbb{N}$. In particular, at $\tau = \pi$, the root is double, while for $\tau = 3\pi, 5\pi, 7\pi, \ldots$ it is only simple. In order to illustrate the invariance property, we compute the Puiseux series near the critical pairs (j,π), $(j,3\pi)$, and $(j,5\pi)$. More precisely, for the critical pair (j,π) and for some sufficiently small $\varepsilon > 0$, the Puiseux series is written as

$$\lambda = j \pm (0.3547 + 0.2594j)(\varepsilon)^{\frac{1}{2}} + o((\varepsilon)^{\frac{1}{2}}).$$

Next, for the critical pair $(j,3\pi)$ the corresponding Puisseux (Taylor) series is

$$\lambda = j - 0.1592j\varepsilon + (-0.0199 + 0.0190j)(\varepsilon)^2 + o((\varepsilon)^2).$$

Finally, for the critical pair $(j,5\pi)$ the Puisseux (Taylor) series is

$$\lambda = j - 0.0796j(\varepsilon) + (-0.0025 + 0.0055j)(\varepsilon)^2 + o((\varepsilon)^2).$$

Proposition 6.4 describes the invariance property. It is important to point out that the first-order analysis was not sufficient since the corresponding coefficients are located on the imaginary axis. Next, we may alternatively employ the frequency-sweeping test as stated in Proposition 6.14 to analyze the asymptotic behavior. The corresponding frequency-sweeping curve is shown in Figure 6.1. It is easy to observe that at the critical frequency $\omega = 1$, the frequency-sweeping curve touches without crossing (it is in fact tangent with) the line with ordinate equal to 1, and thus, from Proposition 6.14, it follows that the asymptotic behavior of the characteristic roots located on the imaginary axis does not cause a change in the number of unstable roots. To further verify the above analysis, we give the root loci near the critical pairs (j,π) and $(j,3\pi)$ in Figure 6.2.

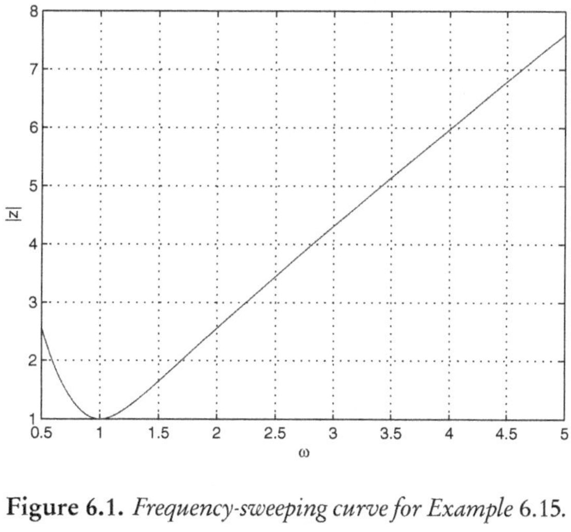

Figure 6.1. *Frequency-sweeping curve for Example 6.15.*

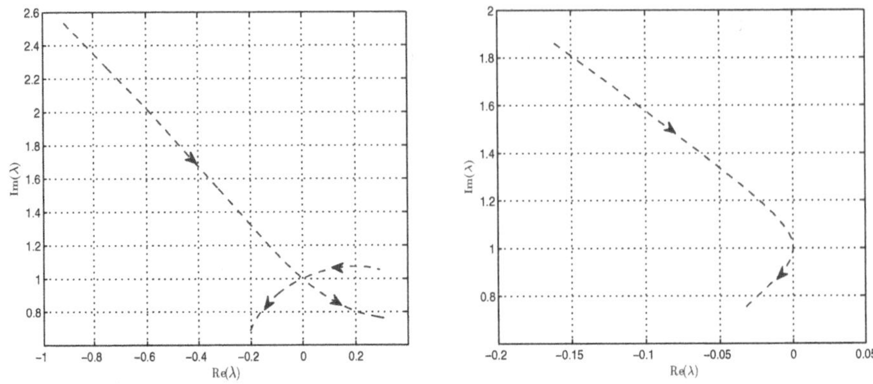

Figure 6.2. *Root loci near (j,π) (left) and $(j,3\pi)$ (right) for Example* 6.15.

Quasi-polynomials with one delay

Consider the following class of quasi-polynomials including one delay:

$$p(\lambda;\ \tau) := Q(\lambda) + P(\lambda)e^{-\lambda\tau}, \tag{6.37}$$

where P and Q are coprime polynomials with real coefficients such that $\deg(Q) > \deg(P)$. Such a form is motivated by some stabilization problems for linear systems using delayed output feedback, as we shall see in Chapter 10. It is easy to see that (6.37) can be rewritten as $p(\lambda;\ \tau) = Q(\lambda)\big(1+P(\lambda)e^{-\lambda\tau}/Q(\lambda)\big)$. Hence, under the assumption that Q has no zeros on the imaginary axis, the stability can be analyzed as outlined in the previous section. In what follows we present an alternative method that does not make such an assumption.

If p has a zero $j\omega_0$, $\omega_0 > 0$, for some delay τ_0, then $Q(j\omega_0) = -P(j\omega_0)e^{-j\omega_0\tau_0}$, which implies that Ω consists of the strictly positive zeros of the function

$$F : \mathbb{R}_+ \to \mathbb{R}, \qquad \omega \mapsto F(\omega) := \mid Q(j\omega)\mid^2 - \mid P(j\omega)\mid^2, \tag{6.38}$$

while the delay crossing set can be derived from the phase information of $Q(j\omega_0) = -P(j\omega_0)e^{-j\omega_0\tau_0}$. Furthermore, in [301] it is shown that under the assumption that the roots of F are simple, the crossing direction information can be obtained from the derivative of F, leading to the following result:

Proposition 6.16. *Assume that the zeros of F are simple. The characteristic function p has a zero $j\omega_0$, $\omega_0 > 0$, for some delay value τ_0 if and only if*

$$F(\omega_0) = 0. \tag{6.39}$$

Furthermore, for any ω_0 satisfying (6.39) *the set of corresponding delay values is given by*

$$\mathcal{T}_{\omega_0} = \left\{ \frac{1}{\omega_0}\left[-j\,\mathrm{Log}\left(-\frac{P(j\omega_0)}{Q(j\omega_0)} \right) + 2\pi l \right] \geq 0,\ l \in \mathbb{Z} \right\}. \tag{6.40}$$

When increasing the delay the corresponding crossing direction of the zero is towards instability (stability) if $F'(\omega_0) > 0$ (< 0).

Remark 6.17. *The methodology considered here for the characterization of the delay crossing set is sometimes called the* direct approach *(see, e.g.,* [130]*). To the best of our knowledge,* *the crossing characterization in terms of the function F was proposed by Cooke and van den Driessche [81], generalizing the second-order characterization proposed in [79].*

Pseudodelay and matrix pencils

Recall that the pseudodelay technique allows us to detect characteristic roots on the imaginary axis using the bilinear transformation

$$z = \frac{1-\lambda T}{1+\lambda T}, \quad T \geq 0,$$

which puts the quasi-polynomial $p(\lambda;\ \tau)$, defined in (6.17), into the form

$$p_3(\lambda;\ T) = \sum_{i=0}^{n_d} p_i(\lambda)(1-T\lambda)^i(1+T\lambda)^{n_d-i} := \sum_{i=0}^{n_d} q_i(\lambda)T^i,$$

with $q_0,\ldots,q_{n_d}$ polynomials. More precisely, Proposition 6.5 establishes a connection between the characteristic function $p(\lambda;\ \tau)$ and the parameter-dependent polynomial $p_3(\lambda;\ T)$ in terms of zeros on the imaginary axis, and, thus, the detection of such zeros for particular delay values is reduced to finding parameters T such that p_3 has zeros on the imaginary axis. We outline how this problem can be solved by computing the generalized eigenvalues of an appropriately defined matrix pencil.

With an arbitrary polynomial $a(\lambda) := \sum_{i=0}^{n_a} a_i \lambda^{n_a-i}$, we can associate a matrix $H(a)$, defined as

$$H(a) := \begin{bmatrix} a_1 & a_3 & a_5 & \cdots & a_{2n_a-1} \\ a_0 & a_2 & a_4 & \cdots & a_{2n_a-2} \\ 0 & a_1 & a_3 & \cdots & a_{2n_a-3} \\ 0 & a_0 & a_2 & \cdots & a_{2n_a-4} \\ \vdots & & & \ddots & \vdots \\ 0 & 0 & 0 & \cdots & a_{n_a} \end{bmatrix} \in \mathbb{R}^{n_a \times n_a}, \tag{6.41}$$

where the coefficients $a_l = 0$ are assumed zero for $l > n_a$. Next, we introduce the matrix pencil

$$\Gamma(z) := zU + V,$$

with U, V given by

$$U = \begin{bmatrix} I & & & \\ & \ddots & & \\ & & I & \\ & & & H(q_{n_d}) \end{bmatrix}, \quad V = \begin{bmatrix} 0 & -I & \cdots & 0 \\ \vdots & \vdots & \ddots & \vdots \\ 0 & 0 & \cdots & -I \\ H(q_0) & H(q_1) & \cdots & H(q_{n_d-1}) \end{bmatrix}, \tag{6.42}$$

where the identity and the zero-block matrices have appropriate dimensions.

The following result gives a characterization of the zeros of p_3 on the imaginary axis as a function of T, and generalizes the matrix pencil solution proposed in [67] in the context of static output feedback:

Proposition 6.18. *Let $0 < \lambda_1 < \lambda_2 < \cdots < \lambda_h$, with $h \leq nn_d$, be the real eigenvalues of the matrix pencil Γ. The parameter-dependent polynomial p_3 has a zero on the imaginary axis if and only if $T \in \{\lambda_1,\ldots,\lambda_h\}$. Furthermore, if there are r unstable zeros for $T = T^*$, $T^* \in (\lambda_i, \lambda_{i+1})$, then there are r unstable roots for all $T \in (\lambda_i, \lambda_{i+1})$. In other words, the* *instability degree of p_3 remains constant as T varies within each interval $(\lambda_i, \lambda_{i+1})$. The same property holds for the intervals $(0, \lambda_1)$ and (λ_h, ∞).*

In conclusion, the generalized eigenvalues of Γ on the real axis define the values of the parameter T for which the polynomial p_3 has zeros on the imaginary axis. For all $T \in \{\lambda_1, \dots, \lambda_h\}$ one can compute these zeros on the imaginary axis, whose imaginary parts define the *frequency crossing set* Ω. Next, for every frequency $\omega \in \Omega$ and corresponding $T \in \{\lambda_1, \dots, \lambda_n\}$, the set $\mathcal{T}_\omega$ can be derived by solving the equation

$$e^{-j\omega\tau} = \frac{1 - j\omega T}{1 + j\omega T}.$$

It is important to point out that this procedure for detection the crossing frequency set Ω is quite distinct to the procedures proposed in [313, 348].

6.3.4 ▪ Extensions to delay-differential algebraic equations

All the ideas and results proposed in the previous section can be extended to time-delay systems described by neutral functional differential equations or by delay-differential algebraic equations. In what follows, we will address only the second class and we will focus on the matrix pencil approach.

Consider now the following class of delay systems:

$$\begin{cases} \dot{x}_1(t) = Ax_1(t) + \sum_{k=1}^{m}(A_k x_1(t-k\tau) + B_k x_2(t-k\tau)), \\ x_2(t) = Cx_1(t) + \sum_{k=1}^{m}(C_k x_1(t-k\tau) + D_k x_2(t-k\tau)), \end{cases} \tag{6.43}$$

where $x_i \in \mathbb{R}^{n_i}$, for $i = 1, 2$, and under some appropriate initial conditions on $(x_{1t}(\cdot), x_{2t}(\cdot)) \in \mathscr{C}([-m\tau, 0], \mathbb{R}^{n_1}) \times \mathscr{C}([-m\tau, 0], \mathbb{R}^{n_2})$, where $x_{it}(\cdot)$ denotes the restriction to the interval $[t - m\tau, t]$, translated to $[-m\tau, 0]$. The corresponding characteristic function is written as $H(\lambda;\ \tau) = \det M(\lambda; \tau)$, where

$$M(\lambda; \tau) = \begin{bmatrix} \lambda I_{n_1} - A - \sum_{k=1}^{m} A_k e^{-k\lambda\tau} & -\sum_{k=1}^{m} B_k e^{-k\lambda\tau} \\ C + \sum_{k=1}^{m} C_k e^{-k\lambda\tau} & I_{n_2} - \sum_{k=1}^{m} D_k e^{-k\lambda\tau} \end{bmatrix}. \tag{6.44}$$

As discussed in Chapter 1, in the stability analysis of delay-differential algebraic equations, one needs explicitly the stability of the *difference operator* $\mathscr{D} : \mathscr{C}([-m\tau, 0], \mathbb{R}^{n_2}) \mapsto \mathbb{R}^{n_2}$ defined by

$$\mathscr{D}(\phi) = \phi(0) - \sum_{k=1}^{m} D_k \phi(-k\tau). \tag{6.45}$$

The stability condition is given by:

Lemma 6.19. *The difference operator $\mathscr{D}$ defined by* (6.45) *is stable if and only if*

$$\rho(D) < 1, \tag{6.46}$$

where

$$D := \begin{bmatrix} D_1 & \cdots & D_{m-1} & D_m \\ I & \cdots & 0 & 0 \\ \vdots & \ddots & \vdots & \vdots \\ 0 & \cdots & I & 0 \end{bmatrix}. \tag{6.47}$$

Furthermore, the stability is guaranteed for all positive delay values τ.

Introduce now the following matrix pencils:

$$\Lambda_i(z) = zU_i + V_i, \qquad i = 1,2, \tag{6.48}$$

where U_i, V_i are given by

$$U_1 = \begin{bmatrix} I & & & \\ & \ddots & & \\ & & I & \\ & & & H_m \end{bmatrix}, \quad V_1 = \begin{bmatrix} 0 & -I & \cdots & 0 & \cdots & 0 \\ \vdots & \vdots & & \ddots & & \vdots \\ 0 & 0 & \cdots & 0 & \cdots & -I \\ H_{-m} & H_{-m+1} & \cdots & H_0 & \cdots & H_{m-1} \end{bmatrix},$$

$$U_2 = \begin{bmatrix} -\sum_{k=1}^{m} A_k & -\sum_{k=1}^{m} B_k \\ -\sum_{k=1}^{m} C_k & -\sum_{k=1}^{m} D_k \end{bmatrix}, \quad V_2 = \begin{bmatrix} -A & 0 \\ -C & I_{n_2} \end{bmatrix},$$

where the identity and zero matrices have appropriate dimensions, and the matrices H_0 and $H_{\pm k}$ $(k = 1, \ldots, m)$ are given as follows:

$$H_0 = \begin{bmatrix} A \oplus A^T & 0 & I_{n_1} \otimes C^T \\ C \otimes I_{n_1} & I_{n_1 n_2} & 0 \\ 0 & 0 & I_{n_1 n_2} \end{bmatrix},$$

$$H_k = \begin{bmatrix} A_k \otimes I_{n_1} & B_k \otimes I_{n_1} & 0 \\ C_k \otimes I_{n_1} & -D_k \otimes I_{n_1} & 0 \\ 0 & 0 & 0 \end{bmatrix}, \quad H_{-k} = \begin{bmatrix} I_{n_1} \otimes A_k^T & 0 & I_{n_1} \otimes C_k^T \\ 0 & 0 & 0 \\ I_{n_1} \otimes B_k^T & 0 & -I_{n_1} \otimes D_k^T \end{bmatrix}$$

for all $1 \le k \le m$. The construction of the matrix pencils Λ_i, $i = 1,2$, leads to the following result, similar to the one developed in [289] in the retarded case:

Lemma 6.20. *Assume that $\rho(D) < 1$. If the original system that is free of delay is asymptotically stable, then the matrix pencils Λ_1 and Λ_2 are both regular.*

Frequency crossing set characterization

Similarly to the retarded case, the following result gives a complete description of the characteristic roots located on the imaginary axis in terms of the generalized eigenvalues distribution of the matrix pencil Λ. The proof was presented in [297] and follows closely the ideas discussed in [289].

Proposition 6.21. *Assume that the difference operator $\mathcal{D}$ defined by* (6.45) *is stable. Assume further that the matrix pencil Λ_1 is regular.*

Then the characteristic function associated to (6.43) *has a root $j\omega_0$, $\omega_0 > 0$, for some positive delay value τ if there exists a complex number*

$$z_0 \in \sigma(\Lambda_1) \cap \mathscr{C}(0,1) \tag{6.49}$$

such that $j\omega_0 \in \sigma(\mathscr{A}(z_0))$, where

$$\mathscr{A}(z_0) := A + \sum_{k=1}^{m} A_k z_0^k + B_k z_0^k \left(I_{n_2} - \sum_{k=1}^{m} D_k z_0^k \right)^{-1} \left(C + \sum_{k=1}^{m} C_k z_0^k \right).$$

Furthermore, the corresponding delay values are given by

$$\mathcal{T}_{\omega_0} = \left\{ \frac{\angle \overline{z_0}}{j\omega_0} + \frac{2\pi \ell}{\omega_0} > 0 : j\omega_0 \in \sigma(\mathscr{A}(z_0)) - \{0\}, \ell \in \mathbb{Z} \right\}. \tag{6.50}$$

As in the retarded case (see, for instance, Definition 6.7), we may define the set σ_g of all crossing generators as well as the frequency crossing set Ω and the delay crossing set $\mathcal{T}$ as $\mathcal{T} := \bigcup_{\omega \in \Omega} \mathcal{T}_{\omega}$.

Crossing direction characterization: Simple characteristic roots

For the sake of brevity, we will consider only the case of *simple* characteristic roots located on the imaginary axis. With the notations and the remarks above, we have:

Proposition 6.22 (crossing direction characterization: Simple roots). *Assume that the difference operator $\mathscr{D}$ defined by* (6.45) *is stable. Assume further that the matrix pencil Λ_1 is regular.*

Under the assumption of simple crossing roots, let $z_0 \in \sigma_g$ be a crossing generator of some root $j\omega_0 \neq 0$ for the critical delay τ_0 of the system (6.43).

Then, the corresponding characteristic root will cross the imaginary axis towards instability (stability) if and only if

$$Re \left\{ -\frac{j\omega_0}{u_1^* u_1} \sum_{k=1}^{m} k z_0^k \begin{bmatrix} v_1 \\ v_2 \end{bmatrix}^* \begin{bmatrix} A_k & B_k \\ C_k & D_k \end{bmatrix} \begin{bmatrix} u_1 \\ u_2 \end{bmatrix} \right\} > 0 (< 0), \tag{6.51}$$

where $v_1^* \in \mathbb{C}^{n_1}$ ($u_1 \in \mathbb{C}^{n_1}$) and $v_2^* \in \mathbb{C}^{n_2}$ ($u_2 \in \mathbb{C}^{n_2}$) represent a partition of the corresponding left (right) eigenvectors u^* (v) of the eigenvalue $j\omega_0$ of the characteristic matrix $M(\lambda_0; \tau_0)$ as defined by (6.44).

Comments and interpretations

For the sake of simplicity, we shall consider only the single delay case ($m = 1$). However, the proposed comparisons, discussions, and comments are also valid for the general case. Our interest lies in seeing how the results above rewrite in the particular retarded and neutral cases, and in establishing the existing connections between the computed matrix pencils and those proposed in the literature [70, 289, 116] for handling the corresponding cases.

Retarded case

It is easy to see that we completely recover the stability conditions for *retarded* linear systems including a single delay, but with a different definition of the matrix pencil Λ_1 [70, 289]. In fact, the corresponding matrix pencil in [70, 289] is derived using a different *strong* linearization of the corresponding matrix polynomial (see also [290]) $\mathscr{P}(z) = B \otimes I_{n_1} z^2 + A \oplus A^T z + I_{n_1} \otimes B^T$, leading to

$$\widetilde{\Lambda}_1(z) = z \begin{bmatrix} I_{n_1^2} & 0 \\ 0 & B \otimes I_{n_1} \end{bmatrix} + \begin{bmatrix} 0 & -I_{n_1^2} \\ I_{n_1} \otimes B^T & A \oplus A^T \end{bmatrix}.$$

In this particular configuration, Λ_1 is rewritten as

$$\Lambda_1(z) = z \begin{bmatrix} 0 & B \otimes I_{n_1} & 0 \\ 0 & 0 & 0 \\ 0 & 0 & -I_{n_1^2} \end{bmatrix} + \begin{bmatrix} A \oplus A^T & 0 & I_{n_1^2} \\ I_{n_1^2} & -I_{n_1^2} & 0 \\ I_{n_1} \otimes B^T & 0 & 0 \end{bmatrix}.$$

Simple computations show that $\det(\Lambda_1(z)) = \det(\widetilde{\Lambda}_1(z))$ and, in conclusion:

Proposition 6.23. *The matrix pencils Λ_1 and $\widetilde{\Lambda}_1$ share the same generalized eigenvalues on the unit circle of the complex plane.*

The approach proposed in the previous paragraphs takes advantage of the form of Λ_1 instead of $\widetilde{\Lambda}_1$, which cannot be used directly in the lossless propagation case. Furthermore, the form of the pencil Λ_1 is particularly adapted to handling (regular) *singular* systems with delays.

Neutral case

Consider now the neutral system

$$\dot{x}(t) - A_{-1}x(t-\tau) = A_0 x(t) + A_1 x(t-\tau). \tag{6.52}$$

As shown in [116], the stability analysis makes use of the matrix pencil

$$\widehat{\Lambda}_1 := \begin{bmatrix} I & 0 \\ 0 & A_1 \otimes I - A_{-1} \otimes A_0^T \end{bmatrix} + \begin{bmatrix} \begin{pmatrix} 0 \\ I \otimes A_1^T \\ -A_0 \otimes A_{-1}^T \end{pmatrix} & \begin{pmatrix} -I \\ A_0 \oplus A_0^T - A_1 \otimes A_{-1}^T \\ -A_{-1} \otimes A_1^T \end{pmatrix} \end{bmatrix}. \tag{6.53}$$

By rewriting (6.52) in the form (6.43), we will have $A = A_0$, $B = A_0 A_{-1} + A_1$, $C = I_n$, and $D = A_{-1}$. Some simple, but tedious, computations of $\det(\Lambda_1(z))$ and $\det(\widehat{\Lambda}_1(z))$ show that

$$\det(\Lambda_1(z)) = \det\left[((I_n - Dz) \otimes I_n)^{-1}\right] \cdot \det(\widehat{\Lambda_1}(z)) \cdot \det\left[\left(I_n \otimes (I_n - D^T \bar{z})\right)^{-1}\right].$$

Next, since $\rho(D) < 1$, it follows that:

Proposition 6.24. *The matrix pencils Λ_1 and $\widehat{\Lambda}_1$ share the same generalized eigenvalues on the unit circle of the complex plane.*

As in the retarded case, the matrix pencil $\widehat{\Lambda}_1$ cannot be directly adapted to the lossless propagation case, but it opens up the perspective of giving a complete characterization of *singular* neutral systems including commensurate delays. For the sake of brevity, such a case study is not discussed here.

Illustrative example: Scalar lossless propagation system

Consider now the scalar (propagation) system ($n_1 = n_2 = 1$). The system (6.43) is rewritten as

$$\begin{cases} \dot{x}_1(t) = a x_1(t) + b x_2(t-\tau), \\ x_2(t) = c x_1(t) + d x_2(t-\tau), \end{cases} \tag{6.54}$$

where $a,b,c,d \in \mathbb{R}$. The matrix pencils Λ_1 and Λ_2 are given by

$$\Lambda_1(z) = z\begin{bmatrix} 0 & b & 0 \\ 0 & d & 0 \\ 0 & 0 & -1 \end{bmatrix} + \begin{bmatrix} 2a & 0 & c \\ c & -1 & 0 \\ b & 0 & d \end{bmatrix}, \quad \Lambda_2(z) = z\begin{bmatrix} 0 & -b \\ 0 & -d \end{bmatrix} + \begin{bmatrix} -a & 0 \\ -c & 1 \end{bmatrix}. \tag{6.55}$$

In this case, Propositions 6.22 and 6.21 lead to a complete characterization of the stability property. More precisely, the scalar lossless propagation system may have not more than one simple characteristic root $j\omega_c$ (with $\omega_c > 0$), and the corresponding crossing direction is towards instability. Here, $\omega_c = \sqrt{(ad-bc)^2 - a^2}/\sqrt{1-d^2}$, and it is easy to see that by an appropriate choice of coefficients (a,b,c,d), we recover both retarded and neutral cases. Finally, under the assumption that the system which is free of delay is asymptotically stable, we have the following result:

Proposition 6.25. *Assume* $a(1-d)+bc < 0$ *and* $\mid d \mid < 1$. *Then the system* (6.54) *is delay-independent asymptotically stable if and only if*

$$-\mid a \mid + ad \leq bc \leq \mid a \mid + ad. \tag{6.56}$$

If not, it is asymptotically stable for all the delays $\tau \in [0, \tau_m)$*, where the delay margin is given by*

$$\tau_m = \sqrt{\frac{1-d^2}{(ad-bc)^2 - a^2}} arcctg\left(\frac{a(1+d^2) - bcd}{\sqrt{((bc-ad)^2 - a^2)(1-d^2)}}\right). \tag{6.57}$$

6.4 ▪ Geometric methods

In what follows, we focus on the particular problem of the characterization of the stability regions and of their boundaries in the delay-parameter space by considering a simplified characteristic function including two independent delays of the form

$$p(\lambda;\ \tau_1, \tau_2) := p_0(\lambda) + p_1(\lambda)e^{-\lambda\tau_1} + p_2(\lambda)e^{-\lambda\tau_2}, \tag{6.58}$$

where $p_l(\lambda) := \sum_{i=1}^{n_l} p_{li}\lambda^i$ for $i = 0,1,2$, with $n_0 > n_1, n_2$. This condition implies that the corresponding system is of *retarded* type. The analysis of the neutral case is omitted, but it can be treated in a similar way under some appropriate assumptions on the difference operator (see, for instance, [132]).

Our first objective is to identify the regions of (τ_1, τ_2) in $\mathbb{R}_+^2$ such that $p(\lambda;\ \tau_1, \tau_2)$ has zeros on the imaginary axis. We first exclude some simple trivial cases and restrict the analysis to the case when $p(\lambda;\ \tau_1, \tau_2)$ satisfies the following conditions:

I. Zero frequency:

$$p_0(0) + p_1(0) + p_2(0) \neq 0. \tag{6.59}$$

II. The polynomials $p_0(\lambda)$, $p_1(\lambda)$, and $p_2(\lambda)$ do not have any common zeros.

It is easy to see that if condition I is not satisfied, then 0 is a zero of $p(\lambda;\ \tau_1, \tau_2)$ for any $(\tau_1, \tau_2) \in \mathbb{R}_+^2$, and therefore it can never be asymptotically stable. Next, condition II is also natural. If it is not satisfied, then there exists a common factor $c(\lambda) \neq$ constant such that $p_l(\lambda) = c(\lambda)q_l(\lambda)$, $l = 0,1,2$. Choose $c(\lambda)$ to be the highest possible order; then $q_l(\lambda)$, $l = 0,1,2$, do not have any common zeros, and the underlying delay-differential equation can be decomposed to an ODE with characteristic polynomial $c(\lambda)$ and a delay-differential equation with characteristic quasi-polynomial

$$q_0(\lambda)+q_1(\lambda)e^{-\lambda\tau_1}+q_2(\lambda)e^{-\lambda\tau_2}, \tag{6.60}$$

which satisfies condition II. Due to the continuity, given $\tau_1=\tau_1^0$ and $\tau_2=\tau_2^0$, in principle, we may find the number of zeros of $p(\lambda;\ \tau_1,\tau_2)$ on $\mathbb{C}_+$ using the following simplified *algorithm*: first, find the number of unstable roots in the case that is free of delays, that is, the polynomial $p(\lambda:0,0)$; second, form a curve in the τ_1-τ_2 plane within $\mathbb{R}_+^2$ initiating from the origin and ending at the desired point $(\tau_1^0,\ \tau_2^0)$; finally, find all the points of (τ_1,τ_2) in the curve such that there are zeros of $p(\lambda;\tau_1,\tau_2)$ crossing the imaginary axis, and find the directions of crossing (from left to right or vice versa) as one moves along the curve. In the remainder of the section we will identify these crossing points and the corresponding stability crossing curves. We will need the following definition [132]:

Definition 6.26. *Let $\mathscr{C}_k:[a,b]\to\mathbb{R}^2$, $k=1,2,\ldots$, be a series of curves satisfying the condition $\mathscr{C}_k(b)-\mathscr{C}_k(a)=A$ ($k=1,2,\ldots$), where $A\in\mathbb{R}^2$ is a constant 2D vector independent of k, and $\mathscr{C}_{k+1}(a)=\mathscr{C}_k(b)$. Then, the curve $\mathscr{C}$ formed by connecting all the curves $\mathscr{C}_k$ ($k=1,2,\ldots$) $\mathscr{C}=\bigcup_{k=1}^{\infty}\mathscr{C}_k$ is known as a* spiral-like curve, *and A is known as its* axis. *If, in addition, $\mathscr{C}_{k+1}(\xi)=\mathscr{C}_k(\xi)+A$ for all $\xi\in[a,b]$, then $\mathscr{C}$ is known as a* spiral.

The definition above simply says that a spiral is formed by connecting identical curves head to tail. It is worth mentioning that the composite curves in a spiral-like curve do not have to be identical. In what follows, in the spiral-like curves case, $\mathscr{C}_{k+1}$ can often be viewed as formed from $\mathscr{C}_k$ with a small deformation, which justifies the term "spiral-like curve."

6.4.1 ▪ Identification of crossing points

Let us adapt the notations from the commensurate and single delay cases to the delay-parameter space defined by τ_1 and τ_2. Thus, $\mathscr{T}$ denotes the set of all the points of (τ_1,τ_2) in $\mathbb{R}_+^2$ such that $p(\lambda;\ \tau_1,\tau_2)$ has at least one zero on the imaginary axis. Any $(\tau_1,\tau_2)\in\mathscr{T}$ is known as a *crossing point*. The set $\mathscr{T}$, which is the collection of all the crossing points, is called the *stability crossing curves.*

Define now $a_l(\lambda)=p_l(\lambda)/p_0(\lambda)$, $l=1,2$, and

$$a(\lambda;\ \tau_1,\tau_2):=1+a_1(\lambda)e^{-\lambda\tau_1}+a_2(\lambda)e^{-\lambda\tau_2}. \tag{6.61}$$

Using an argument similar to the one proposed in the rank one delayed matrix case, for any given τ_1 and τ_2, as long as $p_0(\lambda)$ does not have characteristic roots on the imaginary axis, $p(\lambda;\ \tau_1,\tau_2)$ and $a(\lambda;\ \tau_1,\tau_2)$ *share* all the zeros in some neighborhood of the imaginary axis. Therefore, in general, we may obtain all the crossing points and directions of crossing from

$$a(\lambda;\ \tau_1,\tau_2)=0 \tag{6.62}$$

instead of $p(\lambda;\ \tau_1,\tau_2)=0$.

For each given $\lambda=j\omega$, $\omega>0$, we may consider the three terms in $a(j\omega;\ \tau_1,\tau_2)$ as three vectors in the complex plane, with the magnitudes 1, $|a_1(j\omega)|$, and $|a_2(j\omega)|$, respectively. Furthermore, if we adjust the values of τ_1 and τ_2, we may arbitrarily adjust the directions of the vectors represented by the second and third terms. Equation (6.62) means that if we put these vectors head to tail, they form a triangle as illustrated in Figure 6.3. This allows us to conclude the following proposition (see, for instance, [132] for a complete proof):

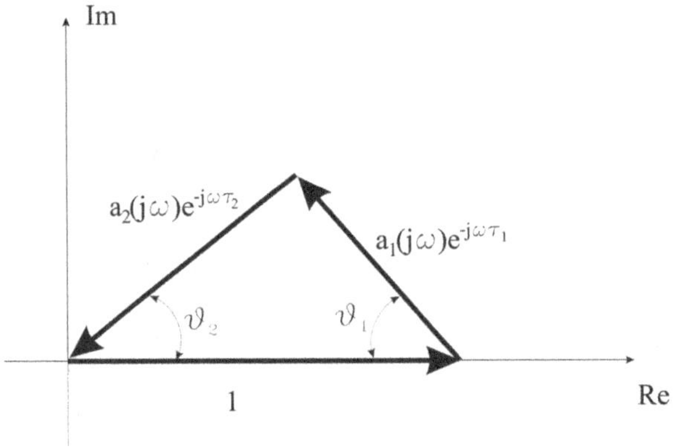

Figure 6.3. *Triangle formed by* 1, $|a_1(j\omega)|$, *and* $|a_2(j\omega)|$ [132].

Proposition 6.27. *For each* ω, $\omega \neq 0$, $p_0(j\omega) \neq 0$, $\lambda = j\omega$ *can be a solution of* $p(\lambda;\ \tau_1, \tau_2) = 0$ *for some* $(\tau_1, \tau_2) \in \mathbb{R}_+^2$ *if and only if*

$$|a_1(j\omega)| + |a_2(j\omega)| \geq 1, \tag{6.63}$$

$$-1 \leq |a_1(j\omega)| - |a_2(j\omega)| \leq 1. \tag{6.64}$$

For $\omega \neq 0$ *satisfying* $p_0(j\omega) = 0$, $\lambda = j\omega$ *can be a zero of* $p(\lambda;\ \tau_1, \tau_2)$ *for some* $(\tau_1, \tau_2) \in \mathbb{R}_+^2$ *if and only if*

$$|p_1(j\omega)| = |p_2(j\omega)|. \tag{6.65}$$

Let Ω be the set of all $\omega > 0$ satisfying (6.63) and (6.64) if $p_0(j\omega) \neq 0$ and (6.65) if $p_0(j\omega) = 0$. Similarly to the commensurate and single delay cases, we will refer to Ω as the *frequency crossing set*. It contains all the ω such that some zero of $p(\lambda;\ \tau_1, \tau_2)$ may cross the imaginary axis at $j\omega$. Then, for any given $\omega \in \Omega$, $p_l(j\omega) \neq 0$, $l = 0, 1, 2$, one may easily find all the pairs of (τ_1, τ_2) satisfying (6.62) as follows:

$$\tau_1 = \tau_1^{u\pm}(\omega) = \frac{\angle a_1(j\omega) + (2u-1)\pi \pm \theta_1}{\omega} \geq 0,\ u = u_0^{\pm}, u_0^{\pm}+1, u_0^{\pm}+2, \ldots, \tag{6.66}$$

$$\tau_2 = \tau_2^{v\pm}(\omega) = \frac{\angle a_2(j\omega) + (2v-1)\pi \mp \theta_2}{\omega} \geq 0,\ v = v_0^{\pm}, v_0^{\pm}+1, v_0^{\pm}+2, \ldots, \tag{6.67}$$

where $\theta_1, \theta_2 \in [0, \pi]$ are the internal angles of the triangle in Figure 6.3 and can be calculated by the law of cosine as

$$\theta_1 = \cos^{-1}\left(\frac{1 + |a_1(j\omega)|^2 - |a_2(j\omega)|^2}{2|a_1(j\omega)|}\right), \tag{6.68}$$

$$\theta_2 = \cos^{-1}\left(\frac{1 + |a_2(j\omega)|^2 - |a_1(j\omega)|^2}{2|a_2(j\omega)|}\right), \tag{6.69}$$

and u_0^+, u_0^-, v_0^+, v_0^- are the smallest possible integers (which may be negative and may depend on ω) such that the corresponding $\tau_1^{u_0^+ +}$, $\tau_1^{u_0^- -}$, $\tau_2^{v_0^+ +}$, $\tau_2^{v_0^- -}$ calculated are nonnegative. Notice $u_0^+ \leq u_0^-$, $v_0^+ \geq v_0^-$. The position in Figure 6.3 corresponds to $(\tau_1^{u+}, \tau_2^{v+})$. The position corresponding to $(\tau_1^{u-}, \tau_2^{v-})$ is its mirror image about the real axis. Next, let $\mathscr{T}_{\omega,u,v}^{+}$ and $\mathscr{T}_{\omega,u,v}^{-}$ be the singletons defined by

$$\mathscr{T}_{\omega,u,v}^{\pm} = \{(\tau_1^{u\pm}(\omega), \tau_2^{v\pm}(\omega))\},$$

and define

$$\mathscr{T}_\omega = \left(\bigcup_{u \geq u_0^+, v \geq v_0^+} \mathscr{T}_{\omega,u,v}^+ \right) \bigcup \left(\bigcup_{u \geq u_0^-, v \geq v_0^-} \mathscr{T}_{\omega,u,v}^- \right),$$

which generalizes the delay crossing set notion encountered in the commensurate or single delay cases. Here, $\mathscr{T}_\omega$ represents the set of all (τ_1, τ_2) such that $p(\lambda;\ \tau_1, \tau_2)$ has a zero at $\lambda = j\omega$. In the following remark, we will discuss the degenerate cases of $p_i(j\omega) = 0$ for at least one i $(= 1, 2)$.

Remark 6.28. *If $p_0(j\omega) = 0$, $\omega \in \Omega$. Then $p(j\omega) = 0$ and assumption* II *imply $|p_1(j\omega)| = |p_2(j\omega)| \neq 0$. In this case, $\mathscr{T}_\omega$ consists of the solutions of*

$$\angle p_1(j\omega) - \omega\tau_1 + 2\pi u = \angle p_2(j\omega) - \omega\tau_2 + 2\pi v + \pi$$

in $\mathbb{R}_+^2$ for integers u, v. Instead of isolated points, $\mathscr{T}_\omega$ now consists of an infinite number of straight lines of slope 1 *of equal distance.*

On the other hand, if $p_0(j\omega) \neq 0$, $\omega \in \Omega$, and $p_1(j\omega) = 0$, then $a_1(j\omega) = 0$ and $|a_2(j\omega)| = 1$, we have $\theta_2 = 0$, θ_1 can assume all the values in $[0, \pi]$, and $\mathscr{T}_{\omega,u,v}^{\pm}$ contains all the points calculated by (6.66) *and* (6.67) *with $\theta_1 \in [0, \pi]$, $\theta_2 = 0$. The corresponding $\mathscr{T}_\omega$ is a series of horizontal lines. Similarly, for $\omega \in \Omega$ satisfying $p_0(j\omega) \neq 0$, $p_2(j\omega) = 0$, the corresponding $\mathscr{T}_{\omega,u,v}^{\pm}$ contains all the points calculated by* (6.66) *and* (6.67) *with $\theta_1 = 0$, $\theta_2 \in [0, \pi]$, and $\mathscr{T}_\omega$ is a series of vertical lines.*

Obviously, $\mathscr{T} = \bigcup_{\omega \in \Omega} \mathscr{T}_\omega$. Since the behavior of the degenerate cases discussed in the above remark is easily understood, for brevity we will exclude these degenerate situations in what follows, by imposing the following last assumption:

III. Nondegeneracy:

$$p_l(j\omega) \neq 0 \quad \forall \omega \in \Omega \text{ and } l = 0, 1, 2. \tag{6.70}$$

6.4.2 ▪ Stability crossing curves

In this section, we will give the complete characterization of the crossing set Ω and of the stability crossing curves $\mathscr{T}$. The presentation follows closely the approach proposed in [132]. We have the following result:

Proposition 6.29. *The crossing set Ω consists of a finite number of intervals of finite length, including the cases which may violate* (6.70).

From Propositions 6.27 and 6.29, it follows that the end intervals frequency points ω of the crossing set Ω should satisfy one of the following conditions:

$$|a_1(j\omega)| + |a_2(j\omega)| = 1, \tag{6.71}$$

or

$$|a_1(j\omega)| - |a_2(j\omega)| = 1, \tag{6.72}$$

or

$$|a_2(j\omega)| - |a_1(j\omega)| = 1. \tag{6.73}$$

Let these intervals be Ω_k, $k = 1, 2, \ldots, N$, arranged in such an order that the left end point of Ω_k increases with increasing k. Then

$$\Omega := \bigcup_{k=1}^{N} \Omega_k. \tag{6.74}$$

It is worth clarifying that $0 \notin \Omega$ by definition even if $\omega = 0$ satisfies (6.63) and (6.64). Indeed, if (6.63) and (6.64) are satisfied for $\omega = 0$ and sufficiently small positive values of ω, then $\Omega_1 = (0, \omega_1^r]$, and we will let $\omega_1^l = 0$ in this case. Otherwise, $\Omega_1 = [\omega_1^l, \omega_1^r]$, $\omega_1^l \neq 0$. For $k \geq 2$, $\Omega_k = [\omega_k^l, \omega_k^r]$. We will subdivide the intervals if necessary so that for any $\omega \in (\omega_k^l, \omega_k^r)$, none of the three equations (6.71), (6.72), and (6.73) is satisfied.

Let

$$\mathcal{T}_{u,v}^{\pm k} = \bigcup_{\omega \in \Omega_k} \mathcal{T}_{\omega,u,v}^{\pm} = \{(\tau_1^{u\pm}(\omega), \tau_2^{v\pm}(\omega)) \mid \omega \in \Omega_k\}$$

and

$$\mathcal{T}^k = \bigcup_{u=-\infty}^{\infty} \bigcup_{v=-\infty}^{\infty} \left(\mathcal{T}_{u,v}^{+k} \cup \mathcal{T}_{u,v}^{-k}\right) \cap \mathbb{R}_+^2 = \bigcup_{\omega \in \Omega_k} \mathcal{T}_\omega. \tag{6.75}$$

Then,

$$\mathcal{T} = \bigcup_{k=1}^{N} \mathcal{T}^k.$$

Note that we allow part of $\mathcal{T}_{u,v}^{+k}$ or $\mathcal{T}_{u,v}^{-k}$ to be outside of $\mathbb{R}_+^2$ in some cases for the convenience of discussion. We should, however, keep in mind that the part of $\mathcal{T}_{u,v}^{+k}$ or $\mathcal{T}_{u,v}^{-k}$ outside of $\mathbb{R}_+^2$ no longer represents the boundary of a meaningful change of the number of RHP zeros of $p(\lambda;\ \tau_1, \tau_2)$. As is well known, $p(\lambda;\ \tau_1, \tau_2)$ has an infinite number of RHP zeros if τ_1 or τ_2 assumes a negative value [23].

We will not restrict $\angle a_l(j\omega)$ to be within a range of 2π but make it a continuous function of ω within each Ω_k. This is always possible due to the way Ω_k is defined. As a result, for a fixed pair of integers (u, v), each $\mathcal{T}_{u,v}^{+k}$ or $\mathcal{T}_{u,v}^{-k}$ is a continuous curve.

To study how each $\mathcal{T}_{u,v}^{+k}$ or $\mathcal{T}_{u,v}^{-k}$ is connected in $\mathcal{T}^k$ at the ends of Ω_k, we make the following observation: under our standing nondegenerate assumption (6.70), the end points of the intervals, ω_k^l, $k = 2, 3, \ldots$, and ω_k^r, $k = 1, 2, \ldots$, must satisfy one and only one of the three equations (6.71), (6.72), and (6.73). Accordingly, we can classify these end points into three types according to which equation $\omega = \omega_k^l$ or $\omega = \omega_k^r$ satisfies. The left end of Ω_1 may have an additional type if $\omega_1^l = 0$. A careful examination of the equations (6.66) and (6.67) allows us to arrive at the following list [132]:

Type 1. (6.72) is satisfied. In this case, $\theta_1 = 0$, $\theta_2 = \pi$, and $\mathcal{T}_{u,v}^{+k}$ is connected to $\mathcal{T}_{u,v-1}^{-k}$ at this end.

Type 2. (6.73) is satisfied. In this case, $\theta_1 = \pi$, $\theta_2 = 0$, and $\mathcal{T}_{u,v}^{+k}$ is connected to $\mathcal{T}_{u+1,v}^{-k}$ at this end.

Type 3. (6.71) is satisfied. In this case, $\theta_1 = \theta_2 = 0$, and $\mathcal{T}_{u,v}^{+k}$ is connected to $\mathcal{T}_{u,v}^{-k}$ at this end.

Type 0. $\omega_k^l = 0$. This requires that $\omega = 0$ satisfy (6.63) and (6.64). In this case, as $\omega \to 0$, $\mathcal{T}_{u,v}^{+k}$ and $\mathcal{T}_{u,v}^{-k}$ approach ∞ with asymptotes passing through the points $(\hat{a}_1 \pm \hat{\theta}_1, \hat{a}_2 \mp \hat{\theta}_2)$ with slopes of

$$\frac{\tau_2^{v\pm}}{\tau_1^{u\pm}} \to x_{u,v}^{\pm} = \frac{\angle a_2(0) + (2v-1)\pi \mp \theta_2(0)}{\angle a_1(0) + (2u-1)\pi \pm \theta_1(0)}, \tag{6.76}$$

where $\theta_1(0)$ and $\theta_2(0)$ are evaluated by (6.68) and (6.69) using $a_1(0)$ and $a_2(0)$, respectively, and

$$\hat{a}_l = \frac{d}{d\omega}[\angle a_l(j\omega)]_{\omega=0}, \tag{6.77}$$

$$\hat{\theta}_l = \frac{d}{d\omega}\theta_l(j\omega)|_{\omega=0}. \tag{6.78}$$

Correspondingly, we say an interval Ω_k (and corresponding curves) are of Type $l\,r$ if its left end of Ω_k is of Type l and its right end is of Type r. There are a total of $4 \times 3 = 12$ possible types of such intervals. Let us consider one of these cases:

Example 6.30. *Consider a second-order system including two delays such that*

$$a_1(\lambda) = \frac{3}{\lambda^2 + 2\lambda + 1}, \qquad a_2(\lambda) = \frac{9\lambda + 1}{\lambda^2 + 2\lambda + 1}. \tag{6.79}$$

In this case, Ω contains two intervals, $\Omega_1 = [0.188, 0.453]$ of Type 12 and $\Omega_2 = [8.532, 9.217]$ of Type 23. Figure 6.4 plots $|a_1(j\omega)| + |a_2(j\omega)|$ and $|a_1(j\omega)| - |a_2(j\omega)|$ against ω.

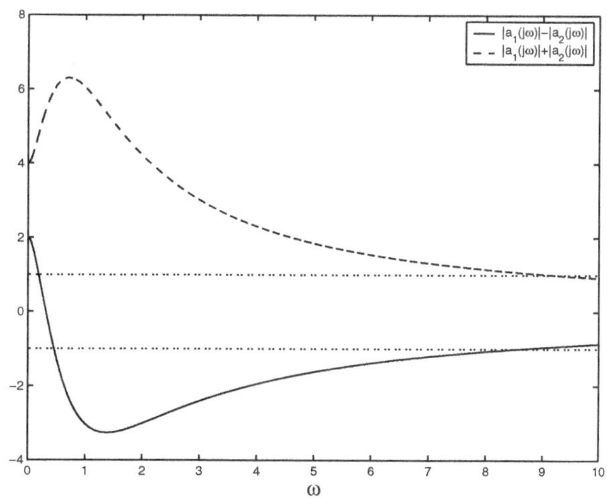

Figure 6.4. $|a_1(j\omega)| \pm |a_2(j\omega)|$ *versus* ω *for system represented by* (6.79) [132].

According to the types of Ω_k, $\mathcal{T}^k$ may have different shapes, as specified in the following proposition.

Proposition 6.31. *Under the standing assumption* (6.70), *the stability crossing curves $\mathcal{T}^k$ corresponding to Ω_k must be an intersection of $\mathbb{R}_+^2$ with a series of curves belonging to one of the following categories:*

A. *A series of closed curves.*

B. *A series of spiral-like curves with axes oriented either horizontally, vertically, or diagonally.*

C. *A series of open ended curves with both ends approaching ∞.*

The rest of this section will be devoted to showing the validity of the above proposition by a detailed list of scenarios (see, for instance, [132] for more details and illustrative examples corresponding to each situation):

Closed curves

As an illustration of this situation, examine first $\mathscr{T}^k$ corresponding to Ω_k of Type 11. In this case, for given u and v, $\mathscr{T}^{+k}_{u,v}$ and $\mathscr{T}^{-k}_{u,v-1}$ are connected on both ends to form a closed curve. As u and v vary, a series of deformed versions of such closed curves are generated along the horizontal and vertical directions. $\mathscr{T}^k$ is the intersection of $\mathbb{R}^2_+$ with this series of closed curves. Similarly, it is easily shown that a $\mathscr{T}^k$ corresponding to Ω_k of Type 22 or Type 33 also forms a similar series of closed curves. In the case of Type 22 (Type 33), a closed curved is formed by connecting both ends of $\mathscr{T}^{+k}_{u,v}$ and $\mathscr{T}^{-k}_{u+1,v}$ ($\mathscr{T}^{+k}_{u,v}$ and $\mathscr{T}^{-k}_{u,v}$).

Spiral-like curves

Several situations occur:

Spiral-like curves with axes oriented diagonally

To illustrate such a case, consider $\mathscr{T}^k$ corresponding to Ω_k of Type 12. In this case, $\mathscr{T}^{+k}_{u,v}$ is connected to $\mathscr{T}^{-k}_{u+1,v}$ at ω^r_k, and the other end of $\mathscr{T}^{-k}_{u+1,v}$ is connected to $\mathscr{T}^{+k}_{u+1,v+1}$ at ω^l_k, which is again connected to $\mathscr{T}^{-k}_{u+2,v+2}$ at ω^r_k, and so on. According to Definition 6.26, with $\mathscr{C}_k = \mathscr{T}^{+k}_{u,v} \cup \mathscr{T}^{-k}_{u+1,v}$, it can be easily verified that this forms a spiral-like curve with the axis

$$\begin{aligned} A &= \left(\tau_1^{u+1-}(\omega^l_k) - \tau_1^{u+}(\omega^l_k), \tau_2^{v-}(\omega^l_k) - \tau_2^{v+}(\omega^l_k)\right) \\ &= \left(\frac{2p}{\omega^l_k}, \frac{2p}{\omega^l_k}\right), \quad \text{independent of } u \text{ (or } v), \end{aligned}$$

forming a 45° angle from the horizontal. This spiral is repeated an infinite number of times in a deformed form as the difference between u and v changes. Shown in Figure 6.5 is $\mathscr{T}^1$ for the system in Example 6.30. We can observe that a $\mathscr{T}^k$ corresponding to Ω_k of Type 21 also form such a series spiral-like curves with axes oriented diagonally. In this case $\mathscr{T}^{+k}_{u,v}$ is connected to $\mathscr{T}^{-k}_{u+1,v}$ at ω^l_k instead, and $\mathscr{T}^{-k}_{u+1,v}$ is connected to $\mathscr{T}^{+k}_{u+1,v+1}$ at ω^r_k, and so on.

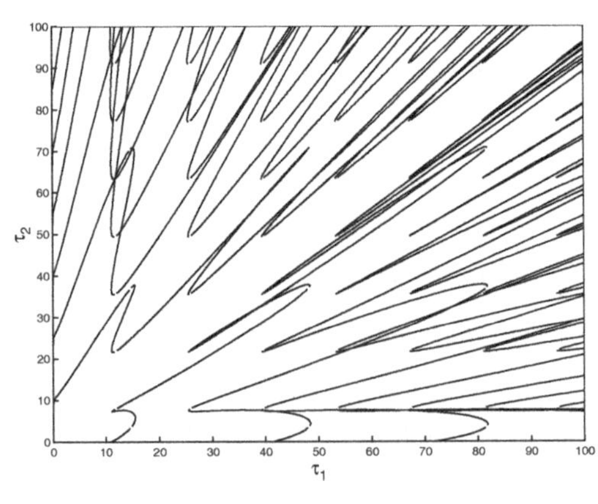

Figure 6.5. $\mathscr{T}^1$ *of the system in Example* 6.30 [132].

Spiral-like curves with vertical axes

To illustrate such a case, consider $\mathcal{T}^k$ corresponding to Ω_k of Type 13. In this case, $\mathcal{T}^{+k}_{u,v}$ is connected to $\mathcal{T}^{-k}_{u,v}$ at ω^r_k, and the other end of $\mathcal{T}^{-k}_{u,v}$ is connected to $\mathcal{T}^{+k}_{u,v+1}$ at ω^l_k, and so on. This forms a spiral-like curve with a vertical axis. This spiral-like curve is repeated in deformed form along the horizontal direction as u changes. It is easily shown that $\mathcal{T}^k$ corresponding to Ω_k of Type 31 is also in the form of a series of vertically oriented spiral-like curves, with $\mathcal{T}^{+k}_{u,v}$ and $\mathcal{T}^{-k}_{u,v}$ connected at ω^l_k, $\mathcal{T}^{-k}_{u,v}$ and $\mathcal{T}^{+k}_{u,v+1}$ connected at ω^r_k, and so on.

Spiral-like curves with horizontal axes

Finally, the curves of $\mathcal{T}^k$ corresponding to Ω_k of Types 23 and 32 are in the forms of a series of *spiral-like curves with horizontal axes.* For Type 23, $\mathcal{T}^{+k}_{u,v}$ is connected to $\mathcal{T}^{-k}_{u+1,v}$ at ω^l_k, and the other end of $\mathcal{T}^{-k}_{u+1,v}$ is connected to $\mathcal{T}^{+k}_{u+1,v}$ at ω^r_k, and so on. For Type 32, $\mathcal{T}^{+k}_{u,v}$ and $\mathcal{T}^{-k}_{u+1,v}$ are connected at ω^r_k, and $\mathcal{T}^{-k}_{u+1,v}$ and $\mathcal{T}^{+k}_{u+1,v}$ are connected at ω^l_k, and so on.

Open-ended curves

Corresponding to $\Omega_1 = (0, \omega^r_1]$, $\mathcal{T}^1$ is a series of *open-ended curves.* For Type 01, $\mathcal{T}^{-1}_{u,v}$ and $\mathcal{T}^{+1}_{u,v+1}$ are connected at ω^r_1. The other end of $\mathcal{T}^{-1}_{u,v}$ extends to infinity with asymptote of a slope $x^-_{u,v}$ passing through the point $(\hat{a}_1 - \hat{\theta}_1, \hat{a}_2 + \hat{\theta}_2)$. The other end of $\mathcal{T}^{+1}_{u,v+1}$ extends to infinity with asymptote of a slope $x^+_{u,v+1}$ passing through the point $(\hat{a}_1 + \hat{\theta}_1, \hat{a}_2 - \hat{\theta}_2)$. This pattern is repeated in a deformed form in both horizontal and vertical directions. Note also that the slopes also change for different u and v. It is easy to show that $\mathcal{T}^1$ corresponding to Ω_1 of Type 02 and Type 03 also forms open-ended curves. For Type 02, $\mathcal{T}^{+1}_{u,v}$ and $\mathcal{T}^{-1}_{u+1,v}$ are connected at ω^r_1. The other ends of $\mathcal{T}^{+1}_{u,v}$ and $\mathcal{T}^{-1}_{u+1,v}$ extend to infinity with slopes $x^+_{u,v}$ and $x^-_{u+1,v}$, respectively. For Type 03, $\mathcal{T}^{+1}_{u,v}$ and $\mathcal{T}^{-1}_{u,v}$ are connected at ω^r_1. The other ends of $\mathcal{T}^{+1}_{u,v}$ and $\mathcal{T}^{-1}_{u,v}$ extend to infinity with slopes $x^+_{u,v}$ and $x^-_{u,v}$, respectively.

Thus far, we have exhausted all 12 types of Ω_k.

6.4.3 ▪ Tangents, smoothness, and crossing direction

We discuss in detail the smoothness of the stability crossing curves, characterize their tangents, and derive expressions for the corresponding direction of crossing.

Tangents and smoothness

For a given k, we will discuss the smoothness of the curves in $\mathcal{T}^k$ and thus in $\mathcal{T}$. For this purpose, we consider τ_1 and τ_2 as implicit functions of $\lambda = j\omega$ defined by (6.62). As λ moves along $j\mathbb{R}$, $(\tau_1, \tau_2) = (\tau_1^{u\pm}(\omega), \tau_2^{v\pm}(\omega))$ moves along $\mathcal{T}^k$. For a given $\omega \in \Omega_k$, let

$$R_0 = \Re\left(\frac{j}{\lambda}\frac{\partial a(\lambda;\, \tau_1, \tau_2)}{\partial \lambda}\right)_{\lambda = j\omega}$$

$$= \frac{1}{\omega}\Re\left(\left[a_1'(j\omega) - \tau_1 a_1(j\omega)\right]e^{-j\tau_1\omega} + \left[a_2'(j\omega) - \tau_2 a_2(j\omega)\right]e^{-j\tau_2\omega}\right), \quad (6.80)$$

$$\begin{aligned} I_0 &= \Im\left(\frac{j}{\lambda}\frac{\partial a(\lambda;\ \tau_1,\tau_2)}{\partial \lambda}\right)_{\lambda=j\omega} \\ &= \frac{1}{\omega}\Im\left(\left[a_1'(j\omega)-\tau_1 a_1(j\omega)\right]e^{-j\tau_1\omega}+\left[a_2'(j\omega)-\tau_2 a_2(j\omega)\right]e^{-j\tau_2\omega}\right), \end{aligned} \tag{6.81}$$

and

$$R_l = -\Re\left(\frac{1}{\lambda}\frac{\partial a(\lambda;\ \tau_1,\tau_2)}{\partial \tau_k}\right)_{\lambda=j\omega} = \Re\left(a_k(j\omega)e^{-j\tau_k\omega}\right), \tag{6.82}$$

$$I_l = -\Im\left(\frac{1}{\lambda}\frac{\partial a(\lambda;\ \tau_1,\tau_2)}{\partial \tau_k}\right)_{\lambda=j\omega} = \Im\left(a_k(j\omega)e^{-j\tau_k\omega}\right) \tag{6.83}$$

for $l = 1,2$. Then, since $a(\lambda;\ \tau_1,\tau_2)$ is an analytic function of λ, τ_1, and τ_2, the implicit function theorem indicates that the tangent of $\mathcal{T}^k$ can be expressed as

$$\begin{pmatrix} \frac{d\tau_1}{d\omega} \\ \frac{d\tau_2}{d\omega} \end{pmatrix} = \begin{pmatrix} R_1 & R_2 \\ I_1 & I_2 \end{pmatrix}^{-1} \begin{pmatrix} R_0 \\ I_0 \end{pmatrix} = \frac{1}{R_1 I_2 - R_2 I_1} \begin{pmatrix} R_0 I_2 - I_0 R_2 \\ I_0 R_1 - R_0 I_1 \end{pmatrix}, \tag{6.84}$$

provided that

$$R_1 I_2 - R_2 I_1 \neq 0. \tag{6.85}$$

It follows from a well-known result [48] that $\mathcal{T}^k$ is smooth everywhere except possibly at the points where either (6.85) is not satisfied, or when

$$\frac{d\tau_1}{d\omega} = \frac{d\tau_2}{d\omega} = 0. \tag{6.86}$$

A careful examination of these cases allows us to conclude with the following [132]:

Proposition 6.32. *Under the standing assumptions including* (6.70), *the curves in* $\mathcal{T}^k$ *are smooth everywhere except possibly at the degenerate points corresponding to* ω *in any one of the following three cases:*

(1) $\lambda = j\omega$ *is a multiple solution of* $a(\lambda;\ \tau_1,\tau_2) = 0$.

(2) ω *is a Type* 3 *end point of* Ω_k, *and*

$$\frac{d}{d\omega}\left(|a_1(j\omega)| + |a_2(j\omega)|\right) = 0.$$

(3) ω *is a Type* 1 *or Type* 2 *end point of* Ω_k, *and*

$$\frac{d}{d\omega}\left(|a_1(j\omega)| - |a_2(j\omega)|\right) = 0.$$

Furthermore, if the point is not among the three cases, then the tangents of the curves in $\mathcal{T}^k$ *can be expressed as*

$$\frac{d\tau_2}{d\tau_1} = \begin{cases} \frac{1/\tan\varphi_0 - 1/\tan\varphi_1}{1/\tan\varphi_0 - 1/\tan\varphi_2}, & \omega \in (\omega_k^l, \omega_k^r), \\ -\frac{|a_1(j\omega)|}{|a_2(j\omega)|}, & \omega \text{ is a Type 3 end point of } \Omega_k, \\ \frac{|a_1(j\omega)|}{|a_2(j\omega)|}, & \omega \text{ is a Type 1 or 2 end point of } \Omega_k, \end{cases} \tag{6.87}$$

where

$$\varphi_0 = \angle\left(\left[a_1'(j\omega) - \tau_1 a_1(j\omega)\right]e^{-j\tau_1\omega} + \left[a_2'(j\omega) - \tau_2 a_2(j\omega)\right]e^{-j\tau_2\omega}\right),$$
$$\varphi_k = \angle\left(a_k(j\omega)e^{-j\tau_k\omega}\right), \quad k = 1, 2.$$

Direction of crossing

Next, we will address the direction in which the solutions of (6.62) cross the imaginary axis as (τ_1, τ_2) deviates from a curve in $\mathcal{T}^k$. We will call the direction of the curve that corresponds to increasing ω the *positive direction*. Notice that, as the curve passes through the points corresponding to the end points of Ω_k, the positive direction is reversed. We will also call the region on the left-hand side, as we head in the positive direction of the curve, the *region on the left*. Again, due to the possible reversion of parametrization, the same region may be considered on the left with respect to one point of the curve, and may be considered as on the right on another point of the curve.

For the purpose of discussing the direction of crossing, we need to consider τ_1 and τ_2 as functions of $\lambda = \sigma + j\omega$, i.e., functions of two real variables σ and ω. Since the tangent of $\mathcal{T}^k$ along the positive direction is $(\partial\tau_1/\partial\omega, \partial\tau_2/\partial\omega)$, the normal to $\mathcal{T}^k$ pointing to the left-hand side of the positive direction is $(-\partial\tau_2/\partial\omega, \partial\tau_1/\partial\omega)$. Also, as a pair of complex conjugate solutions of (6.62) cross the imaginary axis to the RHP, (τ_1, τ_2) moves along the direction $(\partial\tau_1/\partial\sigma, \partial\tau_2/\partial\sigma)$. We can therefore conclude that if the inner product of these two vectors is positive, i.e.,

$$\left[\frac{\partial\tau_1}{\partial\omega}\frac{\partial\tau_2}{\partial\sigma} - \frac{\partial\tau_2}{\partial\omega}\frac{\partial\tau_1}{\partial\sigma}\right]_{\lambda=j\omega} > 0, \tag{6.88}$$

the region on the left of $\mathcal{T}^k$ at ω has two more solutions on the RHP. On the other hand, if the inequality in (6.88) is reversed, then the region on the left of $\mathcal{T}^k$ has two fewer solutions on the right-hand side of the complex plane. We can very easily express, parallel to (6.84), that,

$$\begin{pmatrix} \frac{\partial\tau_1}{\partial\sigma} \\ \frac{\partial\tau_2}{\partial\sigma} \end{pmatrix}_{\lambda=j\omega} = \begin{pmatrix} R_1 & R_2 \\ I_1 & I_2 \end{pmatrix}^{-1} \begin{pmatrix} I_0 \\ -R_0 \end{pmatrix} = \frac{1}{R_1 I_2 - R_2 I_1} \begin{pmatrix} R_0 R_2 + I_0 I_2 \\ -R_0 R_1 - I_0 I_1 \end{pmatrix},$$

where R_l and I_l, $l = 0, 1, 2$, are defined as in (6.80) to (6.83). This allows us to conclude with the following:

Proposition 6.33. *Let $\omega \in (\omega_k^l, \omega_k^r)$ and $(\tau_1, \tau_2) \in \mathcal{T}^k$ such that $j\omega$ is a simple solution of $a(j\omega;\ \tau_1, \tau_2) = 0$, and*

$$a(j\omega';\ \tau_1, \tau_2) \neq 0 \quad \textit{for any } \omega' > 0,\ \omega' \neq \omega. \tag{6.89}$$

Then as (τ_1, τ_2) moves from the region on the right to the region on the left of the corresponding curve in $\mathcal{T}^k$, a pair of solutions of (6.62) cross the imaginary axis to the right if

$$\Im(a_1(j\omega)a_2(-j\omega)e^{j\omega(\tau_2-\tau_1)}) = R_2 I_1 - R_1 I_2 > 0. \tag{6.90}$$

The crossing is in the opposite direction if the inequality is reversed.

The condition (6.89) simply means that (τ_1, τ_2) is not an intersection point of two curves or different sections of a single curve in $\mathcal{T}$. Finally, any given direction, $d =$

(d_1, d_2), with $||d||_2 = 1$, is to the left-hand side of the curve if its inner product with the left-hand side normal $(-\partial\tau_2/\partial\omega, \partial\tau_1/\partial\omega)$ is positive, i.e.,

$$-d_1\partial\tau_2/\partial\omega + d_2\partial\tau_1/\partial\omega > 0, \tag{6.91}$$

from which we have the following:

Corollary 6.34. *Let ω, τ_1, and τ_2 satisfy the same condition as Proposition 6.33. Then as (τ_1, τ_2) crosses the curve along the direction (d_1, d_2), a pair of solutions of (6.62) cross the imaginary axis to the right if*

$$d_1(R_0I_1 - I_0R_1) + d_2(R_0I_2 - I_0R_2) > 0. \tag{6.92}$$

The crossing is in the opposite direction if the inequality is reversed.

6.4.4 ▪ Computing stability radii in the delay-parameter space

Consider now the problem of computing the *stability radii* in the delay-parameter space for the particular case of quasi-polynomials including two independent delays. More precisely, the problem can be formulated as follows: given the nominal delays $\tau_{10} \in \mathbb{R}_+$, $\tau_{20} \in \mathbb{R}_+$ such that the system with the characteristic (nominal) quasi-polynomial

$$p_0(\lambda;\ \tau_{10}, \tau_{20}) = p_0(\lambda) + p_1(\lambda)e^{-\lambda\tau_{10}} + p_2(s)e^{-\lambda\tau_{20}} \tag{6.93}$$

is stable, find the maximum deviation d such that for any $\tau_1 \geq 0$, $\tau_2 \geq 0$, the system with the characteristic quasi-polynomial

$$p(\lambda;\ \tau_1, \tau_2) = p_0(\lambda) + p_1(\lambda)e^{-\lambda\tau_1} + p_2(\lambda)e^{-\lambda\tau_2} \tag{6.94}$$

is stable as long as

$$\sqrt{(\tau_1 - \tau_{10})^2 + (\tau_2 - \tau_{20})^2} < d. \tag{6.95}$$

The special case of $\tau_{10} = \tau_{20} = 0$ corresponds to the computation of the minimum delay to *destabilize* a stable nominal delay-free system. With the notations of the previous sections, an equivalent statement of the problem is to find the minimum distance between (τ_{10}, τ_{20}) and $\mathscr{T}$ (for further details, see also [133]). Since

$$\mathscr{T} = \bigcup_{k=1}^{N} \bigcup_{u=-\infty}^{\infty} \bigcup_{v=-\infty}^{\infty} \left(\mathscr{T}_{u,v}^{+k} \cup \mathscr{T}_{u,v}^{-k}\right) \cap \mathbb{R}_+^2, \tag{6.96}$$

we have

$$d = \min\left\{d_{u,v}^{+k}, d_{u,v}^{-k} \mid (u,v) \in \mathbb{Z}^2,\quad k \in \mathbb{Z}_+^*\right\}, \tag{6.97}$$

where

$$d_{u,v}^{\pm k} = \min\left\{ \sqrt{(\tau_1 - \tau_{10})^2 + (\tau_2 - \tau_{20})^2} \,\middle|\, (\tau_1, \tau_2) \in \left(\mathscr{T}_{u,v}^{\pm k} \cap \mathbb{R}_+^2\right)\right\}. \tag{6.98}$$

In what follows, we will focus on the computation of the "quantity" $d_{u,v}^{\pm k}$ as well as on bounding its corresponding range.

Calculating $d_{u,v}^{\pm k}$

If $\mathcal{T}_{u,v}^{\pm k}$ is smooth, then the minimum distance between (τ_{10}, τ_{20}) and $\mathcal{T}_{u,v}^{\pm k} \cap \mathbb{R}_+^2$ can only be reached at one of the following points:

(i) The point (τ_1, τ_2) in $\mathcal{T}_{u,v}^{\pm k} \cap \mathbb{R}_+^2$ where the tangent of $\mathcal{T}_{u,v}^{\pm k}$ is perpendicular to the vector $(\tau_1 - \tau_{10}, \tau_2 - \tau_{20})$;

(ii) The intersection of $\mathcal{T}_{u,v}^{\pm k}$ with τ_1-axis;

(iii) The intersection of $\mathcal{T}_{u,v}^{\pm k}$ with τ_2-axis;

(iv) The point (τ_1, τ_2) in $\mathcal{T}_{u,v}^{\pm k}$ corresponding to ω_k^l or ω_k^r.

The points in items (ii)–(iv) are independent of (τ_{10}, τ_{20}) and can be easily found. The crucial step in finding $d_{u,v}^{\pm k}$, therefore, is to identify points in item (i). With the expression of the tangent,

$$\frac{d\tau_2}{d\tau_1} = \frac{I_0 R_1 - R_0 I_1}{R_0 I_2 - I_0 R_2}, \tag{6.99}$$

it is easily seen that such points must satisfy

$$f(\omega) = (\tau_1 - \tau_{10})(R_0 I_2 - R_2 I_0) + (\tau_2 - \tau_{20})(R_0 I_1 - R_1 I_0) = 0, \tag{6.100}$$

with $\tau_1 \geq 0$, $\tau_2 \geq 0$. Therefore, in order to find points in case (i) above, one only needs to *sweep* ω through the interval Ω_k to identify the points corresponding to a change of sign of $f(\omega)$. A rather coarse gridding is sufficient in the first sweep. A refined gridding can be introduced near the points where f changes sign. It should be pointed out that the three degenerate cases (corresponding to the case when $\mathcal{T}^k$ is not smooth) discussed in Proposition 6.32 are automatically taken into account in the above procedure.[17]

Bounding the range of $d_{u,v}^{\pm k}$

While there are an infinite number of $d_{u,v}^{\pm k}$ (due to an infinite number of u and v), most of them can be eliminated from consideration. Let

$$\theta_{l\max}^k = \max_{\omega\in\Omega_k} \theta_l, \quad \theta_{l\min}^k = \min_{\omega\in\Omega_k} \theta_l,$$
$$\angle a_{l\max}^k = \max_{\omega\in\Omega_k} \angle a_l(j\omega), \quad \angle a_{l\min}^k = \min_{\omega\in\Omega_k} \angle a_l(j\omega).$$

Then, a bound of $\mathcal{T}_{u,v}^{+k}$ can be easily found as

$$\tau_{1\min} = \frac{\angle a_{1\min}^k + (2u-1)\pi + \theta_{1\min}^k}{\omega_k^r}, \quad \tau_{1\max} = \frac{\angle a_{1\max}^k + (2u-1)\pi + \theta_{1\max}^k}{\omega_k^l},$$
$$\tau_{2\min} = \frac{\angle a_{2\min}^k + (2v-1)\pi - \theta_{2\max}^k}{\omega_k^r}, \quad \tau_{2\max} = \frac{\angle a_{2\max}^k + (2v-1)\pi - \theta_{2\min}^k}{\omega_k^l},$$

[17] Indeed, in case (1) of Proposition 6.32, (6.100) is satisfied, and cases (2) and (3) of Proposition 6.32 belong to case (iv) mentioned above.

and a bound of $\mathcal{T}_{u,v}^{-k}$ can be found as

$$\tau_{1\min} = \frac{\angle a_{1\min}^k + (2u-1)\pi - \theta_{1\max}^k}{\omega_k^r}, \quad \tau_{1\max} = \frac{\angle a_{1\max}^k + (2u-1)\pi - \theta_{1\min}^k}{\omega_k^l},$$
$$\tau_{2\min} = \frac{\angle a_{2\min}^k + (2v-1)\pi + \theta_{2\min}^k}{\omega_k^r}, \quad \tau_{2\max} = \frac{\angle a_{2\max}^k + (2v-1)\pi + \theta_{2\max}^k}{\omega_k^l}.$$

With these bounds, we can eliminate $\mathcal{T}_{u,v}^{\pm k}$ for some u and v from consideration in searching for d. First, if $\tau_{1\max} < 0$ or $\tau_{2\max} < 0$, then $\mathcal{T}_{u,v}^{\pm k}$ does not have to be considered since it is outside of $\mathbb{R}_+^2$. Second, if we already know the distance from (τ_{10}, τ_{20}) to a point (τ_1, τ_2) in $\mathcal{T}$ is d_0, then, obviously, $d \leq d_0$. In this case, for certain $\mathcal{T}_{u_0,v_0}^{+k}$, if $\tau_{1\min} \geq \tau_{10} + d_0$, then $\mathcal{T}_{u,v}^{+k}$ for any $u \geq u_0$ can be eliminated from consideration in searching for d. Similarly, if $\tau_{2\min} \geq \tau_{20} + d_0$, then $\mathcal{T}_{u,v}^{+k}$ for $v \geq v_0$ can be eliminated from consideration. The same idea applies for $\mathcal{T}_{u,v}^{-k}$.

This process typically allows us to consider only very few u and v in searching for d, and the corresponding algorithm can be derived straightforwardly.

6.5 ▪ Delay ratio sensitivity and delay-interference phenomenon

The main objective of this section is to investigate the way in which the stability properties are affected by the ratio between the delays. More precisely, we will consider a particular stability property, the *delay-independent* stability along some ray in the delay-parameter space. As a by-product of this analysis, we will give also a characterization of the so-called delay-interference phenomena. To the best of the authors' knowledge, the notion of delay-interference[18] was first mentioned by MacDonald [222] (see also [221]), where a second-order system including two delays is shown to be subjected to delay-interference if it has the following property: delay-independent stability if the delays are equal, and delay-dependent stability with respect to each delay if the other is equal to zero. For the sake of brevity, throughout the section we will consider only delay systems of retarded type. For the details we refer the reader to [255].

Consider now the system including constant delays ($\tau_i \geq 0$, $i = 1, \ldots, m$):

$$\dot{x}(t) = A_0 x(t) + \sum_{i=1}^{m} A_i x(t - \tau_i), \tag{6.101}$$

where the matrices $A_i \in \mathbb{R}^{n\times n}$, $i = 0, \ldots, m$, are given. We start this section by studying the eigenvalue distribution of some complex matrix-valued functions, derived from the original delay system (6.101) leading to some new necessary and sufficient conditions for *delay-independent stability*. The derived criteria cover both the case where all delays vary independently of each other, and that where they are restricted to *an arbitrary* ray in the delay-parameter space, more exactly,

$$(\tau_1, \ldots, \tau_m) = \tau_0\,(r_1, \ldots, r_m), \quad r_i \in \mathbb{R}_+,\ i = 1, \ldots, m, \tag{6.102}$$

with $\tau_0 \in \mathbb{R}_+$ being the free parameter.

[18] In physics, interference represents the combination of two or more wave motions to form a resultant wave in which the displacement is reinforced or canceled.

6.5.1 ▪ Definitions, assumptions, and preliminary results

Given a direction $(r_1,\ldots,r_m)\in\mathscr{B}_+^m$ in the delay-parameter space, we define the associated *ray*, $\mathscr{T}(\vec{r})$, as follows:

Definition 6.35. *For* $\vec{r}\in\mathscr{B}_+^m$, *let* $\mathscr{T}(\vec{r}):=\{\tau_0\,\vec{r}:\ \tau_0\in\mathbb{R}_+\}$.

As we address delay-independent stability properties for the case where all delays vary independently of each other, and that where they are restricted to a particular ray, from now on we will consequently use the following terminology in order to avoid confusion:

Definition 6.36. *The system* (6.101) *is delay-independent stable if and only if its zero solution is asymptotically stable for all* $\vec{\tau}\in\mathbb{R}_+^m$.

Definition 6.37. *The ray* $\mathscr{T}(\vec{r})$ *is stable if and only if the zero solution of* (6.101) *is asymptotically stable for all* $\vec{\tau}\in\mathscr{T}(\vec{r})$.

Regarding delay-interference, we have:

Definition 6.38. *A stable ray* $\mathscr{T}(\vec{r})$ *is subject to the delay-interference phenomenon if and only if for all* $\epsilon>0$ *there exists a* $\vec{s}\in\mathscr{B}_+^m$ *with* $\|\vec{r}-\vec{s}\|<\epsilon$ *such that the ray* $\mathscr{T}(\vec{s})$ *is not stable.*

Without any loss of generality, the following technical assumption (excluding characteristic roots at the origin) will be made throughout this section:

Assumption 6.39.

$$\det\left(A_0+\sum_{i=1}^{m}A_i\right)\neq 0.$$

Introductory example

Consider the following scalar system including two distinct delays taken from [86]:

$$\dot{x}(t)=-x(t)-x(t-\tau_1)-\frac{1}{2}x(t-\tau_2), \tag{6.103}$$

where $(\tau_1,\tau_2)\in\mathbb{R}_+\times\mathbb{R}_+$. The geometric approach proposed in the previous section for the computation of the stability crossing curves applies to the system (6.103). In this case, the crossing set Ω is given by all the frequencies $\omega\geq 0$ such that $2/3\leq|a(j\omega)|\leq 2$ with $a(\lambda)=1/(\lambda+1)$. More precisely, $\Omega=[0,\sqrt{5}/2]$, where $\tilde{\omega}=\sqrt{5}/2$ is the unique positive solution of $|a(j\tilde{\omega})|=2/3$, and the stability crossing curves characterization follows.

Consider now the analysis of the asymptotic stability on some particular ray $(\tau_1,\tau_2)=\tau(1,2)$, $\tau\geq 0$, in the delay-parameter space, and on some of its perturbations $(\tau_1,\tau_2)=\tau(1,2+\varepsilon)$, for some $\varepsilon>0$, but sufficiently small. In this case, if $(\tau_1,\tau_2)=\tau(1,2)$, the system (6.103) is rewritten as follows:

$$\dot{x}(t)=-x(t)-x(t-\tau)-\frac{1}{2}x(t-2\tau), \tag{6.104}$$

where τ is constant. It is easy to see that the system (6.104) is *delay-independent* asymptotically stable (by using, for instance, the elimination technique). Furthermore, the same property holds if $\tau_1=0$ and $\tau_2\neq 0$ or, if $\tau_1\neq 0$ and $\tau_2=0$, that is, both delay axes are also *stable rays.* Since the system (6.103) is *not* asymptotically stable for *all positive delays* τ_1 and τ_2, the natural question is whether *such a ray will intersect some stability crossing curve in the domain.* This is indeed the case. As shown by Datko [86], there exists a sequence $\{\varepsilon_n\}_{n\geq 1}$, where $\varepsilon_n = \frac{1}{2(2n+1)}$, such that the ray $(\tau_1, \tau_2) = \tau(1, 2+\varepsilon)$ with $\varepsilon = \varepsilon_n$ is not stable. More precisely, for some delay values $\tau > 2(2n+1)\pi$, the system becomes unstable on the ray corresponding to $\epsilon = \varepsilon_n$. Hence, the ray $\tau(1,2)$ is subject to the so-called *delay-interference* phenomenon as specified by Definition 6.38.

Properties of some associated matrix-valued functions

Consider now the following functions:

$$L_1 : [0,\ 2\pi]^m \mapsto \mathbb{C}^{n\times n} \text{ given by } L_1(\vec{\theta}) := L_1(\theta_1, \ldots, \theta_m) = A_0 + \sum_{i=1}^{m} A_i e^{-j\theta_i},$$

$$L_2 : \mathbb{R}_+ \times \mathbb{R}_+^m \mapsto \mathbb{C}^{n\times n} \text{ given by } L_2(\theta, \vec{r}) := L_2(\theta, r_1, \ldots, r_m) = A_0 + \sum_{i=1}^{m} A_i e^{-j\theta r_i}.$$

The following sets and quantities will play a major role in the characterization of delay-independent stability and the interference phenomenon:

Definition 6.40. *Let*

$$\mathbb{W} = \bigcup_{\vec{\theta}\in[0,\ 2\pi]^m} \sigma\left(L_1(\vec{\theta})\right), \qquad \alpha_0 = \sup\{\Re(\lambda) :\ \lambda \in \mathbb{W}\},$$

and for $\vec{r} \in \mathscr{B}_+^m$, *let*

$$\mathbb{V}(\vec{r}) = \bigcup_{\theta\geq 0} \sigma(L_2(\theta, \vec{r})), \qquad \alpha(\vec{r}) = \sup\{\Re(\lambda) :\ \lambda \in \mathbb{V}(\vec{r})\}.$$

It directly follows that

$$\forall \vec{r} \in \mathbb{B}_+^m, \qquad \overline{\mathbb{V}(\vec{r})} \subseteq \mathbb{W}, \qquad \alpha(\vec{r}) \leq \alpha_0. \tag{6.105}$$

As we shall see, the relations (6.105) can be strict ($\subset$, resp., $<$). However, if the components of $\vec{r}$ are rationally independent,[19] then the next proposition applies:

Proposition 6.41. *If the components of* $\vec{r}$ *are rationally independent, then*

$$\overline{\mathbb{V}(\vec{r})} = \mathbb{W}, \qquad \alpha(\vec{r}) = \alpha_0.$$

Sketch of the proof. The proof idea can be resumed as follows: since $\mathbb{V}(\vec{r}) \subseteq \mathbb{W}$ and $\mathbb{W}$ is closed, it remains to prove for the first statement that every element of W can be approximated arbitrarily well with elements from $\mathbb{V}(\vec{r})$. Such a conclusion follows by applying Kronecker's theorem [155, Theorem 444], combined with a Rouché-type result (Theorem A1 in [242]). A complete proof can be found in [255].

Next, with $\mathscr{B}_+^m$ equipped with the Euclidean norm and $\mathscr{P}(\mathbb{C})$ with the Hausdorff metric, we address the continuity of the functions $\mathbb{V}(\cdot)$ and $\alpha(\cdot)$; see Appendix A.3 for

[19] See Appendix A.4 for definitions regarding the interdependency of numbers.

precise definitions. As illustrated in what follows, these functions are in general **not** continuous at each point; however, the following weaker property holds (see [255] for the proof and Appendix A.3 for the semicontinuity definition):

Proposition 6.42. *The function* $\mathbb{V}: \mathscr{B}_+^m \to \mathscr{P}(\mathbb{C}),\ \vec{r} \mapsto \mathbb{V}(\vec{r})$ *is lower semicontinuous at each* $\vec{r} \in \mathscr{B}_+^m$.

Corollary 6.43. *The function* $\alpha: \mathscr{B}_+^m \to \mathbb{R},\ \vec{r} \mapsto \alpha(\vec{r})$ *is lower semicontinuous at each* $\vec{r} \in \mathscr{B}_+^m$.

By combining Propositions 6.41 and 6.42 with relations (6.105) we arrive at a stronger result at rationally independent $\vec{r}$:

Proposition 6.44. *The function* $\mathbb{V}: \mathscr{B}_+^m \to \mathscr{P}(\mathbb{C}),\ \vec{r} \mapsto \mathbb{V}(\vec{r})$ *is continuous at rationally independent* $\vec{r}$. *The function* $\alpha: \mathscr{B}_+^m \to \mathbb{R},\ \vec{r} \mapsto \alpha(\vec{r})$ *is continuous at rationally independent* $\vec{r}$.

We illustrate the above concepts and properties with the equation

$$\dot{x}(t) = -1.3x(t) - x(t-\tau_1) - \frac{1}{2}x(t-\tau_2), \tag{6.106}$$

an example that slightly modifies (6.103). We have

$$\mathbb{W} = \bigcup_{(\theta_1,\theta_2)\in[0,\,2\pi]^2} \left(-1.3 - e^{-j\theta_1} - \frac{1}{2}e^{-j\theta_2}\right) = \left\{\lambda \in \mathbb{C}:\ \frac{1}{2} \le |\lambda + 1.3| \le \frac{3}{2}\right\},$$

$$\mathbb{V}(\vec{r}) = \bigcup_{\theta \ge 0}\left(-1.3 - e^{-jr_1\theta} - \frac{1}{2}e^{-jr_2\theta}\right).$$

In Figure 6.6 we have shown the set $\mathbb{V}(\vec{r})$ for some values of $\vec{r}$ as well as the boundaries of $\mathbb{W}$. When the components of $\vec{r}$ are commensurate, $\mathbb{V}(\vec{r})$ forms a closed curve. In all the cases displayed we have $\alpha(\vec{r}) < \alpha_0$ and $\overline{\mathbb{V}(\vec{r})} \subset \mathbb{W}$ (strict relations). *Discontinuities* of the function $\vec{r} \mapsto \mathbb{V}(\vec{r})$ occur at all rationally dependent $\vec{r}$. To illustrate the mechanism, let us first compare cases (a) and (f) of Figure 6.6. Ratio $r_2/r_1 = 1.9$ can be seen as a perturbation of $r_2/r_1 = 2$ and, although the components of $\vec{r}$ are still commensurate, they are "more independent" since the coprime numbers 19 and 10 (having ratio 1.9) are larger than 2 and 1 (having ratio 2). As a consequence, the curve

$$\theta \ge 0 \mapsto -1.3 - e^{-j\theta} - \frac{1}{2}e^{-j\frac{r_2}{r_1}\theta} \tag{6.107}$$

with $r_2/r_1 = 1.9$ only closes at $\theta = 20\pi$ when θ is increased from zero (instead of $\theta = 2\pi$ for $r_2/r_1 = 2$), and a "larger portion" of $\mathbb{W}$ is covered by $\mathbb{V}(\vec{r})$. This is clearly visible in subplot (f), where we have also depicted the values of (6.107) for $\theta \in [0,\,2\pi]$ (bold). Next, if $r_2/r_1 = 2$ were perturbed instead to an irrational value, such as $2 - \pi/n, n \ge 2$, then the corresponding curve $\mathbb{V}(\vec{r})$ would never close and its points would densely fill $\mathbb{W}$, as follows from Proposition 6.41. Since n can be taken arbitrarily large, the perturbation can be taken arbitrarily small.

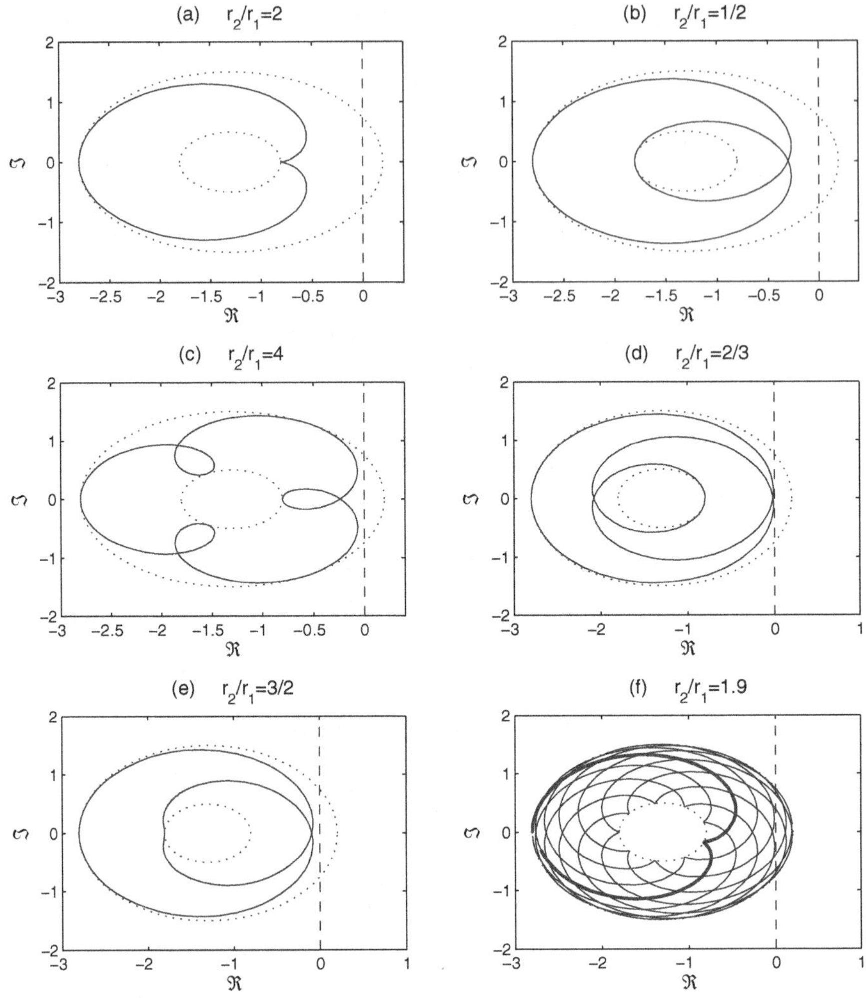

Figure 6.6. *The set* $\mathbb{V}(\vec{r})$ *for the system* (6.106), *for different values of* $\vec{r}$ *(solid curves). The dotted curves are the boundaries of* $\mathbb{W}$ [255].

6.5.2 ▪ Stability and delay-interference characterizations

Necessary and sufficient conditions for delay-independent stability are expressed by the following theorem (see [255] for a complete proof, where both implications are proven by contradiction):

Theorem 6.45. *The system* (6.101) *is delay-independent stable if and only if*

$$\mathbb{W} \subset \left(\mathbb{C}_{-} \cup \{0\}\right). \tag{6.108}$$

Remark 6.46. *For a delay-independent stable system, the condition* $0 \notin \mathbb{W}$, *respectively* $0 \in \mathbb{W}$, *is referred to in the literature as* strong, *respectively* weak, *delay-independent stability (see* [28, 290] *and the references therein), notions which are related to stability properties of some associated 2-D system.*

The stability of a ray $\mathcal{T}(\vec{r})$ can be characterized in an analogous way to Theorem 6.45 and it is omitted:

Proposition 6.47. *The ray $\mathcal{T}(\vec{r})$ is stable if and only if*

$$\mathbb{V}(\vec{r}) \subset \left(\mathbb{C}_- \cup \{0\}\right). \tag{6.109}$$

The next proposition states that if the system (6.101) is not delay-independent stable, then it is always prone to the delay-interference phenomenon:

Proposition 6.48. *Assume that $\mathbb{W} \not\subset \left(\mathbb{C}_- \cup \{0\}\right)$. If a ray $\mathcal{T}(\vec{r})$ is stable, then it is subject to the delay-interference phenomenon.*

Proof. From the assumption $\mathbb{W} \not\subset \left(\mathbb{C}_- \cup \{0\}\right)$ it follows that (6.101) is not delay-independent stable. So there exist delay values $\vec{\tau} \in \mathbb{R}_+^m$ and a frequency $\omega > 0$ such that $H(j\omega;\vec{\tau}) = 0$. For any given $\vec{k} \in \mathbb{N}^m$, $\vec{k} \neq 0$, this implies that $H(j\omega;\vec{\tau}(n)) = 0$ for all $n \in \mathbb{N}$, where

$$\vec{\tau}(n) = \left(\tau_1 + k_1 n \frac{2\pi}{\omega}, \ldots, \tau_m + k_m n \frac{2\pi}{\omega}\right). \tag{6.110}$$

Therefore, the ray $\mathcal{T}\left(\frac{\vec{\tau}(n)}{\|\vec{\tau}(n)\|}\right)$ is unstable for all $n \in \mathbb{N}$. From $\lim_{n\to\infty} \vec{\tau}(n)/\|\vec{\tau}(n)\| = \vec{k}/\|\vec{k}\|$ and the fact that this analysis can be repeated for every nonzero $\vec{k} \in \mathbb{N}^m$, it follows that every ray $\mathcal{T}(\vec{r})$ where $\vec{r}$ has *commensurate* components becomes unstable when applying certain infinitesimal perturbations to $\vec{r}$. Now, if the components of $\vec{r}$ are noncommensurate, then there exist infinitesimal $\delta\vec{r}$ such that $\vec{r} + \delta\vec{r}$ has commensurate components ($\mathbb{Q}^m$ is dense in $\mathbb{R}^m$), and the above arguments can be repeated. This completes the proof. □

Combining the previous results for the case where $\mathbb{W}$ has a nonempty intersection with the open right half plane yields:

Theorem 6.49. *Assume that $\mathbb{W} \cap \mathbb{C}_+ \neq \phi$. Then the following hold:*

1. *If the components of $\vec{r}$ are rationally independent, then the ray $\mathcal{T}(\vec{r})$ is unstable.*

2. *If the ray $\mathcal{T}(\vec{r})$ is stable, then it is subject to the delay-interference phenomenon.*

3. *The set $\left\{\vec{r} \in \mathcal{B}_+^m : \mathcal{T}(\vec{r}) \text{ stable}\right\}$ is nowhere dense in $\mathcal{B}_+^m$.*

Proof. If the components of $\vec{r}$ are rationally independent, then $\overline{\mathbb{V}(\vec{r})} = \mathbb{W}$ by Proposition 6.41. A combination with the assumption of the theorem leads to $\mathbb{V}(\vec{r}) \cap \mathbb{C}_+ \neq \phi$. Applying Proposition 6.47 then yields the first assertion. The second assertion follows from Proposition 6.48, whose proof mainly relies on the invariance property (6.110). Finally, the third assertion follows from Propositions 6.41–6.42 and Proposition 6.47. □

Under the assumption of the previous theorem, we comment on the detection of *stable* rays $\mathcal{T}(\vec{r})$ (if any) *without* the explicit computation of stability/instability regions of (6.101) in the delay-parameter space. From Theorem 6.49, the components of $\vec{r}$ cannot be rationally independent. By Proposition 6.47, we have to search for values for which $\mathbb{V}(\vec{r}) \subset \left(\mathbb{C}_- \cup \{0\}\right)$, whereas we have $\mathbb{W} \cap \mathbb{C}_+ \neq \phi$ and $\mathbb{V}(\vec{r}) \subseteq \mathbb{W}$. Thus, values of $\vec{r}$, for

Table 6.1. *Values of $\alpha(\vec{r})$, where $\vec{r} = (n_1, n_2)/\sqrt{n_1^2 + n_2^2}$, computed for the system* (6.106) [255].

$\alpha(\vec{r})$	0	1	2	3	4	5	6	7	8	$\cdots$ n_1
0		−0.8								
1	−1.8	0.20	−0.27	0.20	0.06	0.20	0.13	0.20	0.16	
2		−0.55		−0.01		0.11		0.15		
3		0.20	−0.08		0.08	0.20		0.20	0.16	
4		−0.06		0.06		0.13		0.16		
5		0.20	0.05	0.20	0.11		0.15	0.20	0.17	
6		0.07				0.14		0.16		
7		0.20	0.11	0.20	0.14	0.20	0.16		0.17	
8		0.13		0.14		0.16		0.17		
$\vdots$										
n_2										

which $\mathbb{V}(\vec{r})$ "doesn't cover $\mathbb{W}$ very well," are good candidates. Following from Propositions 6.41 and 6.42, such values must be characterized by a "large dependence," for instance, $\vec{r} = \vec{n}/||\vec{n}||$, with $\vec{n} \in \mathbb{N}^m$ and $||\vec{n}||$ small. To fix the ideas, we pick up the example (6.106), which satisfies $\mathbb{W} \cap \mathbb{C}_+ \neq \phi$ since $\alpha_0 = 0.2$. Table 6.1 displays the corresponding values of $\alpha(\vec{r})$ for rationally dependent $\vec{r}$. It becomes apparent that stable rays correspond to

$$(n_1, n_2) \in \{(1,0),\ (0,1),\ (1,2),\ (1,4),\ (2,1),\ (2,3),\ (3,2)\}. \tag{6.111}$$

Notice that for five of these cases the set $\mathbb{V}(\vec{r})$ is depicted in Figure 6.6(a)–(e).

Finally, we look at the special case where $\mathbb{W} \not\subset (\mathbb{C}_- \cup \{0\})$ but still lies in the *closed* left half plane. Inspired by the proof of Lemma 2.2 in [151], and using an approximation and a continuation argument, we have the following result (see, for instance, [255] for a complete proof):

Lemma 6.50. *Assume that* (6.101) *has a characteristic root in $\mathbb{C}_+$ for some delay values. Then* $\mathbb{W} \cap \mathbb{C}_+ \neq \phi$.

Finally, the combination of Theorem 6.45, Proposition 6.48, and Lemma 6.50 results in:

Theorem 6.51. *Assume that $\mathbb{W} \cap \mathbb{C}_+ = \phi$ and $\mathbb{W} \not\subset (\mathbb{C}_- \cup \{0\})$. Then there exist delay values for which* (6.101) *has characteristic roots on the imaginary axis, while there are no delay values for which* (6.101) *has characteristic roots in $\mathbb{C}_+$. Every stable ray is subjected to the delay-interference phenomenon.*

The main results of this section are summarized in Table 6.2. The cases $\mathbb{W} \cap \mathbb{C}_+ \neq \phi$ and $\mathbb{W} \subset (\mathbb{C}_- \cup \{0\})$ with $0 \notin \mathbb{W}$ are generic. The other cases, where the rightmost elements of $\mathbb{W}$ are on the imaginary axis, characterize situations where a system is on the edge of either losing or acquiring delay-independent stability. Recall that this analysis has been performed under Assumption 6.39, which can be checked a priori (if the assumption is not satisfied, then there exist no values of $\vec{\tau}$ for which the zero solution of (6.101) is asymptotically stable).

To conclude, we briefly discuss some existing results related to the interference phenomenon. In [216] it is shown that delay-independent stability is equivalent to

Table 6.2. *Characterization of stability of the steady state solution of* (6.101) *along rays in the delay-parameter space, as a function of* $\mathbb{W}$ *and* $\mathbb{V}(\cdot)$ [255].

<table>
<tr><th>$\mathbb{W}\subset(\mathbb{C}_-\cup\{0\})$</th><th>$\mathbb{W}\not\subset(\mathbb{C}_-\cup\{0\})$ and $\mathbb{W}\cap\mathbb{C}_+=\phi$</th><th>$\mathbb{W}\cap\mathbb{C}_+\neq\phi$</th></tr>
<tr><td rowspan="3">Delay-independent stable:
$0\notin\mathbb{W}$: strong
$0\in\mathbb{W}$: weak</td><td colspan="2">Not delay-independent stable;
ray $\mathcal{T}(\vec{r})$ stable if and only if $\mathbb{V}(\vec{r})\subset(\mathbb{C}_-\cup\{0\})$</td></tr>
<tr><td colspan="2">Stable ray (if any) subject to the delay-interference phenomenon</td></tr>
<tr><td>Characteristic roots in $\mathbb{C}_+$ not possible</td><td>Ray $\mathcal{T}(\vec{r})$ unstable for $\vec{r}$ rationally independent</td></tr>
</table>

asymptotic stability for all delay values lying in a nontrivial sector in the delay-parameter space, and to the robustness of stability of a ray, consisting of commensurate delay values, w.r.t. small perturbations of the direction (see [216] for precise formulations). Note that the latter two statements imply the existence of a stable ray, which is not subject to the interference phenomenon. Given a stable ray, the additional conditions of having interference of [222] (delay-dependent stability when one of the (two) delays is set to zero) and of [292] (frequency-sweeping test) are in fact conditions for the presence of characteristic roots on the imaginary axis for some delay values, and thus conditions for *not* having delay-independent stability. In this way, the above cited results are a direct corollary of the fact that *either* a system is delay-independent stable *or* every stable ray is subject to the delay-interference phenomenon; see Table 6.2.

6.5.3 ▪ Illustrative example

Consider the system

$$\dot{x}(t)=A_0x(t)+A_1x(t-\tau_1)+A_2x(t-\tau_2), \tag{6.112}$$

where

$$A_0=\begin{bmatrix} 0 & 1 \\ a-a^2-\frac{5}{4} & 2a-1 \end{bmatrix},\quad A_1=\begin{bmatrix} 0 & 0 \\ \frac{1}{5} & 0 \end{bmatrix},\quad A_2=\begin{bmatrix} 0 & 0 \\ -\frac{4}{5} & 0 \end{bmatrix}, \tag{6.113}$$

and $a\in\mathbb{R}$ is a parameter. The corresponding set $\mathbb{W}$ is depicted in Figure 6.7. A change of parameter a results in a shift of this set along the real axis. For $a<0$ the system is strongly delay-independent stable, and for $a>0$ we have $\mathbb{W}\cap\mathbb{C}_+\neq\phi$; see Table 6.2. For the intermediate value, $a=0$, we are in the special case where the assumptions of Theorem 6.51 are satisfied.

In Figure 6.8, we plot the stability region in the (τ_1,τ_2)-space for $a=1/16$ and $a=0$. In the first case there are three stable rays, determined by $\vec{r}=\vec{n}/\|\vec{n}\|$, with $\vec{n}\in\{(1,0),(1,1),(4,1)\}$. These directions can also be obtained directly by constructing a table similar to Table 6.1. As $a\rightarrow 0+$, all closed curves in the (τ_1,τ_2)-space, corresponding to characteristic roots on the imaginary axis, shrink, and at the limit $a=0$, where Theorem 6.51 applies, they have collapsed to equally spaced points. For such delay values, which can be computed analytically as

$$(\tau_1,\tau_2)=\left(\frac{10\pi}{3\sqrt{3}}+k\frac{4\pi}{\sqrt{3}},\frac{4\pi}{3\sqrt{3}}+l\frac{4\pi}{\sqrt{3}}\right),\quad k,l\in\mathbb{N}, \tag{6.114}$$

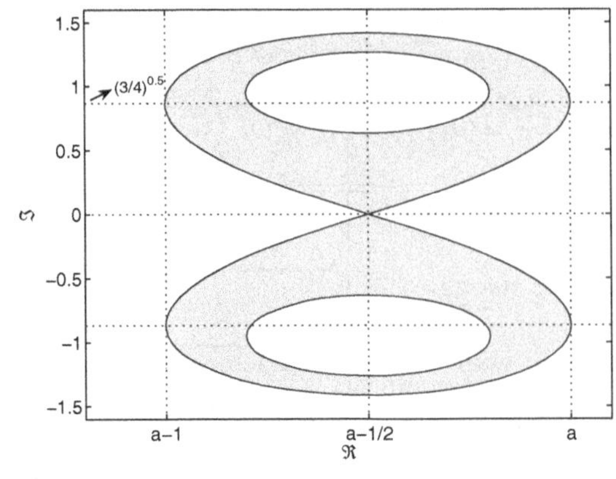

Figure 6.7. *The set* $\mathbb{W}$*, corresponding to the system* (6.112)–(6.113)*, is shaded* [255].

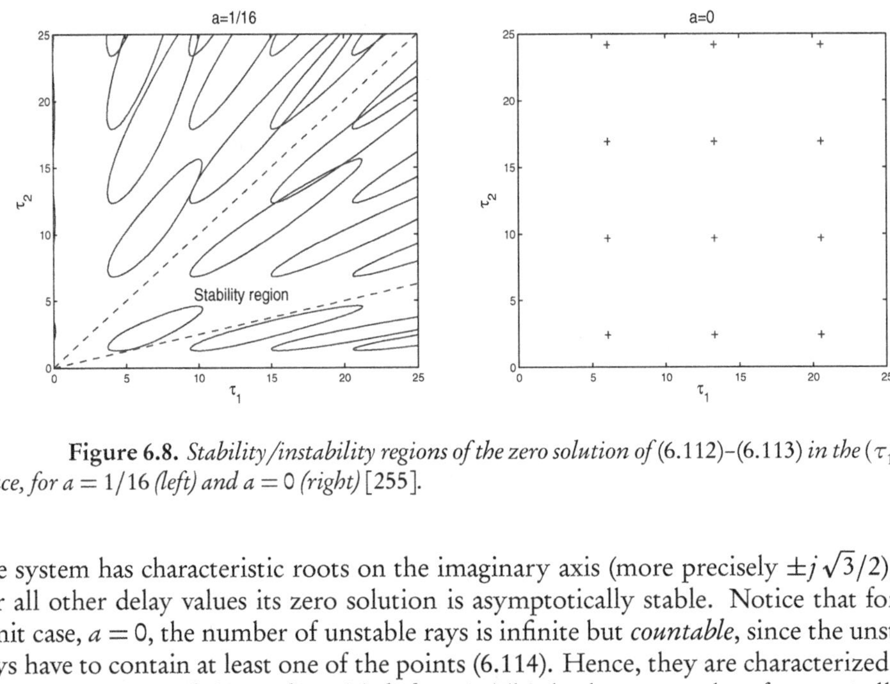

Figure 6.8. *Stability/instability regions of the zero solution of* (6.112)–(6.113) *in the* (τ_1, τ_2)*-space, for* $a = 1/16$ *(left) and* $a = 0$ *(right)* [255].

the system has characteristic roots on the imaginary axis (more precisely $\pm j\sqrt{3}/2$), but for all other delay values its zero solution is asymptotically stable. Notice that for the limit case, $a = 0$, the number of unstable rays is infinite but *countable*, since the unstable rays have to contain at least one of the points (6.114). Hence, they are characterized by a slope $r_2/r_1 = (2+6l)/(5+6k)$ (with $k, l \in \mathbb{N}$). This leads to a paradox: for $a = 0$ all rays $\mathcal{T}(\vec{r})$, with r_2/r_1 irrational, are stable, whereas for any $a > 0$, they are unstable (since $\mathbb{W} \cap \mathbb{C}_+ \neq \phi$ and Theorem 6.49 applies). The explanation is as follows: for all rationally independent $\vec{r}$ and every value of a, we have for the example $\alpha(\vec{r}) = \alpha_0, \Re(\lambda) < \alpha_0$ for all $\lambda \in \mathbb{V}(\vec{r})$. Consequently, $\mathbb{V}(\vec{r})$ intersects the closed right half plane if $\alpha_0 > 0$ but does not if $\alpha_0 = 0$.

6.6 ▪ Notes and references

This chapter addressed the stability problem in the delay-parameter space. More precisely, we discussed and presented various algebraic tests for characterizing the crossing existence together with corresponding delay intervals in the case of linear systems in state-space representation and including commensurate delays. We also presented the geometry of the stability crossing curves for some quasi-polynomials including two independent delays.

A deeper analysis of delay ratio sensitivity as well as delay-interference phenomena completes the presentation.

Among the results we are proposing, it is worth mentioning that Proposition 6.4 (invariance principle for some double and non–semisimple root) was presented in a different form in [178] (see also [207] for the characterization of the invariance principle in the general case). Next, the analysis of the crossing roots and the corresponding stability characterization of linear systems with commensurate delays presented here follows along the lines of [296] with some complements taken from [290, 130]. It is important to point out that, to the best of our knowledge, in the systems and control literature the interpretation of the characteristic equation of a linear system with commensurate delays as an equation including two variables—one on the imaginary axis and one on the unit circle—goes back to the work of Kamen [185, 186] in the 1980s in the context of *delay-independent* (asymptotic) *stability*. Such an idea was largely exploited in the literature in the 1980s (quasi-polynomials) and in the 1990s (state-space representation). A more extensive overview of such results can be found in [290, 130]. The bilinear transformation-based approaches can be found in [290, 130]. For the 2-D polynomial approach, we mention the test proposed by [47, 76]. The pseudodelay technique goes back to the work of Rekasius [331] and Thowsen [370, 371] (delay-independent analysis). The complete characterization of the commensurate delays case can be found in [313], and several extensions of the method for dealing with neutral and multiple delays can be found in [348].

Proposition 6.9 can be found in [296] and represents a natural extension of the root crossing characterization proposed by [81, 79] to the state-space representation of the delay system. The characterization of the crossing direction in the case of semisimple characteristic roots is taken from [115]. The particular cases covering the characterizations of hyperbolicity and of delay-independent stability can be found in [290] and have been revisited in [296]. The rank one delayed matrix presentation is new and will be useful in the stability analysis of some models from the biosciences in the last chapter. A thorough discussion on frequency-sweeping tests can be found in [71] (see also [130]). The stability crossing characterization of quasi-polynomials including one delay was considered first in [81] and [402].

The connection between pseudodelays and the computation of generalized eigenvalues of some appropriate matrix pencils is new and is inspired by some characterization proposed in the context of distributed delays in [282]. Such an idea will be adopted later to the case of delayed output feedback stabilization of SISO systems. The extension of the analysis to systems described by delay-differential algebraic equations follows closely the approach proposed in [297].

Next, a geometric approach was used for the characterization of the stability crossing curves in the delay-parameter space of quasi-polynomials with two delays. The presentation follows closely the approach proposed by [132]. More precisely, we proved that the frequency crossing set can be expressed by three constraints and consists of a finite number of intervals of finite length. Except for in a few degenerate cases, the stability crossing curves are smooth. These curves may be closed, open ended, or spiral-like with the axis in the horizontal, vertical, or diagonal direction. The classification of the curves is determined by the constraints which are violated at the ends points of the frequency crossing set. Next, smoothness properties of the stability crossing curves were investigated and a characterization of the crossing direction was made. We used an approach similar to the one described in Chapter 11 of [93] based on the implicit function theorem. The discussions concerning the computation of the stability radius in the delay-parameter space can be found in [133]. This geometric approach will be particularly exploited in the next chapters to give some insight into the case of a Smith predictor subject to delay uncertainty, the analysis of some immune dynamics models in chronic leukemia, and the delayed output feedback stabilization problem of SISO systems.

Finally, the characterization of the delay-interference phenomenon is based on [255] and Chapter 5 of [256]. In summary, this phenomenon corresponds to the situation where a system with multiple delays is not delay-independent stable if all delays are considered as free parameters, but it is delay-independent stable along particular directions (rays) in the delay space.

Part II

Stabilization and robust fixed-order control

Chapter 7

Stabilization using a direct eigenvalue optimization approach

7.1 ▪ Fixed-order control design

In this chapter and in Chapter 9 we address the design of controllers for linear time-delay systems, which are characterized by a fixed order or a fixed structure, depending on a finite number of parameters $p = (p_1, \dots, p_{n_p})$. This class of systems includes classical static and dynamic state and output feedback controllers, PI and PID controllers, as well as various kinds of prediction-based controllers.

Since time-delay systems are inherently infinite-dimensional and the number of controller parameters is finite, the control design problem is difficult, exhibiting hard limitations on stabilizability, performance, and robustness. In this book we take a direct optimization approach, where stability, robustness, and performance criteria are optimized as a function of the controller or design parameters. Conceptually, these methods can be seen as extensions to infinite-dimensional systems of the recently proposed fixed-order control design approach for LTI system at the basis of the package HIFOO ($\mathscr{H}_\infty - \mathscr{H}_2$ Fixed Order Optimization; see, e.g., [143, 138]).

For reasons of clarity we first treat, in the next sections, the stabilization problem for the case where the closed-loop system can be brought into the form

$$\dot{x}(t) = \sum_{j=0}^{m} A_j(p)\, x(t - \tau_j), \tag{7.1}$$

where $x(t) \in \mathbb{R}^n$ is state, $A_j \in \mathbb{R}^{n \times n}$, $j = 0, \dots, m$, are the system matrices which smoothly depend on the parameters $p = (p_1, p_2, \dots, p_{n_p}) \in \mathbb{R}^{n_p}$, and $\tau_j \geq 0$, $j = 0, \dots, m$, are fixed delays, with $\tau_0 = 0$. In Section 7.4 we treat the extension towards delay-differential algebraic equations (DDAEs), which can be used to model a much broader class of systems and controllers. The method for $\mathscr{H}_\infty$ optimization proposed in Chapter 9 also supports a system description by DDAEs.

The stabilization problem for (7.1) can be recast as the problem of finding parameter values p for which the spectral abscissa

$$c(p) := \max\left\{ \Re(\lambda) :\ \det\left(\lambda I - A_0(p) - \sum_{j=1}^{m} A_j(p) e^{-\lambda \tau_j} \right) = 0 \right\} \tag{7.2}$$

is strictly negative. The stabilization approach consists of directly solving the optimization problem

$$\min_{p} c(p). \tag{7.3}$$

Recall that the system is stabilizable if either $c(p)$ is unbounded from below or a global minimizer p^* exists for which $\alpha(p^*) < 0$. In the latter case, the parameters p^* result in an "optimal spectrum," where the characteristic roots are pushed to the left as far as possible and the trajectories consequently follow a maximal asymptotic decay rate to the steady state. Note that, from a purely stabilization point of view, it is not necessarily to go all the way to the minimum, since it suffices to halt the minimization procedure of α once it is strictly negative.

Because the spectral abscissa is a nonsmooth function, standard optimization methods cannot be used to solve (7.3). Instead, we combine BFGS with a so-called bundle gradient method, specifically the *gradient sampling algorithm* of [58], an optimization method that is able to find minima of general nonsmooth, nonconvex objective functions. In [56, 58] the latter algorithm was successfully used for designing stabilizing low-order static output feedback controllers. The combination of both algorithms is also included in the package HIFOO mentioned above.

7.2 ▪ Stabilization via eigenvalue optimization

We examine some smoothness properties of the function $c(p)$ as defined in (7.2) and explain why it is hard, if not impossible, to optimize with standard optimization tools. Next, we discuss the recently developed gradient sampling algorithm that is able to find local minima for functions satisfying such smoothness properties. Finally, we apply this algorithm to the optimization of the spectral abscissa function (7.2) and briefly discuss the extension towards classes of nonlinear time-delay systems.

7.2.1 ▪ Smoothness properties of the spectral abscissa function

A first important property of the spectral abscissa $c(p)$ regarding its optimization is that it is nonconvex, and therefore may have many local minima. Consider, as a constructed example, the system

$$\ddot{x}(t)+p^2x(t)-\epsilon p^2x(t-1)=0, \tag{7.4}$$

where p is the variable parameter and ϵ is a small, fixed value.

Figure 7.1 shows the real part of the rightmost characteristic roots versus p for $\epsilon = 0.1$ (left) and $\epsilon = 0.01$ (right). It is clear that for this example the spectral abscissa has several local minima. Moreover, by taking ϵ sufficiently small, one can even make the number of minima arbitrarily large, since the function $c(p)/\epsilon$ uniformly converges on a compact interval to $-\frac{1}{2}p\sin(p)$ if ϵ tends to zero. It is clear that with such a behavior the global minimum is very hard to find, and most optimization algorithms will converge to a local minimum, without being able to give any guarantee about whether or not this is also the global minimum.

But even if we are satisfied with finding a local minimum, standard optimization algorithms will still fail, because the spectral abscissa $c(p)$ is also a *nonsmooth* function. Indeed, despite the fact that isolated characteristic roots behave smoothly with respect to changes of the parameters, the spectral abscissa does not. A first reason for this is the presence of the maximum operator appearing in (7.2). If for certain parameter values there is more than one *active* characteristic root, i.e., several characteristic roots satisfy $\Re(\lambda) = c(p)$, it is easy to see that at this point the function $c(p)$ is most likely not differentiable.

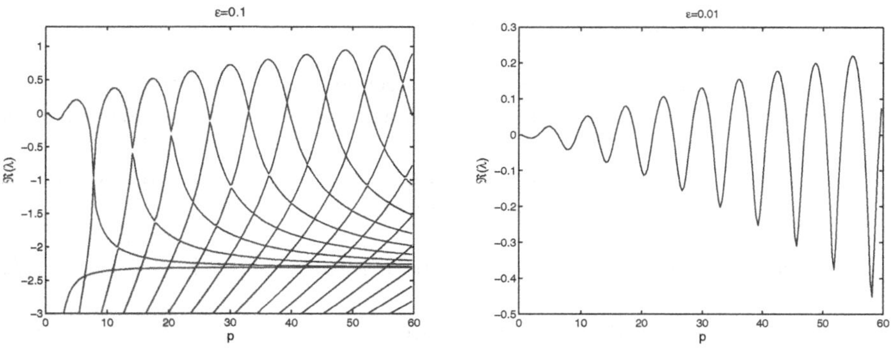

Figure 7.1. *The real part of the rightmost characteristic roots of system* (7.4) *as a function of* p, *for* $\epsilon = 0.1$ *(left) and* $\epsilon = 0.01$ *(right)* [387].

A special situation occurs when an active characteristic root has an algebraic multiplicity larger than the geometric multiplicity (i.e., it is a multiple, non–semisimple root). Then, $c(p)$ is typically not locally Lipschitz even. This is, for example, the case at parameter values where a complex pair of characteristic roots is about to split up into two real characteristic roots, or vice versa. As another example, consider the one-dimensional system $\frac{d^3x}{dt^3}(t) = px(t)$, which yields the spectral abscissa function

$$c(p) = \begin{cases} \sqrt[3]{p}, & p \geq 0, \\ \sqrt[3]{-p}/2, & p < 0. \end{cases} \tag{7.5}$$

Here, α indeed exhibits a non-Lipschitz point at $p = 0$, corresponding to a triple characteristic root $\lambda^\star = 0$. The parameter value that yields this triple root is also the minimizer of the spectral abscissa function. This is actually not a coincidence, as it is a common observation that the solution of a fixed-order stabilization problem occurs at a characteristic root with higher multiplicity; see, for instance, [54].

For ordinary differential equation systems, the number of active characteristic roots (multiplicity taken into account) cannot be larger than the dimension of the system. Typical for the delay case is that, since the number of characteristic roots is infinite, the number of active characteristic roots and even their multiplicity *can* be larger than the system's dimension n. This is, for example, illustrated with the case study of Chapter 8.

7.2.2 ▪ The gradient sampling algorithm

For practical nonsmooth functions, such as the spectral abscissa function, nondifferentiable points will in general only occur on a set with measure zero, meaning that such functions are differentiable *almost everywhere*. Indeed, for randomly selected parameter values p, the gradient of $c(p)$ exists with probability one. This property is exploited by the gradient sampling algorithm, an optimization algorithm recently developed by Burke, Lewis, and Overton that uses a bundle gradient strategy to find a local minimum of a nonsmooth objective function [58]. It consists of the following main steps:

ALGORITHM 7.1. Gradient sampling algorithm for nonsmooth optimization.

Input A function $\phi(p)$, continuous and differentiable almost everywhere, and its gradient, $\vec{\nabla}\phi(p)$ (whenever it exists)

Step 0 Initialize a starting value p_0 arbitrarily.

Step 1 Compute the nonsmooth steepest descent direction at p_k, using a gradient sampling approximation. *If* the norm of this direction is very small, *then* stop (succeed).

Step 2 Perform a line search along the direction computed in the previous step to determine a step size that has a lower function value.

Step 3 *If* Step 2 succeeded, update p_k to p_{k+1} and go back to Step 1, *else* stop (fail).

Essentially, the algorithm has the same outline as the classical steepest descent method, except for using the *nonsmooth steepest descent direction* as search direction. This direction is defined as

$$\arg\max_{\|d\|\leq 1} \min_{z\in\partial_c\phi(p_k)} \langle -z, d\rangle, \tag{7.6}$$

where $\langle\cdot,\cdot\rangle$ is the standard Euclidean inner product, and $\partial_c\phi(p_k)$ denotes the *Clarke subdifferential* (or generalized gradient) at p_k, given by

$$\partial_c\phi(p_k) := \text{conv}\left\{\lim_{p\to p_k} \vec{\nabla}\phi(p) : p \in N\right\}. \tag{7.7}$$

Here, "conv" denotes the convex hull and N is any full-measure subset of a neighborhood around p_k containing differentiable points.

To understand why the nonsmooth descent direction is used in the gradient sampling algorithm, we refer the reader to Figures 7.2–7.3, where some typical contour lines of a nonsmooth function, depending on the configuration of the intersection of smooth manifolds, are plotted. The dashed vectors represent the negative gradients of these different manifolds in a certain nondifferentiable point, and their convex hull can be considered as the "negative" of the Clarke subdifferential. A full vector denotes the nonsmooth steepest descent direction. As seen in the left frame of Figure 7.2, the nonsmooth steepest descent direction lies exactly along the negative gradient when there is no nonsmoothness (the two vectors are drawn slightly shifted for visibility reasons). The other two frames show configurations where two, respectively three, manifolds meet, but without their gradients being in conflict. In both cases, the line of nonsmooth points will simply be traversed and the descent can be proceeded on the lowest manifold without difficulty.

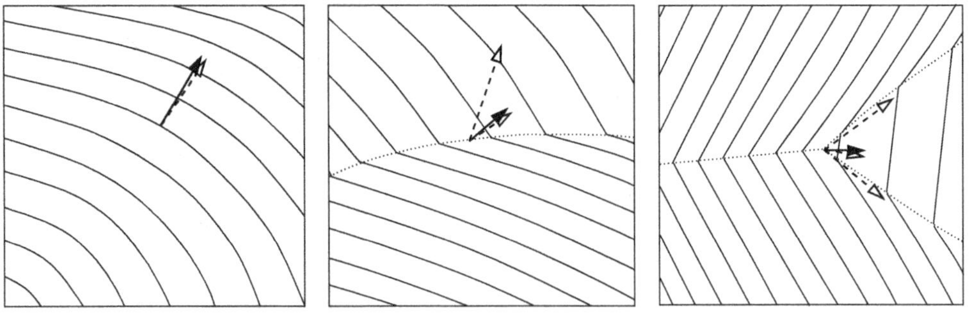

Figure 7.2. *Contours of a nonsmooth function with typical configurations of meeting manifolds. The dotted line consists of nondifferentiable points. The dashed vectors are the negative gradients of the respective manifolds and a full vector shows the nonsmooth steepest descent direction* [387].

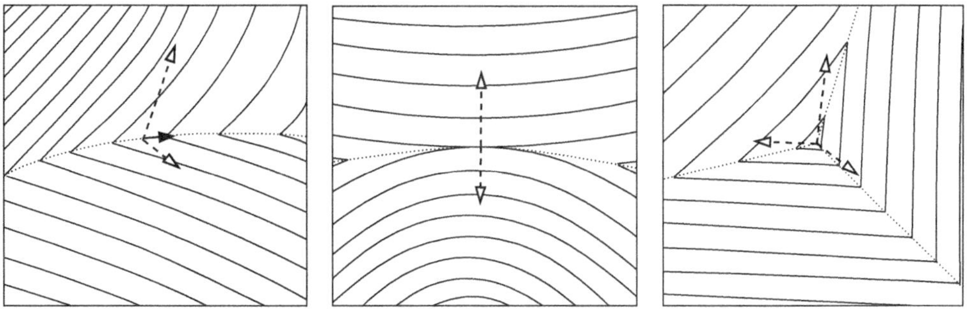

Figure 7.3. *(Left) two conflicting manifolds resulting in a nonsmooth steepest direction lying along the intersection. (Right) two possible configurations in an optimum* [387].

In the left frame of Figure 7.3, another situation occurs. Here, two manifolds come together forming a kind of ridge. A standard optimization algorithm such as the steepest descent method will jump back and forth across this ridge, taking smaller and smaller steps, and eventually become jammed. The gradient sampling algorithm, using the nonsmooth steepest descent direction, will behave quite differently. According to (7.6) the nonsmooth steepest descent direction is in the opposite direction of the vector for which the minimum of the inner product with the subgradients is maximal, which corresponds to the vector in the negative Clarke subdifferential with smallest norm. In case of two intersecting manifolds with conflicting gradients as in Figure 7.3 (left), this vector is tangent to the ridge containing the nonsmooth points. Indeed, in that case the largest decay of the objective function can be achieved by a decrease along the ridge. In the remaining two frames of Figure 7.3, two possible configurations for a nonsmooth local minimum are depicted. It is observed that in such a local minimum the Clarke subdifferential *contains the zero vector*, and hence (7.6) is equal to zero. Note that all points for which (7.6) is zero are *Clarke stationary* [58].

Figure 7.4 shows a typical evolution of the gradient sampling algorithm, compared with the behavior of the classical steepest descent algorithm. It is seen that the latter, denoted with a dashed line, indeed stagnates upon reaching a ridge of nonsmooth points, whereas the full line representing the gradient sampling algorithm, by following the nonsmooth steepest descent direction, is able to proceed until a nonsmooth local minimum is found.

Because the nonsmooth steepest descent direction, to be determined in Step 1 of Algorithm 7.1, is in practice difficult to compute exactly, an approximation is constructed using the following computational procedure:

ALGORITHM 7.2. Computation of nonsmooth steepest descent direction.

Step 1a Sample a number of gradients in a small neighborhood around the current point.

Step 1b Collect these gradients into a bundle, serving as an approximation for the Clarke subdifferential.

Step 1c Compute the vector with smallest norm out of this bundle by solving a quadratic program.

Assume that the current iteration point is p_k. Besides the gradient $\vec{\nabla}\phi(p_k)$, additional gradients are computed in a number of random points sampled in a small neighborhood

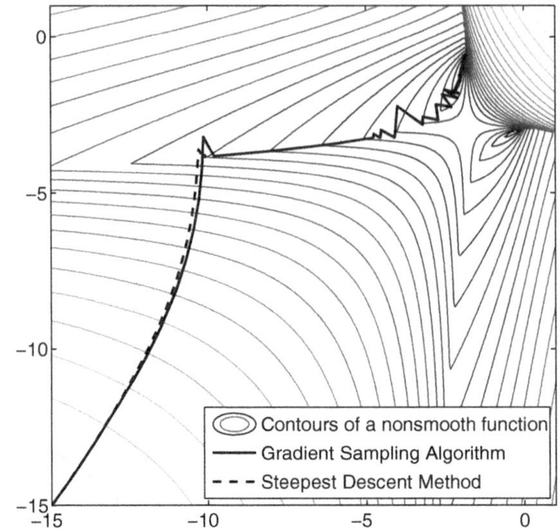

Figure 7.4. *Comparison of the behavior of the classical steepest descent method (dashed line) and the gradient sampling algorithm (solid line) on a typical nonsmooth function* [387].

around p_k. This neighborhood is typically chosen to be the ball with center at p_k and radius ϵ, where ϵ is some small number. If one of the sampled points should happen to be nondifferentiable, it is simply neglected and resampled. Along with the gradient in p_k, these sampled gradients are then regarded as a *bundle* of gradients, and it was shown in [55] that the closure of the convex hull of this bundle can serve as a good approximation to the Clarke subdifferential as long as a sufficiently large amount of samples is taken. The final direction for the line search is then taken as the negative of the vector out of this convex hull that has the smallest norm. This vector can easily be computed by solving a quadratic program [58].

The fact that (7.6) is zero in a (Clarke) stationary point can be used as an indication to decide when to stop the gradient sampling algorithm. In particular, if the norm of the computed search direction is smaller than a threshold value, this point is assumed to be a local minimum (due to the nature of a saddle point and the sampling step in the algorithm, a breakdown in a saddle point is not expected). A critical value for the accuracy of this approximate minimum is the sampling radius ϵ that was used in the computation of the nonsmooth steepest descent. Because it is very hard to determine an appropriate value for this ϵ beforehand, the gradient sampling algorithm is usually conducted several times, beginning with a relatively large ϵ, and repeatedly using a smaller sampling radius, each time starting off at the end point of the previous run. For more practical implementation aspects of the gradient sampling algorithm, and a convergence proof, we refer the reader to [58].

In summary, the gradient sampling procedure can find local minima of a nonsmooth function, under the assumption that it is differentiable almost everywhere. The algorithm relies on an evaluation of the objective function and the corresponding gradient at designated points.

In order to accelerate the optimization procedure, one can use the BFGS method with weak Wolfe line search first (see [206] for the favorable behavior of BFGS applied to nonsmooth problems), and only switch to gradient sampling on the moment of stagnation. This two-stage approach is implemented in our code, as well as in the package HIFOO for fixed-order control of delay free systems.

7.2.3 ▪ Application to linear time-delay systems

As the spectral abscissa function is continuous, as well as differentiable almost everywhere (see Section 7.2.1), Algorithm 7.1 can be applied for its minimization. The algorithm requires the computation of the objective function and its gradient.

The objective function, $c(p)$, can be evaluated using the procedure for determining the rightmost characteristic roots, described in Chapter 2.

The gradient, $\vec{\nabla} c$, can be computed analytically. Indeed, let u and v be the respective left and right null vector of the matrix $\Lambda(\lambda^\star; p)$, with $\lambda^\star$ an isolated characteristic root. Differentiating the equation $\Lambda(\lambda^\star; p)v = 0$ results, after some calculus, in the following formula for the gradient of α:

$$\vec{\nabla} c(p) = \Re\left(\frac{u^*\left(\frac{\partial A_0}{\partial p} + \sum_{j=1}^m e^{-\lambda^\star \tau_j} \frac{\partial A_j}{\partial p}\right)v}{u^*\left(I + \sum_{j=1}^m \tau_j e^{-\lambda^\star \tau_j} A_j\right)v} \right). \tag{7.8}$$

However, tests indicate that in general this formula is not always the most robust choice. Alternatively, the gradient $\vec{\nabla} c$ can be approximated by a finite difference formula as follows. The number of parameters in system (7.1), n_p, equals the dimension of the tangent plane at $c(p)$. Thus, this plane—or, equivalently, the gradient $\vec{\nabla} c(p)$—can be constructed by using values $\alpha(p^i)$ at n_p+1 points p^i close to p. In our implementation, the points p^i are selected as the n_p+1 vertices of a regular polytope in the space of n_p free parameters with center of mass p and radius ε. For our experiments we used $\varepsilon = 10^{-7}$. Thus, in principle, one solves n_p+1 independent eigenvalue problems. However, after computing the active characteristic root $\lambda^\star$ at the center of the polytope p, the active characteristic roots at the vertices p^i can efficiently be computed using Newton's method with starting value $\lambda^\star$.

Finally, we note that the number of optimization parameters n_p affects the number of samples one should take in each iteration to ensure a good approximation of the nonsmooth steepest descent direction, and thus the total number of function evaluations. Practice proved that the double is a safe choice.

7.2.4 ▪ Extension to nonlinear time-delay systems

There are two important issues when adapting the approach of the previous section to a nonlinear delay-differential system of the form

$$\dot{x}(t) = f(x(t), x(t-\tau_1), \ldots, x(t-\tau_m); p), \tag{7.9}$$

where $f(\cdot)$ is continuously differentiable. First, the constant $x(t) \equiv 0$ is in general no stationary solution of (7.9). Moreover, the existence and uniqueness of steady states for fixed system parameters no longer hold; e.g., if a steady state solution $x^\star \equiv x(t)$ exists for certain system parameters, then a nearby point in the parameter space can have a steady state "close" to $x^\star$ or there can be no steady state at all. Hence, the spectral abscissa α depends not only on the parameters, but also on the considered steady state $x^\star$. This implies that for fixed parameter values, the spectral abscissa is not uniquely defined. The investigation of *branches* of solutions of a nonlinear system for varying parameters is called *bifurcation analysis*; see, e.g., [342]. The second difference is the meaning of the objective function that is optimized. Consider the linearization of (7.9) about a particular steady state. One can prove that the characteristic roots of this linearization determine the *local stability* of the original nonlinear system in a neighborhood of this steady state. Hence by optimizing the spectral abscissa of the linearization, we optimize the reaction of the system to small perturbations.

The consequences on the computational procedure are the following. Most importantly, the value of the steady state must be corrected after each change in the system's parameters. For this purpose, we employ Newton's method using the previous solution as initial guess. If for the current point the value of α is negative or zero, then the optimization algorithm will not leave the current branch. Indeed, the values of α along the optimization path decrease monotonically, and, in a bifurcation point, there is at least one characteristic root on the imaginary axis, and hence the value of α is positive or zero. Additional branches that exist for the same parameter range should therefore be independently explored if no stabilizing parameters were found on the initial branch. A second consequence is that, since α also depends on the steady state $x^\star$, we have to adapt formula (7.8) by including a term for the partial derivative w.r.t. $x^\star$. When computing the gradient by using finite differences, the steady state $x^\star$ must be corrected by Newton's method for all vertices p^i of the polytope used in the finite difference formula.

7.3 ▪ Numerical examples

We present two numerical examples.

7.3.1 ▪ Model problem

Consider the system

$$\dot{x}(t)=Ax(t)+Bu(t-\tau), \tag{7.10}$$

with

$$A=\begin{bmatrix} -0.08 & -0.03 & 0.2 \\ 0.2 & -0.04 & -0.005 \\ -0.06 & -0.2 & -0.07 \end{bmatrix},\ B=\begin{bmatrix} -0.1 \\ -0.2 \\ 0.1 \end{bmatrix},\ \tau=5, \tag{7.11}$$

which we control by static state feedback,

$$u(t)=[p_1\ p_2\ p_2]\,x(t).$$

Applying the gradient sampling algorithm yields (nearly) identical numerical values for the optimal parameters, $p=\begin{bmatrix} 0.472 & 0.505 & 0.603 \end{bmatrix}$. Particularities of the gradient sampling approach are the monotonic decrease of the spectral abscissa function with the iteration number of the algorithm, as seen in Figure 7.5, and the fact that the procedure is fully automatic.

7.3.2 ▪ Semiconductor laser

We investigate the stabilization of a system consisting of a delay-differential equation coupled to a partial differential equation. The system models a semiconductor laser subject to conventional optical feedback and lateral carrier diffusion [390]. The system variables are the complex scalar variable $E(t)$, representing the electric field, and the real variable $N(x,t)$, representing the carrier density in the interval $x\in[-0.5,0.5]$, reads as

$$\dot{E}(t)=(1-i\beta)E(t)\zeta(t)+\eta E(t-\tau)e^{-i\phi}-ibE(t), \tag{7.12}$$

$$\begin{aligned} T\frac{\partial N(x,t)}{\partial t}&=d\frac{\partial^2 N(x,t)}{\partial x^2}-N(x,t)+P(x)\\ &\quad -F(x)(1+2N(x,t))|E(t)|^2, \end{aligned} \tag{7.13}$$

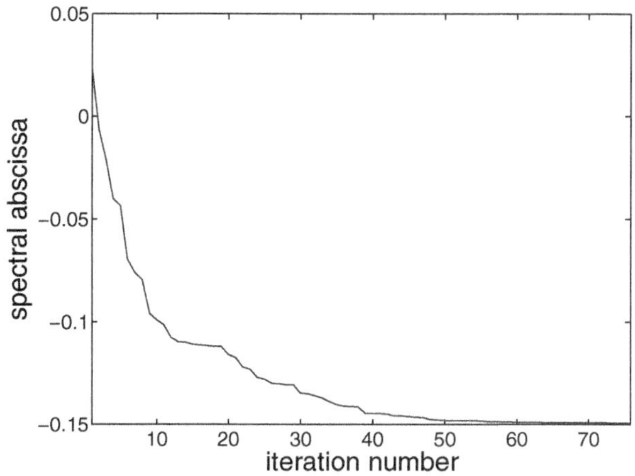

Figure 7.5. *Evolution of the spectral abscissa during the optimization process* [387].

where $\zeta(t)$ is a weighted average of the carrier density $N(x,t)$, specifically,

$$\zeta(t) = \frac{\int_{-0.5}^{0.5} F(x)N(x,t)\,\mathrm{d}x}{\int_{-\infty}^{\infty} F(x)\,\mathrm{d}x}. \tag{7.14}$$

The functions $P(x)$ and $F(x)$ are specified in [390]. We split (7.12) into real and imaginary parts in order to work only with real numbers. The symmetry of the spatial domain about $x = 0$ is exploited by considering only the interval $[0, 0.5]$ and imposing zero Neumann boundary conditions at $x = 0$, i.e., $\partial N(0,t)/\partial x = 0$. We also use zero Neumann boundary conditions at $x = 0.5$.

The partial differential system (7.12)–(7.13) cannot be treated directly by our method. It is first transformed to a system of differential equations of the form (7.9). For this purpose, we use a standard finite difference discretization in space, in particular a second-order central difference formula with constant stepsize $\Delta x = 0.5/128$. Moreover, the integrals in (7.14) are approximated using the trapezoidal quadrature rule. This transformation gives a time-delay system of size 131 in the unknowns $\Re(E(t))$, $\Im(E(t))$, and $N(x_j, t)$ for $x_j := j\Delta x \in [0, 0.5]$, with $j = 0, \ldots, 128$. Note that this transformation introduces a discretization error. However, since the characteristic roots determining the linear stability after space discretization are computed up to a desired accuracy (due to the Newton corrections), the computed rightmost characteristic roots correspond to the original system (7.12)–(7.13) if the error of the *spatial* discretization is sufficiently small, that is, if the system's properties are captured sufficiently well by the spatial discretization. The quality of the characteristic roots is confirmed by our experiments using different numbers of mesh intervals in space.

Due to the form of system (7.12)–(7.13), every solution (E, N) belongs to a family of solutions of the form (cE, N), where c is an arbitrary complex number on the unit circle, i.e., $|c| = 1$. It is said that the system is *rotationally symmetric* in E. The tangent at this continuum of solutions (cE, N) is the eigenvector corresponding to a characteristic root at zero. This characteristic root at zero would not occur if one particular solution were selected from the continuum of solutions by imposing a so-called pinning condition or phase condition. It can therefore be safely ignored. For this reason, our algorithm removes the computed characteristic root that is zero, up to rounding error, before the active (pair of) characteristic root(s) is selected.

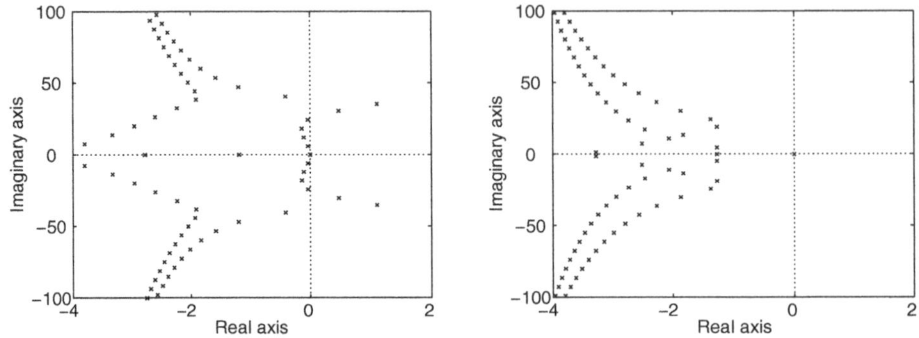

Figure 7.6. *The rightmost characteristic roots corresponding to the starting point of the optimization (left) and the obtained minimum of the spectral abscissa function (right). The characteristic root at zero is induced by the rotational symmetry of the system* [387].

We fix the parameters to the values $\beta = 3$, $\phi = 0$, $T = 1000$ and delay $\tau = 1000$; and we optimize w.r.t. parameters η and d, representing the feedback strength and diffusion coefficient, respectively. To increase the numerical stability of the computations, the time variable and the time delay are rescaled by a factor of $1/1000$, and the parameters η and d are rescaled by a factor of 1000. Recall that a nonlinear system can have multiple steady state solutions for a given set of parameters; see Section 7.2.4. In this case, one can show that for η between 0 and 7×10^{-3} and arbitrary d there are four branches that contain *stable* steady state solutions. This is illustrated in [390, Figure 3].

The starting point of the optimization algorithm is an unstable steady state with $\eta = 6.4051 \times 10^{-3}$ and $d = 1.68 \times 10^{-2}$. This initial steady state lies on one of the branches mentioned above. The algorithm returned a local minimum $\alpha^\star = -1.2713$ on this branch at $\eta^\star \approx 1.9926 \times 10^{-3}$ and $d^\star \approx 2.3203 \times 10^{-2}$. Figure 7.6 shows the rightmost characteristic roots for the starting point of the optimization (left) and for the computed local minimum (right).

Figure 7.7 shows the computed local minimum and some contours of the spectral abscissa α (spaced 0.02 units apart) in the parameter space η–d. It also shows the path followed by the optimization algorithm (dashed line and tick markers) and the points of nondifferentiability (dotted lines). In this figure one can clearly see how the optimization algorithm tends to follow these lines of nonsmoothness to converge to the local minimum.

7.4 ▪ Systems described by delay-differential algebraic equations

We now consider equations of the form

$$E\dot{x}(t) = A_0(p)x(t) + \sum_{i=1}^{m} A_i(p)x(t-\tau_i), \tag{7.15}$$

where E is possibly singular. The flexibility of the system description (7.15) has been discussed in Section 1.3. Summarizing, a broad class of linear time-delay systems and controllers can be brought into this form by simply augmenting the equations describing systems, controllers, and interconnections. Hence, elimination of inputs and outputs, which is not always possible in the presence of delays, is not necessary, and the closed-loop matrices remain linearly dependent on the controller parameters. Moreover, neutral-type systems also can be addressed in this framework. For more advantages in the presence of

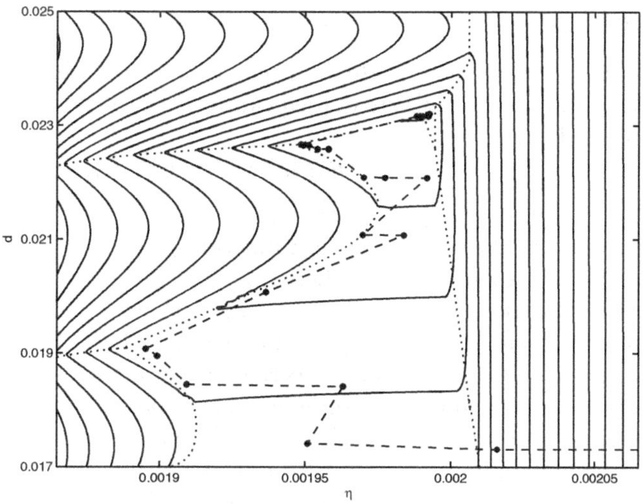

Figure 7.7. *Contour lines of the spectral abscissa function α w.r.t. parameters η and d about the optimum $\approx (1.9926\times 10^{-3}, 2.3203\times 10^{-2})$, the path followed by the optimization algorithm (dashed line and tick markers), and the points of nondifferentiability (dotted lines)* [387].

extra inputs and outputs to characterize performance and robustness, we refer the reader to Chapter 9.

7.4.1 ▪ An optimization point of view

As before, exponential stability of the null solution of (7.15) can be imposed by finding values of p for which the spectral abscissa is strictly negative. However, we know from Chapter 1 that stability may not be robust against arbitrarily small delay perturbations. If the achieved stability is needed to be robust against small delay perturbations, the requirement of a negative spectral abscissa must be strengthened to the negativeness of the robust spectral abscissa. This brings us to the optimization problem

$$\inf_p C(\vec{\tau};\, p). \tag{7.16}$$

Strongly stabilizing values of p exist if the objective function can be made strictly negative. By Proposition 1.54 the latter can be evaluated as

$$C_D(\vec{\tau};\, p) = \max(c(\vec{\tau};\, p), C_D(\vec{\tau};\, p)). \tag{7.17}$$

An alternative approach consists of solving the constrained optimization problem

$$\inf_p c(\vec{\tau};\, p) \quad \text{subject to } \gamma_0(p) < \gamma, \tag{7.18}$$

where $\gamma < 1$. If the objective function is strictly negative, then the satisfaction of the constraint implies strong stability.

The advantage of solving problem (7.16) is that for all values of p a bound on the exponential growth rate of the solutions is ensured, which takes into account delay perturbations. The advantage of solving (7.18) instead is that evaluating $\gamma_0(p)$ is less computationally demanding than evaluating $C_D(\vec{\tau};\, p)$; see Remark 2.15.

7.4.2 ▪ Algorithms

We start with problem (7.16). The spectrum of the DDAE (7.15) behaves similarly to the spectrum of a delay-differential equation of retarded type whenever the parameters are restricted to the set

$$\{p \in \mathbb{R}^{n_p} : c(\vec{\tau};\, p) > C_D(\vec{\tau};\, p)\}. \tag{7.19}$$

Also the spectral properties for retarded systems carry over whenever (7.19) is satisfied. In particular, the spectral abscissa function $p \mapsto c(\vec{\tau};\, p)$ may not be everywhere differentiable, and even not everywhere Lipschitz continuous. A lack of differentiability may occur when there is more than one *active* characteristic root, i.e., a characteristic root whose real part equals the spectral abscissa. A lack of Lipschitz continuity may occur when an active characteristic roots is multiple and non–semisimple. On the contrary, the spectral abscissa function is differentiable at points where there is only one active characteristic root with multiplicity one. Since this is the case with probability one when randomly sampling parameters from the set (7.19), the spectral abscissa is smooth almost everywhere. The function f in Proposition 1.51 is a maximum eigenvalue function, similarly to the spectral abscissa function. Therefore, the above properties also hold for the function

$$p \mapsto C_D(\vec{\tau};\, p),$$

and they are preserved by the maximum operator in (7.17). We conclude that the objective function in (7.16), the robust spectral abscissa, is not everywhere differentiable, and not everywhere Lipschitz continuous, but it is differentiable almost everywhere.

The properties of the problem (7.16) preclude the use of standard optimization methods developed for smooth problems. Instead, similarly to the retarded case, we use a combination of BFGS with weak Wolfe line search and gradient sampling. The overall algorithm only requires the evaluation of the objective function, as well as its derivatives with respect to the controller parameters, *whenever* it is differentiable.

When the robust spectral abscissa is differentiable we can express

$$\frac{\partial C}{\partial p_k}(\vec{\tau};\, p) = \begin{cases} \frac{\partial c}{\partial p_k}(\vec{\tau};\, p), & c(\vec{\tau};\, p) > C_D(\vec{\tau};\, p), \\ \frac{\partial C_D}{\partial p_k}(\vec{\tau};\, p), & c(\vec{\tau};\, p) < C_D(\vec{\tau};\, p) \end{cases}$$

for $k = 1, \ldots, n_p$. In case $c > C_D$, the derivative of the robust spectral abscissa is inferred from the sensitivity of an individual characteristic root. More precisely, we can express

$$\frac{\partial C}{\partial p_k}(\vec{\tau};\, p) = \Re\left(\frac{w^* \left(\frac{\partial A_0}{\partial p_k}(p) + \sum_{i=1}^{m} \frac{\partial A_i}{\partial p_k}(p) e^{-\lambda \tau_i} \right) z}{w^* \left(E + \sum_{i=1}^{m} A_i \tau_i e^{-\lambda \tau_i} \right) z} \right), \tag{7.20}$$

where the tuple (λ, w, z) satisfies

$$w^* \Delta(\lambda) = 0, \quad \Delta(\lambda) z = 0, \quad z \neq 0, \quad w \neq 0,$$

and λ corresponds to the rightmost characteristic roots. If $c < C_D$, then the robust spectral abscissa is differentiable in the generic case, where for $\zeta = C_D(\vec{\tau};\, p)$ the maximum in (2.61) is isolated. We can then express

$$\frac{\partial C}{\partial p_k}(\vec{\tau};\, p) = \frac{\Re\left(\bar{\lambda} u^* \left(\frac{\partial H_1}{\partial p_k}(p) e^{-\zeta \tau_1} + \sum_{i=2}^{m} \frac{\partial H_i}{\partial p_k}(p) e^{-\zeta \tau_i} e^{j\theta_i} \right) v \right)}{\Re\left(\bar{\lambda} u^* \left(H_1(p) \tau_1 e^{-\zeta \tau_1} + \sum_{i=2}^{m} H_i(p) \tau_i e^{-\zeta \tau_i} e^{j\theta_i} \right) v \right)}, \tag{7.21}$$

where $(\zeta, \lambda, u, v, \vec{\theta})$ refers to the corresponding solution of (2.62). An alternative to the analytic formulae (7.20) and (7.21) consists of computing the derivatives by finite differences.

Finally, we come back to the constrained problem (7.18). It can be solved using the barrier method proposed in [396], which is in its turn inspired by interior point methods for solving convex optimization problems; see, e.g., [38]. The first step consists of finding a feasible point, i.e., a set of values for p satisfying the constraint. If the feasible set is nonempty such a point can be found by solving

$$\min_{p} \gamma_0(p). \tag{7.22}$$

Once a feasible point $p = p_0$ has been obtained, one can solve in the next step the unconstrained optimization problem

$$\min_{p} \{c(p) - r \log(\gamma - \gamma_0(p))\}, \tag{7.23}$$

where $r > 0$ is a small number and γ satisfies

$$\gamma_0(p) < \gamma \leq 1.$$

The second term (the barrier) ensures that the feasible set cannot be left when the objective function is decreased in a quasi-continuous way (because the objective function will go to infinity when $\gamma_0 \to \gamma$). If (7.23), with $\gamma = 1$, is repeatedly solved for decreasing values of r and with the previous solution as starting value, a solution of (7.18) is obtained. Strong *exponential* stability can be imposed by setting γ strictly smaller than one.

In our implementation we use HANSO [317] for solving both (7.22) and (7.23). Derivatives of γ_0 exist for almost all values of p and can be computed using the formula

$$\frac{\partial \gamma_0(p)}{\partial p_k} = \frac{1}{|\lambda|} \Re \left(\bar{\lambda} u^* \left(\frac{\partial H_1}{\partial p_k}(p) + \sum_{i=2}^{m} \frac{\partial H_i}{\partial p_k}(p) e^{j\theta_i} \right) v \right), \quad k = 1, \ldots, n_p,$$

with the quadruple $(\lambda, u, v, \vec{\theta})$ solving (2.62) (where ζ is set to zero and one equation is dropped; see Remark 2.15). Note that the objective function in (7.23) still contains an implicit constraint in the sense that it is defined only on the feasible set of (7.18). This problem can be solved by explicitly setting the objective function to infinity outside the feasible set [396].

7.4.3 ▪ Numerical examples

We illustrate the approach with several case studies. A user-friendly MATLAB implementation of the control design algorithms, as well as all data corresponding to the presented examples, are publicly available from

http://twr.cs.kuleuven.be/research/software/delay-control/stab/

A stabilization problem with input delay

As a first example we reconsider system (7.10)–(7.11). The uncontrolled system is unstable, characterized by the spectral abscissa $c = 0.108$. We design a dynamic controller of the form

$$\begin{cases} \dot{x}_c(t) = A_c x_c(t) + B_c y(t), \\ u(t) = C_c x_c(t) + D_c y(t), \quad x_c(t) \in \mathbb{R}^{n_c}, \end{cases} \tag{7.24}$$

Table 7.1. *Results of minimizing the spectral abscissa of* (7.10) *and* (7.24) [241].

Controller	Minimum c
Output feedback, $n_c = 0$	-0.1489
Output feedback, $n_c = 1$	-0.2293
Output feedback, $n_c = 2$	-0.2682
Output feedback, $n_c = 3$	-0.4575

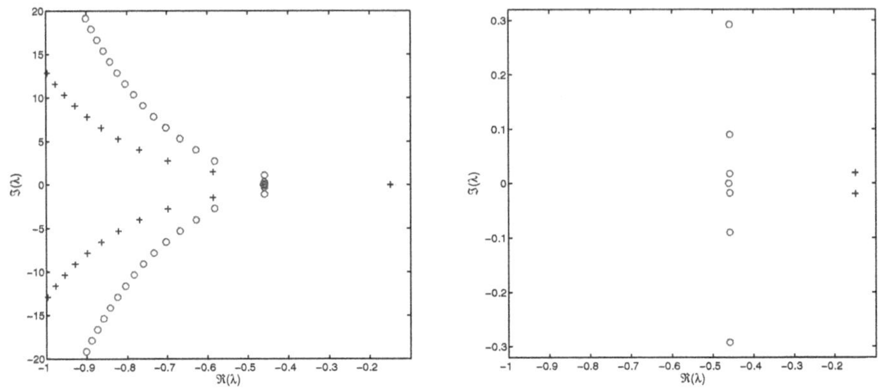

Figure 7.8. *Characteristic roots of* (7.10) *and* (7.24), *corresponding to a minimum of the spectral abscissa, for $n_c = 0$ (pluses) and $n_c = 3$ (circles). The left and right panes correspond to different scalings of the imaginary axis* [241].

where we set

$$p = \text{vec}\left(\begin{bmatrix} A_c & B_c \\ C_c & D_c \end{bmatrix}\right).$$

Since the transfer function from u to y is strictly proper, the robust spectral abscissa equals the spectral abscissa, and the optimization problems (7.16) and (7.18) reduce to the (unconstrained) minimization of the spectral abscissa. The results for different controller orders are displayed in Table 7.1. The case $n_c = 0$ refers to static feedback, already treated. In Figure 7.8 the rightmost characteristic roots of the closed-loop system are shown for $n_c = 0$ and $n_c = 3$.

For the second example we assume that the measured output of system (7.10)–(7.11) is instead given by

$$\tilde{y}(t) = x(t) + \begin{bmatrix} 3 \\ 4 \\ 1 \end{bmatrix} u(t-2.5) + \begin{bmatrix} 2/5 \\ -2/5 \\ -2/5 \end{bmatrix} u(t-5), \tag{7.25}$$

and we design a static controller,

$$u(t) = D_c \tilde{y}(t). \tag{7.26}$$

In this case there is a high-frequency feedthrough term in the control loop. Solving the optimization problem (7.16) leads to

$$C = -0.0309, \qquad D_c = [0.0409\ \ 0.0612\ \ 0.3837]. \tag{7.27}$$

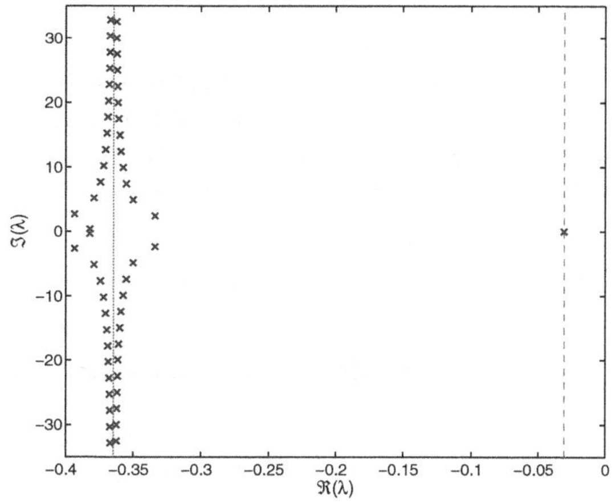

Figure 7.9. *Characteristic roots corresponding to the minimum of the robust spectral abscissa of (7.10) and (7.25) for the control law (7.26). The rightmost characteristic singular root, $\lambda \approx -0.0309$, has multiplicity three* [241].

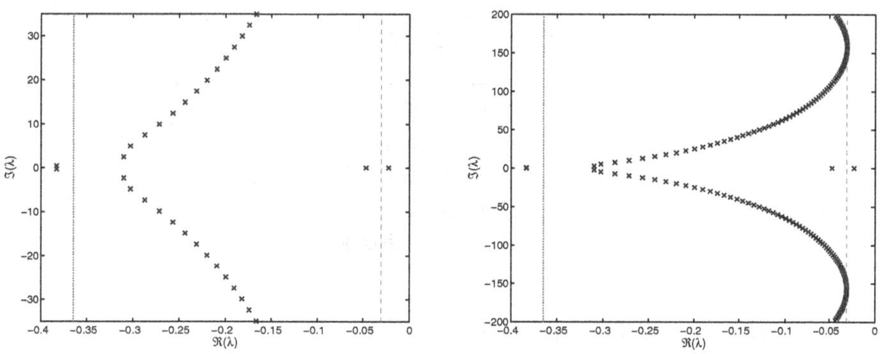

Figure 7.10. *Effect on the characteristic roots of a perturbation of the delays* (2.5, 5) *in* (7.25) *to* (2.51, 5)*. The difference between the left and right panes lies in the scaling of the imaginary axis* [241].

In Figure 7.9 we show the rightmost characteristic roots corresponding to the minimum of the robust spectral abscissa (7.27). The dotted line corresponds to $\Re(\lambda) = c_D$, and the dashed line to $\Re(\lambda) = C_D$. The minimum of C is characterized by an equality between C_D and the spectral abscissa c, the latter induced by a rightmost characteristic root with multiplicity three. This is compatible with the three degrees of freedom in the controller. In order to illustrate that we indeed have $c = C_D$, we depict in Figure 7.10 the rightmost characteristic roots after perturbing the delay value 2.5 in (7.25) to 2.51.

Solving the optimization problem (7.23) with $r = 10^{-3}$ and $\gamma = 10^{-3}$ yields

$$c = -0.0345, \quad C_D = -0.00602, \quad D_c = [0.0249\ 0.1076\ 0.3173].$$

Compared to (7.27), where we had $C = c = C_D$, a further reduction of the spectral abscissa has been achieved, at the price of an increased value of C_D. This is expected because the constraint $\gamma_0 < 1$ imposes robustness of stability, yet no bound on the exponential decay rate of the solutions.

Heating system

In [398] a linear model of an experimental heat transfer setup at the Czech Technical University in Prague is proposed, consisting of 10 delay-differential equations. The inclusion of an integrator to achieve a zero steady state error to a set-point of one of the state variables eventually leads to equations of the form

$$\dot{x}(t)=A_0x(t)+\sum_{i=1}^{5}A_ix(t-\tau_i)+Bu(t-\tau_6), \quad x(t)\in\mathbb{R}^{11\times 11}, \quad u(t)\in\mathbb{R}, \tag{7.28}$$

see [398] for the corresponding matrices and delay values. We consider two outputs

$$y(t)=\begin{bmatrix} x_5(t)+x_6(t) \\ -x_{10}(t)+x_{11}(t) \end{bmatrix}. \tag{7.29}$$

The spectral abscissa of the uncontrolled system is equal to zero. In Table 7.2 we show the result of optimizing the spectral abscissa using static state feedback, $u(t)=Kx(t)$, $K\in\mathbb{R}^{1\times 11}$, and dynamic output feedback (7.24).

Table 7.2. *Results of minimizing the spectral abscissa of* (7.28)–(7.29) *for static state feedback and for dynamic output feedback* (7.24) [241].

Controller	Minimum c
Static state feedback,	−0.0577
Output feedback, $n_c=0$	−0.0187
Output feedback, $n_c=1$	−0.0218
Output feedback, $n_c=2$	−0.0237

7.5 ▪ Notes and references

We presented a stabilization method for time-delay systems based on tuning a finite number of parameters. Stabilization was achieved by optimizing the spectral abscissa of the system. For this, recently developed nonsmooth, nonconvex optimization algorithms were combined with algorithms for the computation of the rightmost characteristic roots of time-delay systems.

The direct eigenvalue approach has several advantages w.r.t. the other stabilization methods for time-delay systems. First, the optimization procedure induces no conservatism in the sense that stabilizing parameters can in principle always be found when they exist. Second, preliminary model transformations, which may introduce additional dynamics [131, 218], can easily be dealt with. This was illustrated with the semiconductor laser example, where a nonphysical characteristic root at zero, due to the rotational symmetry, was known a priori and explicitly ignored by the optimization procedure. Here, we notice that from a mathematical point of view an introduction of an unstable characteristic root renders the system unstable. Hence, sufficient stability conditions for the transformed systems become automatically infeasible. Third, the algorithm is fully automatic and requires no interactions with the user, in contrast with the continuous pole placement procedure of [243], where the number of controlled characteristic roots needs to be increased manually on stagnation. In the final stage, where gradient sampling is used, the algorithm yields a monotonic decrease of the spectral abscissa as a function of the number of iterations until a Clarke stationary point is reached. Finally, the method is very general in the sense that there are no limitations neither on the number of time-delays in state, inputs, and outputs, nor on the type of delays. The extension to periodically varying delays and distributed delays is straightforward, since the method only requires that a procedure to compute the stability determining eigenvalues be at hand.

In Section 7.4 we discussed the extension to delay-differential algebraic equations. This allowed us to apply to approach to a much broader class of systems and controllers. By optimizing the robust spectral abscissa, the potential sensitivity of stability with respect to arbitrarily small perturbations is explicitly taken into account.

The approach has been implemented in generic software which is publicly available, see Appendix A.5. Even though the software has been written for systems with pointwise delays, it should be noted that with minimal adaptations it can handle a broad class of systems with distributed delays, via a model transformation. To sketch the main idea, consider the equation

$$\dot{x}(t) = Ax(t) + \int_{t-\tau}^{t} Bx(s)ds.$$

When rewriting the equation as

$$\dot{x}(t) = Ax(t) + By(t), \quad y(t) = \int_{t-\tau}^{t} x(s)ds,$$

differentiation of the "algebraic" equation yields the following system with pointwise delays,

$$\begin{cases} \dot{x}(t) = Ax(t) + By(t), \\ \dot{y}(t) = x(t) - x(t-\tau). \end{cases}$$

This idea can be generalized to distributed delays with a so-called gamma-distribution kernel (combination of polynomials and exponentials); see [278]. The model transformation will in general create additional, nonphysical characteristic roots (zeros due to differentiation in the above example), yet as they are known beforehand, they can be removed after a characteristic roots computation.

Finally, we note that in a practical control design the stabilization phase is usually only a first step in the overall design procedure. The (subsequent) fixed-order $\mathcal{H}_\infty$ control design is addressed in Chapter 9.

The chapter is based on [387, 58, 243, 241] and the references therein.

Chapter 8

Stabilizability with delayed feedback: A numerical case study

8.1 ▪ Introduction

In the previous chapter we proposed a stabilization method for linear time-delay systems. Among others, in Section 7.3.1 the method was illustrated by means of the stabilization problem of a linear controllable single input system in the presence of an input delay using static state feedback. This results in a closed-loop system of the form

$$\dot{x}(t) = Ax(t) + BK^T x(t-\tau), \tag{8.1}$$

where K is the feedback gain. To shed light on the restrictions of delayed feedback, in this chapter we completely characterize the class of all second-order systems, which are stabilizable with delayed state feedback.

Notice that the closed-loop system (8.1) is of the form

$$\dot{x}(t) = Ax(t) + Mx(t-\tau). \tag{8.2}$$

where the "control matrix" $M = BK^T$ is of rank 1. We also consider the case where all the elements of M can be chosen independently. Then (8.2) can be interpreted as the feedback controlled *multiple input* system,

$$\dot{x}(t) = Ax(t) + I\mathbf{u}(t-\tau), \quad \mathbf{u}(t) = Mx(t). \tag{8.3}$$

Even though the controllability assumption of (A, B) implies that A is cyclic and generically a rank 1 feedback is sufficient for pole assignment in the ODE case, we will show that the use of two inputs allows us to extend the class of stabilizable systems considerably. Furthermore, the availability of two inputs causes no restriction since it is shown in [261] that it is always possible to construct a stabilizing controller for the single input case if the stabilization problem (8.3) can be solved. More precisely, the observer-predictor-based controller

$$\begin{aligned} \dot{\hat{x}}(t) &= A\hat{x}(t) + Bu(t) + M(\hat{x}(t-\tau) - x(t)), \\ u(t) &= K^T \hat{x}(t) \end{aligned}$$

results in the closed-loop system

$$\begin{cases} \dot{x}(t) = (A + BK^T)x(t) + BK^T e(t), \\ \dot{e}(t) = Ae(t) + Me(t-\tau), \end{cases} \tag{8.4}$$

where $e(t) = \hat{x}(t-\tau) - x(t)$ is the prediction error. Because of the triangular structure, the closed-loop system is stable if and only if $(A+BK^T)$ is Hurwitz and the e-subsystem of (8.4) is asymptotically stable. Notice that a similar type of control laws is used in the context of predictive synchronization; see, for instance, [310] and the references therein.

Although the above stabilizability problems are independent of practical numerical methods for calculating the feedback gain, the analysis in this chapter starts from a repeated application of the spectral abscissa optimization, because this reveals some important properties of the spectrum which form the starting point of the stabilizability analysis. The followed approach further relies on methods and tools from the field of numerical *bifurcation analysis* [342].

In Section 8.2 we characterize all second-order systems which are stabilizable for a fixed, given value of the time delay. In Section 8.3 we discuss the simultaneous stabilizability with the same controller gain for all delays in a given interval including zero, that is, the stabilizability problem for the corresponding problem without delay, with a guaranteed delay margin. Finally, in Section 8.4 we briefly comment on a similar case study for Pyragas-type time-delayed feedback.

8.2 ▪ Characterization of stabilizable systems

We consider equation (8.2) with $A, M \in \mathbb{R}^{2\times 2}$. We simultaneously treat both the case where M is of rank 1, for instance, $M = BK^T$, and the case where the components of M can be chosen independently. For simplicity we first assume that the time-delay τ is equal to one. The case $\tau \neq 1$ is treated separately. At the end of this section, we also consider the special case where the system matrix A is noncyclic.

8.2.1 ▪ System representation

In the second-order case and for $\tau = 1$, system (8.2) can be transformed into the canonical form

$$\dot{x} = \begin{bmatrix} 0 & 1 \\ -a_2 & -a_1 \end{bmatrix} + \begin{bmatrix} -k_4 & -k_3 \\ -k_2 & -k_1 \end{bmatrix} x(t-1). \tag{8.5}$$

Furthermore, when $M = BK^T$, with (A,B) controllable, it is always possible to obtain the form (8.5) with $k_3 = k_4 = 0$, which then reduces to the so-called control canonical form. The characteristic function of (8.5) is given by

$$H(\lambda) = \lambda^2 + (a_1 + \bar{k}_1 e^{-\lambda})\lambda + (a_2 + \bar{k}_2 e^{-\lambda}) + \bar{k}_3 e^{-2\lambda}, \tag{8.6}$$

where

$$\begin{cases} \bar{k}_1 = k_1 + k_4, \\ \bar{k}_2 = k_2 + a_1 k_4 - a_2 k_3, \\ \bar{k}_3 = k_1 k_4 - k_2 k_3. \end{cases} \tag{8.7}$$

In the remainder of this section, we study the stability of the quasi-polynomial (8.6) as a function of the three control parameters $\bar{k}_1, \bar{k}_2, \bar{k}_3$. In this way we eliminate the redundancy in (8.5). However, it turns out that $\bar{k}_1, \bar{k}_2, \bar{k}_3$ cannot always be considered as free parameters, since for any given value of these parameters, corresponding values for k_1, k_2, k_3, k_4 in (8.5) should be found (using the relations (8.7)). This is not always possible.

When M is of rank 1, we have $\bar{k}_3 = 0$, and $\bar{k}_1$ and $\bar{k}_2$ can be freely assigned by an appropriate choice of k_1 and k_2.

When the control term is of full rank, i.e., $\bar{k}_3 \neq 0$, we have a constraint on the possible values of the control parameters $\bar{k}_1, \bar{k}_2, \bar{k}_3$. Indeed, eliminating k_1 and k_2 from the first two equations of (8.7) and substituting into the last one leads to the quadratic form

$$k_4^2 + a_2 k_3^2 - a_1 k_3^2 - a_1 k_3 k_4 + \bar{k}_1 k_4 - \bar{k}_2 k_3 - \bar{k}_3 = 0.$$

When $a_2 - a_1^2/4 \neq 0$, this expression can be written as

$$Ax^2 + y^2 - B = 0,$$

where

$$x = k_3 + \frac{\bar{k}_1 a_1/2 - \bar{k}_2}{2(a_2 - a_1^2/4)}, \quad y = k_4 - a_1 k_3/2 + \bar{k}_1/2,$$

$$A = a_2 - a_1^2/4, \qquad B = \bar{k}_3 + \bar{k}_1^2/4 + \frac{\left(\bar{k}_1 a_1/2 - \bar{k}_2\right)^2}{4(a_2 - a_1^2/4)}.$$

When $A > 0$ this equation can only have real solutions if $B \geq 0$. Hence, the constraint on the control parameters is given by

$$\bar{k}_3 \geq -\left(\bar{k}_1^2/4 + \frac{\left(\bar{k}_1 a_1/2 - \bar{k}_2\right)^2}{4(a_2 - a_1^2/4)}\right) \tag{8.8}$$

whenever $a_2 - \frac{a_1^2}{4} > 0$. When $a_2 = a_1^2/4$, one can show that no constraints on the control parameters are present, except for a degenerate case, which does not occur in the procedure followed in this chapter.

8.2.2 ▪ Class of stabilizable systems for the unit delay

As follows from the canonical form (8.5) and the characteristic equation (8.6), stabilizability conditions on the system can be expressed as a function of its parameters a_1 and a_2. Note that $\lambda^2 + a_1\lambda + a_2 = 0$ is the characteristic polynomial of the (uncontrolled) system.

Define the *optimized* spectral abscissa

$$c_{\text{opt}}(a_1, a_2) := \min_M F(M, a_1, a_2), \tag{8.9}$$

where

$$F(M, a_1, a_2) := \sup\left\{\Re(\lambda) : H(\lambda) = \det\left(\lambda I - A - Me^{-\lambda}\right) = 0\right\}. \tag{8.10}$$

Then in the (a_1, a_2)-plane, stabilizable and unstabilizable systems are separated by curves on which $c_{\text{opt}}(a_1, a_2) = 0$. We refer to these curves as the *stabilizability boundary*.

In Figure 8.1, the stabilizability boundary is outlined and its computation depicted. Before discussing its properties, we outline its computation.

An exhaustive approach would consist of optimizing the spectral abscissa, using the approach of the previous chapter, for a large number of values of the plant parameters (a_1, a_2), chosen on a fine grid, to check whether the system is stabilizable. The stabilizability boundary then separates the regions where the system is stabilizable and where it is not. However, a more efficient calculation is possible by taking into account specific properties of the optimization problem (8.9). This is now explained in detail for the case where M is of rank 1. Similar ideas apply to the case where M is of full rank.

From the analysis in Chapter 7 we recall that the objective function of the optimization problem (8.9) is not everywhere differentiable not even everywhere Lipschitz

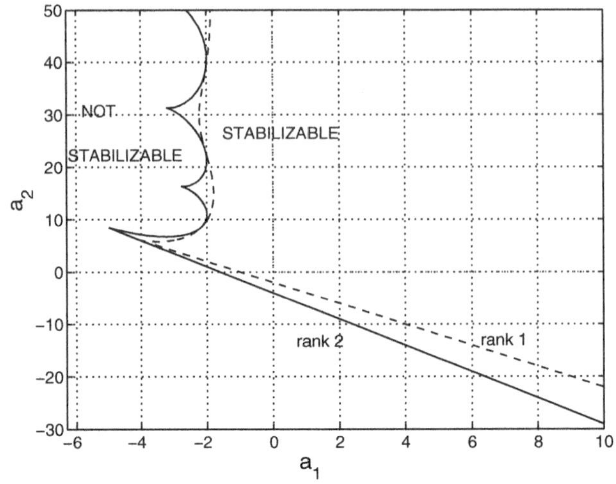

Figure 8.1. *Stabilizability boundary of* (8.2) *and* (8.5) *for* $\tau = 1$ *when* M *is of rank* 1 *(dotted line) and* M *is of rank* 2 *(solid line). The characteristic equation of the uncontrolled system is given by* $\lambda^2 + a_1\lambda + a_2 = 0$*. As we shall see in Section* 8.2.3 *the depicted curves remain valid for* $\tau \neq 1$*, provided that the axes are relabeled as* $a_1\tau$ *and* $a_2\tau^2$ [262].

continuous. Moreover, we have observed from our numerical experiments that the minimum is typically characterized by characteristic roots with multiplicity larger than one, and hence the optimum cannot be calculated directly using the relations $\frac{\partial F}{\partial m_i} = 0$, where m_i is a component of M. The following general result (which is easily shown to hold also when M has rank 2) can be found in [262]:

Proposition 8.1. *When the function* $M \to F(M, a_1, a_2)$ *with* $M = BK^T$ *is minimal, there are at least three characteristic roots with real part equal to* $c_{\text{opt}}(a_1, a_2)$.

As a consequence, the possible configurations of the rightmost characteristic roots at the global minimum of (8.10) can be reduced to the four basic situations shown in Table 8.1, where we also show the mathematical relations characterizing the position of the rightmost characteristic roots. In our numerical experiments only situations I and II occur. Theoretically situations I′ and II′ are possible, but are less generic than situations I and II.

Indeed, in cases I and II, the mathematical relations allow a direct computation of the minimal value c of (8.10) and the corresponding control parameters $\bar{k}_1$ and $\bar{k}_2$ when good starting values are available. However, in cases I′ and II′ there is one extra unknown variable, and therefore the mathematical relations in Table 8.1 define curves in their unknowns, through the point corresponding to the optimal parameter values. In such situations, a suitable small parameter change generically reduces the value of c, meaning that the situation does not correspond to the global minimum, unless an extra condition is satisfied (e.g., a turning point on the curve, or another characteristic root becoming the rightmost one). Our numerical experiments indicate that this extra condition is characterized by $\omega = 0$ for case I′, which then reduces to case I, and by $\omega_2 = \omega_1$ for case II′, which reduces to case II. These extra conditions are in some sense natural because situation II occurs when the uncontrolled system is highly oscillatory, with ω approximating the natural frequency, whereas situation I occurs when it is highly damped, and in these cases an extra "dominant frequency" in the controlled system is not expected.

Table 8.1. *According to Proposition* 8.1, *four configurations of the rightmost characteristic roots are possible in the global minimum of* (8.10) *when* M *is of rank* 1. *Only situation* I *and* II *occur. In the second column the mathematical relations, which characterize each configuration, are displayed.* $H^{(i)}$ *refers to the ith derivative of* H *w.r.t.* λ [262].

Poss.	RM eigenvalues (multiplicity)	Equations	Unknowns
I	c (3—f)	$\begin{cases} H(c)=0 \\ H^{(1)}(c)=0 \\ H^{(2)}(c)=0 \end{cases}$	$c,\bar{k}_1,\bar{k}_2$
I′	c $c \pm j\omega$	$\begin{cases} H(c)=0 \\ \Re(H(c+j\omega))=0 \\ \Im(H(c+j\omega))=0 \end{cases}$	$c,\omega,\bar{k}_1,\bar{k}_2$
II	$c \pm j\omega$ (2—f)	$\begin{cases} \Re(H(c+j\omega))=0 \\ \Im(H(c+j\omega))=0 \\ \Re\left(H^{(1)}(c+j\omega)\right)=0 \\ \Im\left(H^{(1)}(c+j\omega)\right)=0 \end{cases}$	$c,\omega,\bar{k}_1,\bar{k}_2$
II′	$c \pm j\omega_1$ $c \pm j\omega_2$	$\begin{cases} \Re(H(c+j\omega_1))=0 \\ \Im(H(c+j\omega_1))=0 \\ \Re(H(c+j\omega_2))=0 \\ \Im(H(c+j\omega_2))=0 \end{cases}$	$c,\omega_1,\omega_2,\bar{k}_1,\bar{k}_2$

Using the information contained in Table 8.1, the calculation of the stabilizability boundary only requires us to optimize the spectral abscissa for plant parameters chosen on a *coarse* grid in the (a_1,a_2)-plane. Indeed, suppose that we have applied the method for the plant parameters $(a_1^{(i)},a_2^{(i)})$, $i=1,2$, leading to $c_{\mathrm{opt}}(a_1^{(1)},a_2^{(1)})>0$ and $c_{\mathrm{opt}}(a_1^{(2)},a_2^{(2)})<0$. Then in the (a_1,a_2)-plane these two pairs are separated by the stabilizability boundary. Moreover, when $c_{\mathrm{opt}}(a_1^{(i)},a_2^{(i)})$ is sufficiently close to zero, information is available about the type of the minimum (I or II) to be expected on the stabilizability boundary in that neighborhood. This allows us to compute points on the stabilizability boundary directly using the mathematical relations displayed in Table 8.2. Compared to Table 8.1, here the extra condition $c=0$ is required, while the parameters a_1 and a_2 are freed. Hence, the mathematical relations in Table 8.2 define a branch which can be numerically continued in an efficient way. Good starting values are obtained from a spectral abscissa optimization for plant parameters $(a_1^{(i)},a_2^{(i)})$. The emanating branch forms part of the stabilizability boundary, until it intersects with another branch.

Based on information in Table 8.2, the components of the stabilizability boundary are displayed in Figure 8.2. The frequency ω on branch II approximates the natural frequency of the open-loop system and tends to zero along the branch when approaching the intersection with branch I.

When the control term is of full rank, there are three control parameters $\bar{k}_1,\bar{k}_2,\bar{k}_3$ in (8.6). The configurations of the rightmost characteristic roots on the stabilizability boundary are analogous to the previous case but, since there is an extra control parameter, one extra condition on the rightmost characteristic roots needs to be fulfilled; see Table 8.2. Four cases can be distinguished, and accordingly the stabilizability boundary can be decomposed into four components; see Figure 8.2. The presence of an extra

Table 8.2. *Positions of the rightmost characteristic roots on the stabilizability boundary of the system* (8.1)–(8.2) *with* $\tau = 1$*, and determining systems.* $H^{(i)}$ *refers to the ith derivative of H w.r.t.* λ*. Situations* I *and* II *occur when M is of rank* 1*; situations* III, IV, V, *and* VI *occur when M is of rank* 2*. The corresponding mathematical relations define branches, of which the stabilizability boundary is composed* [262].

Branch	RM eigenvalues (multiplicity)	Equations	Unknowns (Analytic solution)
I	0 (3–f)	$\begin{cases} H(0)=0 \\ H^{(1)}(0)=0 \\ H^{(2)}(0)=0 \end{cases}$	$a_1, a_2, \bar{k}_1, \bar{k}_2$ $(a_2 = -2a_1 - 2)$
II	$\pm j\omega$ (2–f)	$\begin{cases} H(j\omega)=0 \\ H^{(1)}(j\omega)=0 \end{cases}$	$a_1, a_2, \bar{k}_1, \bar{k}_2, \omega$
III	0 (4–f)	$\begin{cases} H(0)=0 \\ H^{(1)}(0)=0 \\ H^{(2)}(0)=0 \\ H^{(3)}(0)=0 \end{cases}$	$a_1, a_2, \bar{k}_1, \bar{k}_2, \bar{k}_3$ $(a2 = -2.5a_1 - 4)$
IV	0 $\pm j\omega$ (2–f)	$\begin{cases} H(0)=0 \\ H(j\omega)=0 \\ H^{(1)}(j\omega)=0 \end{cases}$	$a_1, a_2, \bar{k}_1, \bar{k}_2, \bar{k}_3, \omega$
V	$\pm j\omega$ (2–f)	$\begin{cases} H(j\omega)=0 \\ H^{(1)}(j\omega)=0 \\ J_{\bar{k}_1,\bar{k}_2,\bar{k}_3,\omega} X = 0 \\ X^T X = 1 \end{cases}$	$a_1, a_2, \bar{k}_1, \bar{k}_2, \bar{k}_3, \omega$
VI	$\pm j\omega$ (2–f)	$\begin{cases} H(j\omega)=0 \\ H^{(1)}(j\omega)=0 \\ \bar{k}_3 = -\left(\bar{k}_1^2/4 + \frac{(\bar{k}_1 a_1/2 - \bar{k}_2)^2}{4(a_2 - a_1^2/4)}\right) \end{cases}$	$a_1, a_2, \bar{k}_1, \bar{k}_2, \bar{k}_3, \omega$

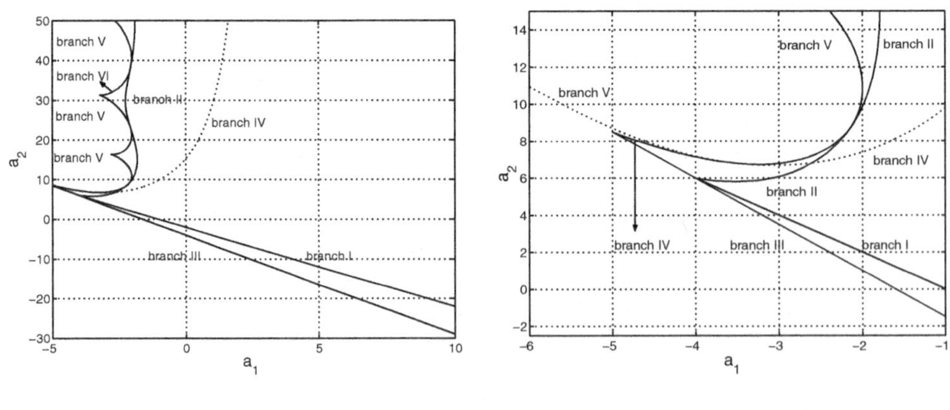

Figure 8.2. *Components of the stabilizability boundary (left) and detail (right) for system* (8.1)–(8.2) *with* $\tau = 1$*. The different branches refer to Table* 8.2 [262].

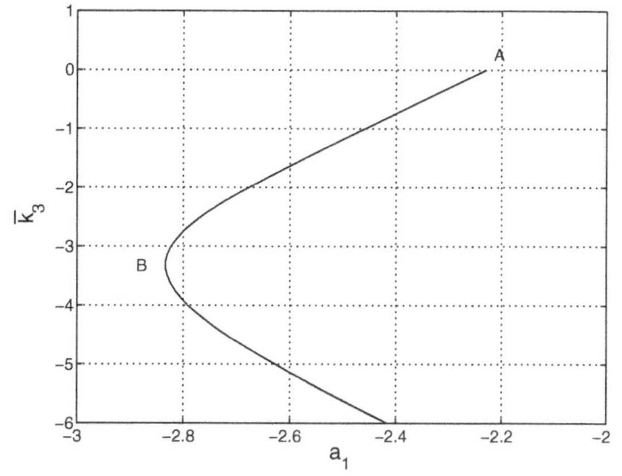

Figure 8.3. *When fixing a_2 and continuing the solution of* (8.11) *from point* A *on branch* II *as a function of parameter a_1, a turning point* B *occurs. Branch* V *is composed of such turning points. In point* A *we have* $(a_1, a_2) = (-2.223, 30)$ [262].

control parameter leads to four coinciding real characteristic roots on branch III and two coinciding pairs of complex conjugate characteristic roots and one real characteristic root on branch IV. At their intersection we have five characteristic roots equal to zero. The calculation of branches V and VI deserves further attention. Therefore, consider Figure 8.3. Point A lies on branch II, and hence $\bar{k}_3 = 0$ and its parameters $\bar{k}_1, \bar{k}_2, \omega, a_1, a_2$ satisfy

$$\left\{ \begin{array}{l} \Re(H(j\omega)) = 0, \\ \Im(H(j\omega)) = 0, \\ \Re\left(H^{(1)}(j\omega)\right) = 0, \\ \Im\left(H^{(1)}(j\omega)\right) = 0. \end{array} \right. \tag{8.11}$$

We now free parameter $\bar{k}_3$, and while keeping a_2 constant, we continue the solution of (8.11) as a function of a_1. The turning point B, where a minimum of a_1 is reached, defines a point of branch V. Let $J_{\bar{k}_1,\bar{k}_2,\bar{k}_3,\omega}(H, H^{(1)})$ be the Jacobian matrix of (8.11) as a function of parameters $\bar{k}_1, \bar{k}_2, \bar{k}_3, \omega$. Since $J_{\bar{k}_1,\bar{k}_2,\bar{k}_3,\omega}(H, H^{(1)})$ is singular in a turning point [342], we can express the extra condition to be fulfilled along branch V as

$$\det\left(J_{\bar{k}_1,\bar{k}_2,\bar{k}_3,\omega}(H, H^{(1)})\right) = 0, \tag{8.12}$$

or as

$$\left\{ \begin{array}{l} J_{\bar{k}_1,\bar{k}_2,\bar{k}_3,\omega}(H, H^{(1)})X = 0, \\ X^T X = 1, \end{array} \right. \tag{8.13}$$

which is better suited for numeral calculations [342]. By computing the branch of turning points, i.e., by continuing the solution of (8.11)–(8.13), branch V is obtained. The sharp edges on Figure 8.2 are caused by the projection on the (a_1, a_2)-plane.

Notice that constraint (8.8) forms a lower bound on $\bar{k}_3$. Hence, it may be that this constraint becomes active before the turning point is reached. In that case the stabilizability boundary is determined by (8.11) and

$$\bar{k}_3 = -\left(\bar{k}_1^2/4 + \frac{\left(\bar{k}_1 a_1/2 - \bar{k}_2\right)^2}{4(a_2 - a_1^2/4)}\right),$$

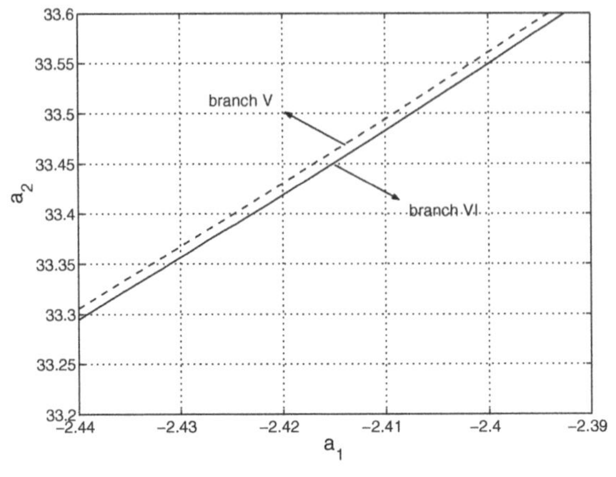

Figure 8.4. *For $32.4 \leq a_2 \leq 34.8$, the stabilizability boundary is not formed by branch* V *(dashed line), but by branch* VI *(solid line). On branch* VI, (8.8) *is an equality constraint* [262].

which expresses that constraint (8.8) is active. This is the case for $32.4 \leq a_2 \leq 34.8$, and here the stabilizability boundary coincides with branch VI. See Figure 8.4, where the relevant part of Figure 8.2 (left) is enlarged.

8.2.3 ▪ Class of stabilizable systems for arbitrary delay values and discussion on stabilizability

We now consider the computation of the stabilizability boundary of (8.1)–(8.2) when the delay $\tau \neq 1$. With the substitution

$$\bar{\lambda} = \tau\lambda,$$

the characteristic equation of (8.2),

$$\det\left(\lambda I - A - Me^{-\lambda\tau}\right) = 0, \tag{8.14}$$

becomes

$$\det\left(\bar{\lambda} I - \tau A - \tau M e^{-\bar{\lambda}}\right) = 0.$$

This can be seen as the characteristic equation of

$$\dot{x} = \bar{A}x + \bar{M}x(t-1), \qquad \bar{A} = \tau A, \ \bar{M} = \tau M. \tag{8.15}$$

Because of the transformation, the stabilizability boundary of system (8.15) can be computed as outlined in the previous paragraphs, but it is expressed as a function of the parameters of $\bar{A}$. As follows from the definition of $\bar{A}$, it can be expressed in the normalized coordinates $a_1\tau$ and $a_2\tau^2$, where $\lambda_2 + a_1\lambda + a_2 = 0$ is the characteristic equation of the uncontrolled system. Hence, the stabilizability region in the (a_1, a_2, τ) space is simply given by Figure 8.1, where the axes are *relabeled* as $a_1\tau$ and $a_2\tau^2$. By means of this figure, we now summarize some interesting properties of delayed state feedback:

- The uncontrolled system is only asymptotically stable when $a_1 > 0$ and $a_2 > 0$. With delayed state feedback, the class of stabilizable systems is determined by the stabilizability boundary. This class becomes larger and grows unbounded in all directions in the (a_1, a_2)-plane as the delay τ is reduced. Contrary to the ODE case, where the whole spectrum can be controlled with rank 1 feedback, the class of stabilizable systems is considerably larger with rank 2 feedback in the DDE case, a consequence of the infinite-dimensional nature of delay-differential equations.

- From Figure 8.1 one can deduce stabilizability information of a plant with fixed parameters (a_1,a_2) as a function of the delay τ. When the delay changes, the normalized plant parameters $(\hat{a}_1,\hat{a}_2) = (a_1\tau,\ a_2\tau^2)$ move on a (half) parabola. When the plant has a characteristic root in the open right half plane, i.e., either $a_1 < 0$ or $a_2 < 0$, this parabola always intersects the stabilizability boundary. Hence, when the uncontrolled system is (exponentially) unstable, it cannot be stabilized for large values of the time delay. On the other hand, when the rightmost characteristic root of the open-loop system lies on the imaginary axis, stabilization is always possible, whatever the value of the time delay. This result also follows from [298, 236]. Furthermore, the gain can be chosen arbitrarily small [236].

- When τ is fixed and $a_2 \to \infty$, the stabilizability boundary does not converge to the a_2-axis. Consequently, the set of stabilizable systems includes a class of exponentially unstable systems ($a_1 < 0$) with arbitrarily large natural frequencies (i.e., imaginary parts of the characteristic roots). For such systems with highly oscillatory behavior, stabilization with delayed feedback is intuitively possible because tuning of the matrix M in (8.2) allows us to give the feedback signal Me precisely the necessary phase-shift, thereby compensating (only mod 2π) the phase-shift introduced by the delay. But this implies that the larger a_2 is, the more sensitive is the achieved stability w.r.t. changes in the system parameters, and in particular the delay. For instance, the achieved asymptotic stability property will only hold in a small delay interval around the nominal delay.

8.3 ▪ Simultaneous stabilization over a delay interval

We reconsider the stabilizability of the second-order systems (8.1). We characterize the class of systems which are *simultaneously* stabilizable for all delay values in an interval $[0,\ \tau_{\max})$. For the sake of conciseness, we restrict ourselves to the case of rank-1 feedback, where the characteristic equation simplifies to

$$\lambda^2 + (a_1 + k_1 e^{-\lambda\tau})\lambda + (a_2 + k_2 e^{-\lambda\tau}) = 0. \tag{8.16}$$

The approach is based on a qualitative study of the evolution of the stability region in the (k_1,k_2)-plane as the delay is increased from zero, and a quantitative characterization of the situations where either stabilizability is lost or the intersection of the stability region with the stability region for $\tau = 0$ becomes empty.

For any pair of system parameters (a_1,a_2) and delay τ, the set S_τ of stabilizing feedback gains in the (k_1,k_2)-plane can be found by searching for characteristic roots on the imaginary axis, i.e., by substituting $\lambda = j\omega$ in (8.16). Therefore, S_τ is determined by the line $k_2 = -a_2$ and by the spiral curve Γ_τ,

$$\Gamma_\tau :\ \omega \in [0,\ \infty) \to \left\{ \begin{array}{l} k_1(\omega,\tau), \\ k_2(\omega,\tau), \end{array} \right. \tag{8.17}$$

where

$$\begin{aligned} k_1(\omega,\ \tau) &= \left(\omega^2 \sin(\omega\tau) - a_1\omega\cos(\omega\tau) - a_2\sin(\omega\tau)\right)/\omega, \\ k_2(\omega,\tau) &= \omega^2\cos(\omega\tau) + a_1\omega\sin(\omega\tau) - a_2\cos(\omega\tau). \end{aligned}$$

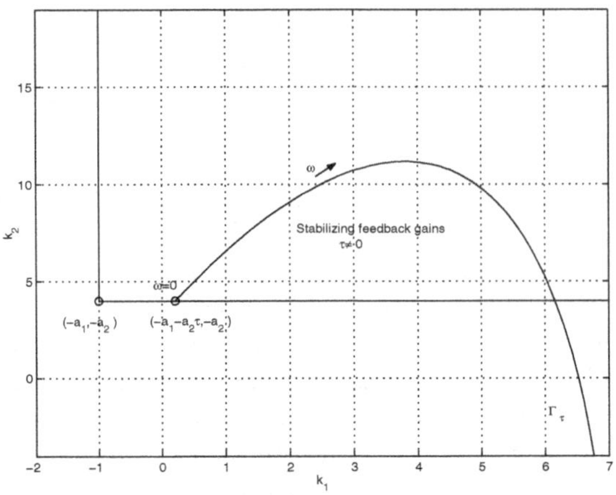

Figure 8.5. *Stabilizing values of* (k_1,k_2) *for equation* (8.16) *with* $(a_1,a_2)=(1,-4)$, *when* $\tau=0$ *and* $\tau=0.3$.

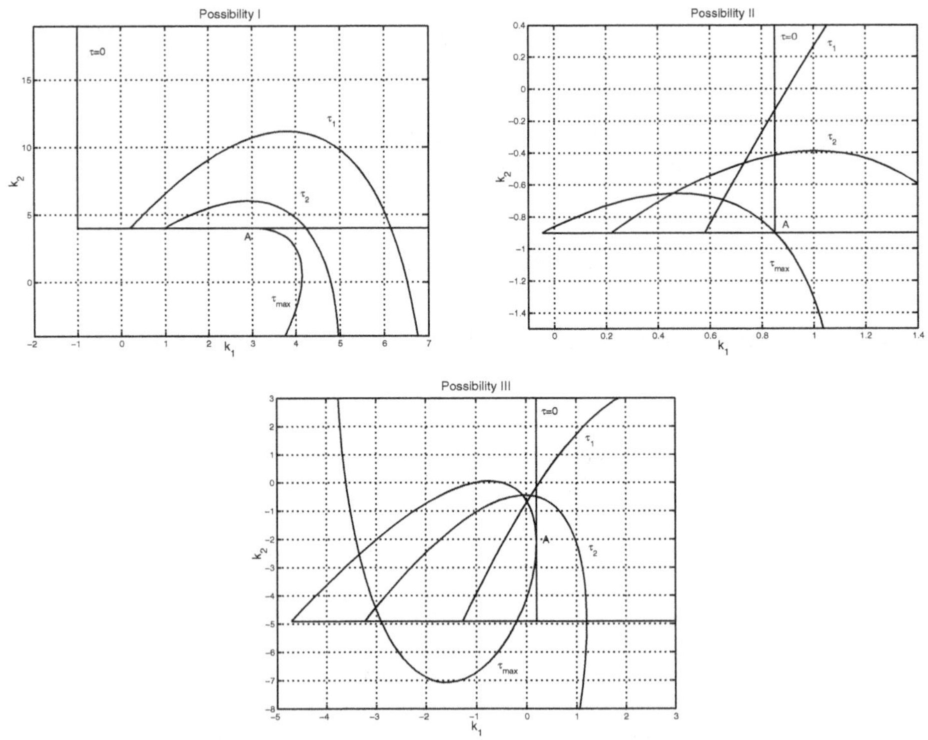

Figure 8.6. *The three possible cases for the evolution of the set* S_τ *as a function of the delay*, $0<\tau_1<\tau_2<\tau_{\max}$. *The feedback gain corresponding to point* A *and the delay* $\tau_{\max}$ *solve the optimization problem* (8.18).

In Figure 8.5 we show the set S_τ for small values of the time delay. For $\tau=0$, it is given by

$$k_1>-a_1,\quad k_2>-a_2,$$

and for small delay values, it is enclosed by the line $k_2=-a_1$ and the curve Γ_τ. Depending on the system parameters (a_1,a_2), there are qualitatively three possible situations as the delay is further increased, which are sketched in Figure 8.6. For these three cases, the

Table 8.3. *Mathematical relations which characterize the three possible situations shown in Figure 8.6. When $\tau_{\max}$ is fixed, these relations define branches in the (a_1,a_2)-plane. For $\tau_{\max}=1$, these branches are shown in Figure 8.7.*

Case	Equations	Unknowns	Analytic solution
I	$\lim_{\omega\to 0+}\frac{\frac{\partial k_2}{\partial\omega}(\omega,\tau_{\max})}{\frac{\partial k_1}{\partial\omega}(\omega,\tau_{\max})}=0$	$a_1,a_2,\tau_{\max}$	$a_2\tau_{\max}^2=-2a_1\tau_{\max}-2$
II	$\begin{cases} k_1(\omega,\tau_{\max})=-a_1 \\ k_2(\omega,\tau_{\max})=-a_2 \end{cases}$	$a_1,a_2,\omega,\tau_{\max}$	
III	$\begin{cases} k_1(\omega,\tau_{\max})=-a_1 \\ \frac{\partial k_1(\omega,\tau_{\max})}{\partial\omega}=0 \end{cases}$	$a_1,a_2,\omega,\tau_{\max}$	

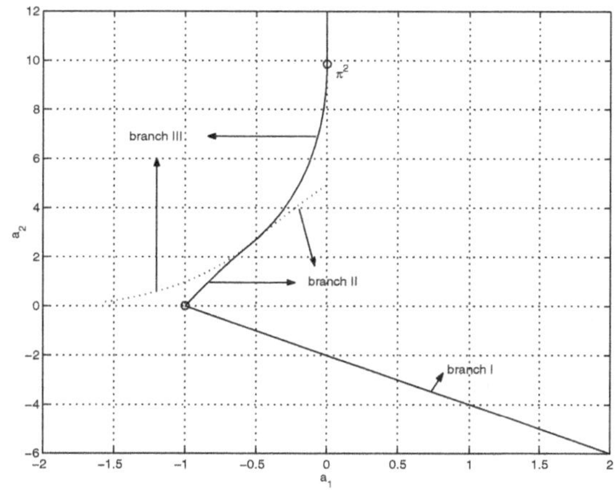

Figure 8.7. *The branches in the (a_1,a_2)-plane, corresponding to the mathematical relations of Table 8.3, for $\tau_{\max}=1$. The set of systems, simultaneously stabilizable in the delay interval $[0,\ \tau_{\max})$, is determined by the parameter values lying to the right of the solid line, which is composed from these branches.*

values of the feedback gain (k_1,k_2) in point A, and the corresponding delay, $\tau=\tau_{\max}$, solve the following optimization problem:

$$\begin{array}{c}\max_{k_1,k_2,\tau}\ \tau\ \text{ subject to}\\ \max_{\theta\in[0,\ \tau]}\left\{\sup\Re(\lambda):\ \lambda^2+(a_1+k_1e^{-\lambda\theta})+(a_2+k_2e^{-\lambda\theta})=0\right\}<0.\end{array}\tag{8.18}$$

In other words, the system can be made globally asymptotically stable for all $\tau\in[0,\ \tau_{\max})$, but not in a larger delay interval. Note that in case I, the set S_τ is empty for $\tau=\tau_{\max}$. This is not so for cases II and III, which means that stabilization is possible for some fixed delay values $\tau>\tau_{\max}$, while it is not possible over the whole interval $[0,\ \tau]$.

In Table 8.3 we display the mathematical relations describing the three cases depicted in Figure 8.6. When $\tau_{\max}$ is fixed, these relations define branches in the (a_1,a_2)-plane, which determine the class of systems, simultaneously stabilizable in the delay interval $[0,\ \tau_{\max})$. In Figure 8.7 we display these branches for $\tau_{\max}=1$. Because of the arguments spelled out in Section 8.2.3, the branches corresponding to other values of $\tau_{\max}$

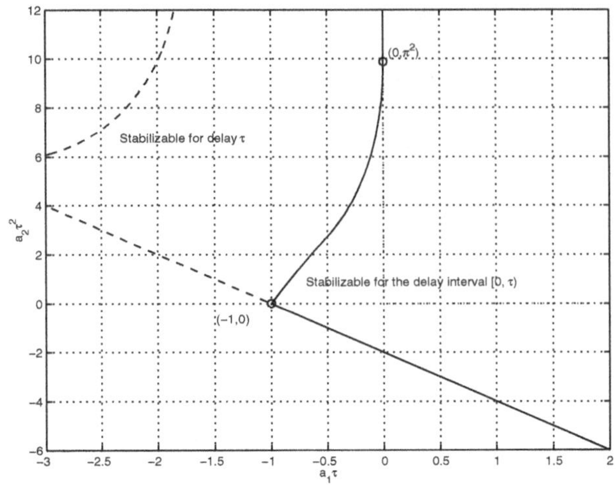

Figure 8.8. *In case of rank 1 feedback, the class of second-order systems (simultaneously) stabilizable over the delay interval $[0,\ \tau)$ is determined by the solid line, while the dashed line, the stabilizability boundary, determines the class of systems which are stabilizable for the (fixed) delay τ.*

coincide with the branches for $\tau_{\max} = 1$, when normalizing the system parameters to $(a_1\tau_{\max}, a_2\tau_{\max}^2)$.

In Figure 8.8, we compare, for a rank 1 feedback, the stabilizable systems for a fixed delay value τ (characterized in Section 8.2) with the systems simultaneously stabilizable over the delay interval $[0,\ \tau)$. In the latter case, stabilization is *not* possible for high values of a_2. For example, if the uncontrolled systems have characteristic roots on the imaginary axis ($a_1 = 0$), stabilization in the delay interval $[0,\ \tau)$ is only possible if $a_2\tau^2 \leq \pi^2$, that is, if $\tau \leq T/2$, with T the period of the natural oscillation.

8.4 ▪ Stabilizability using Pyragas-type feedback

In [171] a stabilizability study similar to (8.1) was performed. The system under consideration was a second-order linear single input controllable system

$$\dot{x}(t) = Ax(t) + Bu(t).$$

The controllers under consideration were based on static state and output feedback, depending on the difference $x(t) - x(t-\tau)$, with τ a design parameter, i.e., Pyragas-type feedback [287, 329], also called time-delayed feedback. We note that this type of feedback is commonly used in the context of chaos control of nonlinear systems. In that context its main advantage is that the presence and shape of a periodic orbit with period τ is not affected by the control, in contrast to its stability properties. As a consequence, Pyragas-type feedback can be used to stabilize unstable periodic orbits. Pyragas-type controllers are also used to stabilize unstable equilibria (which are special cases of periodic orbits), without changing their location.

In order to illustrate the main similarities and differences between Pyragas-type feedback and delayed feedback addressed in the previous sections, we consider in what follows the case of Pyragas-type state feedback, which leads us to the closed-loop system

$$\dot{x}(t) = Ax(t) + BK^T(x(t) - x(t-\tau)).$$

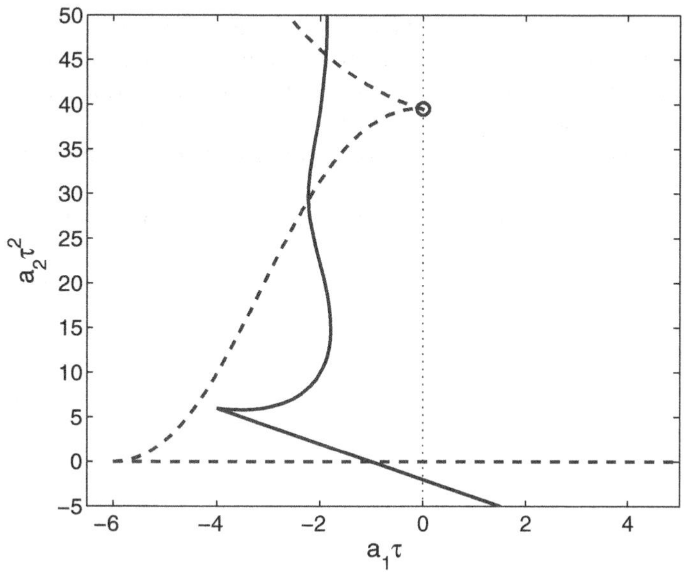

Figure 8.9. *Stabilizability region for delayed state feedback (bounded by the full curves) and for Pyragas-type feedback (bounded by the dashed curves).*

Once more, we assume that A is a 2-by-2 matrix with characteristic polynomial $\lambda^2+a_1\lambda+a_2$, and we look at the stabilizability with the elements of K as controller parameters and with τ given. In Figure 8.9 we compare the stabilizability region in the (a_1,a_2,τ) space, with τ fixed, for Pyragas-type state feedback and for delayed state feedback. We make the following observations:

- Using Pyragas-type feedback the system is not stabilizable whenever $a_2 < 0$. The reason is that for $a_2 < 0$, the uncontrolled systems have one characteristic root (and thus an odd number) in the closed right half plane. With Pyragas feedback an odd number of characteristic roots cannot be moved from the right to the left half plane, since roots cannot cross the imaginary axis at zero (note that $(1-\exp(-\lambda\tau))$ vanishes at $\lambda = 0$), and hence they can only cross the imaginary axis in pairs. This limitation is closely related to the so-called odd-number limitation in stabilizing unstable periodic orbits [287, 181].
- Using Pyragas-type feedback the stabilizability boundary touches the stability region of the uncontrolled systems at
$$a_1 = 0, \quad a_2 = (2\pi\ell)^2, \quad \ell = 1,2,\ldots.$$
The reason is that the uncontrolled system is oscillatory in such a case, and for the above choice of a_2, the frequency of the oscillation is matching, in the sense that the control law vanishes along the solutions.

8.5 ▪ Notes and references

We explored the limits of stabilizability with delayed state feedback by completely characterizing the class of stabilizable second-order systems. Due to the infinite-dimensional nature of time-delay systems, this is no trivial problem, despite the low dimension, and a combination of both analytical and numerical methods was required. The case study demonstrated a fruitful interaction of methods and tools from control theory and numerical bifurcation analysis. An application of the approach to the analysis of Pyragas-type feedback was briefly discussed. Further discussions on the use of Pyragas-type controllers in avoiding the existence of triple zero singularity of an inverted pendulum on a cart can be found in [34].

This chapter is based on [262, 171], [240, Appendix A] and the references therein.

Chapter 9

Optimization of $\mathcal{H}_\infty$ norms

9.1 ▪ Introduction

In many control applications, robust controllers are desired to achieve stability and performance requirements under model uncertainties and exogenous disturbances [414]. The design requirements are usually defined in terms of $\mathcal{H}_\infty$ norms of closed-loop transfer functions including the plant, the controller, and weights for uncertainties and disturbances. There are robust control methods to design the optimal $\mathcal{H}_\infty$ controller for linear finite-dimensional multi-input–multi-output (MIMO) systems based on Riccati equations and linear matrix inequalities (LMIs); see, e.g., [96, 117] and the references therein. The order of the controller designed by these methods is typically greater than or equal to the order of the plant. This is a restrictive condition for high-order plants, since low-order controllers are desired in a practical implementation. The design of fixed-order or low-order $\mathcal{H}_\infty$ controllers can be translated into a nonsmooth, nonconvex optimization problem. Recently, fixed-order $\mathcal{H}_\infty$ controllers have been successfully designed for finite-dimensional linear-time-invariant (LTI) MIMO plants using a direct optimization approach [143]. This approach allows the user to choose the controller order and tunes the parameters of the controller to minimize the $\mathcal{H}_\infty$ norm under consideration. An extension to a class of retarded time-delay systems has been described in [139].

In this chapter we design a fixed-order or fixed-structure $\mathcal{H}_\infty$ controller in a feedback connection with a time-delay system. The closed-loop system is a delay-differential algebraic system, and its state-space representation is written as

$$\begin{cases} E\dot{x}(t) = A_0 x(t) + \sum_{i=1}^{m} A_i x(t-\tau_i) + Bw(t), \\ z(t) = Cx(t). \end{cases} \tag{9.1}$$

The time-delays τ_i, $i = 1, \ldots, m$, are positive real numbers and the uppercase letters are real-valued matrices with appropriate dimensions. The input w and output z are disturbances and signals to be minimized to achieve design requirements, and some of the system matrices include the controller parameters.

The system with the closed-loop equations (9.1) represents all interesting cases of the feedback connection of a time-delay plant and a controller. The transformation of the closed-loop system to this form can be easily done by first augmenting the system equations of the plant and controller. As we shall see, this augmented system can subsequently be brought to the form (9.1) by introducing slack variables to eliminate input/output delays and direct feedthrough terms in the closed-loop equations. Hence, the resulting system of the form (9.1) is obtained directly without complicated elimination techniques that may not even be possible in the presence of time delays.

As we shall see, the $\mathcal{H}_\infty$ norm of DDAEs may be sensitive to arbitrarily small delay changes. Since small modeling errors are inevitable in any practical design we are interested in the smallest upper bound of the $\mathcal{H}_\infty$ norm that is insensitive to small delay changes. Inspired by the concept of strong stability of neutral equations [152], we are led to the introduction of the concept of *strong $\mathcal{H}_\infty$ norms* for DDAEs. Several properties of the strong $\mathcal{H}_\infty$ norm are shown and a computational formula is obtained. The theory derived can be considered as the dual of the theory of strong stability as elaborated in Chapter 1 (see also [109, 152, 211, 242, 264, 266] and the references therein).

In addition, a level set algorithm for computing strong $\mathcal{H}_\infty$ norms is presented, which is inspired by the algorithm for $\mathcal{H}_\infty$ computation for time-delay systems of retarded type as described in Chapter 4. However, a main difference lies in the fact that the robustness w.r.t. small delay perturbations needs to be explicitly addressed.

The numerical algorithm for the norm computation is subsequently applied to the design of $\mathcal{H}_\infty$ controllers by a direct optimization approach. In the context of control of LTI systems it is well known that $\mathcal{H}_\infty$ norms are in general nonconvex functions of the controller parameters which arise as elements of the closed-loop system matrices. They are typically not even everywhere smooth, although they are differentiable almost everywhere [143]. These properties carry over to the case of strong $\mathcal{H}_\infty$ norms of DDAEs under consideration. Therefore, special optimization methods for nonsmooth, nonconvex problems are required. We will use a combination of BFGS, whose favorable properties in the context of nonsmooth problems have been reported in [206], and bundle and gradient sampling methods, as implemented in the MATLAB code HANSO.[20] The overall algorithm only requires the evaluation of the objective function, i.e., the strong $\mathcal{H}_\infty$ norm, as well as its derivatives with respect to the controller parameters whenever it is differentiable. The computation of the derivatives is also discussed in this chapter.

The presented method is frequency domain–based and builds on the eigenvalue-based framework developed in the previous chapters. Time-domain methods for the $\mathcal{H}_\infty$ control of DDAEs have been described in, e.g., [113] and the references therein, based on the construction of Lyapunov–Krasovskii functionals.

The structure of the chapter is as follows. In Section 9.2 we illustrate the generality of the system description (9.1). Preliminaries and assumptions are given in Section 9.3. The definition and properties of the strong $\mathcal{H}_\infty$ norm of DDAEs are given in Section 9.4. The numerical algorithm to compute the strong $\mathcal{H}_\infty$ norm is described in detail in Section 9.5. Fixed-order $\mathcal{H}_\infty$ controller design is addressed in Section 9.6. Section 9.7 is devoted to the numerical examples. Section 9.8 discusses the computation of stability radii. In Section 9.9 some concluding remarks are presented.

9.2 ▪ Motivating examples

With some simple examples we illustrate the generality of the system description (9.1).

Example 9.1. *Consider the feedback interconnection of the system*

$$\begin{cases} \dot{x}(t) = Ax(t) + B_1 u(t) + B_2 w(t), \\ y(t) = Cx(t) + D_1 u(t), \\ z(t) = Fx(t) \end{cases}$$

[20] Hybrid Algorithm for Nonsmooth Optimization; see [317].

and the controller

$$u(t)=Ky(t-\tau).$$

For $\tau=0$ it is possible to eliminate the output and controller equation, which results in the closed-loop system

$$\begin{cases} \dot{x}(t)=Ax(t)+B_1K(I-D_1K)^{-1}Cx(t)+B_2w(t), \\ z(t)=Fx(t). \end{cases} \tag{9.2}$$

This approach is, for instance, taken in the software package HIFOO [53]. If $\tau\neq 0$, then the elimination is no longer possible. However, if we let $X=[x^T\ u^T y^T]^T$, we can describe the system by the equations

$$\begin{cases} \begin{bmatrix} I & 0 & 0 \\ 0 & 0 & 0 \\ 0 & 0 & 0 \end{bmatrix} \dot{X}(t)=\begin{bmatrix} A & B_1 & 0 \\ C & D_1 & -I \\ 0 & I & 0 \end{bmatrix} X(t)-\begin{bmatrix} 0 & 0 & 0 \\ 0 & 0 & 0 \\ 0 & 0 & K \end{bmatrix} X(t-\tau)+\begin{bmatrix} B_2 \\ 0 \\ 0 \end{bmatrix} w(t), \\ z(t)=\begin{bmatrix} F & 0 & 0 \end{bmatrix} X(t), \end{cases}$$

which are of the form (9.1). Furthermore, the dependence of the matrices of the closed-loop system on the controller parameters, K, is still linear, unlike in (9.2).

Example 9.2. *The presence of a direct feedthrough term from w to z, as in*

$$\begin{cases} \dot{x}(t)=Ax(t)+A_1x(t-\tau)+Bw(t), \\ z(t)=Fx(t)+D_2w(t), \end{cases} \tag{9.3}$$

can be avoided by introducing a slack variable. If we let $X=[x^T\ \gamma_w^T]^T$, where γ_w is the slack variable, we can bring (9.3) to the form (9.1):

$$\begin{cases} \begin{bmatrix} I & 0 \\ 0 & 0 \end{bmatrix} \dot{X}(t)=\begin{bmatrix} A & 0 \\ 0 & -I \end{bmatrix} X(t)+\begin{bmatrix} A_1 & 0 \\ 0 & 0 \end{bmatrix} X(t-\tau)+\begin{bmatrix} B \\ I \end{bmatrix} w(t), \\ z(t)=[F\ D_2]\,X(t). \end{cases}$$

Example 9.3. *The system*

$$\begin{cases} \dot{x}(t)=Ax(t)+B_1w(t)+B_2w(t-\tau), \\ z(t)=Cx(t) \end{cases}$$

can also be brought to the standard form (9.1) by a slack variable. Letting $X=[x^T\gamma_w^T]^T$ we can express

$$\begin{cases} \dot{X}(t) = \begin{bmatrix} A & B_1 \\ 0 & -I \end{bmatrix} X(t)+\begin{bmatrix} 0 & B_2 \\ 0 & 0 \end{bmatrix} X(t-\tau)+\begin{bmatrix} 0 \\ I \end{bmatrix} w(t), \\ z(t) = [C\ 0]\,X(t). \end{cases}$$

In a similar way one can deal with delays in the output z.

Using the techniques illustrated with the above examples, a broad class of interconnected systems with delays can be brought to the form (9.1), where the external inputs w and outputs z stem from the performance specifications expressed in terms of appropriately defined transfer functions. As a more realistic illustration, the feedback interconnection of any retarded-type time-delay system G and controller K with the following state-space representations can be written in the form of (9.1) using similar techniques in the previous examples:

$$G: \begin{cases} \dot{x}_G(t) = \sum_{i=0}^{m_a} A^i x_G(t-\tau_i^a) + \sum_{i=0}^{m_{b_1}} B_1^i w(t-\tau_i^{b_1}) + \sum_{i=0}^{m_{b_2}} B_2^i u(t-\tau_i^{b_2}), \\ z(t) = \sum_{i=0}^{m_{c_1}} C_1^i x_G(t-\tau_i^{c_1}) + \sum_{i=0}^{m_{d_{11}}} D_{11}^i w(t-\tau_i^{d_{11}}) + \sum_{i=0}^{m_{d_{12}}} D_{12}^i u(t-\tau_i^{d_{12}}), \\ y(t) = \sum_{i=0}^{m_{c_2}} C_2^i x_G(t-\tau_i^{c_2}) + \sum_{i=0}^{m_{d_{21}}} D_{21}^i w(t-\tau_i^{d_{21}}) + \sum_{i=0}^{m_{d_{22}}} D_{22}^i u(t-\tau_i^{d_{22}}), \end{cases} \tag{9.4}$$

$$K: \begin{cases} \dot{x}_K(t) = \sum_{i=0}^{m_{a_k}} A_K^i x_K(t-\tau_i^{a_k}) + \sum_{i=0}^{m_{b_k}} B_K^i y(t-\tau_i^{b_k}), \\ u(t) = \sum_{i=0}^{m_{c_k}} C_K^i x_K(t-\tau_i^{c_k}) + \sum_{i=0}^{m_{d_k}} D_K^i u(t-\tau_i^{d_k}). \end{cases} \tag{9.5}$$

The price to pay for the generality of the framework is the increase of the dimension of the system, n, which affects the efficiency of the numerical methods. However, this is a minor problem in most applications because the delay-difference equations or algebraic constraints are related to inputs and outputs, and the number of inputs and outputs is usually much smaller than the number of state variables.

9.3 ▪ Preliminaries

Assumptions

As in Section 1.3 we let rank$(E) = n - \nu$, with $\nu \leq n$, and let the columns of matrix $U \in \mathbb{R}^{n\times\nu}$, respectively $V \in \mathbb{R}^{n\times\nu}$, be a (minimal) basis for the left, respectively right, nullspace, that is,

$$U^T E = 0, \qquad EV = 0.$$

Throughout this chapter we make the following assumption:

Assumption 9.4. *Matrix $U^T A_0 V$ is nonsingular.*

In order to motivate Assumption 9.4, we note that the equations (9.1) can be separated into coupled delay-differential and delay-difference equations. When we define

$$\mathbf{U} = \begin{bmatrix} U^\perp & U \end{bmatrix}, \qquad \mathbf{V} = \begin{bmatrix} V^\perp & V \end{bmatrix},$$

a premultiplication of (9.1) with $\mathbf{U}^T$ and the substitution

$$x = \mathbf{V}\,[x_1^T\ x_2^T]^T,$$

with $x_1(t) \in \mathbb{R}^{n-\nu}$ and $x_2(t) \in \mathbb{R}^\nu$, yield the coupled equations

$$\begin{cases} E^{(11)}\dot{x}_1(t) = \sum_{i=0}^m A_i^{(11)} x_1(t-\tau_i) + \sum_{i=0}^m A_i^{(12)} x_2(t-\tau_i) + B_1 w(t), \\ \quad 0 = A_0^{(22)} x_2(t) + \sum_{i=1}^m A_i^{(22)} x_2(t-\tau_i) \\ \qquad\qquad + \sum_{i=0}^m A_i^{(21)} x_1(t-\tau_i) + B_2 w(t), \\ z(t) = C_1 x_1(t) + C_2 x_2(t), \end{cases} \tag{9.6}$$

where

$$A_i^{(11)} = U^{\perp T} A_i V^\perp, \quad A_i^{(12)} = U^{\perp T} A_i V,$$
$$A_i^{(21)} = U^T A_i V^\perp, \quad A_i^{(22)} = U^T A_i V, \qquad i = 0, \ldots, m,$$

and

$$E^{(11)} = U^{\perp^T} E V^{\perp}, \quad B_1 = U^{\perp^T} B, \quad B_2 = U^T B, \quad C_1 = C V^{\perp}, \quad C_2 = C V.$$

Matrix $E^{(11)}$ in (9.6) is invertible, following from

$$n - \nu = \operatorname{rank}(E) = \operatorname{rank}(\mathbf{U}^T E \mathbf{V}) = \operatorname{rank}(E^{(11)}).$$

In addition, matrix $A_0^{(22)}$ is invertible, following from Assumption 9.4.

The equations (9.6), with $w \equiv 0$, are semi-explicit delay-differential algebraic equations of index 1, because delay-differential equations are obtained by differentiating the second equation. This precludes the occurrence of impulsive solutions [113]. Moreover, the invertibility of $A_0^{(22)}$ prevents the equations from being of *advanced* type and, hence, noncausal. This further motivates why Assumption 9.4 is natural in the delay case considered, although it restricts the index to one (for a general treatment in the delay free case see, for instance, [364] and the references therein).

We further make the following assumption:

Assumption 9.5. *The zero solution of system* (9.1), *with* $w \equiv 0$, *is strongly exponentially stable.*

Recall that strong exponential stability refers to the fact that the asymptotic stability of the null solution is robust against small delay perturbations [152, 266]. Due to modeling errors and uncertainty, the delays in the model are usually not exact, and strong exponential stability is required in practice. The stability of the closed-loop system (9.1) is a necessary assumption for $\mathcal{H}_\infty$ norm optimization since this norm is finite only for stable systems. We assume that parameters of a controller are available such that the closed-loop system of the form (9.1) is strongly exponentially stable. These parameters can, for instance, be found by minimizing the robust spectral abscissa; see Chapter 7.

Transfer functions

From (9.6) we can write the transfer function of the system (9.1) as

$$T(\lambda) := C\left(\lambda E - A_0 - \sum_{i=1}^{m} A_i e^{-\lambda \tau_i}\right)^{-1} B, \tag{9.7}$$

$$= [C_1 \ C_2]\begin{bmatrix} \lambda E^{(11)} - A_{11}(\lambda) & -A_{12}(\lambda) \\ -A_{21}(\lambda) & -A_{22}(\lambda) \end{bmatrix}^{-1} \begin{bmatrix} B_1 \\ B_2 \end{bmatrix}, \tag{9.8}$$

with

$$A_{kl}(\lambda) = \sum_{i=0}^{m} A_i^{(kl)} e^{-\lambda \tau_i}, \quad k, l \in \{1, 2\}.$$

The *asymptotic* transfer function of the system (9.1) is defined as

$$T_a(\lambda) := -CV\left(U^T A_0 V + \sum_{i=1}^{m} U^T A_i V e^{-\lambda \tau_i}\right)^{-1} U^T B \tag{9.9}$$

$$= -C_2 A_{22}(\lambda)^{-1} B_2.$$

The terminology stems from the fact that the transfer function T and the asymptotic transfer function T_a converge to each other for high frequencies. This is precisely stated in the following proposition:

Proposition 9.6. *For all* $\gamma > 0$, $\exists \Omega > 0$ $\sigma_1(T(j\omega) - T_a(j\omega)) < \gamma$ for all $\omega > \Omega$.

Proof. The assertion follows from the explicit expression for the inverse of the 2-by-2 block matrix in (9.8), combined with the property that

$$\sup_{\Re(\lambda)\geq 0} \left\| (A_{22}(\lambda))^{-1} \right\|_2 \tag{9.10}$$

is finite. The latter is due to Assumption 9.5. □

The $\mathcal{H}_\infty$ norm of the transfer function T of the *stable* system (9.1) is defined as

$$\|T(j\omega)\|_{\mathcal{H}_\infty} := \sup_{\omega\in\mathbb{R}} \sigma_1(T(j\omega)).$$

Similarly, we can define the $\mathcal{H}_\infty$ norm of T_a.

9.4 ▪ The strong $\mathcal{H}_\infty$ norm of time-delay systems

We now analyze continuity properties of the $\mathcal{H}_\infty$ norm of the transfer function T with respect to the delay parameters. The function

$$\vec{\tau} \in (\mathbb{R}_0^+)^m \mapsto \|T(j\omega, \vec{\tau})\|_{\mathcal{H}_\infty} \tag{9.11}$$

is, in general, not continuous, which is inherited from the behavior of the asymptotic transfer function, T_a, more precisely, the function

$$\vec{\tau} \in (\mathbb{R}_0^+)^m \mapsto \|T_a(j\omega, \vec{\tau})\|_{\mathcal{H}_\infty}. \tag{9.12}$$

We start with a motivating example:

Example 9.7. *Let the transfer function T be defined as*

$$T(\lambda, \vec{\tau}) = \frac{\lambda + 2.1}{(\lambda + 0.1)(1 - 0.25e^{-\lambda\tau_1} + 0.5e^{-\lambda\tau_2}) + 1}, \tag{9.13}$$

where $(\tau_1, \tau_2) = (1, 2)$. The transfer function T is stable, its $\mathcal{H}_\infty$ norm is 2.5788, achieved at $\omega = 1.6555$, and the maximum singular value plot is given in Figure 9.1. The high-frequency behavior is described by the asymptotic transfer function

$$T_a(\lambda, \vec{\tau}) = \frac{1}{(1 - 0.25e^{-\lambda\tau_1} + 0.5e^{-\lambda\tau_2})}, \tag{9.14}$$

whose $\mathcal{H}_\infty$ norm is equal to 2.0320, which is less than $\|T(j\omega, \vec{\tau})\|_{\mathcal{H}_\infty}$. However, when the first time delay is perturbed to $\tau_1 = 0.99$, the $\mathcal{H}_\infty$ norm of the transfer function T is 3.9993, reached at $\omega = 158.6569$; see Figure 9.2. The $\mathcal{H}_\infty$ norm of T is quite different from that for $(\tau_1, \tau_2) = (1, 2)$. A closer look at the maximum singular value plot of the asymptotic transfer function T_a in Figure 9.3 shows that the sensitivity is due to the transfer function T_a.

Even if the first delay is perturbed slightly to $\tau_1 = 0.999$, the problem is not resolved, indicating that the functions (9.11) and (9.12) are discontinuous at $(\tau_1, \tau_2) = (1, 2)$. The $\mathcal{H}_\infty$

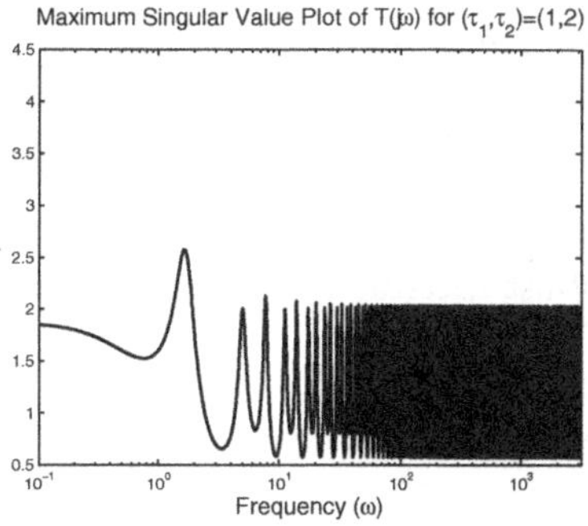

Figure 9.1. *The maximum singular value plot of $T(j\omega,\vec{\tau})$ for $(\tau_1,\tau_2)=(1,2)$ as a function of ω* [141].

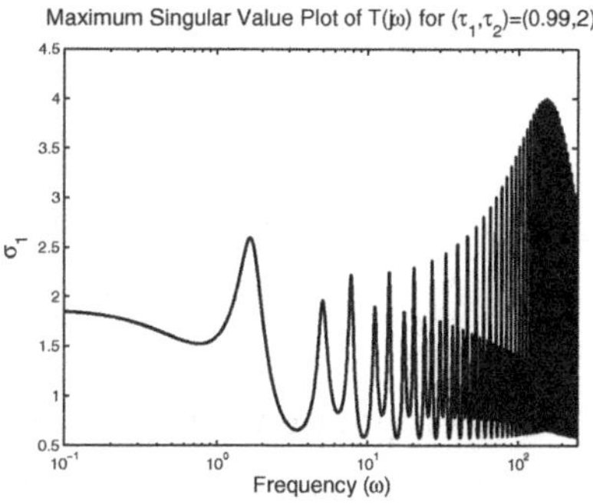

Figure 9.2. *The maximum singular value plot of $T(j\omega,\vec{\tau})$ for $(\tau_1,\tau_2)=$ (0.99, 2) as a function of ω* [141].

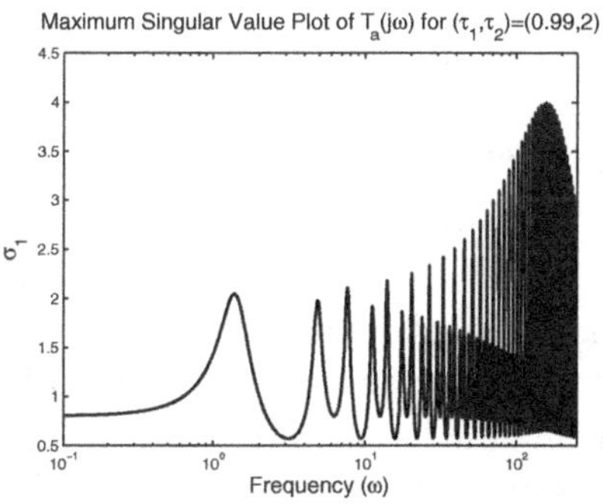

Figure 9.3. *The maximum singular value plot of $T_a(j\omega,\vec{\tau})$ for $(\tau_1,\tau_2)=$ (0.99, 2) as a function of ω* [141].

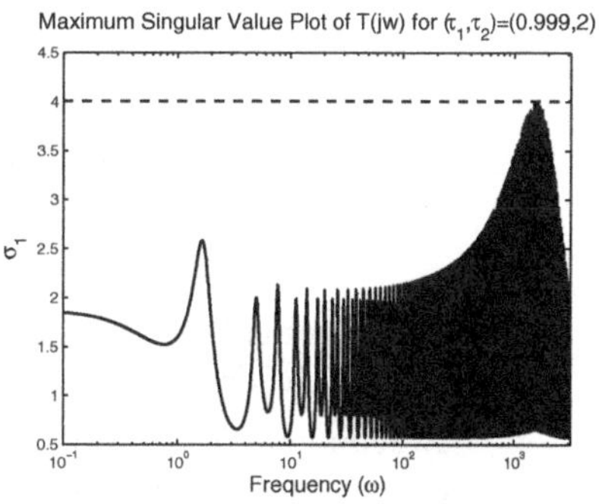

Figure 9.4. *The maximum singular value plot of $T_a(j\omega,\vec{\tau})$ for $(\tau_1,\tau_2)=$ (0.999, 2) as a function of ω* [141].

norm of the transfer function T for $(\tau_1,\tau_2)=(0.999,2)$ is given by 3.9944, and the peak value is reached at $\omega=1515.8091$. The corresponding asymptotic transfer function T_a is shown in Figure 9.4. When the delay perturbation tends to zero, the frequency where the maximum in the singular value plot of the asymptotic transfer function T_a is achieved moves towards infinity.

The above example illustrates that the $\mathcal{H}_\infty$ norm of the transfer function T may be sensitive to *infinitesimal* delay changes. On the other hand, for any $\omega_{\max}>0$, the function

$$\vec{\tau}\mapsto \max_{[0,\,\omega_{\max}]}\sigma_1(T(jw,\vec{\tau})),$$

where the maximum is taken over a compact set, is continuous, because a discontinuity would be in contradiction with the continuity of the maximum singular value function of a matrix. Hence, the sensitivity of the $\mathcal{H}_\infty$ norm is related to the behavior of the transfer function at high frequencies and, hence, the asymptotic transfer function T_a. Accordingly we start by studying the properties of the function (9.12).

Since small modeling errors and uncertainty are inevitable in a practical design, we wish to characterize the smallest upper bound for the $\mathcal{H}_\infty$ norm of the asymptotic transfer function T_a, which is *insensitive* to small delay changes:

Definition 9.8. *For $\vec{\tau} \in (\mathbb{R}_0^+)^m$, let the strong $\mathcal{H}_\infty$ norm of T_a, $|||T_a(j\omega,\vec{\tau})|||_{\mathcal{H}_\infty}$, be defined as*

$$|||T_a(j\omega,\vec{\tau})|||_{\mathcal{H}_\infty} := \lim_{\epsilon\to 0+} \sup\{\|T_a(j\omega,\vec{\tau}_\epsilon)\|_{\mathcal{H}_\infty} : \vec{\tau}_\epsilon \in \mathcal{B}(\vec{\tau},\epsilon)\cap(\mathbb{R}^+)^m\}.$$

Some important properties of this upper bound on $\|T_a(j\omega,\vec{\tau})\|_{\mathcal{H}_\infty}$ are listed below, which are similar to properties of the robust spectral abscissa studied in Chapter 1. For the proof we refer to [141]:

Proposition 9.9. *The following assertions hold:*

1. *For every $\vec{\tau} \in (\mathbb{R}_0^+)^m$, we have*

$$|||T_a(j\omega,\vec{\tau})|||_{\mathcal{H}_\infty} = \max_{\vec{\theta}\in[0,\,2\pi]^m} \sigma_1\left(\mathbb{T}_a(\vec{\theta})\right), \tag{9.15}$$

where

$$\mathbb{T}_a(\vec{\theta}) = -CV\left(U^T A_0 V + \sum_{i=1}^{m} U^T A_i V e^{-j\theta_i}\right)^{-1} U^T B. \tag{9.16}$$

2. $|||T_a(j\omega,\vec{\tau})|||_{\mathcal{H}_\infty} \geq \|T_a(j\omega,\vec{\tau})\|_{\mathcal{H}_\infty}$ *for all delays $\vec{\tau}$.*

3. $|||T_a(j\omega,\vec{\tau})|||_{\mathcal{H}_\infty} = \|T_a(j\omega,\vec{\tau})\|_{\mathcal{H}_\infty}$ *for rationally independent $\vec{\tau}$.*[21]

Formula (9.15) in Proposition 9.9 shows that the strong $\mathcal{H}_\infty$ norm of T_a is independent of the delay values. The formula further leads to a computational scheme based on sweeping on $\vec{\theta}$ intervals. This approximation can be corrected by solving a set of nonlinear equations. Numerical computation details are presented in Section 9.5.1.

We now come back to the properties of the transfer function (9.11) of the system (9.1). As we have illustrated with Example 9.7, a discontinuity of the function (9.12) may carry over to the function (9.11). Therefore, we define the strong $\mathcal{H}_\infty$ norm of the transfer function T in a similar way:

Definition 9.10. *For $\vec{\tau} \in (\mathbb{R}_0^+)^m$, the strong $\mathcal{H}_\infty$ norm of T, $|||T(j\omega,\vec{\tau})|||_{\mathcal{H}_\infty}$, is given by*

$$|||T(j\omega,\vec{\tau})|||_{\mathcal{H}_\infty} := \lim_{\epsilon\to 0+} \sup\{\|T(j\omega,\vec{\tau}_\epsilon)\|_{\mathcal{H}_\infty} : \vec{\tau}_\epsilon \in \mathcal{B}(\vec{\tau},\epsilon)\cap(\mathbb{R}^+)^m\}.$$

The following main theorem describes the desirable property that, in contrast to the $\mathcal{H}_\infty$ norm, the strong $\mathcal{H}_\infty$ norm *continuously* depends on the delay parameters. It also presents an explicit expression that is at the basis of the algorithm for computing the strong $\mathcal{H}_\infty$ norm of a transfer function, presented in the next section. For the proof we refer once more to [141].

Theorem 9.11. *The strong $\mathcal{H}_\infty$ norm of the transfer function of the DDAE* (9.1) *satisfies*

$$|||T(j\omega,\vec{\tau})|||_{\mathcal{H}_\infty} = \max\left(\|T(j\omega,\vec{\tau})\|_{\mathcal{H}_\infty}, |||T_a(j\omega,\vec{\tau})|||_{\mathcal{H}_\infty}\right), \tag{9.17}$$

where T and T_a are the transfer function (9.7) *and the asymptotic transfer function* (9.9). *In addition, the function*

$$\vec{\tau} \in (\mathbb{R}_0^+)^m \mapsto |||T(j\omega,\vec{\tau})|||_{\mathcal{H}_\infty} \tag{9.18}$$

is continuous.

[21] The m components of $\vec{\tau} = (\tau_1,\ldots,\tau_m)$ are rationally independent if and only if $\sum_{k=1}^m z_k\tau_k = 0$, $z_k \in \mathbb{Z}$, implies $z_k = 0$ for all $k = 1,\ldots,m$. For instance, two delays τ_1 and τ_2 are rationally independent if their ratio is an irrational number.

Note that the above theorem is the counterpart of Proposition 1.54.

Example 9.12. *We come back to Example 9.7. The $\mathcal{H}_\infty$ norm of T, as defined by (9.13), is 2.6422, and the strong $\mathcal{H}_\infty$ norm of the corresponding asymptotic transfer function T_a is 4. From property (9.17), we conclude that the strong $\mathcal{H}_\infty$ norm of T (9.13) is 4.*

Remark 9.13. *In contrast to delay perturbations, the $\mathcal{H}_\infty$ norm of T is continuous with respect to changes of the system matrices $A_i, \ldots, A_m$, B, and C.*

9.5 ▪ Computation of strong $\mathcal{H}_\infty$ norms

The algorithm for computing the strong $\mathcal{H}_\infty$ norm of the transfer function of (9.1) is based on property (9.17). Therefore, we first outline in Section 9.5.1 the strong $\mathcal{H}_\infty$ norm computation of the asymptotic transfer function T_a, before presenting the algorithm in Section 9.5.2.

9.5.1 ▪ Strong $\mathcal{H}_\infty$ norm of the asymptotic transfer function

The computation of $|||T_a(j\omega, \vec{\tau})|||_{\mathcal{H}_\infty}$ is based on expression (9.15) in Proposition 9.9. We obtain an approximation by restricting $\vec{\theta}$ in (9.15) to a grid,

$$|||T_a(j\omega, \vec{\tau})|||_{\mathcal{H}_\infty} \approx \max_{\vec{\theta}\in\Theta_h} \sigma_1\left(\mathbb{T}_a(\vec{\theta})\right), \tag{9.19}$$

where Θ_h is an m-dimensional grid over the hypercube $[0,\ 2\pi]^m$ and $\mathbb{T}_a(\vec{\theta})$ is defined by (9.16). If a high accuracy is required, then the approximate results may be corrected by solving the nonlinear equations

$$\left\{\begin{array}{l} \left[\begin{array}{cc} \mathbb{A}_{22}(\vec{\theta}) & -\xi^{-1}B_2B_2^T \\ \xi^{-1}C_2^TC_2 & -(\mathbb{A}_{22}(\vec{\theta}))^* \end{array}\right]\left[\begin{array}{c} u_a \\ v_a \end{array}\right] = 0, \\ n(u_a, v_a) = 0, \\ \Re(e^{-j\theta_i}(v_a^* A_i^{(22)} u_a)) = 0,\ i = 1, \ldots, m, \end{array}\right. \tag{9.20}$$

where

$$\mathbb{A}_{22}(\vec{\theta}) = -U^TA_0V - \sum_{i=1}^{m} U^TA_iVe^{-j\theta_i} \tag{9.21}$$

and $n(u_a, v_a) = 0$ is a normalization constraint. The first equation in (9.20) implies that ξ is a singular value of $\mathbb{T}_a(\vec{\theta})$. The last equation of (9.20) expresses that the derivatives of the singular value ξ with respect to the elements of $\vec{\theta}$ are zero. In our implementation we solve (9.20) using the Gauss–Newton method, which exhibits quadratic convergence because the (overdetermined) equations have an exact solution; see Section 10.2 of [305].

In most practical problems, the number of delays to be considered in $\mathbb{T}_a(\vec{\theta})$ and $\mathbb{A}_{22}(\vec{\theta})$ is much smaller than the number of system delays, m, because most of the terms in (9.21) are zero. This significantly reduces the computational cost of the sweeping in (9.19). Note that in a control application a nonzero term in (9.21) corresponds to a high-frequency feedthrough over the control loop. We illustrate this with the following example.

Example 9.14. *Consider the time-delay system*

$$\begin{cases} \dot{x}(t) = \sum_{i=2}^{m} M_i x(t-\tau_i) + B_1 w(t), \\ z(t) = Px(t) + w(t) + N_1 w(t-\tau_1). \end{cases} \tag{9.22}$$

When defining $X = [x^T \ \gamma_d^T \gamma_w^T]^T$, where γ_d and γ_w are slack variables, the system can be described by equations of the form (9.1) *as*

$$\begin{cases} \underbrace{\begin{bmatrix} I & 0 & 0 \\ 0 & 0 & 0 \\ 0 & 0 & 0 \end{bmatrix}}_{E} \dot{X}(t) = \underbrace{\begin{bmatrix} 0 & 0 & 0 \\ 0 & -I & I \\ 0 & 0 & -I \end{bmatrix}}_{A_0} X(t) + \underbrace{\begin{bmatrix} 0 & 0 & 0 \\ 0 & 0 & N_1 \\ 0 & 0 & 0 \end{bmatrix}}_{A_1} X(t-\tau_1) \\ \qquad + \sum_{i=2}^{m} \underbrace{\begin{bmatrix} M_i & 0 & 0 \\ 0 & 0 & 0 \\ 0 & 0 & 0 \end{bmatrix}}_{A_i} X(t-\tau_i) + \underbrace{\begin{bmatrix} B_1 \\ 0 \\ I \end{bmatrix}}_{B} w(t), \\ z(t) = \underbrace{\begin{bmatrix} P & I & 0 \end{bmatrix}}_{C} X(t). \end{cases}$$

The asymptotic transfer function (9.9) *is given by*

$$T_a(\lambda) = -CV\left(U^T A_0 V + \sum_{i=1}^{m} U^T A_i V e^{-\lambda\tau_i}\right)^{-1} U^T B,$$

where

$$U = V = \begin{pmatrix} 0 & 0 \\ I & 0 \\ 0 & I \end{pmatrix}.$$

Since $U^T A_i V = 0$ for $i = 2, \dots, m$, $T_a(\lambda)$ reduces to

$$\begin{aligned} T_a(\lambda) &= -CV\left(U^T A_0 V + U^T A_1 V e^{-\lambda\tau_1}\right)^{-1} U^T B, \\ &= -\begin{bmatrix} I & 0 \end{bmatrix} \begin{bmatrix} -I & -(I + N_1 e^{-\lambda\tau_1}) \\ 0 & -I \end{bmatrix}^{-1} \begin{bmatrix} 0 \\ I \end{bmatrix} \\ &= I + N_1 e^{-\lambda\tau_1}, \end{aligned}$$

which readily follows from (9.22)*. Although the original system has m delays, the asymptotic transfer function has only one delay τ_1. Accordingly, the grid Θ_h in the approximation* (9.19) *reduces to a grid on the interval* $[0,\ 2\pi]$.

In the numerical implementation, we compute the matrix norm of $U^T A_i V$ for $i = 1, \dots, m$ and omit the corresponding time delays if their norms are less than a tolerance value.

9.5.2 • Algorithm

From (9.17) the following implication can be derived:

$$|||T(j\omega, \vec{\tau})|||_{\mathcal{H}_\infty} > |||T_a(j\omega, \vec{\tau})|||_{\mathcal{H}_\infty} \Rightarrow |||T(j\omega, \vec{\tau})|||_{\mathcal{H}_\infty} = ||T(j\omega, \vec{\tau})||_{\mathcal{H}_\infty}.$$

Moreover, it can be shown that, given a level

$$\xi > |||T_a(j\omega, \vec{\tau})|||_{\mathcal{H}_\infty}, \tag{9.23}$$

there are only *finitely* many frequencies $\omega \geq 0$ for which a singular value of $T(j\omega, \vec{\tau})$ is equal to ξ. These properties allow an adaptation of the standard level set algorithm for $\mathcal{H}_\infty$ computations for finite-dimensional systems as described in [49]. The differences are twofold. First, one has to be restricted to the situation where (9.23) holds. This is possible by a preliminary computation of the strong $\mathcal{H}_\infty$ norm of T_a, as outlined in Section 9.5.1, and setting the initial level such that (9.23) is satisfied. Second, the Hamiltonian eigenvalue problem, from which intersections of singular value curves with level sets are computed, is infinite-dimensional, carrying over from the case of retarded time-delay systems discussed in [247]. Therefore, a discretization is necessary, which brings us to a predictor-corrector approach. In the predictor step, an approximation of the strong $\mathcal{H}_\infty$ norm of T (provided it exceeds $|||T_a(j\omega, \vec{\tau})|||_{\mathcal{H}_\infty}$) is obtained by computing

$$\|T_N(j\omega)\|_{\mathcal{H}_\infty}$$

using the level set method presented in [49]. Here,

$$T_N(\lambda) := \mathbf{C}_N(\lambda \mathbf{E}_N - \mathbf{A}_N)^{-1}\mathbf{B}_N \tag{9.24}$$

is the transfer function of the system

$$\begin{cases} \mathbf{E}_N \dot{z}(t) = \mathbf{A}_N z(t) + \mathbf{B}_N u(t), \\ y(t) = \mathbf{C}_N z(t), \end{cases} \tag{9.25}$$

obtained by a spectral discretization of (9.1) on the grid (2.5), resulting in $\mathbf{E}_N$ given by (2.47), $\mathbf{A}_N$ by (2.6)–(2.7), $\mathbf{B}_N$ and $\mathbf{C}_N$ by (4.53). We refer the reader to [141] for the details.

The correction step serves to remove the discretization error on the result. It is based on solving a system of nonlinear equations that characterize extrema in the singular value curves. The initial conditions are generated in the prediction step, ensuring that the algorithm converges to the right peak value. The overall algorithm for the strong $\mathcal{H}_\infty$ norm computation is as follows.

ALGORITHM 9.1.

Input: system data, N, candidate critical frequency $\{\omega_1, \dots, \omega_l\}$ *if available, tolerance tol for the prediction step,* $|||T_a(j\omega, \vec{\tau})|||_{\mathcal{H}_\infty}$

1. Prediction step:

   (a) calculate the first level,

   $$\xi_l = \max\Big(|||T_a(j\omega, \vec{\tau})|||_{\mathcal{H}_\infty}, \sigma_1(T(j\omega_1)), \dots, \sigma_1(T(j\omega_l))\Big)$$

   (b) repeat until break

      i. set $\xi := \xi_l(1 + 2\text{tol})$

      ii. compute all $\omega^{(i)} \in \mathbb{R}$ satisfying $\sigma_k\left(T_N(j\omega^{(i)})\right) = \xi$. By [118, Proposition 12], this can be done by computing generalized eigenvalues of the pencil

      $$\lambda \begin{bmatrix} \mathbf{E}_N & 0 \\ 0 & \mathbf{E}_N^T \end{bmatrix} - \begin{bmatrix} \mathbf{A}_N & \xi^{-1}\mathbf{B}_N\mathbf{B}_N^T \\ -\xi^{-1}\mathbf{C}_N^T\mathbf{C}_N & -\mathbf{A}_N^T \end{bmatrix}, \tag{9.26}$$

      whose imaginary axis eigenvalues are given by $\lambda = j\omega^{(i)}$.

iii. ***if*** no generalized eigenvalues $j\omega^{(i)}$ of (9.26) exist, ***then***

if $\xi_l = |||T_a(j\omega,\vec{\tau})|||_{\mathcal{H}_\infty}$, ***then***

set $|||T(j\omega,\vec{\tau})|||_{\mathcal{H}_\infty} = |||T_a(j\omega,\vec{\tau})|||_{\mathcal{H}_\infty}$

quit

else

let $\omega^{(i)} \in \mathbb{R}$ satisfying $\sigma_k\left(T_N(j\omega^{(i)})\right) = \xi_l$,

set $\tilde{\xi} = (\xi + \xi_l)/2$, $\tilde{\omega}^{(i)} = \omega^{(i)}$, $i = 1,2,\ldots$

break, go to the correction step 2.

endif

else

calculate $\mu^{(i)} := \sqrt{\omega^{(i)}\omega^{(i+1)}}$, $i = 1,2,\ldots$

set

$$\xi_l := \max_i \max\left(\sigma_1\left(T_N(j\mu^{(i)})\right), |||T_a(j\omega,\vec{\tau})|||_{\mathcal{H}_\infty}\right).$$

endif

2. *Correction step:*

(a) Solve the nonlinear equations

$$\left\{\begin{array}{l} H(j\omega,\xi)\left[\begin{array}{c} u \\ v \end{array}\right] = 0, \\ n(u,v) = 0, \\ \Im\{v^*(E + \sum_{i=1}^m A_i\tau_i e^{-j\omega\tau_i})u\} = 0 \end{array}\right. \tag{9.27}$$

using the Gauss–Newton method, where

$$H(j\omega,\xi) = \left[\begin{array}{cc} j\omega E - A_0 - \sum_{i=1}^m A_i e^{-j\omega\tau_i} & -\xi^{-1}BB^T \\ \xi^{-1}C^TC & j\omega E^T + A_0^T + \sum_{i=1}^m A_i^T e^{j\omega\tau_i} \end{array}\right]$$

and $n(u,v) = 0$ is a normalizing condition, with the starting values

$$\omega = \tilde{\omega}^{(i)}, \quad \xi = \tilde{\xi}, \quad \left[\begin{array}{c} u \\ v \end{array}\right] = \arg\min \|H(j\tilde{\omega}^{(i)},\tilde{\xi})\zeta\|/\|\zeta\|;$$

denote the solutions with $(\hat{u}^{(i)},\hat{v}^{(i)},\hat{\omega}^{(i)},\hat{\xi}^{(i)})$ for $i = 1,2,\ldots,$

(b) set $|||T(j\omega)|||_{\mathcal{H}_\infty} := \max_{1\le i\le p}\hat{\xi}^{(i)}$

The first and second equations in (9.27) describe the presence of a singular value ξ of matrix $T(j\omega,\vec{\tau})$. The third equation expresses that the derivative of this singular value with respect to ω is equal to zero; see [247]. Hence, Equations (9.27) can be used to correct approximate peak values. Note that the correction step is performed only if

$$|||T(j\omega,\vec{\tau})|||_{\mathcal{H}_\infty} > |||T_a(j\omega,\vec{\tau})|||_{\mathcal{H}_\infty}.$$

For details on the choice of the number of discretization points, N, and the tolerance, tol, we refer the reader to [247].

The main ideas behind Algorithm 9.1 are clarified with two examples:

Example 9.15. *We apply Algorithm* 9.1 *to the transfer function T, as specified by* (9.13). *The strong $\mathcal{H}_\infty$ norm of its asymptotic transfer function T_a* (9.14) *satisfies $|||T_a(j\omega,\vec{\tau})|||_{\mathcal{H}_\infty}=4$. Therefore, the first level is equal to $\xi_l=4$ in step* (a) *of the prediction step of Algorithm* 9.1, *provided that no candidate frequencies are given. In step* (b)ii, *there is no intersection for the level ξ as shown in Figure* 9.5. *Therefore the strong $\mathcal{H}_\infty$ norm of T* (9.13) *is equal to* 4 *and the correction step is not carried out.*

Example 9.16. *We consider the transfer function*

$$T(\lambda,\vec{\tau}):=\frac{\lambda+2}{\lambda(1-1/16e^{-\lambda\tau_1}+1/2e^{-\lambda\tau_2})+1}, \tag{9.28}$$

with $\vec{\tau}=(1,2)$, and its asymptotic transfer function

$$T_a(\lambda,\vec{\tau}):=\frac{1}{(1-1/16e^{-\lambda\tau_1}+1/2e^{-\lambda\tau_2})}. \tag{9.29}$$

The first level ξ_l is set to the strong $\mathcal{H}_\infty$ norm of T_a (9.29),

$$|||T_a(j\omega,\vec{\tau})|||_{\mathcal{H}_\infty}=2.2857.$$

In step (b)ii, *there are two intersections for the level $\xi=2.2903$, $\omega^{(1)}=1.6600$ and $\omega^{(2)}=1.8786$, as shown in Figure* 9.6. *We can see the details of the next iterations in Figure* 9.7. *Step* (b)iii *calculates the middle frequency $\mu^{(1)}=1.7656$ and $\xi_l=2.3855$ for the next level. In the second iteration, the level is set to $\xi=2.3903$ and the corresponding intersections are $\omega^{(1)}=1.7656$ and $\omega^{(2)}=1.7786$. Since there is no intersection in the third iteration due to the chosen tolerance in the prediction step,* $\mathrm{tol}=10^{-3}$, *we compute the approximate strong $\mathcal{H}_\infty$ norm of T* (9.28) *and the frequencies as $\tilde{\xi}=2.3879$ and $(\tilde{\omega}_1,\tilde{\omega}_2)=\{1.7657,1.7786\}$. In the correction step, these values are corrected and the strong $\mathcal{H}_\infty$ norm of T* (9.28) *and the corresponding frequency are computed as $\hat{\xi}=2.3859$ and $\hat{\omega}=1.7721$.*

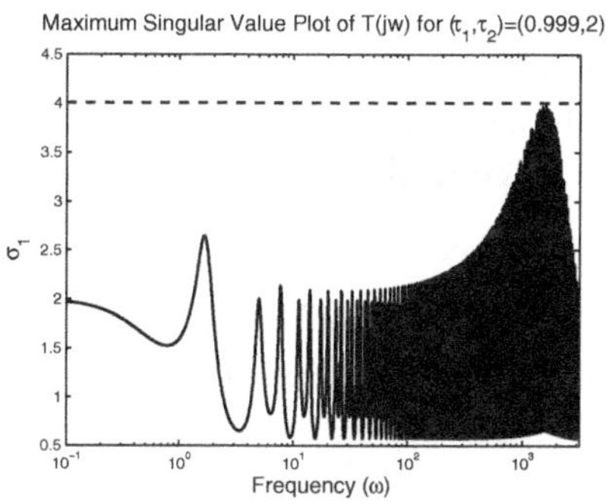

Figure 9.5. *Algorithm* 9.1 *for the maximum singular value plot of T* (9.13)*: no intersection case* [141].

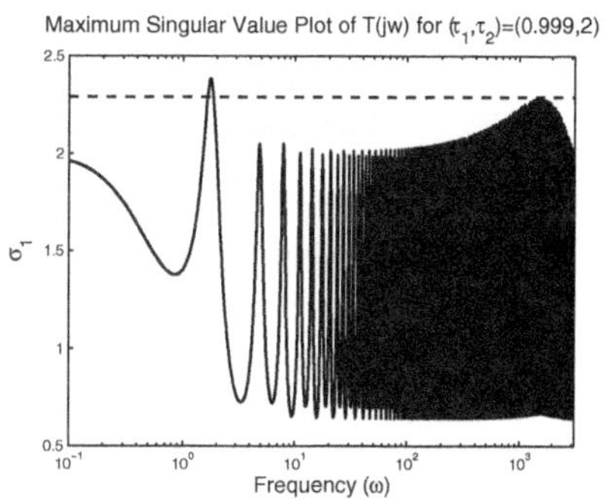

Figure 9.6. *Algorithm* 9.1 *for the maximum singular value plot of T* (9.28)*: with intersections case* [141].

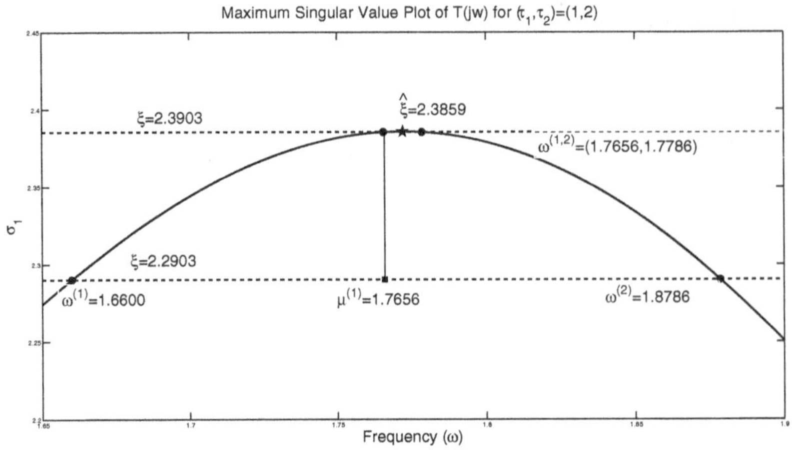

Figure 9.7. *Steps of Algorithm 9.1 for the maximum singular value plot of T (9.28): with intersections case (zoomed)* [141].

9.6 ▪ Fixed-order $\mathcal{H}_\infty$ controller design

We consider the equations

$$\begin{cases} E\dot{x}(t) = A_0(p)x(t) + \sum_{i=1}^{m} A_i(p)x(t-\tau_i) + Bw(t), \\ z = Cx(t), \end{cases} \tag{9.30}$$

where the system matrices smoothly depend on parameters p. As illustrated in Section 9.2, a broad class of interconnected systems can be brought into this form, where the parameters p can be interpreted in terms of a parametrization of a controller. For example, in the feedback interconnection of (9.4) and (9.5) they may correspond to the elements of the matrices of the controller K. Note that, by fixing some elements of these matrices, additional structure can be imposed on the controller, e.g., a proportional-integrative-derivative (PID)-like structure.

The proposed method for designing fixed-order/fixed-structure $\mathcal{H}_\infty$ controllers is based on a direct minimization of the strong $\mathcal{H}_\infty$ norm of the closed-loop transfer function T from w to z as a function of the parameters p. The overall optimization algorithm requires the evaluation of the objective function and its gradients with respect to the optimization parameters, whenever it is differentiable.

The strong $\mathcal{H}_\infty$ norm of the transfer function T can be computed by Algorithm 9.1. The derivatives of the norm with respect to controller parameters exist whenever there are unique values $\vec{\hat{\tau}}$ or $\vec{\hat{\theta}}$ such that

$$|||T(j\omega,\vec{\tau})|||_{\mathcal{H}_\infty} = \hat{\xi} = \begin{cases} \sigma_1\left(\mathbb{T}_a\left(\vec{\hat{\theta}}\right)\right) & \text{if } \hat{\xi} = |||T_a\left(\vec{\hat{\theta}},\vec{\tau}\right)|||_{\mathcal{H}_\infty}, \\ \sigma_1(T(j\hat{\omega})) & \text{if } \hat{\xi} > |||T_a\left(\vec{\hat{\theta}},\vec{\tau}\right)|||_{\mathcal{H}_\infty} \end{cases}$$

holds and, in addition, the largest singular value $\hat{\xi}$ has multiplicity one. We compute the derivative of the strong $\mathcal{H}_\infty$ norm of T with respect to the parameter p_k as

$$\frac{\partial \xi}{\partial p_k} = \begin{cases} -2\xi^2 \left. \dfrac{\Re\left(v_a^* \frac{\partial \mathbb{A}_{22}(\vec{\theta})}{\partial p_k} u_a\right)}{v_a^* B_2 B_2^T v_a + u_a^* C_2^T C_2 u_a} \right|_{(\xi,\vec{\theta})=\left(\hat{\xi},\hat{\vec{\theta}}\right)} & \text{if } \hat{\xi} = |||T_a\left(\hat{\vec{\theta}},\vec{\tau}\right)|||_{\mathcal{H}_\infty}, \\ -2\xi^2 \left. \dfrac{\Re\left(v^* \frac{\partial A(j\omega)}{\partial p_k} u\right)}{v^* B B^T v + u^* C^T C u} \right|_{(\xi,\omega)=(\hat{\xi},\hat{\omega})} & \text{if } \hat{\xi} > |||T_a\left(\hat{\vec{\theta}},\vec{\tau}\right)|||_{\mathcal{H}_\infty}, \end{cases}$$

where given $\xi = \hat{\xi}$, u_a, v_a, and u, v are vectors in (9.20) and (9.27) for $\vec{\theta} = \hat{\vec{\theta}}$ and $\omega = \hat{\omega}$, respectively. For more details on the computation of derivatives we refer the reader to [139].

The overall design procedure is fully automated and does not require any interaction with the user. The computational cost of the optimization algorithm is dominated by the evaluation of the strong $\mathcal{H}_\infty$ norm of the closed-loop transfer function T for the parameters p at each iteration. The first main part in this computation is to find the strong $\mathcal{H}_\infty$ norm of the asymptotic transfer function by computing the maximum singular value of $\mathbb{T}_a$ at $p_a^{m_a-1}$ points spanning the grid Θ_h in (9.19), where p_a is the number of grid points in the interval $[0,\ 2\pi]$ (the default value is 20 in our implementation) and m_a is the number of actual delays appearing in $\mathbb{A}_{22}(\vec{\theta})$; see (9.21). Note that the number of delays m_a is usually much smaller than the number of system delays; see the arguments at the end of Section 9.5.1. Therefore the computational cost for sweeping is usually not very high. It is even completely skipped if there is no high-frequency feedthrough in the control loop (which results in $m_a = 0$). The second main part is the computation of the generalized eigenvalues of the pencil (9.26) in the prediction step of Algorithm 9.1. This computation requires solving a generalized eigenvalue problem with dimensions $2nN$ where the default value for N is 20 in our implementation. The number of iteration steps of the optimization algorithm heavily depends on the optimization problem under consideration. In most cases, satisfactory results are already obtained in the first phase of the optimization algorithm where the BFGS algorithm is used. For the behavior of BFGS, applied to nonsmooth problems, we refer the reader to [206].

Recall that the feedback interconnection of system (9.4) and controller (9.5) can be rewritten in the form (9.30) in such a way that the closed-loop matrices depend affinely on the matrices of the controller. This property improves the performance of the optimization method. Note that in the existing work for systems without delay (see, e.g., [143]) the dependency is in general nonlinear, due to the use of elimination for handling a nontrivial feedthrough, as illustrated in Example 9.1.

9.7 ▪ Examples

In Section 9.7.1 we illustrate some aspects of the proposed approach on a motivating example. In Section 9.7.2 apply the approach to benchmark examples collected from the literature.

9.7.1 ▪ Motivating example

We consider the design of a static controller for the case where the standard $\mathcal{H}_\infty$ norm and the strong $\mathcal{H}_\infty$ norm of the closed-loop systems are different. The transfer function T, as defined by (9.13), can be interpreted as the transfer function of the closed-loop system

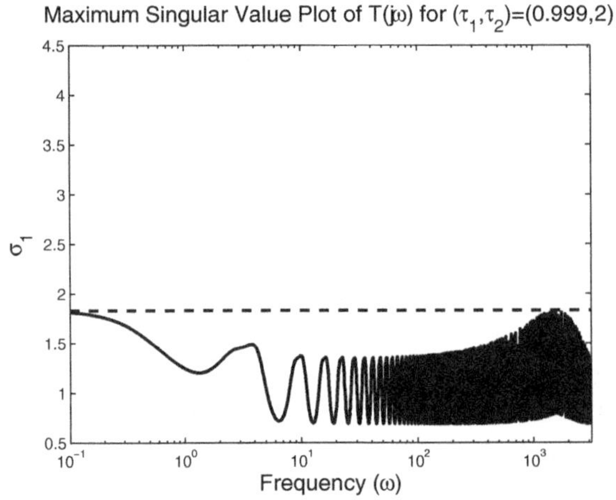

Figure 9.8. *The maximum singular value plot of the closed-loop system $T(j\omega)$ (9.13) for $(\tau_1, \tau_2) = (0.999, 2)$ with the controller K_{opt} [141].*

formed by the plant

$$\begin{cases} \begin{bmatrix} 1 & 0 \\ 0 & 0 \end{bmatrix} \dot{x}(t) = \begin{bmatrix} -0.1 & -1 \\ 1 & -1 \end{bmatrix} x(t) + \begin{bmatrix} 0 \\ 1 \end{bmatrix} u(t) + \begin{bmatrix} 0 \\ 1 \end{bmatrix} w(t), \\ z(t) = \begin{bmatrix} 2 & -1 \end{bmatrix} x(t), \\ y(t) = \begin{bmatrix} 0 & 1 \\ 0 & 0 \end{bmatrix} x(t-\tau_1) + \begin{bmatrix} 0 & 0 \\ 0 & 1 \end{bmatrix} x(t-\tau_2), \end{cases}$$

where $(\tau_1, \tau_2) = (1, 2)$ and the controller is

$$u(t) = Ky(t),$$

where $K = K_{\text{init}} = \begin{bmatrix} 0.25 & -0.5 \end{bmatrix}$. In Example 9.15, we computed the standard $\mathcal{H}_\infty$ norm 2.5788 and the strong $\mathcal{H}_\infty$ norm 4, as illustrated in Figure 9.5 with slightly perturbed delay values. An optimization of the strong $\mathcal{H}_\infty$ norm results in

$$K = K_{opt} = \begin{bmatrix} -0.3533 & -0.1012 \end{bmatrix},$$

and the corresponding optimal value is given by 1.8333. As shown in Figure 9.8, the optimization method pushes the strong $\mathcal{H}_\infty$ norm of the asymptotic transfer function until it is equal to the standard $\mathcal{H}_\infty$ norm. Hence, the minimum is characterized by a balance between low- and high-frequency behavior of the transfer function.

9.7.2 ▪ A collection of examples from the literature

We collected benchmark examples for the $\mathcal{H}_\infty$ optimization of time-delay systems from the literature, whose system description can be found in Table 9.1.

We considered two types of problems: the $\mathcal{H}_\infty$ optimization with *state* and *output* feedback controllers. Our results are given in Table 9.2 and 9.3, respectively.

The designed controllers for the examples in Table 9.3 have a state-feedback-observer structure. The observer is a time-delay system and estimates the states of the original

Table 9.1. *Data for the numerical examples.*

Example 4 in [112]

$$\dot{x}_1(t) = -x_1(t-0.999) - x_2(t-0.999) + w(t),$$
$$\dot{x}_2(t) = x_2(t) - 0.9x_2(t-0.999) + w(t) + u(t),$$
$$z_1(t) = x_2(t), \quad z_2(t) = 0.1u(t), \quad y_1(t) = x_1(t), \quad y_2(t) = x_2(t).$$

Example 1 in [110]

$$\dot{x}_1(t) = 2x_1(t) + x_2(t) - x_1(t-0.1) - 0.5w(t) + 3u(t),$$
$$\dot{x}_2(t) = -x_2(t) - x_1(t-0.1) + x_2(t-0.1) + w(t) + u(t),$$
$$z_1(t) = x_1(t) - 0.5x_2(t), \quad z_2(t) = u(t), \quad y_1(t) = x_1(t), \quad y_2(t) = x_2(t).$$

Example 2 in [113]

$$\dot{x}_1(t) = -x_1(t-1.2) + w(t) - 0.5u(t),$$
$$0 = x_1(t-1.2) - x_2(t-1.2) + w(t) + u(t),$$
$$z_1(t) = x_1(t) + 0.2x_2(t) + 0.1u(t), \quad y_1(t) = x_1(t), \quad y_2(t) = x_2(t).$$

Example 3 in [113]

$$\dot{x}_1(t) = -x_1(t-1.2) + w_1(t),$$
$$0 = x_2(t) + x_1(t-1.2) - x_2(t-1.2) + w_1(t) + u(t),$$
$$z_1(t) = x_1(t) + 0.2x_2(t) + 0.1u(t), \quad y_1(t) = x_1(t) + 0.1w_2(t).$$

Example 4 in [113]

$$\dot{x}_1(t) = -x_1(t-h) - x_2(t-h) + w_1(t),$$
$$\dot{x}_2(t) = x_2(t) - 0.9x_2(t-h) + w_1(t) + u(t),$$
$$z_1(t) = x_2(t), \quad z_2(t) = 0.1u(t), \quad y_1(t) = x_2(t) + 0.1w_2(t).$$

Example in [265]

$$\dot{x}(t) = A_0 + \textstyle\sum_{i=1}^{5} A_i x(t-h_i) + I_{11}w(t) + Bu(t-7),$$
$$z(t) = I_{11}x(t), \quad y(t) = I_{11}x(t)$$

where the time delays are $h_1 = 3$, $h_2 = 5$, $h_3 = 15$, $h_4 = 23$, $h_5 = 29$.

The system matrices A_i for $i = 0, \ldots, 6$ are real-valued 11×11 matrices.

The (k,l) nonzero element of A_ith matrix is denoted by $A_i^{(k,l)}$, and the numerical values are $A_0^{(1,1),(3,3),(5,5),(9,9)} = -0.2$, $A_0^{(4,7),(4,8),(8,3),(8,4)} = 0.1417$,
$A_0^{(2,2)} = -0.04$, $A_0^{(6,6)} = -0.0588$, $A_0^{(6,6)} = -1$, $A_0^{(10,10)} = -0.0667$,
$A_0^{(4,3)} = A_0^{8,7} = 0.1917$, $A_0^{(4,4)} = A_0^{(8,8)} = -0.04$, $A_1^{(5,4)} = 0.195$,
$A_2^{(3,2)} = 0.1966$, $A_2^{(6,5)} = 0.0529$,
$A_2^{(9,8)} = 0.194$, $A_2^{(10,9)} = 0.0613$,
$A_3^{1,6} = 0.1946$, $A_4^{2,1} = 0.0384$, $A_5^{7,7} = -0.0159$.

Example 2 in [139]

$$\dot{x}(t) = \begin{bmatrix} -4.4656 & -0.4271 & 0.4427 & -0.1854 \\ -0.8601 & -5.6257 & 0.8577 & -0.5210 \\ 0.9001 & -0.7177 & -6.5358 & 0.0417 \\ -0.6836 & 0.0242 & 0.4997 & -3.5618xx \end{bmatrix} x(t)$$
$$+ \begin{bmatrix} 0.6848 & -0.0618 & 0.5399 & 0.5057 \\ 0.3259 & -0.3810 & 0.6592 & -0.0066 \\ 0.6325 & 0.3752 & 0.4122 & 0.7303 \\ 0.5878 & 0.9737 & 0.1907 & -0.8639 \end{bmatrix} x(t-3.2)$$
$$+ \begin{bmatrix} 0.9371 & -0.7859 & 0.1332 & 0.7429 \\ -0.8025 & 0.4483 & 0.6226 & 0.0152 \\ 0.0940 & 0.2274 & 0.1536 & 0.5776 \\ -0.1941 & 0.5659 & 0.8881 & -0.0539 \end{bmatrix} x(t-3.4)$$
$$+ \begin{bmatrix} 1 & 0 \\ -1.6 & 1 \\ 0 & 0 \\ 0 & 0 \end{bmatrix} w(t) + \begin{bmatrix} 0.6576 & -0.8543 & -0.3460 & 0.6415 \\ -0.3550 & 0.5024 & 0.6081 & 0.9038 \\ 0.9523 & 0.6624 & 0.0765 & -0.8475 \\ -0.4436 & 0.8447 & -0.0734 & 0.4173 \end{bmatrix} x(t-3.9)$$
$$+ \begin{pmatrix} 0.2 \\ -1 \\ 0.1 \\ -0.4 \end{pmatrix} u(t-0.2)$$

$$z(t) = \begin{bmatrix} 1 & 0 & 0 & -1 \\ 0 & -1 & 1 & 0 \end{bmatrix} x(t) + \begin{bmatrix} 0.1 & 1 \\ -1 & 0.2 \end{bmatrix} w(t) + \begin{bmatrix} 1 \\ -1 \end{bmatrix} u(t)$$

$$y(t) = \begin{bmatrix} 1 & 0 & -1 & 0 \end{bmatrix} x(t) + \begin{bmatrix} -2 & 0.1 \end{bmatrix} w(t) + 0.4u(t-0.2)$$

Table 9.2. *The achieved $\mathcal{H}_\infty$ performances by state-feedback controllers* [141].

Problem	Other methods	Results	Computed controller
Ex. 4, [112]	1.8822, [108] 0.2284, [111] 0.1287, [112]	0.1000	$[-2.3273,\ -9.5004\ 10^3]$
Ex. 1, [110]	0.4215, [110]	0.4005	$[-17.8065,\ 9.5915]$
Ex. 2, [113]	21, [113]	2.9091	$[-1.1151\ 10^3,\ -1.6189\ 10^4]$

Table 9.3. *The achieved $\mathcal{H}_\infty$ performances by output-feedback controllers* [141].

Problem	Other methods	Results	Computed controller
Ex. 3, [113]	2.4, [113]	3.7654	$[-8.6961]$
		1.2618	$\left[\begin{array}{c\|c} -7.1827 & -37.3389 \\ \hline 18.6767 & 90.4893 \end{array}\right]$
		1.2428	$\left[\begin{array}{cc\|c} -2.6837 & -15.1028 & -6.2101 \\ 0.3607 & 1.2086 & 3.6959 \\ \hline 0.1379 & -3.9720 & 10.4548 \end{array}\right]$
Ex. 4, [113] ($h = 0.999$)	0.8600, [112] 11, [113]	0.1617	$[-16.1692]$
Ex. 4, [113] ($h = 1.28$)	20, [113]	0.1617	$[-16.1692]$

plant. Example 3 in the tables is described by a DDAE and Example 4 is a retarded time-delay system. The strong $\mathcal{H}_\infty$ norms of the closed-loop system in Example 4 of [113] for $h = 0.999$ and $h = 1.28$ with respect to controller parameters are given in Figure 9.9 and 9.10. The optimal controller gain does not change for two different delays. In both cases, the optimization method reaches the optimal value.

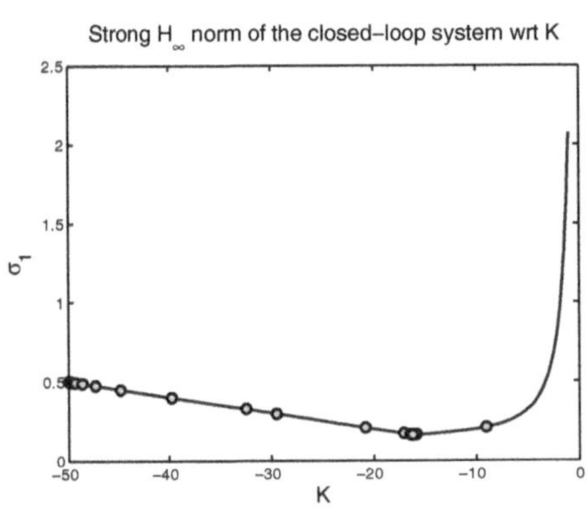

Figure 9.9. *The strong $\mathcal{H}_\infty$ norm of the closed-loop system in Example 4 ($h = 0.999$) of* [113] *with respect to the controller parameter* [141].

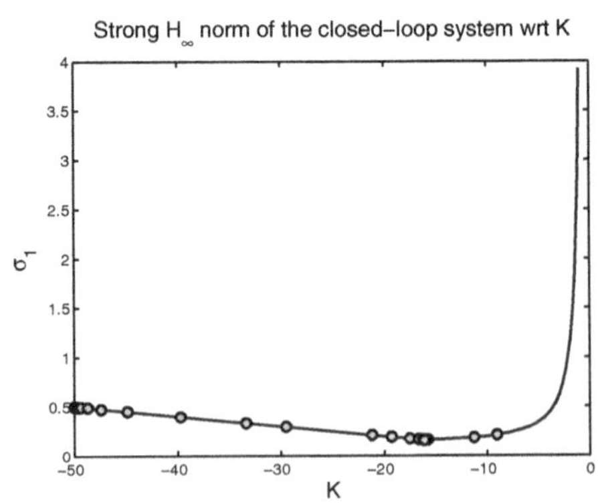

Figure 9.10. *The strong $\mathcal{H}_\infty$ norm of the closed-loop system in Example 4 ($h = 1.28$) of* [113] *with respect to the controller parameter* [141].

In [265] a state-feedback controller is designed for time-delay systems based on quasi-direct pole placement. This approach allows us to assign a number of fixed rightmost poles while shifting the remaining part of the spectrum as far to the left as possible. We designed a static $\mathcal{H}_\infty$ controller for the experimental heat transfer setup described in Section 3 of [265]. This is an 11th-order retarded time-delay system with 5 state delays and 1 input delay,

$$\begin{cases} \dot{x}(t) = A_0 x(t) + \sum_{i=1}^{5} A_i x(t-\tau_i) + w(t) + Bu(t-\tau_6), \\ z(t) = x(t), \quad y(t) = x(t), \end{cases}$$

where the system matrices and delays, A_0, A_i and τ_i for $i = 0, \ldots, 5$ are also specified in Table 9.1. The performance channels are set to identity matrices. The controller parameters of four static controllers designed using quasi-direct pole placement are given in Table 1 of [265]. The closed-loop strong $\mathcal{H}_\infty$ norms with these controllers are 779.1600, 1881.3944, 1155.7140, and 2113.8085. We achieved a minimal closed-loop strong $\mathcal{H}_\infty$ norm 386.3491 by a static $\mathcal{H}_\infty$ controller, $u(t) = Kx(t)$, where

$$\begin{aligned} K = [-1.3414 \;\; -5.7544 \;\; 1.0440 \;\; 0.5181 \;\; -29.9649 \;\; -5.0182 \\ -12.4284 \;\; 0.6694 \;\; 4.7125 \;\; -23.6380 \;\; 2.3902]. \end{aligned}$$

Finally, in [139] a direct optimization approach is applied to the design of fixed-order $\mathcal{H}_\infty$ controllers for a class of retarded time-delay systems where the controller has no feedthrough term. The second example in [139] is a 4th-order time-delay system with 4 delays. The system is stable and its $\mathcal{H}_\infty$ norm is 1.3907. In Table 9.4, we present our results for different controller orders n_K for this example, without any additional restriction on the controller.

Table 9.4. *The achieved $\mathcal{H}_\infty$ performances for the time-delay system in [139] by dynamic controllers [141].*

n_K	Results	Computed controller
1	1.2513	$\left[\begin{array}{c\|c} -0.3068 & 0.9590 \\ \hline 0.0166 & 0.0186 \end{array}\right]$
2	1.2508	$\left[\begin{array}{cc\|c} -0.0959 & -0.0624 & -0.0982 \\ -0.0024 & -0.1984 & 0.0883 \\ \hline -0.0756 & 0.0347 & 0.0234 \end{array}\right]$
3	1.2493	$\left[\begin{array}{ccc\|c} -0.0861 & -0.0673 & -0.0953 & -0.0519 \\ 0.0046 & -0.2170 & -0.0233 & 0.1083 \\ -0.0016 & 0.0010 & -0.2973 & 0.1995 \\ \hline -0.1734 & -0.1040 & -0.0475 & 0.0362 \end{array}\right]$

9.8 ▪ Optimization of stability radii

The optimization of complex stability radii can be turned into an $\mathcal{H}_\infty$ synthesis problem that can be solved using the approach developed in the previous sections. As an illustration we consider the system

$$\dot{x}(t) = Ax(t) + Bu(t-\tau), \quad u = K^T x(t). \tag{9.31}$$

We assume that the controlled system (9.31) is asymptotically stable and consider the stability of the perturbed system

$$\dot{x}(t) = (A + \delta A)x(t) + (B + \delta B)(K + \delta K)^T x(t - \tau) \tag{9.32}$$

under various classes of perturbations on A, B, and K. The nominal characteristic matrix is given by

$$F(\lambda) := \lambda I - A - BK^T e^{-\lambda\tau}.$$

As a perturbation of A corresponds to an additive perturbation of F,

$$\delta F_A = -\delta A,$$

an application of Theorem 3.15 yields

$$r_{\mathbb{C}}(F;\ \mathbb{C}_-, \mathbf{\Delta}_A) = \left(\sup_{\omega \geq 0} \left\| (j\omega I - A - BK^T e^{-j\omega\tau})^{-1} \right\|_2 \right)^{-1}, \tag{9.33}$$

with $\mathbf{\Delta}_A = \mathbb{C}^{3\times 3}$. Similarly, if B or K is perturbed, then the additive perturbations are

$$\delta F_B = -\delta B\, K^T, \qquad \delta F_K = -B\, \delta K^T,$$

and we get

$$\begin{array}{l} r_{\mathbb{C}}(F;\ \mathbb{C}_-, \mathbf{\Delta}_B) = \left(\sup_{\omega \geq 0} \left\| K^T (j\omega I - A - BK^T e^{-j\omega\tau})^{-1} \right\|_2 \right)^{-1}, \\ r_{\mathbb{C}}(F;\ \mathbb{C}_-, \mathbf{\Delta}_K) = \left(\sup_{\omega \geq 0} \left\| (j\omega I - A - BK^T e^{-j\omega\tau})^{-1} B \right\|_2 \right)^{-1}, \end{array} \tag{9.34}$$

with $\mathbf{\Delta}_B = \mathbb{C}^{3\times 1}$ and $\mathbf{\Delta}_K = \mathbb{C}^{1\times 3}$. The expressions (9.33)–(9.34) have the form

$$\left(\sup_{\omega \geq 0} \| M_K(j\omega) \|_2 \right)^{-1} = \left(\| M_K(j\omega) \|_{\mathcal{H}_\infty} \right)^{-1}, \tag{9.35}$$

where $M_K(j\omega)$ is given by

$$M_K(j\omega) = X \left(j\omega I - A - BK^T e^{-j\omega\tau} \right)^{-1} Y,$$

with the constant matrices X and Y depending on which matrix is perturbed.

In what follows we present a numerical example where system matrices A, B and delay τ are given by (7.11) and where perturbations on K are considered. To simplify the notation we denote $r_{\mathbb{C}}(F;\ \mathbb{C}_-, \mathbf{\Delta}_K)$ by $r_{\mathbb{C}}^K$. In Chapter 7 the spectral abscissa of (9.31),

$$c(K) := \sup\{\Re(\lambda) :\ \det\left(\lambda I - A - BK^T\right) = 0\}, \tag{9.36}$$

was minimized, yielding $K = [0.471\ 0.504\ 0.607]^T$ and $c(K) = -0.15$. Starting from these values, we optimize $r_{\mathbb{C}}^K$ as a function of K. The optimized stability radius is 2.287 and the corresponding state-feedback controller is defined by

$$K = \begin{bmatrix} 0.7763 & 1.1119 & 0.5433 \end{bmatrix}.$$

Table 9.5 illustrates the trade-off between performance, in the sense of a fast exponential decay of solutions, and robustness, in the sense of a large complex stability radius $r_{\mathbb{C}}^K$ (i.e., $1/\|M_K(j\omega)\|_{\mathcal{H}_\infty}$).

Table 9.5. *Trade-off between performance, in terms of exponential decay rate and robustness in terms of a complex stability radius.*

Optimal	$\alpha(K)$	$r_{\mathbb{C}}^K$
$\alpha(K)$	-0.15	0.0938
$r_{\mathbb{C}}^K$	-0.073	0.287

9.9 ▪ Notes and references

We considered the fixed-order and fixed-structure $\mathcal{H}_\infty$ controller design problem for systems described by delay-differential algebraic equations. We showed that a very broad class of interconnected systems can be brought to the standard form (9.1) in a systematic way. Input/output delays and direct feedthrough terms can be dealt with by introducing slack variables. The dependence of the closed-loop matrices on the controller parameters always remains linear. We demonstrated the sensitivity of the $\mathcal{H}_\infty$ norm w.r.t. small delay perturbations and introduced the concept of *strong* $\mathcal{H}_\infty$ *norm*, in analogy with the notion of strong stability, and we analyzed its properties. We presented a predictor-corrector–based algorithm for the (strong) $\mathcal{H}_\infty$ norm computation. Based on the numerical algorithm for the strong $\mathcal{H}_\infty$ norm and its gradient computation with respect to controller parameters, we applied nonsmooth, nonconvex optimization methods for designing controllers with a fixed order or structure. We also illustrated how the approach can be used to optimize stability robustness expressed in terms of complex stability.

The proposed method for computing optimal $\mathcal{H}_\infty$ controllers assumes that the initial feedback gain is stabilizing. Such a stabilizing value can be calculated with the method of Chapter 7, which ultimately relies on the same optimization algorithms.

In the controller design problem for large-scale problems, calculating the strong $\mathcal{H}_\infty$ norm is a bottleneck from a computational point of view. A promising direction is taken in [246], albeit in the context of the related problem of calculating pseudospectral abscissa. Here we exploit the property that the boundary of complex pseudospectra can be reached by applying rank one perturbations to the systems matrices, leading to an iteration on a manifold of low rank matrices. The approach for optimizing complex stability radii can be extended to systems with structured perturbations, which results in expressions of stability radii in terms of structured singular values (see Theorem 3.15). It is important to mention that in the cases where only upper bounds on the structured singular value can be computed, an optimization involving such upper bounds eventually yields suboptimal values of the stability radius to be optimized. Multiplicative uncertainty, as occurs for instance in the problem (9.32) with combined uncertainty on A, B, and K, can be "linearized" with the descriptor transformation outlined at the end of Section 3.3.2. Finally, we note that an algorithm for the optimization of real stability radii of time-delay systems was described in [259].

For the optimization of $\mathcal{H}_2$ norms we refer the reader to [385] and the references therein.

This chapter is mainly based on [141], where another set of benchmarks for the proposed controller synthesis methods can be found. It is also based on work reported in [263, 259, 245, 401].

Part III

Applications

Chapter 10

Output feedback stabilization using delays as control parameters

10.1 ▪ Introduction

The existence of a time-delay at the actuating input in a feedback control system is usually known to cause *instability* or poor performance for the closed-loop schemes as largely presented and discussed in the literature (see, for instance, [226, 196, 290, 130] and the references therein). In what follows, we address the opposite problem: characterizing the situations where a delay has a *stabilizing effect*. In other words, we consider the situation where the delay-free feedback system is unstable but becomes asymptotically stable due to the presence of an appropriate delay in the actuating input. More precisely, we study the stabilization of the SISO system with the following state-space representation ($A \in \mathbb{R}^{n\times n}$, $B, C \in \mathbb{R}^{n\times 1}$):

$$\dot{x}(t) = Ax(t) + Bu(t), \qquad y = C^T x(t), \tag{10.1}$$

using the control law

$$u(t) = -ky(t-\tau). \tag{10.2}$$

The *stabilizing delay effect* problem can be defined as follows:

Problem 10.1. *Find explicit conditions on the delay block defined by the pair* (k,τ) *such that the controller* (10.2) *stabilizes the system* (10.1), *but the closed-loop system would be unstable if the delay* τ *is set to zero.*

In the case when one delay block is *not sufficient* for stabilizing the corresponding system, a natural question is to see whether a controller of the form

$$u(t) = -\sum_{i=1}^{l} k_i\, y(t-\tau_i), \tag{10.3}$$

including $l > 1$ delay blocks, may guarantee the *stability* of the closed-loop scheme.

As we shall see below, in the case of one delay block, the conditions derived will lead to an explicit construction of the controller. Furthermore, for each stabilizing pair, we may define a stabilizing delay interval, which can be seen as a *robustness measure* of the corresponding control law if the delay is subject to parametric uncertainty. Furthermore, a geometric characterization of the stability regions in the parameter space (k,τ) will complete the presentation.

The interest in solving Problem 10.1 is twofold: first, the resulting design procedure is rather simple and the controller is easy to implement; second, it allows us to explore the potential of using such a controller (using the delay as a *design parameter*, defining thus a "wait-and-act" strategy) in situations where it is not easy to design or implement a controller without delay (see, for instance, the congestion controllers in high-speed networks [174, 188, 291] or the discussion in Chapter 13 devoted to the stability analysis of some congestion control algorithms in networks).

The structure of the chapter is as follows. In Section 10.2.1, we address the single delay case. A motivating second-order example is presented and the construction of the corresponding closed-loop stability domain boundaries is completely described. Next, necessary stabilizability conditions are explicitly derived. Such conditions, easy to check, are expressed in terms of the Hurwitz stability of some parameter-dependent polynomials. Finally, in order to construct *all stabilizing controllers*, an explicit algorithm is proposed and largely discussed. In Section 10.2.5, the geometry of the stability regions in the space (gain, delay) is proposed. The multiple delay case is presented in Section 10.3, where the stabilization of a chain of integrators by using delays is largely discussed by using the following two approaches in constructing control laws: (i) numerical differentiation and (ii) exact pole placement and low-gain design. Section 10.4 is devoted to the output-feedback stabilization of the SISO systems with input/output network-induced delays. Some notes and references end the chapter.

10.2 ▪ Single delay case

10.2.1 ▪ Characterization of all stabilizable second-order systems

In order to illustrate the potential and limitations of the control law (10.2), we completely characterize the output-feedback stabilizability of the second-order system,

$$\frac{p(\lambda)}{q(\lambda)}=\frac{c_1\lambda+c_2}{\lambda^2+a_1\lambda+a_2}, \tag{10.4}$$

as a function of its parameters (a_1,a_2,c_1,c_2). It is easy to check that with time-invariant output feedback, $u(t)=-ky(t)$, the stabilizability condition is given by

$$\begin{cases} a_2>\frac{c_2}{c_1}a_1 & \text{if} \quad c_1c_2<0, \\ a_1>0 & \text{if} \quad c_1=0, c_2\neq 0, \\ a_2>0 & \text{if} \quad c_2=0, c_1\neq 0, \\ a_1,a_2\in\mathbb{R} & \text{if} \quad c_1c_2>0. \end{cases}$$

In order to check the stabilizability when the delay is also used as a controller parameter, we have to solve an optimization problem for the parameters k and τ, since the stabilizability condition is given by $c(a_1,a_2,c_1,c_2)<0$, with

$$c(a_1,a_2,c_1,c_2):=\min_{k,\tau}\max_{\lambda\in\mathbb{C}}\left\{\Re(\lambda):\ \lambda^2+(a_1+kc_1e^{-\lambda\tau})\lambda+(a_2+kc_2e^{-\lambda\tau})=0\right\}. \tag{10.5}$$

We distinguish between several cases:

Case $c_1=0, c_2=1$. We take a "step-by-step" approach. First, we characterize stabilizability for a *fixed* delay $\tau=1$ using the approach of Chapter 8. Next, we consider the case where τ is also a controller parameter.

Stabilizability region for a fixed delay. For $\tau = 1$ the optimization problem (10.5) simplifies to the determination of

$$c(a_1,a_2) = \min_k \alpha(k),$$

where

$$\alpha(k) = \max\left\{\Re(\lambda):\ h(\lambda) := \lambda^2 + a_1\lambda + (a_2 + ke^{-\lambda}) = 0\right\}. \tag{10.6}$$

For different values of (a_1,a_2), chosen on a coarse grid, the rightmost characteristic roots are computed as a function of k, to check whether or not the system is stabilizable (i.e., $c(a_1,a_2) < 0$) ($c(a_1,a_2) \geq 0$). This allows us to sketch the stability domain boundary, which separates stabilizable and nonstabilizable (a_1,a_2)-pairs, and to characterize the configurations of the rightmost characteristic roots, which occur near the minimum of (10.6). It turns out that, close to the stability domain boundary, there are qualitatively the three following possibilities:

(a) a smooth minimum of (10.6), where the real part of a complex conjugate pair of characteristic roots is minimal;

(b) a nonsmooth but Lipschitz minimum, where a smooth branch of real characteristic roots and a smooth branch of complex conjugate characteristic roots are involved; and

(c) a nonsmooth, non-Lipschitz minimum, where a complex conjugate pair of characteristic roots bifurcates into two real characteristic roots.

As an illustration, we show in Figure 10.1 such configurations of the characteristic roots for (a_1,a_2)-pairs on the stability domain boundary, together with the transitions. The three possible configurations in the minimum are mathematically characterized in Table 10.1. Since there is one degree of freedom in the mathematical relations after freeing a_1 and a_2, the latter define branches, which can be continued efficiently in the (a_1,a_2)-space. The stability domain boundary is composed from segments of these branches; see Figure 10.1.

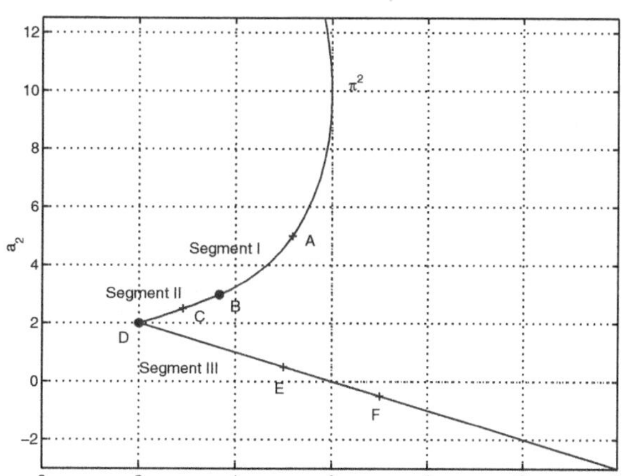

Figure 10.1. *Stability boundary in the (a_1,a_2)-plane for $\tau = 1$ and $(c_1,c_2) = (0,1)$. The different segments refer to Table* 10.1. *For parameter values corresponding to the uppercase letters, the characteristic roots are shown as a function of k in Figure* 10.2 [257].

Table 10.1. *The three possible configurations of the rightmost characteristic roots on the stability domain boundary in the minimum of* (10.6) *and their mathematical description. These relations define branches in the* (a_1, a_2)*-plane and allow for a fast computation of the stability boundary* [257].

Segment	Roots configuration	Mathematical description
I	$\begin{cases} h(j\omega)=0 \\ \left.\frac{\partial\lambda}{\partial k}\right\|_{\lambda=j\omega}=0 \end{cases}$	$\begin{cases} a_2+k\cos(\omega)-\omega^2=0 \\ a_1\omega-k\sin(\omega)=0 \\ a_1\cos(\omega)-2\omega\sin(\omega)-k=0 \end{cases}$
II	$\begin{cases} h(j\omega)=0 \\ h(0)=0 \end{cases}$	$\begin{cases} a_2+k\cos(\omega)-\omega^2=0 \\ a_1\omega-k\sin(\omega)=0 \\ a_2+k=0 \end{cases}$
III	$\begin{cases} h(0)=0 \\ h'(0)=0 \end{cases}$	$\begin{cases} a_2+k=0 \\ a_1-k=0 \end{cases}$

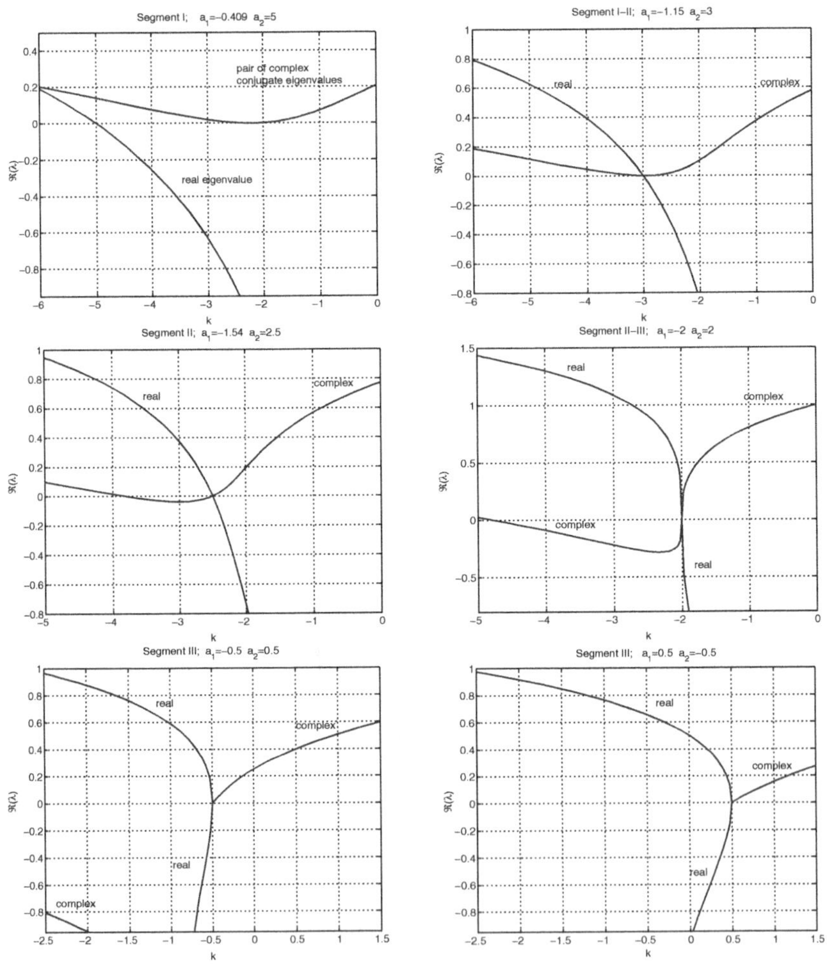

Figure 10.2. *Real parts of the rightmost characteristic roots as a function of k, corresponding to different* (a_1, a_2)*-pairs lying on the stability domain boundary (for* $(c_1, c_2) = (0, 1)$ *and* $\tau = 1$*). The (qualitatively) different configurations of the rightmost characteristic roots in the minimum of* (10.6) *are described in Table* 10.1 [257].

Stabilizability regions with k and τ as parameters. For $\tau \neq 1$, the characteristic equation of the closed-loop system is given by $\lambda^2 + a_1\lambda + a_2 + c_2 k e^{-\lambda\tau} = 0$, which is equivalent to

$$\bar{\lambda}^2 + (a_1\tau)\,\bar{\lambda} + (a_2\tau^2) + c_2 k\tau^2\, e^{-\bar{\lambda}} = 0, \tag{10.7}$$

where $\bar{\lambda} = \lambda\tau$. Therefore, the stability boundary for $\tau \neq 1$ can be directly computed from the one for $\tau = 1$ by normalizing the coefficients (a_1, a_2). In Figure 10.3 the stability domain boundary is shown for different values of τ. Hence, when both k and τ are controller parameters, the stabilizability region extends towards the curves characterized by $\{a_1 = 0,\ a_2 \leq 0\}$ and $\{a_1 \leq 0,\ a_2 = a_1^2/2\}$. Notice that the former curve coincides with a part of the stability domain boundary for $\tau = 0$ and that the latter curve is characterized by a *zero-characteristic root with multiplicity* 3 (point D, intersection of segments II and III in Figure 10.1).

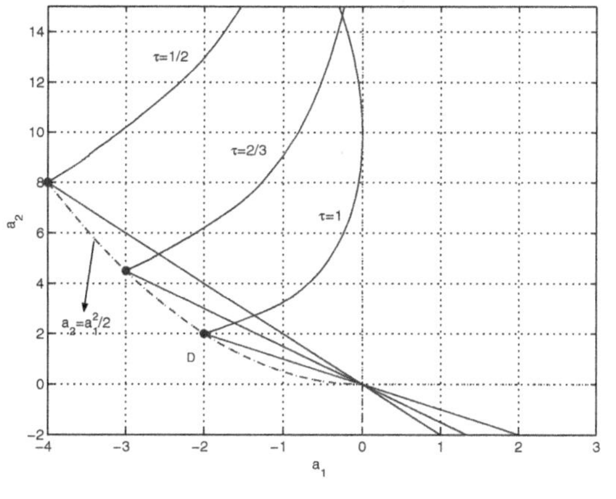

Figure 10.3. *Stability boundary for different values of τ (solid line). When τ is also a controller parameter, the stabilizability region consists of the union of the stabilizability regions for all fixed τ. This region is bounded by the dashed line, determined by the conditions $\{a_1 = 0,\ a_2 \leq 0\}$ and $\{a_1 \leq 0,\ a_2 = a_1^2/2\}$ [257].*

Case $c_1 c_2 < 0$. Although the characteristic function

$$h(\lambda) = \lambda^2 + \left(a_1 + c_1 k e^{-\lambda\tau}\right)\lambda + \left(a_2 + c_2 k e^{-\lambda\tau}\right)$$

cannot be rescaled as (10.7), numerical experiments reveal that the stabilizability boundary is also given partially by the stabilizability boundary for $\tau = 0$, $a_2 = (c_2 a_1)/c_1 - (c_1/c_2)^2$, $a_1 \geq 0$, and partially by a curve, characterized by a triple characteristic root at zero (for $a_1 \leq 0$). The conditions

$$h(0) = h'(0) = h''(0) = 0 \Leftrightarrow \left\{ \begin{array}{l} a_2 + c_2 k = 0, \\ a_1 + c_1 k - c_2 k\tau = 0, \\ -2c_1 k\tau + c_2 k\tau^2 + 2 = 0, \end{array} \right.$$

are equivalent to

$$a_2 = -\frac{c_2}{c_1}\left(\frac{c_2}{c_1} + \sqrt{a_1^2 + \left(\frac{c_2}{c_1}\right)^2}\right).$$

Case $c_2 = 0$. An approach similar to that in the previous case yields that no improvements with respect to time-invariant output feedback can be achieved. An indication of this result is given by the fact that for $a_2 = 0$, there is a characteristic root at zero, independently of the controller parameters.

Finally, combining the three above cases completes the stabilizability study of (10.4) with delayed output feedback. The results are summarized in Table 10.2.

Table 10.2. *Necessary and sufficient conditions for the output feedback stabilizability of the second-order system* (10.4).

$c_1 c_2 < 0$ (relative degree 1, nonminimum phase)	
Time-invariant, $u(t) = -ky(t)$	$a_2 > \frac{c_2}{c_1} a_1$
Delayed, $u(t) = -ky(t-\tau)$	$\left\{a_2 > \frac{c_2}{c_1} a_1\right\} \cup \left\{a_2 > -\frac{c_2}{c_1}\left(\frac{c_2}{c_1} + \sqrt{a_1^2 + \left(\frac{c_2}{c_1}\right)^2}\right)\right\}$
$c_1 = 0,\ c_2 \neq 0$ (relative degree 2)	
Time-invariant	$a_1 > 0$
Delayed	$\{a_1 > 0\} \cup \left\{a_2 > \frac{a_1^2}{2}\right\}$
$c_2 = 0,\ c_1 \neq 0$ (relative degree 1, weakly minimum phase)	
Time-invariant	$a_2 > 0$
Delayed	no improvement
$c_1 c_2 > 0$ (relative degree 1, minimum phase)	
Time-invariant	always stabilizable

10.2.2 ▪ Necessary conditions for stabilizability

Let the transfer function of the system (10.1) be given by

$$\frac{P(\lambda)}{Q(\lambda)} := C^T(\lambda I - A)^{-1}B, \tag{10.8}$$

where P and Q are coprime polynomials satisfying $\deg(P(\lambda)) = m$, $\deg(Q(\lambda)) = n$.

The following necessary stabilizability condition corresponds to [191, Proposition III.3] and is based on an extension of the Lucas Theorem to classes of entire functions [274]:

Proposition 10.2. *A necessary condition for the stabilizability of the system* (10.1) *with the control law* (10.2) *is the Hurwitz stability of the polynomial*

$$\gamma(\lambda;\ \tau) := \left(\frac{d}{d\lambda} + \tau\right)^{m+1} Q(\lambda). \tag{10.9}$$

Example 10.3. *We apply Proposition* 10.2 *to the second-order example* (10.4). *We distinguish between two cases:*

Case $c_1 = 0$, $c_2 \neq 0$. We have $m = 0$ and $\gamma(\lambda;\ \tau) = \tau\lambda^2 + (a_1\tau + 2)\lambda + (a_2\tau + a_1)$. This polynomial is Hurwitz if and only if

$$a_1\tau + 2 > 0, \qquad a_2\tau + a_1 > 0. \tag{10.10}$$

Condition (10.10) is violated for all $\tau \geq 0$ if and only if $a_1 \leq 0$ and $a_2 \leq a_1^2/2$. Hence, a necessary condition for the stabilizability with a control law of the form (10.2) is given by $a_2 > a_1^2/2$ or $a_1 > 0$. As shown in Table 10.2, this condition is also sufficient.

Case $c_1 \neq 0$. We have $m = 1$ and $\gamma(\lambda;\ \tau) = \tau^2\lambda^2 + (a_1\tau^2 + 4\tau)\lambda + (a_2\tau^2 + 2a_1\tau + 2)$, which is Hurwitz if $\tau = 0$, for all a_1 and a_2. This is expected because the necessary condition from Proposition 10.2 does not depend on the parameters of the nominator polynomial $c_1\lambda + c_2$, while the closed-loop stability can always be achieved by using full state feedback.

A direct corollary of Proposition 10.2 is the following:

Corollary 10.4. *If the polynomial $Q(\lambda)$ has at least one unstable root with multiplicity $\geq m + 2$, then the system (10.1) can not be stabilized by the control law (10.2).*

Notice that, to some extent, Proposition 10.2 concerns the stabilizability for a fixed value of τ with the gain k as control parameter, since the necessary condition explicitly depends on the delay (unlike the condition of Corollary 10.4 which is independent of both controller parameters). In the following sections, we will address the dual problem of characterizing the stabilizability for a fixed gain with the delay as controller parameter.

10.2.3 ▪ Controller construction

Using (10.8), the characteristic function of the closed-loop system (10.1)–(10.2) can be written in the form

$$H(\lambda;\ k, \tau) := Q(\lambda) + kP(\lambda)e^{-\lambda\tau}. \tag{10.11}$$

Without any loss of generality we will assume that the polynomials P and Q are coprime, that is, do not have common zeros. Two quantities will play an important role in the controller design procedure:

1. card($\mathcal{U}_+$), where $\mathcal{U}_+$ is the set of roots of $H(\lambda;\ k, 0) = Q(\lambda) + kP(\lambda)$, located in the *closed right half plane*, that is, the instability degree of the closed-loop system free of delays.

2. card($\mathcal{S}_+$), where $\mathcal{S}_+$ the set of strictly positive roots of the polynomial

$$F(\omega;\ k) = \mid Q(j\omega)\mid^2 - k^2 \mid P(j\omega)\mid^2. \tag{10.12}$$

   As we shall see later, $\mathcal{S}_+$ and F characterize the existence of crossing roots on the imaginary axis w.r.t. the delay parameter τ.

Both quantities depend explicitly on the gain parameter only, and we now clarify this dependence in more depth.

The quantity card($\mathcal{U}_+$). Its characterization as a function of the gain corresponds to the static (undelayed) output feedback stabilizability problem. The difficulty of this problem is well known (see, for instance, [366] and the references therein). However, in the SISO system case, the problem is reduced to a *one-parameter problem*, which is relatively easy. The approach below is inspired by Chen's characterization [67] for systems without delay.

We now introduce the following Hurwitz matrix associated to the denominator polynomial $Q(\lambda)=\sum_{i=0}^{n} q_i \lambda^{n-i}$ of the transfer function:

$$H(Q)=\begin{bmatrix} q_1 & q_3 & q_5 & \cdots & q_{2n-1} \\ q_0 & q_2 & q_4 & \cdots & q_{2n-2} \\ 0 & q_1 & q_3 & \cdots & q_{2n-3} \\ 0 & q_0 & q_2 & \cdots & q_{2n-4} \\ \vdots & & & \ddots & \vdots \\ 0 & 0 & 0 & \cdots & q_n \end{bmatrix} \in \mathbb{R}^{n\times n}, \tag{10.13}$$

where the coefficients $q_l = 0$ for all $l > n$. Next, corresponding to the polynomial $P(\lambda)$ of the transfer function, we construct $H(P)$ as an $n \times n$ matrix by the same procedure as (10.13) with the understanding that $p_l = 0$ for all $l > m$. The following result is a slight modification and generalization of Theorem 2.1 by Chen [67]:

Lemma 10.5. *Let $\lambda_1 < \lambda_2 < \cdots < \lambda_h$, with $h \leq n$, the real eigenvalues of the matrix pencil*

$$\Lambda(\lambda) = \det(\lambda H(P) + H(Q)).$$

Then the system (10.8) *cannot be stabilized by the controller $u(t) = -ky(t)$ for any $k = \lambda_i$, $i = 1,2,\dots,h$. Furthermore, if there are "r" unstable closed-loop roots ($0 \leq r \leq n$) for $k = k^*$, $k^* \in (\lambda_i, \lambda_{i+1})$, then there are "$r$" unstable roots for any $k \in (\lambda_i, \lambda_{i+1})$. In other words,* $\text{card}(\mathcal{U}_+)$ *remains constant as k varies within each interval $(\lambda_i, \lambda_{i+1})$. The same holds for the intervals $(-\infty, \lambda_1)$ and (λ_h, ∞).*

Proof. First, we need to show that as k varies, there are closed-loop roots on the imaginary axis if and only if $k = \lambda_i$, $i = 1,2,\dots,h$. The proof follows the same step as proposed by Chen in [67], and therefore will be omitted.

The above implies that for any gain $k \in (\lambda_i, \lambda_{i+1})$, the corresponding closed-loop system has no roots crossing the imaginary axis. Based on the *continuous dependence* of the roots of the polynomial $Q(s)+kP(s)$ on the parameter k, if there exists a k^* in the interval $(\lambda_i, \lambda_{i+1})$ such that $Q(\lambda)+k^*P(\lambda)$ has exactly r unstable roots, then the property is valid for any $k \in (\lambda_i, \lambda_{i+1})$ since the roots cannot jump from $\mathbb{C}_-$ to $\mathbb{C}_+$, or from $\mathbb{C}_+$ to $\mathbb{C}_-$, without crossing the imaginary axis. □

Thus by computing the generalized eigenvalues of the matrix pencil $\Lambda(\lambda)$, yielding the critical gain values, and computing $\mathcal{U}_+$ for intermediate gain values, the function $k \to \text{card}(\mathcal{U}_+)(k)$ is completely determined.

The quantity $\text{card}(\mathcal{S}_+)$. Without any loss of generality assume that $P(0) \neq 0$.[22] Its dependence on k is expressed in the following proposition:

Proposition 10.6. *Assume that* $\text{card}(\mathcal{S}_+)$ *changes at a gain value k^*. Then there exists a frequency $\omega^* \geq 0$ such that for $\omega = \omega^*$,*

$$|Q(j\omega)|^2 - k^{*2}|P(j\omega)|^2 = 0 \tag{10.14}$$

and

$$|P(j\omega)|^2 \frac{d}{d\omega}|Q(j\omega)|^2 - |Q(j\omega)|^2 \frac{d}{d\omega}|P(j\omega)|^2 = 0. \tag{10.15}$$

[22] If not, then $Q(0) \neq 0$, and the results shown still hold by defining the dependence in terms of $\frac{1}{k}$, etc.

Proof. For any k, F cannot have a root ω where $P(j\omega)=0$, because this would imply that also $Q(j\omega)=0$. The roots of F therefore coincide with roots of

$$G(\omega;\, k) := \frac{|Q(j\omega)|^2}{|P(j\omega)|^2} - k^2. \tag{10.16}$$

A change of $\mathrm{card}(\mathcal{S}_+)$ at $k=k^*$ implies that $G(\omega;\, k^*)$ has a roots with multiplicity larger than one at some frequency ω^*, i.e., $G(\omega^*;\, k^*) = G'(\omega^*;\, k^*) = 0$. This leads to (10.14) and (10.15). □

Proposition 10.6 allows computing systematically the behavior of $\mathrm{card}(\mathcal{S}^+)$ as a function of the gain. First, one has to determine the real roots of the polynomial (10.15). Then the critical values of the gain k follow from (10.14). The characterization is complete when computing $\mathcal{S}_+$ for intermediate gain values, which again corresponds to finding the roots of some polynomial.

Remark 10.7. *From the symmetry of the function* (10.16), *it follows that the pairs* $(\omega,k) = (0,\pm\frac{Q(0)}{P(0)})$ *always satisfy* (10.14)–(10.15), *and furthermore, at these k-values,* $\mathrm{card}(\mathcal{S}_+)$ *always changes with* ± 1. *It also follows from* (10.16) *that at other k-values,* $\mathrm{card}(\mathcal{S}_+)$ *can only change with* ± 2.

10.2.4 ▪ Stabilization using the delay parameter

For a given value of the gain k we derive conditions for the existence of stabilizing delay values. We make the following technical assumptions:

Assumption 10.8. *Let the gain* $k \in \mathbb{R}$ *be such that* (1) *all the roots of F are simple,* (2) $0 \notin \mathcal{U}_+$, *and* (3) $\mathrm{card}(\mathcal{U}_+) \neq 0$.

Notice that the first condition is satisfied for almost all k. The second condition is *necessary* for stabilization because it excludes a characteristic root at zero, the latter being *invariant* w.r.t. delay changes. The third assumption excludes the trivial case where the system is asymptotically stable with the choice $\tau = 0$.

An important result is the following:

Theorem 10.9. *The characteristic equation has a root* $j\omega$ *for some delay value* τ_0 *if and only if*

$$\omega \in \mathcal{S}_+. \tag{10.17}$$

Furthermore, for any ω *satisfying* (10.17) *the set of corresponding delay values is given by*

$$\mathcal{T}_\omega = \left\{ \frac{1}{\omega}\left[\angle\left(-\frac{kP(j\omega)}{Q(j\omega)}\right) + 2\pi l\right] \geq 0,\ l \in \mathbb{Z} \right\}. \tag{10.18}$$

When increasing the delay the corresponding crossing direction of characteristic roots is towards instability (stability) when $F'(\omega) > 0\ (<0)$.

Proof. The result is a direct consequence of Proposition 6.16 (Chapter 6) applied to the corresponding characteristic function $H(\lambda;\, k,\tau)$. □

Theorem 10.9, combined with the continuous dependence of the characteristic roots w.r.t. the delay, allows us to characterize completely stability/instability regions in the delay parameter. Indeed, the set $\mathcal{T} = \bigcup_{\omega \in \mathcal{S}_+} \mathcal{T}_\omega$ makes a partition of the delay space ($\mathbb{R}_+$) into intervals in which the number of strictly unstable roots is constant.

Remark 10.10 (crossing direction). *Assume that $\mathcal{S}_+$ is not empty and denote its elements in descending order with $\omega_1 > \omega_2 > \cdots$. Since $\lim_{\omega \to \infty} F(\omega) = +\infty$ and the roots $\{\omega_1, \omega_2, \ldots\}$ of F are simple, the sign of F' at these roots alternates, with $F'(\omega_1) > 0$. As a consequence, as the delay is monotonically increased, all root crossings for delay values $\mathcal{T}_{\omega_1}$ are towards instability, all root crossings for delay values $\mathcal{T}_{\omega_2}$ are towards stability, etc.*

Taking into account the number of unstable roots for $\tau = 0$ and the crossing direction in each point of $\mathcal{T}$, the number of unstable roots for each delay value can be determined. Based on Theorem 10.9 and its underlying ideas, we now focus on the delay *stabilization* problem. We have the following results:

Proposition 10.11. *Assume that* card($\mathcal{U}_+$) *is an odd number. Then the delay stabilization problem has no solution.*

Proof. The proof is by contradiction. Assume that the closed-loop system is asymptotically stable for some delay value τ_s. Because the number of roots in the closed right half plane changes *from odd to even* when increasing the delay from zero (number of closed half plane roots equal to card($\mathcal{U}_+$)) to τ_s (number of closed half plane roots equal to zero), a characteristic root at *zero* must occur for some $\tau_0 \in [0, \tau_s]$. But $H(0; k, \tau_0) = 0$ implies $H(0; k, \tau) = 0$ for all $\tau \geq 0$, which contradicts the asymptotic stability at $\tau = \tau_s$. □

Remark 10.12. *Such a result is relatively simple and proves the existence of a strictly real positive root of the characteristic equation for all delay values if such a root exists for the system free of delay. Finally, we would like to point out that similar results, with slightly different formulations, and similar (or different) proofs already have been proposed in the literature (see, e.g., [359, 123] to cite only a few).*

Proposition 10.13. *If either* card($\mathcal{S}_+$) $= 0$ *or* card($\mathcal{S}_+$) $= 1$, *then the delay stabilizing problem has no solution.*

Proof. When card($\mathcal{S}_+$) $= 0$, characteristic roots cannot cross the imaginary axis, as the delay is varied and the instability for $\tau = 0$ persists for all delay values. When card($\mathcal{S}_+$) $=$ 1, there is one crossing frequency, and from Remark 10.10 the crossing direction is always towards instability as the delay is increased. Combining this fact with the instability for $\tau = 0$ yields the statement of the proposition. □

Notice that the condition card($\mathcal{S}_+$) $= 0$ corresponds to the *delay-independent hyperbolicity* property (fixed number of unstable roots for all positive delay values), as defined in [150] (see also [151]). For the remaining cases, one needs to count the roots crossing the imaginary axis towards stability/instability, and to define the corresponding delay intervals (see also [290, Chapters 4 and 7]). More explicitly, it follows that the first case when the delay has a stabilizing effect may appear if card($\mathcal{U}_+$) $= 2$, and card($\mathcal{S}_+$) $\in \{2, 3\}$, and if the first crossing is towards stability. Furthermore, as proved in what follows, such a condition is also necessary:

Proposition 10.14. *Assume that* $\mathrm{card}(\mathscr{S}_+)=2$ *or* $\mathrm{card}(\mathscr{S}_+)=3$. *Then the delay stabilizing problem has a solution if and only if* $\mathrm{card}(\mathscr{U}_+)=2$ *and* $\tau_- < \tau_+$, *where*

$$\tau_- = \min \bigcup_{\omega\in\mathscr{S}_+,\ F'(\omega)<0} \mathscr{T}_\omega, \qquad \tau_+ = \min \bigcup_{\omega\in\mathscr{S}_+,\ F'(\omega)>0} \mathscr{T}_\omega \setminus \{0\}.$$

If stabilizable, all delay values $\tau \in (\tau_-,\ \tau_+)$ *are stabilizing.*

Proof. "$\Rightarrow$" Consider first the case where $\mathrm{card}(\mathscr{S}_+)=2$. Let $\mathscr{S}_+=\{\omega_1,\omega_2\}$, with $\omega_1 > \omega_2$. Then from Remark 10.10 we have $F'(\omega_1)>0$ (associated crossings for $\mathscr{T}_{\omega_1}$ towards instability when increasing the delay) and $F'(\omega_2)<0$ (associated crossings for $\mathscr{T}_{\omega_2}$ towards stability). The set $\mathscr{T}_{\omega_1}$ consists of delay values, equally spaced with $2\pi/\omega_1$, whereas the elements of $\mathscr{T}_{\omega_2}$ are equally spaced with $2\pi/\omega_2 > 2\pi/\omega_1$. As a consequence, between two stability crossings, an instability crossing must occur, i.e., the number of unstable roots in the closed right half plane cannot be reduced by more than two by increasing the delay. Thus the occurrence of a stabilizing delay value necessarily implies that the number of roots in the closed right half plane is two for $\tau=0$ (i.e., $\mathrm{card}(\mathscr{U}_+)=2$) and that the first[23] crossing is towards stability, mathematically expressed by $\tau_- < \tau_+$. If $\mathrm{card}(\mathscr{S}_+)=3$, then $\mathscr{S}_+=\{\omega_1,\omega_2,\omega_3\}$ with $\omega_1>\omega_2>\omega_3$ and $F'(\omega_1)>0, F'(\omega_2)<0, F'(\omega_3)>0$. Compared to the previous case, there is one additional crossing frequency ω_3 where additional crossings towards *in*stability occur, and the argument remains the same.

"$\Leftarrow$" The condition $\tau_- < \tau_+$ implies that the first crossing is towards stability when the delay in increased from zero. Since $\mathrm{card}(\mathscr{U}_+)=2$, the closed-loop system is asymptotically stable for any $\tau\in(\tau_-,\ \tau_+)$. □

Remark 10.15. *The importance of this result lies in the fact that, in order to check stabilizability in the delay parameter, one only has to investigate the* first *root crossing of the imaginary axis as the delay is increased from zero. This is particularly useful when one determines stabilizability by numerically computing the rightmost characteristic roots as a function of the delay. After the first root crossing one can stop the computations.*

In the case when $\mathrm{card}(\mathscr{S}_+)=2$, the set of *all* stabilizing delay values can be expressed analytically:

Corollary 10.16. *Assume that the conditions of Proposition* 10.14 *are satisfied, and in addition* $\mathrm{card}(\mathscr{S}_+)=2$. *Then all the stabilizing delay values are defined by* $\tau \in (\underline{\tau_l},\ \overline{\tau_l})$, $l=0,1,2,\dots,l_m$, *where*

$$\underline{\tau_l} = \tau_- + \frac{2\pi l}{\omega_-}, \qquad \overline{\tau_l} = \tau_+ + \frac{2\pi l}{\omega_+},$$

$\mathscr{S}_+=\{\omega_+,\omega_-\}$ *with* $\omega_+>\omega_-$, *and* l_m *is the largest integer for which* $\underline{\tau_l} < \overline{\tau_l}$, *which can be expressed as*

$$l_m = \max_{l\in\mathbf{Z}} \left\{ l < \frac{\omega_+\omega_-}{2\pi} \cdot \frac{\tau_+-\tau_-}{\omega_+-\omega_-} \right\}. \tag{10.19}$$

Proposition 10.17. *Assume that* $\mathrm{card}(\mathscr{S}_+)=2n$ *or* $\mathrm{card}(\mathscr{S}_+)=2n+1$, *with* $n\geq 1$. *Assume further that* $\mathrm{card}(\mathscr{U}_+)>2n$. *Then the delay stabilizing problem has no solution.*

[23]When $0\in\mathscr{T}_{\omega_1}$, the crossing at $\tau=0$ is not counted since $\mathrm{card}(\mathscr{U}_+)$ does not change.

Proof. Let $\mathcal{S}_+ = \{\omega_1, \omega_2, \ldots\}$ with $\omega_1 > \omega_2 > \cdots$. From Corollary 10.10 we have the following alternating sequence: $F'(\omega_1) > 0, F'(\omega_2) < 0, F'(\omega_3) > 0, \ldots$. Next, consider the pair (ω_1, ω_2). By the same arguments as used in the proof of Proposition 10.14, there must be an element of $\mathcal{T}_{\omega_1}$ between two elements of $\mathcal{T}_{\omega_2}$. When the delay is increased from zero, the root crossings at $j\omega_1$ and $j\omega_2$ therefore cannot contribute to reduction of closed half plane roots by more than two. When $n > 1$ the same argument can be used for the pairs $(\omega_3, \omega_4), \ldots, (\omega_{2n-1}, \omega_{2n})$. Thus taking into account the root crossings at $j\omega_1, j\omega_2, \ldots, j\omega_{2n}$, no more than $2n$ unstable roots can be shifted to the left half plane, and the proof is complete for $\text{card}(\mathcal{S}_+) = 2n$. Finally, if $\text{card}(\mathcal{S}_+) = 2n+1$, the argument remains the same since $F'(\omega_{2n+1}) > 0$. □

Define now the following quantities:

$$n_+(\tau) = \sum_{\omega \in \mathcal{S}_+,\ F'(\omega) > 0} \text{card}\{\mathcal{T}_\omega \cap (0, \tau]\}, \tag{10.20}$$

$$n_-(\tau) = \sum_{\omega \in \mathcal{S}_+,\ F'(\omega) < 0} \text{card}\{\mathcal{T}_\omega \cap [0, \tau]\} \tag{10.21}$$

for some positive $\tau > 0$. Furthermore, introduce the sets $\mathcal{T}^+$ and $\mathcal{T}^-$, which represent a partition of $\mathcal{T}$ function of the sign of the derivative F' evaluated at the corresponding crossing frequency, that is,

$$\mathcal{T}^+ = \bigcup_{\omega \in \mathcal{S}_+,\ F'(\omega) > 0} \mathcal{T}_\omega \setminus \{0\}, \qquad \mathcal{T}^- = \bigcup_{\omega \in \mathcal{S}_+,\ F'(\omega) < 0} \mathcal{T}_\omega.$$

Based on the conditions and the notations above, we conclude with the following:

Proposition 10.18. *For a given gain k, the stabilizing control problem has a solution of the form $u(t) = -ky(t-\tau)$ if and only if the following conditions hold simultaneously:*

(i) $\text{card}(\mathcal{U}_+)$ *is a strictly positive even integer, which satisfies the inequality* $\text{card}(\mathcal{U}_+) \leq \text{card}(\mathcal{S}_+)$, *and*

(ii) *there exists at least one delay value $\hat{\tau} \in \mathcal{T}$ such that the following equality is verified:*

$$2n_-(\hat{\tau}) = 2n_+(\hat{\tau}) + \text{card}(\mathcal{U}_+). \tag{10.22}$$

Then all delay values $\tau \in (\hat{\tau}, \hat{\tau}_+)$, with

$$\hat{\tau}_+ = \min\left\{\mathcal{T}^+ \cap (\hat{\tau}, +\infty)\right\}, \tag{10.23}$$

guarantee the closed-loop asymptotic stability.

Proof. While the condition (i) is clear, the condition (10.22) in (ii) simply characterizes the existence of crossings such that there are no more unstable rightmost roots for delays $\tau = \hat{\tau} + \varepsilon$, for sufficiently small $\varepsilon > 0$, and the definition of the delay interval follows straightforwardly. □

Remark 10.19. *Proposition* 10.18 *above can be reformulated by defining an appropriate function* [29] $N : \mathbb{R}_+ \mapsto \mathbb{N}$ *defined for some positive delay value τ. Such a function computes explicitly the number of strictly unstable roots in $\mathbb{C}_+$ of the characteristic function H by taking into account all the crossing roots together with their crossing direction for each critical delay inside the interval* $[0, \tau]$. *A similar formula was proposed by* [313] *but the characterization of the critical delays for which crossing roots exist was derived using a different argument.*

Table 10.3. *Output-feedback stabilizability conditions when using the delay as controller parameter (necessary and sufficient conditions given by Proposition* 10.18*; if* $\text{card}(\mathscr{U}_+) = 2$ *and* $\text{card}(\mathscr{S}_+) \in \{2,3\}$, *Proposition* 10.14 *applies).*

	0	1	2	3	4	5	6	7	8	9	$\text{card}(\mathscr{S}_+)$
1	/	/	/	/	/	/	/	/	/	/	
2	/	/	$\tau_- < \tau_+$								
3	/	/	/	/	/	/	/	/	/	/	
4	/	/	/	/							
5	/	/	/	/	/	/	/	/	/	/	
6	/	/	/	/	/	/					
7	/	/	/	/	/	/	/	/	/	/	
8	/	/	/	/	/	/	/	/			
$\text{card}(\mathscr{U}_+)$											

Stabilization using the delay and gain parameter

The main results of the previous section are displayed in Table 10.3. Recall that $\text{card}(\mathscr{U}_+)$ and $\text{card}(\mathscr{S}_+)$ depend *only* on the gain k. The first quantity can be efficiently determined as a function of k by computing the generalized eigenvalues of a matrix pencil (Proposition 10.5), and the second quantity by computing the roots of a polynomial (Proposition 10.6).

The procedure for deriving a stabilizing pair (k,τ) (if any) can be resumed as follows:

- First, compute $\text{card}(\mathscr{S}_+)$ and $\text{card}(\mathscr{U}_+)$ as functions of the gain parameter k, and next select possible gain intervals $(\underline{k},\overline{k})$ such that condition (i) of Proposition 10.18 holds;
- second, for a given k, search for stabilizing delay values. In the special case where $\text{card}(\mathscr{S}_+) = 2$ or $\text{card}(\mathscr{S}_+) = 3$ and $\text{card}(\mathscr{U}_+) = 2$, Proposition 10.14 can be applied; i.e., it is sufficient to investigate only whether the *first* root crossing of the imaginary axis is towards stability as the delay is increased from zero. In general, a more complete characterization of stability/instability regions becomes necessary, as we shall illustrate in Section 10.2.6. According to condition (ii) of Proposition 10.18 this is, however, a systematic task. Indeed, one has to compute first the set $\mathscr{T}$, and next the partition $\mathscr{T}^+$ and $\mathscr{T}^-$, which gives the root crossings (function of the delay values) towards instability and stability, respectively. Condition (10.22) together with (10.23) will define the corresponding stabilizing delay intervals.

Finally, we note that the set of all stabilizing pairs (k,τ) is bounded if the delay-free stabilization problem is not solvable:

Proposition 10.20. *Assume that* $\text{card}(\mathscr{U}_+)(k) > 0$ *for all* $k \in \mathbb{R}$. *Then the set of all stabilizing pairs* (k,τ) *is bounded.*

Proof. For sufficiently large $|k|$ the polynomial $F(\omega;\, k)$ has only one zero. From Theorem 10.9 and, in particular, the crossing direction characterization, it follows that the corresponding delay stabilizing problem has no solution. So, there exists an upper bound on the possible gain values of the stabilizing controllers. The same holds for the corresponding delay values, since for any fixed value of k, the function $n_+(\tau) - n_-(\tau)$ tends to infinity as τ tends to infinity. ☐

10.2.5 ▪ Geometry of stability regions

The difference w.r.t. the approach of the previous section lies in the way in which the stability regions in the (k,τ)-space are computed, or the corresponding stabilization problem is solved. Essentially, the approach of the previous section corresponds to the following three steps: (1) choose and fix the gain k, (2) compute the corresponding crossing frequencies ω, and (3) compute the corresponding delay values τ. On the contrary, the approach presented below corresponds to the following steps: (1) choose possible values of ω and (2) compute corresponding pairs (k,τ). Hence, ω is explicitly treated as a variable which parametrizes the stability crossing curves in the (k,τ)-plane. This allows one to make a geometric characterization of the stability crossing curves.

Identification of crossing points

Let $\mathcal{T}_c$ denote the set of all $(k,\tau)\in\mathbb{R}\times\mathbb{R}_+$ such that the characteristic function $H(\lambda;\,k,\tau)$ has at least one zero on the imaginary axis. Any $(k,\tau)\in\mathcal{T}_c$ is known as a *crossing point*. The set $\mathcal{T}_c$, which is the collection of all crossing points, forms the so-called *stability crossing curves*. Consider also the set Ω of all $\omega\in\mathbb{R}$ such that $j\omega$ is a root of H for at least one pair $(k,\tau)\in\mathbb{R}\times\mathbb{R}_+$. We refer to Ω as the *frequency crossing set*.

Remark 10.21. *If ω is a real number and $(k,\tau)\in\mathbb{R}\times\mathbb{R}_+$, then*

$$Q(-j\omega)+kP(-j\omega)e^{j\omega\tau}=\overline{Q(j\omega)+kP(j\omega)e^{-j\omega\tau}}.$$

Therefore, we only need to consider $\omega\in\mathbb{R}_+$. For the sake of brevity, we make the *additional assumption* that $P'(j\omega)\neq 0$ whenever $P(j\omega)=0$. We have:

Proposition 10.22. *Given any $\omega>0$, $\omega\in\Omega$ if and only if it satisfies*

$$|P(j\omega)|>0, \tag{10.24}$$

and all the corresponding pairs (k,τ) can be calculated as

$$k(\omega)=\pm\left|\frac{Q(j\omega)}{P(j\omega)}\right|, \tag{10.25}$$

$$\tau_m(\omega)=\frac{1}{\omega}\left(-\angle\left(\frac{Q(j\omega)}{P(j\omega)}\right)+(2m+\epsilon_k+1)\pi\right), \tag{10.26}$$

where $m=0,\pm1,\pm2,\dots$ and

$$\epsilon_k=\begin{cases}0 & k\geq 0,\\ -1 & k<0.\end{cases}$$

Proof. For the necessity of (10.24), let ω be a crossing frequency in Ω. Hence, there exists a pair (k,τ) such that $Q(j\omega)+kP(j\omega)e^{-j\omega\tau}=0$. This implies that $|Q(j\omega)|=|k||P(j\omega)|$ is satisfied. It becomes clear that $|P(j\omega)|>0$ is necessary. Otherwise, $P(j\omega)=0$, which implies $Q(j\omega)=0$ for all the gains k, which contradicts the technical assumption that P and Q do not have common zeros. For the sufficiency of (10.24), we only need to recognize that the pair (k,τ) given by (10.25)–(10.26) makes $\lambda=j\omega$ a root of the characteristic function of the closed-loop system. □

Because of Proposition 10.20 and our prior interest in characterizing *stability* regions, we restrict the gain k in what follows to some *finite interval* $[\alpha,\beta]$. From (10.25), this allows restricting the *crossing set* Ω to the set of the frequencies $\omega > 0$ satisfying *simultaneously* the inequalities

$$\min(|\alpha|,|\beta|)\,|P(j\omega)|\,|\leq |Q(j\omega)| \leq \max(|\alpha|,|\beta|)\,|P(j\omega)|, \qquad |P(j\omega)|>0. \quad (10.27)$$

Due to the form of (10.27), and from the standard assumptions considered (strictly proper transfer function, P and Q coprime), Ω *consists of a finite number of intervals*. If such an interval contains a frequency where $|Q(j\omega)| = 0$, which is the case if the uncontrolled systems exhibits oscillatory modes, we first subdivide such an interval into two intervals with a common end point. This will be motivated later on. Next, we denote the resulting intervals with $\Omega_1, \Omega_2, \ldots, \Omega_N$, thus, $\Omega = \cup_{k=1}^{N}\Omega_k$. In what follows, we let $\Omega_i = [\omega_i^l,\ \omega_i^r]$ for all $i = 1,2,\ldots,N$. Without any loss of generality, we can order these intervals from left to right; i.e., for any $\omega_1 \in \Omega_{i_1}$, $\omega_2 \in \Omega_{i_2}$, $i_1 < i_2$, we have $\omega_1 < \omega_2$. Note that ω_1^l can be 0, and in this case Ω_1 is open to the left.

It is clear that $k(\omega_i^l), k(\omega_i^r) \in \{\alpha,\beta\}$ for all $i = 1,\ldots,N$ if $\omega_1^l \neq 0$. In the remainder of this section, we will not restrict $\angle Q(j\omega)/P(j\omega)$ to a 2π range. Rather, we let it vary *continuously* within each interval Ω_i. Note that this is always possible because the continuity of $\angle(Q(j\omega)/P(j\omega))$ can be guaranteed by a suitable choice of the phase angle at every frequency, except at those frequencies where $Q(j\omega) = 0$. But if such a frequency occurs, then it corresponds to an end point of the interval due to the subdivision mentioned above. In summary, for each fixed m, (10.25) and (10.26) give us two continuous curves.

Condition (10.25) and k finite imply that $P(j\omega) \neq 0$ for all $\omega \in \Omega$. We denote the curves defined by (10.25) and (10.26) by $\mathcal{T}_i^{m\pm}$, with the $\pm$ referring to the choice of the sign in (10.25) and i referring to the corresponding interval Ω_i. Therefore, we have an infinite number of continuous stability crossing curves $\mathcal{T}_i^{m\pm}$, $m = 0,\pm1,\pm2,\ldots$, associated to Ω_i.

Finally, it should be noted that, for some m, either part of the curve or the entire curve may be outside of the range $\mathbb{R}\times\mathbb{R}_+$, and therefore may not be physically meaningful. The collection of all the points in $\mathcal{T}_c$ corresponding to Ω_i may be expressed as

$$\mathcal{T}_i := \bigcup_{m=-\infty}^{+\infty} \left[(\mathcal{T}_i^{m+} \cap (\mathbb{R}\times\mathbb{R}_+)) \cup (\mathcal{T}_i^{m-} \cap (\mathbb{R}\times\mathbb{R}_+)) \right].$$

Obviously, $\mathcal{T}_c = \cup_{i=1}^{N}\mathcal{T}_i$. Also it is easy to see that, for each Ω_i, we define two curves, one to the right of the $O\tau$ axis and the other to the left. According to the fixed limits α,β of the interval where k varies, we can eliminate some of these curves.

Classification of stability crossing curves, smoothness, and crossing directions

Stability crossing curves

Similarly to Chapter 6, we give a classification of the crossing curves with respect to their shape. We first classify the ends of the crossing curves confined to $k \in [\alpha,\ \beta]$. Each end point ω_i^l or ω_i^r must belong to one of the following types:

Type 1. It satisfies $k(\omega) = \alpha,\ \omega \neq 0$.

Type 2. It satisfies $k(\omega) = \beta,\ \omega \neq 0$.

Type 3. It equals 0, $Q(0) \neq 0$ and $Q(0)/P(0) \in [\alpha, \beta]$.

Type 4. It equals 0, $Q(0) \neq 0$, and $-Q(0)/P(0) \in [\alpha, \beta]$.

Type 5. It equals 0 and $Q(0) = 0$.

Obviously, only ω_1^l can be of Types 3–5. We note that all the crossing curves are situated in the vertical strip $\mathscr{D}$ between the lines $k = \alpha$ and $k = \beta$. Now, let ω_* be an end point of the interval Ω_i. We already said that each $\mathscr{T}_i^{m\pm}$ is a continuous curve, so $(k(\omega_*), \tau_m(\omega_*))$ is an end point of $\mathscr{T}_i^{m\pm}$. In the case $\alpha \geq 0$, it can be characterized as follows:

- If ω_* is of Type 1, then $k(\omega_*) = \alpha$, $\tau(\omega_*)$ are finite. More precisely, $\mathscr{T}_i^{m+}$, $m \in \mathbb{Z}$, intersects the vertical line $k = \alpha$, which is the left bound of the strip $\mathscr{D}$.
- If ω_* is of Type 2, then $k(\omega_*) = \beta$, $\tau(\omega_*)$ are finite. In this case, $\mathscr{T}_i^{m+}$, $m \in \mathbb{Z}$, intersects the vertical line $k(\omega) = \beta$, which is the right bound of the strip $\mathscr{D}$.
- If ω_* is of Type 3, then k approaches $Q(0)/P(0)$ and τ approaches ∞ for all values of m. In other words, $\mathscr{T}_i^{m+}$, $m \in \mathbb{Z}$, has a vertical asymptote given by $k = Q(0)/P(0)$.
- If ω_* is of Type 4, then k approaches $-Q(0)/P(0)$ and $\mathscr{T}_i^{0+}$ approaches $(Q'(0)/Q(0))–(P'(0)/P(0))$, whereas, $\mathscr{T}_i^{m+}$, $m \in \mathbb{Z}_0$, has a vertical asymptote given by $k = -Q(0)/P(0)$.
- If ω_* is of Type 5, then k approaches 0 and τ approaches ∞ for all m. In other words, $\mathscr{T}_i^{m+}$, $m \in \mathbb{Z}$, has a vertical asymptote given by $k = 0$.

We say that an interval Ω_k is of Type $l\,r$ if its left end is of Type l and its right end is of Type r. We may divide accordingly these intervals into the following nine types:

Type 11. In this case, $\mathscr{T}_i^{m+}$, $m \in \mathbb{Z}$, starts at a point on the vertical line $k = \alpha$ and ends at another point on the vertical line $k = \alpha$.

Type 12. In this case, $\mathscr{T}_i^{m+}$, $m \in \mathbb{Z}$, starts at a point on the vertical line $k = \alpha$ and ends at a point on the vertical line $k = \beta$.

Type 21. This is the reverse of Type 12. $\mathscr{T}_i^{m+}, m \in \mathbb{Z}$, starts at a point on the vertical line $k = \beta$ and ends at a point on the vertical line $k = \alpha$.

Type 22. In this case, $\mathscr{T}_i^{m+}$, $m \in \mathbb{Z}$, starts at a point on the vertical line $k = \beta$ and ends at another point on the vertical line $k = \beta$.

Type 31. In this case, $\mathscr{T}_i^{m+}$, $m \in \mathbb{Z}$, begins at ∞ with a vertical asymptote $k = Q(0)/P(0)$. The other end is on the vertical line $k = \alpha$.

Type 32. In this case, $\mathscr{T}_i^{m+}$, $m \in \mathbb{Z}$, again begins at ∞ with a vertical asymptote $k = Q(0)/P(0)$. The other end is on the vertical line $k = \beta$.

Type 41. In this case, $\mathscr{T}_i^{m+}$, $m \in \mathbb{Z}_0$, begins at ∞ with a vertical asymptote $k = -Q(0)/P(0)$. The other end is on the vertical line $k = \alpha$. The curve $\mathscr{T}_i^{0+}$ begins in $(k, \tau) = (-Q(0)/P(0), Q'(0)/Q(0) - P'(0)/P(0))$ and ends on the line $k = \alpha$. Here it is important to mention that the special case corresponds to $m = 0$ only if the convention of the phase angle is such that

$$\lim_{\omega \to 0+} \angle(Q(j\omega)/P(j\omega)) = \begin{cases} 0, & Q(0)/P(0) \geq 0, \\ \pi, & Q(0)/P(0) < 0. \end{cases} \tag{10.28}$$

This is relevant because, as mentioned before, the classical convention for $\angle(\cdot)$ is sometimes given up because of continuity arguments regarding the curves defined by (10.25)–(10.26).

Type 42. This is the reverse of Type 41. Under the assumption (10.28), $\mathcal{T}_i^{m+}$, $m \in \mathbb{Z}_0$, begins at ∞ with a vertical asymptote $k = -Q(0)/P(0)$. The other end is on the vertical line $k = \beta$. The curve $\mathcal{T}_i^{0+}$ begins in

$$(k,\tau) = (-Q(0)/P(0), Q'(0)/Q(0) - P'(0)/P(0))$$

and ends on the line $k = \beta$.

Type 52. In this case, $\mathcal{T}_i^{m+}$, $m \in \mathbb{Z}$, begins at ∞ with the vertical asymptote $k = 0$. The other end is on the vertical line $k = \beta$.

Remark 10.23. *In case $\beta \le 0$, the previous description remains valid provided $\mathcal{T}_i^{m+}$ is replaced with $\mathcal{T}_i^{m-}$, and Type 52 is replaced with Type 51 (the difference w.r.t. Type 52 lies in the fact that the curves end on the vertical line $k = \alpha$).*

Tangents and smoothness

For a given i, the smoothness properties of the curves in $\mathcal{T}_i^{m\pm}$ and thus in

$$\mathcal{T}_c = \bigcup_{m=-\infty}^{+\infty} \left[\left(\mathcal{T}_i^{m+} \cap (\mathbb{R} \times \mathbb{R}_+)\right) \cup \left(\mathcal{T}_i^{m-} \cap (\mathbb{R} \times \mathbb{R}_+)\right) \right]$$

can be investigated by adopting the approach proposed in Chapter 6 (crossing curves in delay-parameter spaces). We have the following result (see also Chapter 6):

Proposition 10.24. *The curve $\mathcal{T}_i^{m\pm}$ is smooth everywhere except possibly at the point corresponding to $\lambda = j\omega$ in either of the following cases:*

(1) *$\lambda = j\omega$ is a multiple root of the characteristic function $H(\lambda;\, k,\tau)$, and*

(2) *$Q(j\omega) = 0$, which is equivalent to $k = 0$.*

Directions of crossing

Some appropriate quantities need to be introduced. For a given pair $(k,\tau) \in \mathcal{T}_i$ and corresponding $\omega \in \Omega_i$, let

$$R_1 = \Re\left(\frac{\partial H(\lambda;\, k,\tau)}{\partial k}\right)_{\lambda=j\omega} = \Re\left[P(j\omega)e^{-j\omega\tau}\right],$$

$$I_1 = \Im\left(\frac{\partial H(\lambda;\, k,\tau)}{\partial k}\right)_{\lambda=j\omega} = \Im\left[P(j\omega)e^{-j\omega\tau}\right],$$

$$R_2 = \Re\left(\frac{\partial H(\lambda;\, k,\tau)}{\partial \tau}\right)_{\lambda=j\omega} = \Im\left(k\omega P(j\omega)e^{-j\omega\tau}\right),$$

$$I_2 = \Im\left(\frac{\partial H(\lambda;\, k,\tau)}{\partial \tau}\right)_{\lambda=j\omega} = -\Re\left(k\omega P(j\omega)e^{-j\omega\tau}\right).$$

In order to characterize the direction in which the roots of the characteristic function $H(\lambda;\, k,\tau)$ cross the imaginary axis as (k,τ) deviates from the curve $\mathcal{T}_i^{m\pm}$, we call the direction of the curve that corresponds to increasing ω the *positive direction*. We also call the region on the left-hand side as we head in the positive direction of the curve the *region on the left*. Similarly to the characterization of the crossing direction in Chapter 6, we have the following:

Proposition 10.25. *Let $\omega \in (\omega_i^l, \omega_i^r)$ and $(k,\tau) \in \mathcal{T}_i$ such that $j\omega$ is a simple root of the characteristic function $H(\lambda;\ k,\tau)$.*

If (k,τ) moves from the region on the right to the region on the left of a stability crossing curve, then a pair of roots of the characteristic function H crosses the imaginary axis to the right, through $\lambda = \pm j\omega$, whenever $R_2 I_1 - R_1 I_2 > 0$. The crossing is to the left if the inequality is reversed.

Discussions and interpretations

Under standard regularity assumptions, Proposition 10.25 above gives the explicit *crossing direction* towards stability (*reversals*) or instability (*switches*) as a function of some quantity evaluated at the corresponding crossing point. Such a result, together with Lemma 10.5 and Propositions 10.22–10.24, leads to the following simplified procedure: first, compute card($\mathcal{U}_+$) (using Lemma 10.5); second, derive the crossing set Ω, and related dependence $(k(\omega),\tau(\omega))$ (using Proposition 10.22, and its corresponding remarks and derivations); next, determine the crossing direction using Proposition 10.25 above. Finally, in order to find explicitly the (closed-loop) *stability regions*, one can count the number of crossing roots (towards stability/instability) corresponding to each region whose boundaries are given by the crossing curves, starting from the crossing direction of the roots and the fact that for $\tau = 0$, card($\mathcal{U}_+$) gives the complete (stability/instability) information in terms of k (closed-loop free of delays).

10.2.6 ▪ Illustrative example: Second-order system

We consider the system with transfer function

$$\frac{P(\lambda)}{Q(\lambda)} = \frac{1}{\lambda^2 - \gamma\lambda + 2}, \tag{10.29}$$

where $\gamma > 0$ a real parameter. The polynomial $Q(\lambda) = \lambda^2 - \gamma\lambda + 2$ is unstable, and for all $k \in \mathbb{R}$, the polynomial $Q(\lambda) + kP(\lambda)$ has at least one unstable root. Furthermore, if $\gamma = 0$, then (10.29) corresponds to an oscillator (the characteristic equation has two roots on the imaginary axis). Now, with the controller (10.2), the characteristic equation of the closed-loop system is given by

$$\lambda^2 - \gamma\lambda + 2 + ke^{-\lambda\tau} = 0, \tag{10.30}$$

and polynomial $F(\omega;\ k)$ by

$$F(\omega) = |\,Q(j\omega)\,|^2 - |\,P(j\omega)\,|^2 = \omega^4 - (4-\gamma^2)\omega^2 + (4-k^2). \tag{10.31}$$

For $\gamma > 2$ the quantities card($\mathcal{S}_+$) and card($\mathcal{U}_+$) behave as a function of the gain k as displayed in the following table:

k	<-2	$\in(-2,2)$	>2
card($\mathcal{S}_+$)	1	0	1
card($\mathcal{U}_+$)	1	2	2

while for $\gamma < 2$ we have

k	<-2	$\in(-2,-k^*)$	$\in(-k^*,k^*)$	$\in(k^*,2)$	>2
$\mathrm{card}(\mathscr{S}_+)$	1	2	0	2	1
$\mathrm{card}(\mathscr{U}_+)$	1	2	2	2	2

where

$$k^* = 2\sqrt{1-\left(1-\frac{\gamma^2}{4}\right)^2}.$$

According to the results of the previous section, summarized in Table 10.3, a necessary condition for asymptotic stability of the closed-loop system is therefore given by

$$\gamma < 2, \qquad |k| \in (k^*,\, 2). \tag{10.32}$$

Furthermore, for a gain satisfying (10.32) the existence of a stability region in the delay parameter is determined by the condition $\tau^- < \tau^+$. Summarizing we have:

Proposition 10.26. *The system* (10.29) *can be stabilized with a controller of the form* $u(t) = -ky(t-\tau)$ *if and only if the pair* (γ, k) *satisfies*

$$\gamma \in [0,2), \ |k| \in \left(2\sqrt{1-\left(1-\frac{\gamma^2}{4}\right)^2},\, 2\right), \tag{10.33}$$

and in addition $\tau_- < \tau_+$, *where*

$$\begin{array}{ll} \tau_\pm = \dfrac{1}{\omega_\pm}\left[\mathrm{Log}\left(\dfrac{\omega_\pm^2 - 2}{k} - j\dfrac{\gamma\omega_\pm}{k}\right) + (1+\mathrm{sign}\, k)\pi\right], & \gamma \neq 0, \\ \left\{\begin{array}{l} \tau_+ = (3+\mathrm{sign}\, k)\frac{\pi}{2\omega_+}, \\ \tau_- = (1+\mathrm{sign}\, k)\frac{\pi}{2\omega_-}, \end{array}\right. & \gamma = 0, \end{array} \tag{10.34}$$

and

$$\omega_\pm = \sqrt{1-\frac{\gamma^2}{4}} \cdot \sqrt{2\left(1 \pm \sqrt{1-\frac{1-\frac{k^2}{4}}{\left(1-\frac{\gamma^2}{4}\right)^2}}\right)}. \tag{10.35}$$

A stabilizing controller is then defined by the gain k *and* $\tau \in (\tau_-, \tau_+)$.

Remark 10.27. *If* $\gamma = 0$ *and* $k \in (-2, 0)$, *we recover the results proposed in* [1, 294]: $\tau_- = 0$, $\tau_+ = \pi/\sqrt{2+|k|}$. *Furthermore, the number of delay intervals is given by*

$$\max_{l\in\mathbb{Z}} \left\{ l \le \frac{1}{2} \cdot \frac{1}{\sqrt{\frac{2+|k|}{2-|k|}} - 1} \right\}.$$

Thus, the smaller the gain is, the larger the number of stabilizing delay intervals is, a property coherent with the graphical representation in [1].

Let us now apply the geometric approach of the previous section to the case $\gamma = 0$. If we consider the interval $[\alpha, \beta] = [-2, 0]$, then we get $\Omega_1 = (0, \sqrt{2}]$, $\Omega_2 = [\sqrt{2}, 2]$. The curves $\mathcal{T}_1^{m-}$ and $\mathcal{T}_2^{m-}$, $m \in \mathbb{Z}$, are of Types 42 and 21, respectively (notice that $\mathcal{T}_1^{0-}$ is the curve $\tau = 0$). In Figure 10.4 these stability crossing curves are plotted. The stability regions are indicated with the letter S.

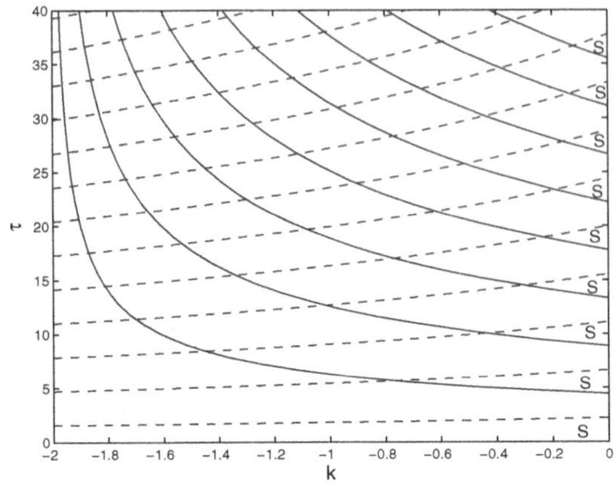

Figure 10.4. *Stability crossing curves for the controlled system* (10.29) *with* $\gamma = 0$. *The solid curves correspond to* $\mathcal{T}_1^{m-}$, *the dashed curves to* $\mathcal{T}_2^{m-}$.

10.3 ▪ Multiple delays case

10.3.1 ▪ Necessary conditions for stabilizability

The following result of [191] generalizes Proposition 10.2 to the multiple delay case:

Proposition 10.28. *Let the transfer function of the system* (10.1) *be given by*

$$H(\lambda) := \frac{P(\lambda)}{Q(\lambda)}, \tag{10.36}$$

where P *and* Q *are coprime polynomials satisfying* $\deg(P) = n$ *and* $\deg(Q) = m$.

A necessary condition for the stabilizability of (10.36) *with the control law* (10.3) *is the Hurwitz stability of the polynomial*

$$\gamma(\lambda;\, \tau_1, \ldots, \tau_l) := \left(\frac{d}{d\lambda} + \tau_1\right)^{m+1} \left(\frac{d}{d\lambda} + \tau_2\right)^{m+1} \cdots \left(\frac{d}{d\lambda} + \tau_l\right)^{m+1} Q(\lambda).$$

Example 10.29. *Consider the example* (10.4), *where* $c_1 = 0$ *and* $c_2 \neq 0$. *With* $l > 1$, *the necessary condition of Proposition* 10.28 *is satisfied when taking all but one delay equal to zero, and the nonzero delay sufficiently small (whatever the value of* a_1 *and* a_2*). In fact, two blocks are* always *sufficient for stabilization (unlike one block; see* Example 10.3), *because the control law*

$$u(t) = -k_1 y(t) - k_2 \frac{y(t) - y(t-\tau)}{\tau},$$

with $a_1 + k_2 c_2 > 0$, $a_2 + k_1 c_2 > 0$, *is stabilizing for sufficiently small values of* τ. *For this, notice that the control law*

$$u(t) = -k_1 y(t) - k_2 \dot{y}(t)$$

is stabilizing. According to [195] *asymptotic stability is preserved when replacing the output* *derivative, $\dot{y}(t)$ with the difference approximation $(y(t)-y(t-\tau))/\tau$, where τ is sufficiently small.*

Similarly to the single delay case, we have:

Corollary 10.30. *Assume that $l(m+1) < n$. If the polynomials $Q(\lambda)$ have at least one unstable root with multiplicity $> l(m+1)$, then the system* (10.36) *cannot be stabilized by the controller* (10.3).

10.3.2 ▪ Stabilization of multiple integrators

We address the stabilization of a chain including n integrators, whose transfer function is given by

$$H(\lambda) := \frac{1}{\lambda^n}. \tag{10.37}$$

Since (10.37) has a pole at zero with multiplicity n, an application of Corollary 10.30 yields that *at least* n terms in the control law (10.3) are *necessary* for stabilization.

In what follows, we show that n terms are also *sufficient* by explicitly constructing stabilizing control laws. In the derivation we will employ the following scaling property, which expresses a natural trade-off between "gain" and "delay" in the controller:

Property 10.31. *The control law*

$$u(t) = -\sum_{j=1}^{n} k_j x(t-\tau_j) \tag{10.38}$$

is asymptotically stabilizing if and only if

$$u(t) = -\sum_{j=1}^{n} \frac{k_j}{\rho^n} x(t-\rho\tau_j), \qquad \rho > 0, \tag{10.39}$$

is asymptotically stabilizing.

Proof. The transformation from (10.38) to (10.39) involves a scaling of the characteristic roots of the closed-loop system by $1/\rho$. □

Note that an analogous scaling property was the basis for the construction of state feedback controllers in the presence of input constraints in [260, 234, 235] and also played a crucial role in the study of the so-called peaking phenomena; see [365, 341] and the references therein.

Control laws based on numerical differentiation with backwards differences

This construction is inspired by [195] and consists of approximating output derivatives with (delayed) output measurements, as we have done in Example 10.29. This initially leads to control laws with small delays, but by Property 10.31 control laws with arbitrary delays can be directly derived. The system (10.37) can be stabilized with the feedback law

$$u(t) = -q_0\, y(t) - q_1\, y'(t) - \cdots - q_{n-1}\, y^{(n-1)}(t), \tag{10.40}$$

where the polynomial $q(\lambda) = \lambda^n + \sum_{k=0}^{n-1} q_k \lambda^k$ is Hurwitz. The latter implies that $q_k > 0$, $k = 1, \dots, n$. Hence, all the derivatives of the output, up to order $(n-1)$, are needed in the control law.

The key idea in the controller construction consists of approximating the output derivatives in (10.40) with (delayed) output measurements. For instance, we have

$$y'(t) \approx \frac{y(t)-y(t-\epsilon)}{\epsilon} \tag{10.41}$$

for small ϵ, which corresponds to an approximation $\lambda \approx (1-e^{-\lambda\epsilon})/\epsilon$ in the frequency domain. Note that the right-hand side of (10.41) is the derivative of the linear approximation of y through the points $(t, y(t))$ and $(t-\epsilon, y(t-\epsilon))$. We now outline how this idea can be generalized to approximate higher-order derivatives of y also.

Choose a set of n delays satisfying $0 \leq \tau_1 < \tau_2 < \cdots < \tau_n$. We may approximate the output $y(t)$ around any time $t = t_0$ with the polynomial

$$y_p(t) = c_0 + c_1(t-t_0) + c_2(t-t_0)^2 + \cdots + c_{n-1}(t-t_0)^{n-1},$$

which interpolates $y(t)$ at the n past instants $t_0 - \epsilon\tau_1, \ldots, t_0 - \epsilon\tau_n$, i.e.,

$$y_p(t_0 - \epsilon\tau_i) = y(t_0 - \epsilon\tau_i), \qquad i = 1, \ldots, n. \tag{10.42}$$

Here $\epsilon > 0$ is a small scaling parameter. Since the Vandermonde matrix

$$T(\tau) := \begin{bmatrix} 1 & \tau_1 & \tau_1^2 & \cdots & \tau_1^{n-1} \\ \vdots & & & & \vdots \\ 1 & \tau_n & \tau_n^2 & \cdots & \tau_n^{n-1} \end{bmatrix} \tag{10.43}$$

is invertible when τ_i are different, the conditions (10.42) can be written as

$$\begin{bmatrix} c_0 \\ c_1 \\ \vdots \\ c_{n-1} \end{bmatrix} = \begin{bmatrix} 1 & & & \\ & \frac{1}{(-\epsilon)} & & \\ & & \ddots & \\ & & & \frac{1}{(-\epsilon)^{n-1}} \end{bmatrix} T(\tau)^{-1} \begin{bmatrix} y(t_0-\epsilon\tau_1) \\ y(t_0-\epsilon\tau_2) \\ \vdots \\ y(t_0-\epsilon\tau_n) \end{bmatrix}, \tag{10.44}$$

and we may approximate

$$y^{(i)}(t_0) \approx y_p^{(i)}(t_0) = i!\, c_i, \qquad i = 1, \ldots, n. \tag{10.45}$$

In this way, the control law (10.40) at $t = t_0$ can be approximated with

$$u(t_0) = -q_0\, y_p(t_0) - q_1\, y_p'(t_0) - \cdots - q_{n-1}\, y_p^{(n-1)}(t_0).$$

Substituting (10.44) and (10.45) into this expression and applying the same principle for all $t_0 > 0$ leads to the control law

$$u(t) = -\begin{bmatrix} q_0 & \frac{1}{(-\epsilon)} q_1 & \frac{2!}{(-\epsilon)^2} q_2 & \cdots & \frac{(n-1)!}{(-\epsilon)^{n-1}} q_{n-1} \end{bmatrix} T(\tau)^{-1} \begin{bmatrix} y(t-\epsilon\tau_1) \\ \vdots \\ y(t-\epsilon\tau_n) \end{bmatrix}. \tag{10.46}$$

When $\epsilon \to 0+$ the approximation of (10.40) becomes better and we have:

Proposition 10.32. *Assume that the polynomial* $q(\lambda) := \lambda^n + q_{n-1}\lambda^{n-1} + \cdots + q_0$ *is Hurwitz. Assume further that* $0 \leq \tau_1 < \tau_2 < \cdots < \tau_n$ *and let* $T(\tau)$ *be defined by* (10.43). *Then the* *control law* (10.46) *achieves asymptotic stability of* (10.37) *for small values of* ϵ. *Moreover, if* $\epsilon \to 0+$, *then the* n *rightmost characteristic roots of the closed-loop system converge to the* n *zeros of* $q(\lambda)$.

Proof. With the control law (10.40) the characteristic equation of the closed-loop system is given by $q(\lambda)=0$, while the control law (10.46) yields $q_\epsilon(\lambda)=0$, where

$$q_\epsilon(\lambda)=\lambda^n+\left[q_0 \ \frac{1}{(-\epsilon)}q_1 \ \frac{2!}{(-\epsilon)^2}q_2 \ \cdots \ \frac{(n-1)!}{(-\epsilon)^{n-1}}q_{n-1}\right]T(\tau)^{-1}\left[\begin{array}{c} e^{-\epsilon\tau_1\lambda} \\ \vdots \\ e^{-\epsilon\tau_n\lambda}\end{array}\right]. \tag{10.47}$$

We first establish a relation between $q(\lambda)$ and $q_\epsilon(\lambda)$ as $\epsilon \to 0+$. Therefore, consider an arbitrary $\lambda \in \mathbb{C}$. Using a Taylor expansion we have

$$e^{-\epsilon\tau_i\lambda}=1+\frac{(-\epsilon\tau_i\lambda)}{1!}+\cdots+\frac{(-\epsilon\tau_i\lambda)^{n-1}}{(n-1)!}+O((\epsilon\lambda)^n), \quad i=1,\ldots,n,$$

which can be written as

$$\left[\begin{array}{c} e^{-\epsilon\tau_1\lambda} \\ e^{-\epsilon\tau_2\lambda} \\ \vdots \\ e^{-\epsilon\tau_n\lambda}\end{array}\right]=T(\tau)\left[\begin{array}{c} 1 \\ \frac{(-\epsilon\lambda)}{1!} \\ \vdots \\ \frac{(-\epsilon\lambda)^{n-1}}{(n-1)!}\end{array}\right]+\left[\begin{array}{c} O((\epsilon\lambda)^n) \\ O((\epsilon\lambda)^n) \\ \vdots \\ O((\epsilon\lambda))^n\end{array}\right]. \tag{10.48}$$

Substituting (10.48) into (10.47) leads to

$$q_\epsilon(\lambda)=q(\lambda)+O(\epsilon\lambda^n). \tag{10.49}$$

Define a *compact* subset S of the complex plane, which contains all the zeros of $q(s)$. From the expression (10.49) it follows that the analytic function $q_\epsilon(\lambda)$ *uniformly* converges to $q(\lambda)$ on S as $\epsilon \to 0_+$. Therefore, both functions have the same number of zeros in S when ϵ is sufficiently small and, moreover, as $\epsilon \to 0_+$, the n zeros of $q_\epsilon(\lambda)$ in S converge to n corresponding zeros of $q(\lambda)$. These statements follow from Rouché's theorem (see also the appendix). The proof is complete when we also show that in any right half plane, $q_\epsilon(\lambda)$ has at most n zeros when ϵ is sufficiently small. This follows from the scaling property, Property 10.31: the condition $q_\epsilon(\lambda)=0$ is equivalent to

$$\bar{\lambda}^n+\left[q_0\epsilon^n \ \frac{\epsilon^{n-1}}{(-1)}q_1 \ \frac{2!\epsilon^{n-2}}{(-1)^2}q_2 \ \cdots \ \frac{(n-1)!\epsilon}{(-1)^{n-1}}q_{n-1}\right]T(\tau)^{-1}\left[\begin{array}{c} e^{-\tau_1\bar{\lambda}} \\ \vdots \\ e^{-\tau_n\bar{\lambda}}\end{array}\right]=0,$$

where $\bar{\lambda}=\epsilon\lambda$. This equation can be interpreted as the characteristic equation of a feedback controlled multiple integrator with fixed feedback delays, where the gain can be made arbitrarily small. As proven in [260], n characteristic roots converges to 0 as the gain tends to 0, while the real parts of the other characteristic roots move off to $-\infty$. This implies that for any $r \in \mathbb{R}$, $q_\epsilon(\lambda)$ has as most n zeros in the half plane $\Re(\bar{\lambda}) \geq r$ or, equivalently, in $\Re(\lambda) \geq r/\epsilon$, provided ϵ is sufficiently small. □

Remark 10.33. *When the complex variable* λ *in the characteristic equation of the closed-loop system with control law* (10.40) *is formally replaced with* $(1-e^{-\lambda\epsilon})/\epsilon$ *(except for the term*

λ^n) and the resulting expression is developed in powers of $e^{-\lambda\epsilon}$, the characteristic equation of a system with proportional+delay compensator with $(n-1)$ commensurate delays is obtained. This is exactly the controller of Proposition 10.32 when taking one delay equal to zero and the other delays commensurate, i.e., $\tau_i = (i-1)$, $i = 1,\dots,n$.

Using Property 10.31, Proposition 10.32 can be rephrased as follows:

Theorem 10.34. *Assume that $0 \le \tau_1 < \cdots < \tau_n$ and $q(\lambda)$ Hurwitz. Then the control law*

$$u(t) = -\begin{bmatrix} \epsilon^n q_0 & \frac{\epsilon^{n-1}}{(-1)} q_1 & \frac{2!\,\epsilon^{n-2}}{(-1)^2} q_2 & \cdots & \frac{(n-1)!\,\epsilon}{(-1)^{n-1}} q_{n-1} \end{bmatrix} T(\tau)^{-1} \begin{bmatrix} y(t-\tau_1) \\ \vdots \\ y(t-\tau_n) \end{bmatrix} \tag{10.50}$$

achieves asymptotic stability of (10.37) for small values of ϵ. As $\epsilon \to 0_+$, the n rightmost characteristic roots converge to $\epsilon\lambda_i$, $i = 1,\dots,n$, with λ_i the zeros of $q(\lambda)$.

Control laws based on exact pole placement and low-gain design

This approach is inspired by [260, 235, 210, 339] and consists of a placement of the n rightmost characteristic roots of the closed-loop by means of low-gain control laws. In the control law $u(t) = -\sum_{j=1}^{n} k_j x(t-\tau_j)$, there are n degrees of freedom, which allow us to place n characteristic roots at prescribed values. This way stability cannot be ensured in general because the number of characteristic roots is infinite and only n of them are controlled. However, this conflict can be solved when using the low-gain approach, developed in [260, 235] in the context of the stabilization of integrators with an input delay and input constraints.

The basic idea is as follows. When the controlled characteristic roots are placed close to zero, it is expected that the gains are low. But when the gains tend to zero, all characteristic roots, except for n, are shifted far away in the left half plane, because the governing delay-differential equation behaves as an ODE with a vanishing (delayed) perturbation.

Theorem 10.35. *Assume $0 \le \tau_1 < \tau_2 \cdots < \tau_n$ and let $T(\tau)$ be defined by (10.43). Then the control law*

$$u(t) = (-1)^n \begin{bmatrix} \epsilon^n & n\epsilon^{n-1} \dots n!\,\epsilon \end{bmatrix} T(\tau)^{-1} \begin{bmatrix} e^{-\epsilon\tau_1} & & \\ & \ddots & \\ & & e^{-\epsilon\tau_n} \end{bmatrix} \begin{bmatrix} y(t-\tau_1) \\ \vdots \\ y(t-\tau_n) \end{bmatrix} \tag{10.51}$$

achieves asymptotic stability of (10.37) for small values of ϵ. Moreover, there the closed-loop system has a characteristic root at $\lambda = -\epsilon$, with multiplicity n.

Proof. The characteristic equation of the closed-loop system is given by $p(\lambda) := \lambda^n + \sum_{j=1}^{n} k_j e^{-\lambda\tau_j} = 0$. Assigning n characteristic roots to $\lambda = \bar{\lambda}$ yields the conditions $p(\bar{\lambda}) = 0, \dots, p^{n-1}(\bar{\lambda}) = 0$, or

$$\begin{bmatrix} e^{-\bar{\lambda}\tau_1} & \cdots & e^{-\bar{\lambda}\tau_n} \\ -\tau_1 e^{-\bar{\lambda}\tau_1} & \cdots & -\tau_n e^{-\bar{\lambda}\tau_n} \\ \vdots & & \vdots \\ (-\tau_1)^{n-1} e^{-\bar{\lambda}\tau_1} & \cdots & (-\tau_n)^{n-1} e^{-\bar{\lambda}\tau_n} \end{bmatrix} \begin{bmatrix} k_1 \\ k_2 \\ \vdots \\ k_n \end{bmatrix} = -\begin{bmatrix} \bar{\lambda}^n \\ n\bar{\lambda}^{n-1} \\ \vdots \\ n(n-1)\cdots 2\bar{\lambda} \end{bmatrix}.$$

This can be written as

$$\begin{bmatrix} 1 & & & \\ & (-1) & & \\ & & \ddots & \\ & & & (-1)^{n-1} \end{bmatrix} T(\tau)^T \begin{bmatrix} e^{-\bar{\lambda}\tau_1} & & \\ & \ddots & \\ & & e^{-\bar{\lambda}\tau_n} \end{bmatrix} \begin{bmatrix} k_1 \\ k_2 \\ \vdots \\ k_n \end{bmatrix} = -\begin{bmatrix} \bar{\lambda}^n \\ n\bar{\lambda}^{n-1} \\ \vdots \\ n(n-1)\cdots 2\bar{\lambda} \end{bmatrix},$$

and therefore,

$$\begin{bmatrix} k_1 \\ k_2 \\ \vdots \\ k_n \end{bmatrix} = -\begin{bmatrix} e^{\bar{\lambda}\tau_1} & & \\ & \ddots & \\ & & e^{\bar{\lambda}\tau_n} \end{bmatrix} T(\tau)^{-T} \begin{bmatrix} \bar{\lambda}^n \\ (-1)n\bar{\lambda}^{n-1} \\ \vdots \\ (-1)^{n-1}n(n-1)\cdots 2\bar{\lambda} \end{bmatrix}.$$

Choosing $\bar{\lambda} = -\epsilon$ leads to the control law (10.51). When we let $\epsilon \to 0+$, we have $K(\epsilon) = [k_1(\epsilon) \cdots k_n(\epsilon)]^T \to 0$. Hence, the n characteristic roots at zero of the uncontrolled system are shifted to $-\epsilon$, while the other characteristic roots cannot cause instability if ϵ (that is, $K(\epsilon)$) is sufficiently small. □

Remark 10.36. *For* $q(\lambda) = (\lambda + 1)^n$ *the control law* (10.50) *reduces to*

$$u(t) = -\begin{bmatrix} \epsilon^n & \frac{n\epsilon^{n-1}}{(-1)} & \frac{n(n-1)\epsilon^{n-2}}{(-1)^2} & \cdots & \frac{n!\,\epsilon}{(-1)^{n-1}} \end{bmatrix} T(\tau)^{-1} \begin{bmatrix} y(t-\tau_1) \\ \vdots \\ y(t-\tau_n) \end{bmatrix}. \quad (10.52)$$

This control law doesn't coincide with (10.51) *because it is based on an asymptotic* approximation *of* $q(\lambda)$, *while* (10.51) *is based on an* exact *placement of* n *characteristic roots.*

Remark 10.37. *Both Theorems* 10.34 *and* 10.35 *guarantee asymptotic stability for sufficiently small values of* ϵ. *A threshold can be computed by performing a numerical continuation of the characteristic roots of the closed-loop systems as a function of the parameter* ϵ, *as illustrated in the next section. Even when the structure of* (10.50) *or* (10.51) *is not explicitly used, a stabilizing feedback law may still be synthesized by means of the approaches outlined in Chapter* 7.

10.3.3 ▪ Illustrative example: Controlling triple integrators

For the *triple integrator* the control law (10.50) with $q(\lambda) = (\lambda + 1)^3$ and $\tau_i = (i-1)$, $i = 1, \dots, 3$, takes the form

$$u(t) = \left(-3\epsilon - \frac{9}{2}\epsilon^2 - \epsilon^3\right) y(t) + (6\epsilon + 6\epsilon^2)\, y(t-1) + \left(-3\epsilon - \frac{3}{2}\epsilon^2\right) y(t-2). \quad (10.53)$$

In Figure 10.5 (left) the rightmost characteristic roots of the closed-loop system are displayed as a function of the parameter ϵ. For $\epsilon < \bar{\epsilon}$, indicated in the figure, the closed-loop

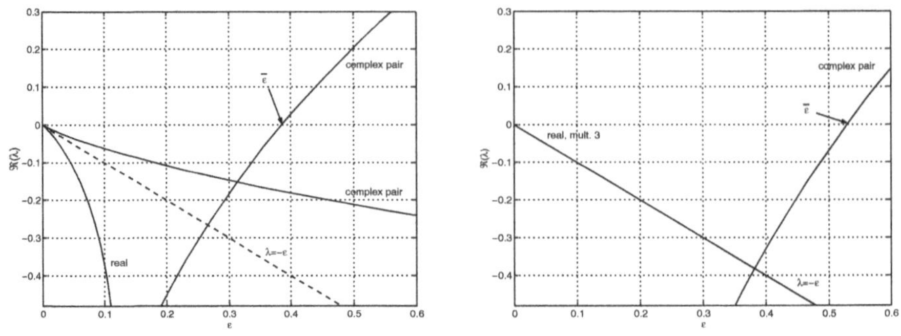

Figure 10.5. *Real parts of the characteristic roots of the triple integrator, controlled with* (10.53) *(left) and* (10.54) *(right)* [300].

system is asymptotically stable. According to Theorem 10.34, the three rightmost characteristic roots converge to $\lambda = -\epsilon$ as $\epsilon \to 0_+$.

The control law (10.51) with $\tau_i = (i-1)$, $i = 1,\ldots,3$, is given by

$$u(t) = \left(-3\epsilon + \frac{9}{2}\epsilon^2 - \epsilon^3\right) y(t) + (6\epsilon - 6\epsilon^2)e^{-\epsilon}\, y(t-1) + \left(-3\epsilon + \frac{3}{2}\epsilon^2\right) e^{-2\epsilon}\, y(t-2) \quad (10.54)$$

and the closed-loop characteristic roots are shown in Figure 10.5 (right). Following Theorem 10.35 three characteristic roots lie at $\lambda = -\epsilon$ for all values of ϵ. Note that when ϵ is small, the spectrum is similar to the previous case, which is not surprising since the dominant terms in (10.53) and (10.54) (the terms $\sim \epsilon$) are equal.

Recall that when ϵ is fixed to a value smaller than $\bar{\epsilon}$, (10.53) and (10.54) actually define a whole one-parameter family of stabilizing feedback laws by making use of Property 10.31.

10.4 ▪ Network-induced delays case

In what follows, we will focus on the output-feedback stabilization problem of a class of SISO networked control systems (NCS) with periodic sampling and constant delays and, as in the previous paragraphs, our main interest concerns the effects induced by the delay presence. In this context, except for the "gain" and the "delay" we will have a further parameter (*"sampling"*) to be taken into account. As reflected in some of the proposed examples, the relationship between the delay and the sampling is particularly important in the stability analysis. For the sake of brevity, we will discuss only a few problems such as, for example, the crossing set characterization, the computation of the stabilizing gains, and the corresponding crossing directions, as well as the sensitivity w.r.t. the delay parameter. For further details on such topics, as well as for complete proofs of the proposed results, the reader is referred to [239].

10.4.1 ▪ Problem formulation and motivating example

Consider the continuous-time linear SISO system

$$\begin{cases} \dot{x}(t) = Ax(t) + b\widehat{u}(t), & t \in [\ell h + \widetilde{\tau}, (\ell+1)h + \widetilde{\tau}), \\ y(t) = cx(t), \end{cases} \quad (10.55)$$

and the control law

$$\widehat{u}(t^+) = -ky(t-\widetilde{\tau}), \quad t \in \{\ell h + \widetilde{\tau}, \ell \in \mathbb{N}\}, \tag{10.56}$$

where $\widehat{u}(t^+)$ is piecewise continuous and only changes its value at $\ell h + \widetilde{\tau}$, where $\widetilde{\tau}$ is the induced network delay satisfying $\widetilde{\tau} = (r-1)h + \tau$ for $0 < \tau \leq h$, $x \in \mathbb{R}^n$, $k, y \in \mathbb{R}$, and $r \in \mathbb{N}$. We are interested in *finding all parameters* (k, h, τ) *such that the controller* (10.56) *(asymptotically) stabilizes the closed-loop SISO system* (10.55). Since the system (10.55) includes both continuous- and discrete-time dynamics, the classical analysis consists in discretizing the entire system in order to homogenize the state variables (it is worth mentioning that, to the best of the authors' knowledge, this approach was promoted for the first time in [146]). Towards this end, we apply ideas similar to those proposed by [10, 411], leading to the representation

$$x[\ell+1] = \Phi(h)x[\ell] + \Phi(h-\tau)\Gamma(\tau)b\widehat{u}[\ell-r] + \Gamma(h-\tau)b\widehat{u}[\ell-r+1], \tag{10.57}$$

where $\Gamma(t) := \int_0^t e^{As}ds$, $\Phi(t) := e^{At}$, $\widetilde{\tau} = (r-1)h + \tau$ with $r \in \mathbb{N}$ and $0 < \tau \leq h$. Define the augmented state vector as $z[\ell] \triangleq \left[x^T[\ell], u[\ell-r], u[\ell-r+1], \ldots, u[\ell-1]\right]^T$, leading to the "augmented" closed-loop system

$$z[\ell+1] = \widetilde{\Phi}(p)z[\ell]. \tag{10.58}$$

Smaller delay case

If the induced network delay $\widetilde{\tau}$ satisfies the condition $0 < \widetilde{\tau} \leq h$, then $\widetilde{\tau} = \tau$, $r = 1$ in (10.57), and the corresponding transfer matrix can be written as

$$\widetilde{\Phi}(p) := \begin{bmatrix} \Phi(h) - k\Gamma(h-\tau)bc & \Phi(h-\tau)\Gamma(\tau)b \\ -kc & 0 \end{bmatrix}. \tag{10.59}$$

In this case, the transfer function can be written as

$$H_{yu}(z; p_o) := \frac{N(z; p_o)}{D(z; p_o)} = c(zI - \Phi(h))^{-1}\left[\frac{\Phi(h-\tau)\Gamma(\tau)b}{z} + \Gamma(h-\tau)b\right]. \tag{10.60}$$

By taking into account the control law (10.56), the characteristic function of the closed-loop system is obtained as

$$F(z; p) = D(z; p_o) + kN(z; p_o). \tag{10.61}$$

Larger delay case

If the induced network delay $\widetilde{\tau}$ satisfies $\widetilde{\tau} = (r-1)h + \tau$ for some positive integer $r > 1$, with $\tau \leq h$, then the corresponding transfer matrix becomes

$$\widetilde{\Phi}(p) := \begin{bmatrix} \Phi(h) & \Phi(h-\tau)\Gamma(\tau)b & \Gamma(h-\tau)b & \cdots & 0 \\ 0 & 0 & 1 & \cdots & 0 \\ \vdots & \vdots & \vdots & \ddots & \vdots \\ 0 & 0 & 0 & \cdots & 1 \\ -kc & 0 & 0 & 0 & 0 \end{bmatrix}. \tag{10.62}$$

Finally, the corresponding transfer function is given by

$$\begin{aligned}\widetilde{H}_{yu} &= c(zI-\Phi(h))^{-1}\left[\Phi(h-\tau)\Gamma(\tau)bz^{-r}+\Gamma(h-\tau)bz^{1-r}\right],\\ \widetilde{H}_{yu} &\equiv z^{1-r}H(z;p_o)\equiv \frac{N(z;p_o)}{z^{r-1}D(z;p_o)},\end{aligned} \tag{10.63}$$

leading to the closed-loop characteristic function

$$F_{r-1}(z;p)=z^{r-1}D(z;p_o)+kN(z;p_o). \tag{10.64}$$

From (10.64), we have that $F_0(z;p)\equiv F(z;p)$. Note also that for $\widetilde{\tau}>h$ the characteristic function of the closed-loop system contains the term z^{r-1}. We write $D_{r-1}(z;p_o):=z^{r-1}D(z;p_o)$.

Motivating example: Chain of integrators

Consider the chain of n integrators

$$y^{(n)}(t)=\widehat{u}(t^+), \qquad t\in[\ell h+\tau,(\ell+1)h+\tau), \tag{10.65}$$

where $\widehat{u}(t^+)$ is given by

$$\widehat{u}(t^+)=-ky(t-\tau), \quad t\in\{\ell h+\tau,\ \ell\in\mathbb{N}\}. \tag{10.66}$$

In the case of NCS stabilization of one integrator ($n=1$), [411] derived the stability region in the parameter space $(h(k),\tau)$ as follows:

$$\max\left\{\frac{1}{2}hk-1,0\right\}<k\tau<\min\{1,hk\}. \tag{10.67}$$

Unfortunately, as found by [238, Proposition 1], the above result is *not longer valid* for higher-order systems ($n\geq 2$). In other words, a *small gain* is always destabilizing. In order to better understand this property, consider the triple-integrator case ($n=3$). For small values of the parameter gain, i.e., $k=\varepsilon$, the closed-loop characteristic roots behave as

$$z_1(\varepsilon)=w_{0,3}\varepsilon+\frac{w_{0,3}\left(-w_{1,3}^3+3h^2w_{0,1}+3!hw_{0,2}\right)}{3!}\varepsilon^2+\mathcal{O}\left(\varepsilon^3\right), \tag{10.68}$$

$$z_{\ell+1}(\varepsilon)=1+w_{1,\ell}\varepsilon^{\frac{1}{3}}+\frac{2w_{1,3}^3+h^2w_{0,1}-w_{1,\ell}^3}{3w_1(\ell)}\varepsilon^{\frac{2}{3}}+\mathcal{O}(\varepsilon), \tag{10.69}$$

where, $w_{0,\ell}:=\tau^\ell/\ell!$, $w_{1,\ell}:=e^{\frac{\ell\pi}{3}j}h$ and $\ell=1,2,3$. Figure 10.6 illustrates the result for $k<0$. It is worth mentioning that the same conclusion was reached in [3, 4] by using a different approach.

10.4.2 ▪ Spectral radius properties

Consider now the continuity property of the spectra of the augmented closed-loop system (10.58) w.r.t. the delay parameter. Toward this end, we have the following:

Definition 10.38. *Let (h^*,k^*) be fixed; the spectral radius (function) $\rho:\mathbb{R}_+\mapsto\mathbb{R}_+$ is defined by*

$$\rho(\tau):=\max\left\{|z|:z\in\sigma\left(\widetilde{\Phi}(p_{h^*,k^*})\right)\right\}.$$

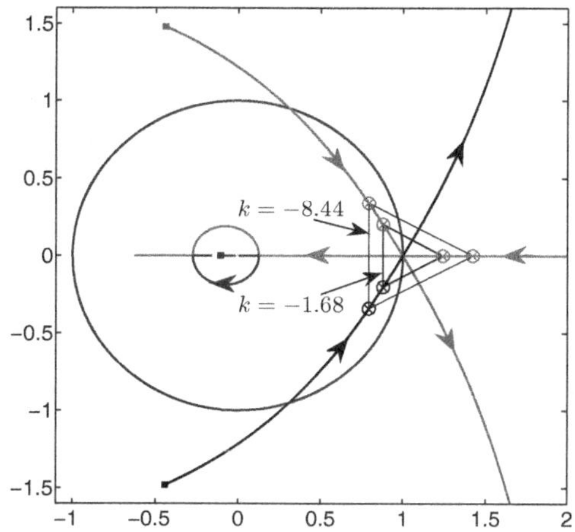

Figure 10.6. *Triple-integrator* ($n = 3$) *subject to* $k < 0$ [239].

Property 10.39. *The spectral radius function has the following properties:* (i) *it always exists,* (ii) *it is finite, and* (iii) *it is continuous.*

Remark 10.40. *Please note that the above result simply says that the spectral radius function is continuous w.r.t. the delay parameter, and in particular when the delay parameter varies from* $\mathscr{I}_\ell$ *to* $\mathscr{I}_{\ell+1}$. *Such a property represents nothing else that the discrete-type version of the continuity of the spectral abscissa (see Chapter 1).*

10.4.3 ▪ Crossing set characterization

Denote by $F(z;p)$ the characteristic function of the closed-loop system. As mentioned in the previous section, for smaller/larger network delays, F can be either $F_{r-1}(z;p) = \sum_{i=0}^{n+r} a_i^{(r-1)}(p)z^i$ (larger delay) or $F_0(z;p)$ (smaller delay). Inspired by [132], we introduce the following notion.

Definition 10.41. *The crossing set* Θ *is defined as the collection of all frequencies* $\theta \in [0, 2\pi)$ *such that there exists a triplet* p^* *of parameters such that* $F_{r-1}(e^{j\theta};p^*) = 0$. *The collection of all triplets* p^* *corresponding to crossing frequencies in* Θ *is defined as the stability crossing surfaces* $\mathscr{S} \subset \mathbb{R}_+^2 \times \mathbb{R}$. *Finally, for a fixed sampling* h_0, *the restriction* $\mathscr{S}|_{p=p_{h_0}}$ *denotes the corresponding stability crossing curves.*

The critical values can be computed in the following way. Associate now to F_{r-1} the parameter-dependent matrices $M_1, M_2 \in \mathbb{R}^{(n+r)\times(n+r)}$:

$$M_1(p) := \begin{bmatrix} a_0(p) & 0 & \cdots & 0 \\ a_1(p) & a_0(p) & \cdots & 0 \\ \vdots & \vdots & \ddots & \vdots \\ a_{n+r-1}(p) & a_{n+r-2}(p) & \cdots & a_0(p) \end{bmatrix}, \tag{10.70}$$

$$M_2^T(p) := \begin{bmatrix} a_{n+r}(p) & a_{n+r-1}(p) & \cdots & a_1(p) \\ 0 & a_{n+r}(p) & \cdots & a_2(p) \\ \vdots & \vdots & \ddots & \vdots \\ 0 & 0 & \cdots & a_{n+r}(p) \end{bmatrix}. \tag{10.71}$$

Introduce the matrix

$$W_F(p) := M_2(p)M_2^T(p) - M_1(p)M_1^T(p). \tag{10.72}$$

Proposition 10.42. *Consider the system* (10.55) *with the control law* (10.56). *The following statements hold:*

(a) p^* *is a crossing point* ($p^* \in \mathscr{S}$) *if and only if* (i) $0 \in \sigma(W_F(p^*))$, *and* (ii) $\sigma\left(\widetilde{\Phi}(p^*)\right) \cap \mathscr{C}(0,1) \neq \emptyset$.

(b) *If for some fixed pair* (h^*, k^*), $F_\ell(z; p^*_{h^*,k^*})|_{\tau=0}$ *is Schur-stable, then* $(h^*, \tau_{min}, k^*) \in \mathscr{S}$, *where* τ_{min} *is the minimal value of the set*

$$\Lambda_\ell(h^*, k^*) := \left\{\tau \in \mathbb{R}_+ : 0 \in \sigma\left(W_{F_\ell}\left(p^*_{h^*,k^*}\right)\right)\right\}.$$

Remark 10.43. *It is important to point out that* (i) *is* necessary *but* not sufficient *for the existence of a crossing since, except for the real crossing points, the determinant* det $W_F(p)$ *vanishes also for symmetric points (of* $F(z; p^*)$*) w.r.t. the unit circle* $\mathscr{C}(0,1)$[24] *(see, e.g.,* [239] *for further discussions and illustrative examples).*

Remark 10.44. *Note that the above result also can be applied to the computation of the stability crossing curves* $\mathscr{S}|_{h=h_0}$ *for some sampling* h_0.

Proof. (a) According to [228], $F_{r-1}(z; p^*)$ and $F^*_{r-1}(z; p^*) := z^{n+r}\overline{F_{r-1}(\frac{1}{\bar{z}}; p^*)}$ has some common root if and only if the determinant of the resultant matrix,

$$R_{F_{r-1},F^*_{r-1}} := \begin{bmatrix} M_1(p) & M_2^T(p) \\ M_2(p) & M_1^T(p) \end{bmatrix}, \tag{10.73}$$

has the property det $R_{F_{r-1},F^*_{r-1}} = 0$. Now, a common root of $F_{r-1}(z; p^*)$ and $F^*_{r-1}(z; p^*)$ means that $F_{r-1}(z; p^*)$ can be factored as $F_{r-1}(z; p^*) = \psi(z)f(z)$, with

$$\psi(z) = \prod_j^{\nu}(z - \rho_j e^{i\phi_j})\left(z - \frac{1}{\rho_j}e^{-i\phi_j}\right) \tag{10.74}$$

and $f(z)$ collect the rest of the roots belonging to the unit circle. On the other hand, the fact that $M_1(p)$ and $M_2(p)$ commute implies that det $R_{F_{r-1},F^*_{r-1}} \equiv -\det W_{F_{r-1}}(p)$. Finally, observe that the condition $\sigma\left(\widetilde{\Phi}(p^*)\right) \cap \mathscr{C}(0,1) \neq \emptyset$ excludes all solutions of the form $(z - \rho_j e^{i\phi_j})(z - \frac{1}{\rho_j}e^{i\phi_j})$, which completes the proof.

(b) Let's prove this property by contradiction, i.e., assume that $(h^*, \tau_{min}, k^*) \notin \mathscr{S}$; then there exist $z^{(0)} \in \mathbb{C}$ such that $F_\ell(z^{(0)}; h^*, \tau_{\min}, k^*) = F_\ell(\frac{1}{\bar{z}^{(0)}}; h^*, \tau_{\min}, k^*) = 0$, where $z^{(0)}$ is symmetric with respect to the unit circle. But this implies that $F_\ell(z^{(0)}; h^*, \tau_{\min}, k^*)$ is unstable, contradicting the fact that the roots of $F_\ell(z; h^*, \tau, k^*)$ are continuous with respect to the delay argument. □

[24]Here, we say that two points z_1 and $z_2 \in \mathbb{C}$ are *symmetric* w.r.t. a circle of radius R and center located at z_0 if $|z_1 - z_0| \cdot |z_2 - z_0| = R^2$, etc.

10.4.4 ▪ Stabilizing gains and crossing directions

Assume now that the parameters (h,τ) are fixed. Then we are interested in developing an algorithm to compute the *set of all stabilizing gains* $k \in \mathbb{R}$. By an abuse of notation, we will construct the $(n+r)\times(n+r)$ matrices (10.70)–(10.71) for lower-order polynomials by setting the coefficient of higher-order terms as zeros. In order to simplify the notations, we will define the resultant of any polynomial $F_{r-1}(z;p)$ with its associated polynomial $F^*_{r-1}(z;p)$ by $R_{F_{r-1}}(p) := R_{F_{r-1},F^*_{r-1}}(p)$. Then, we have the following result:

Proposition 10.45. *Assume (h,τ) is a fixed and known pair $p_o := (h_0,\tau_0)$. Introduce the matrix pencil $\lambda R_N(p_o)+R_{D_{r-1}}(p_o)$ and denote by Λ the set of all its real generalized eigenvalues. Define*

$$\Lambda_s := \left\{\xi \in \Lambda : \quad \exists \eta \in \Lambda,\ \xi\eta = 1, \sigma\left(\widetilde{\Phi}(p_o,\xi)\right)\cap \mathscr{C}(0,1) = \right\}.$$

Let $\lambda_1 < \cdots < \lambda_\ell$, with $\ell \le n+r$ and $\lambda_i \in \Lambda - \Lambda_s$. Then, the system (10.57) cannot be stabilized for any $k=\lambda_i$, $i=1,2,\ldots,\ell$. Furthermore, the number of unstable roots remains invariant for all $k \in (\lambda_i,\lambda_{i+1})$. The same holds for the intervals $(-\infty,\lambda_1)$ and (λ_ℓ,∞).

Proof. First, by construction, the set Λ_s includes the real spectrum of the matrix pencil $R_F(p_o,\lambda) = R_{D_{r-1}}(p_o)+\lambda R_{N_{r-1}}(p_o)$ having the property of symmetry with respect to $\mathscr{C}(0,1)$, but without being characteristic roots of the closed-loop system. Thus, the set Λ– Λ_s collects all the gains for which some crossing w.r.t. $\mathscr{C}(0,1)$ exists. Then, the remaining proof is a direct consequence of Proposition 10.42. □

Remark 10.46. *The result above can be simply interpreted as the* discrete-time *version of the matrix pencil approach proposed by* [67] *in solving the output-feedback stabilization problem of SISO LTI systems.*

Remark 10.47. *Notice that the above characterization for the gain controllers can also be obtained by applying Neimark's D-decomposition method (see, e.g.,* [288, 97]*). Thus,*

$$k(\theta) = -\frac{e^{i(r-1)\theta}D_{r-1}\left(e^{i\theta},p_o^*\right)}{N(e^{i\theta},p_o^*)}, \qquad Im(k(\theta)) = 0, \qquad \theta \in [0,\pi].$$

Example 10.48. *In order to illustrate the above results, consider the following second-order oscillatory system:*

$$\begin{cases} \dot{x}(t) = \begin{bmatrix} 0 & 1 \\ -\omega_0^2 & 0 \end{bmatrix} x(t) + \begin{bmatrix} 0 \\ 1 \end{bmatrix} \hat{u}(t),\ t \in [\ell h+\tau, (\ell+1)h+\tau), \\ y(t) = \begin{bmatrix} \omega_0 & 0 \end{bmatrix} x(t). \end{cases} \tag{10.75}$$

Assume $\omega_0 = 3$, and consider the induced-network delay as $\tilde{\tau} = h+\tau$. The corresponding discretized open-loop transfer function is

$$H_{yu}(z,p_o) = \frac{n_2(p_o)z^2 + n_1(p_o)z + n_0(\tau)}{3z^3 - 6\cos(3h)z^2 + 3z}, \tag{10.76}$$

where $n_2(h,\tau) = 1-\cos(3h-3\tau)$, $n_1(h,\tau) = \cos(3h-3\tau) - 2\cos(3h) + \cos(3\tau)$, and $n_0(\tau) = 1 – \cos(3\tau)$. Then, a direct application of Proposition 10.45 *with $p_o = (19/25, 16/25)$ leads to the results summarized in Table* 10.4.

Table 10.4. *Generalized eigenvalues for* $(h,\tau) = (19/25, 16/25)$.

	λ_1	λ_2	λ_3	λ_4	λ_5	
	−3	0	2.1061	2.7302	4.271	
gain interval*	$(-\infty,-3)$	$(-3,0)$	$(0,2.10)$	$(2.11,2.73)$	$(2.74,4.27)$	$(4.28,\infty)$

* Here, the proposed gain intervals are subject to numerical rounding.

Denote by $\mathscr{D}(\ell, n+r-\ell)$ *the interval with* ℓ*-stable zeros and* $(n+r-\ell)$*-unstable zeros. Then, according to Remark* 10.47, *we apply Neimark's D-decomposition method to compute* $k(\theta)$, *leading to Figure* 10.7. *A simple inspection of the figure shows that we obtain the same results as the ones given in Table* 10.4.

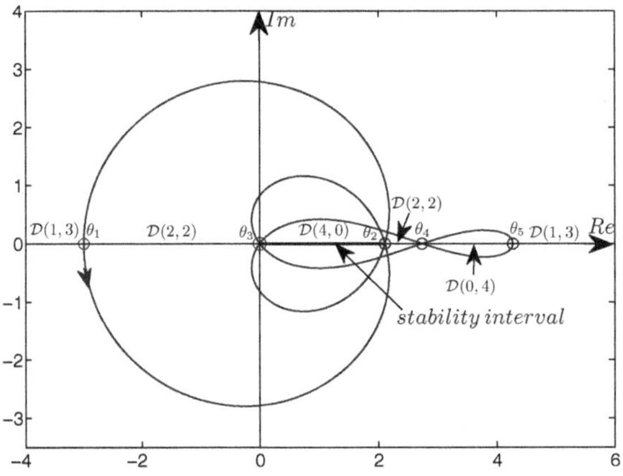

Figure 10.7. *Stability intervals* $\mathscr{D}(\ell, n+r-\ell)$ *for* (10.76), *where* $k(\theta_i) = \lambda_j$ *for some appropriate coefficients* i *and* j [239].

Assume now that the controller gain parameter is fixed to some critical gain value k^* for which there exists at least one critical characteristic root on $\mathscr{C}(0,1)$. For the sake of brevity, we will assume that such critical roots are simple. Then the characterization of the corresponding crossing directions is given as follows:

Proposition 10.49. *Assume that the sampling* h *and the delay* τ *are known and fixed at* h_0 *and* τ_0, *respectively. Let* $k = k^*$ *be a critical gain for the crossing frequency* $\theta = \theta^*$ *(where* $z^* = e^{j\theta^*}$*). Under the assumption that the critical characteristic roots of* F *are simple, the following conditions are equivalent:*

$$\left.\frac{d\,|z|}{dk}\right|_{k=k^*} > 0 \quad (<0), \tag{10.77}$$

$$\Re\left\{\frac{\left.\frac{dF(e^{j\theta^*};k,p_o)}{dk}\right|_{k=k^*}}{\left.\left(z\frac{dF(z,p^*)}{dz}\right)\right|_{z=e^{j\theta^*}}}\right\} < 0 \quad (>0) \tag{10.78}$$

for any k *sufficiently close to and larger than* k^*.

Proof. First, express z as $z = re^{j\theta}$. Next, by deriving it w.r.t. k, we get

$$\frac{dz}{dk} = \frac{z}{r}\frac{dr}{dk} + jz\frac{d\theta}{dk} \Rightarrow \left(\frac{r}{z}\right)\frac{dz}{dk} = \frac{dr}{dk} + jr\frac{d\theta}{dk}. \tag{10.79}$$

Since both dr/dk and $d\theta/dk$ are real numbers, from (10.79) we have

$$\frac{dr}{dk} = \Re\left\{\left(\frac{r}{z}\right)\frac{dz}{dk}\right\}. \tag{10.80}$$

Now, according to the implicit function theorem, we have that

$$\frac{dz}{dk} = -\frac{\frac{dF(z;p)}{dk}}{\frac{dF(z;p)}{dz}}. \tag{10.81}$$

Since $|z| = 1$, equations (10.80)–(10.81) complete the proof. □

10.4.5 ▪ Sensitivity w.r.t. the delay parameter

In what follows, we are interested in better understanding the existing links between the (constant) sampling period h and the network-induced delay τ as well as their effects on the stability. Since the proofs are technical and take advantage of the properties of the corresponding operators, they are omitted (see, for instance, [239] for more details). Thus, consider now the case when the delay is a multiple of the sampling period: $\tilde{\tau} = rh$, with the integer $r > 1$. Then the open-loop system has the transfer function

$$H(z;p_o) = c\,(zI - \Phi(h))^{-1}\Gamma(h)bz^{-r} = \frac{N(z,h)}{z^r D(z,h)}. \tag{10.82}$$

We have the following result:

Proposition 10.50. *Assume that for some fixed parameter $p^* = (h^*, \tau^*, k^*)$ with $\tau^* = h^*$, the following polynomial*

$$P_r(z;p^*) = z^r D(z,p_o^*) + k^* N(z,p_o^*) \tag{10.83}$$

is Schur-stable. Then, there exists some sufficiently small $\epsilon > 0$ such that the same property holds for the perturbed polynomials $F_{r-1}(z;p^*_{\tau=h-\epsilon})$ and $F_r(z;p^*_{\tau=h+\epsilon})$. Furthermore,

$$\sigma(F_r(z;p_{\tau=h+})) = \sigma(F_{r-1}(z;p_{\tau=h-})) \cup \{0\}, \tag{10.84}$$

where $\tau = h+$ ($\tau = h-$) defines the corresponding right (left) limit.

Sketch of proof. The proof idea can be resumed as follows: Without any loss of generality, assume that the closest eigenvalue to $\mathscr{C}(0,1)$, denoted in what follows as $\lambda^{(0)}$, is simple:

$$\lambda^{(0)} := \left\{ \lambda \in \mathbb{C} : \quad |\lambda| = \max_{\tilde{\lambda} \in \sigma(\tilde{\Phi}(p^*))} |\tilde{\lambda}| \right\}.$$

Without any loss of generality, assume that $r = 1$ (i.e., $\tau = h^*$). Then, since the characteristic polynomial associated to the closed-loop system has different degrees for larger or smaller network delays, we need to consider two independent cases: (i) $\tau^* \to \tau^* - \epsilon$ and (ii) $\tau^* \to \tau^* + \epsilon$, respectively. In both situations, a continuity-type argument for the corresponding spectral radius will allow us to conclude.

The following result is a direct consequence of Proposition 10.50:

Proposition 10.51. *Assume $p^* = (h^*, \tau^*, k^*)$ fixed such that $F_{\ell-1}(z; p^*)$ is Schur-stable. Let $\overline{\tau} := \tau - \tau^*$,*

$$\Lambda_{\ell-1}(h^*,k^*) := \left\{\overline{\tau} \in \mathbb{R}_+ : \sigma\left(W_{F_{\ell-1}}(\overline{\tau})\right) = 0\right\},$$
$$\Lambda_{\ell}(h^*,k^*) := \left\{\overline{\tau} \in \mathbb{R}_+ : \sigma\left(W_{F_{\ell}}(\overline{\tau})\right) = 0\right\},$$

and define the minimum elements of $\Lambda_{\ell-1}(h^*,k^*)$ and $\Lambda_\ell(h^*,k^*)$ by

$$\tau_{\ell-1}^- := \inf \Lambda_{\ell-1}(h^*,k^*), \qquad \tau_\ell^- := \inf \Lambda_\ell(h^*,k^*).$$

Then, the system (10.55) *is asymptotically stable for all $\tau \in [(\ell-1)h^* + \tau^*, \ell h^* + \tau^*]$ if and only if the following inequality holds:*

$$(h^* - \tau^*)\tau^* < \tau_{\ell-1}^- \tau_\ell^- . \tag{10.85}$$

Finally, in the case of a simple root, we have the following crossing direction characterization (see, e.g., [239] for a complete proof):

Proposition 10.52. *Let $\lambda_o^* = e^{j\theta^*}$ be a simple eigenvalue of $\Phi(p_{\tau=\tau^*})$. Then for any τ sufficiently close to τ^*, the characteristic zero corresponding to λ_o^* can be expanded by the power series*

$$\lambda_o^* + \left(r_1^T T'(0) q_1\right)(\tau - \tau^*) + o(\tau - \tau^*). \tag{10.86}$$

Thus, for τ sufficiently close to and larger than τ^*, but $\tau > \tau^*$, the corresponding characteristic zero is going outside (inside) the unit circle $\mathscr{C}(0,1)$ if the following condition is satisfied:

$$\cos(\theta_\ell - \theta^*) > 0 \ (<0), \qquad \ell = 1, \ldots, m-1, \tag{10.87}$$

where $T'(0)$ is given by

$$T'(0) = \begin{bmatrix} k\Phi(h^* - \tau^*)bc & (\Phi(h^*) - A\Phi(h^* - \tau^*)\Gamma(\tau^*))b \\ 0 & 0 \end{bmatrix}. \tag{10.88}$$

10.4.6 ▪ Illustrative example: Still second-order systems

Recall the second-order oscillatory system (10.75):

$$\begin{cases} \dot{x}(t) = \begin{bmatrix} 0 & 1 \\ -\omega_0^2 & 0 \end{bmatrix} x(t) + \begin{bmatrix} 0 \\ 1 \end{bmatrix} \hat{u}(t), \ t \in [\ell h + \tau, (\ell+1)h + \tau), \\ y(t) = \begin{bmatrix} \omega_0 & 0 \end{bmatrix} x(t). \end{cases} \tag{10.89}$$

Assume now $\omega_0 = 3$ and the control gain $k = 2$. Then the corresponding stability regions are depicted in Figure 10.8. As shown in Figure 10.8 some pairs (h^*, τ^*) preserve the stability property for $\widetilde{\tau} = \tau + h$. However, it is important to point out that, even in the case when $F_r(z, p_o^*)$ and $F_{r+1}(z, p_o^*)$ are stable, not all $\tau \in [rh + \tau, (r+1)h + \tau]$ preserves this property. To illustrate such a situation, consider $(h, \tau) = (6.1, 2.5)$. Figure 10.9 depicts the root trajectory when τ is increased from $\tau_0 = 2.5$ to $\tau_f = 8.60$. The behavior for the first five critical delays is given in Table 10.5.

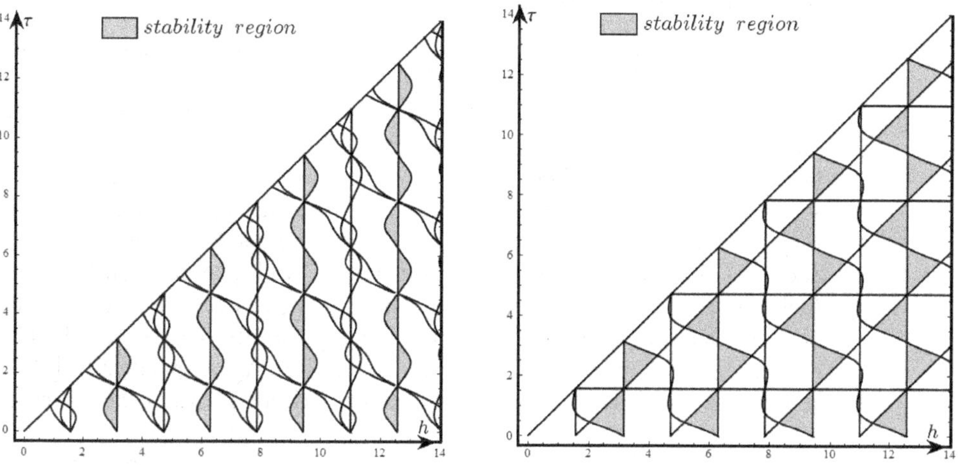

Figure 10.8. *Stability crossing curves $\mathscr{S}_{k^*}$ for $k^* = 2$ and $\omega_o = 3$: $F_{r+1}(z,p)$ (left) and $F_r(z,p)$ (right) with $r = 0$ [239].*

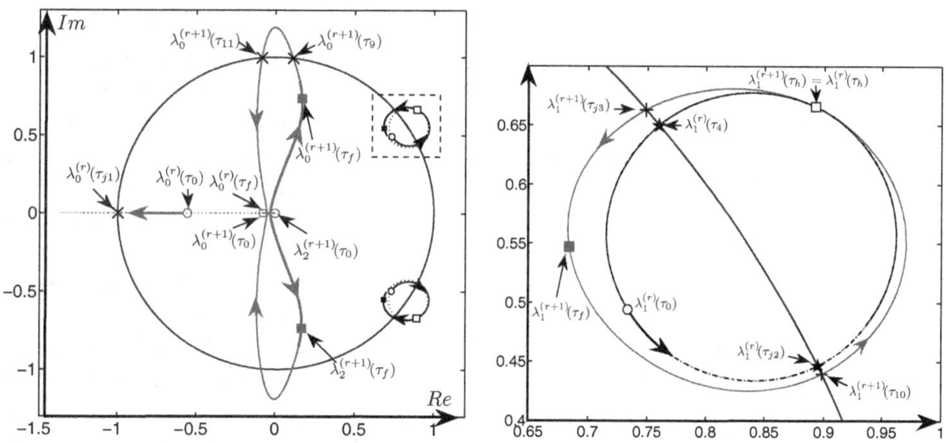

Figure 10.9. *Roots trajectories of the second-order system* (10.75) *with $\omega_0 = 3$, $k = 2$ for $\tilde{\tau} \in [\tau^*, h^* + \tau^*)$ (left) and zoom of the dashed region (right)* [239].

Table 10.5. *Critical delays and crossing directions for the system* (10.75) *with $\omega_0 = 3$ and $k = 2$* [239].

	Critical delay values									
	$\tau_1^{(\ell)}$				$\tau_2^{(\ell)}$		$\tau_3^{(\ell)}$		τ_4	τ_5
	$\ell = 1$	2	3	4	$\ell = 1$	2	$\ell = 1$	2	$\ell = 1$	$\ell = 1$
	2.69	3.4	4.78	5.5	2.98	5.22	6.35	8.44	4.27	6.66
Crossing direction	+	−	+	−	+	+	−	−	−	+

10.5 ▪ Notes and references

This chapter was devoted to the stabilization problem of a class of linear SISO systems subject to delayed output feedback. More precisely, we considered the problem where the delay in the control law may induce a *stabilizing* effect, that is, the closed-loop stability is guaranteed due to the presence of the delay existence. Some results in this direction have been considered in [1, 294], but without any attempt to treat the problem in the general setting. For instance, a Nyquist criterion is used in [1] to prove that a pair (gain, delay) may stabilize some simple second-order linear oscillatory systems. Next, the paper [294] addresses the general static delayed output-feedback problem, and some *existence results* (delay-independent, delay-dependent, instability persistence) are derived in terms of generalized eigenvalues distribution of some appropriate matrix pencils, but without any *explicit construction* of the (stabilizing) controllers. More specifically, [294] compares the stability of the closed-loop schemes with or without delays in the corresponding control laws.

Although only strictly proper SISO systems are considered above, most of the ideas still work for more general SISO systems, such as a restricted class of (not necessarily strictly) proper systems, or systems with internal delays in addition to the feedback input delay. For the sake of conciseness, we do not present such extensions here. Finally, it is important to note that the techniques proposed also can be used for the closed-loop stability analysis of a dead-time plant subject to a proportional controller. We feel that our methodology gives a simpler answer to the corresponding problem (see, for instance, [345] and the references therein, for different frequency-domain approaches).

In Section 10.3 of the chapter, we addressed the output-feedback stabilization problem using control laws of the form (10.3), which involve several delay blocks. First, *necessary stabilizability* conditions were given for the general case. Next, it was shown that a chain of n integrators can be stabilized by a proportional+delay controller including $(n-1)$ delays, or by a chain of n delay blocks. Two constructive approaches were presented and applied to a numerical example. The first approach makes use of some numerical differentiation scheme combined with backward differences, and the second approach is based on an exact pole placement method and a low-gain design.

Necessary conditions for the output-feedback stabilizability of linear systems with control law of the form (10.3) were given in [191]. The problem of stabilizing a chain of integrators with state feedback and input constraints was treated in [369] (without delay) and in [235, 260] (with a single delay in the control loop). Concerning the multiple delay case, some particular cases (single integrator, double integrator with one or two delays) were considered [290, 75, 291, 260], while the general case was treated in [300]. A further example where the pole placement idea is used can be found in Chapter 13, devoted to the analysis of some control congestion algorithms. Finally, note that the use of delays, in particular delay-difference feedback, is well known in the context of stabilizing unstable orbits in nonlinear systems; see, for instance, [329].

Section 10.4 of the chapter is devoted to the output-feedback control problem for some class of SISO NCS with *periodic sampling* and *constant* input/output network-induced delays. For a deeper discussion on NCS, we refer the reader to [161, 209, 411]. Our intention was to show that the eigenvalue-based approach can be used also for analyzing the effects induced by the delays on the corresponding systems' dynamics. Particular attention was paid to the corresponding crossing set characterization as well as to the computation of the stabilizing gains and related crossing directions.

The results presented in this chapter are based on [301] (single delay), [278, 279] (single delay, geometric approach), [300, 191, 260, 257] (multiple delays), [239] (network-induced delays) and some of the references therein.

Chapter 11

Smith predictor for stable systems: Delay sensitivity analysis

11.1 ▪ Introduction

We consider a SISO stable system with a (discrete) input delay, described by the transfer function

$$H(\lambda) = H_0(\lambda)e^{-\lambda\tau}, \tag{11.1}$$

where the rational function $H_0(\lambda)$ is the transfer function of the system free of delay. One of the simplest method for controlling such a system if H_0 is a stable transfer function was proposed by Smith in the 1960s [353].

The idea behind *Smith predictors* is to use a controller structure which takes the delay out of the control loop and allows a feedback design based only on $H_0(\lambda)$. More precisely, in the classical Smith predictor [353], a controller of the form

$$C'(\lambda) = \frac{C(\lambda)}{1 + C(\lambda)(H_0(\lambda) - H_0(\lambda)e^{-\lambda\tau})} \tag{11.2}$$

is used, which gives the following closed-loop transfer function from an external reference signal r to the plant output y:

$$H_{y,r}(\lambda) = \frac{C(\lambda)H_0(\lambda)}{1 + C(\lambda)H_0(\lambda)}e^{-\lambda\tau},$$

and hence $C(\lambda)$ can be designed based on $H_0(\lambda)$.

The advantage in using such an approach in controlling input delay systems lies in its simplicity. To the best of the authors' knowledge, three main problems of the Smith predictor schemes have been considered in the control literature during the last 20 years: (a) robustness (plant and/or delay) [106, 340, 410], (b) disturbance rejection characteristics [406], and (c) extension of the Smith principle idea to the case of integrative plants [9, 233, 306].

This chapter deals with the first problem. Therefore, we assume that the delay of the plant is given by $\tau + \delta$, with the real constant δ modeling the *delay uncertainty*, and that the controller design is based on the nominal delay value τ. Then the closed-loop transfer function becomes [295]

$$H_{y,r}^{\delta}(\lambda) = \frac{C(\lambda)H_0(\lambda)e^{-\lambda(\tau+\delta)}}{1 + C(\lambda)H_0(\lambda) + C(\lambda)H_0(\lambda)e^{-\lambda\tau}(e^{-\lambda\delta} - 1)}. \tag{11.3}$$

When the delay-free plant and the controller are factorized as

$$H_0(\lambda)=\frac{B_1(\lambda)}{A_1(\lambda)},\quad C(\lambda)=\frac{B_2(\lambda)}{A_2(\lambda)},$$

we have

$$H^{\delta}_{y,r}(\lambda)=\frac{B(\lambda)e^{-\lambda(\tau+\delta)}}{A(\lambda)+B(\lambda)e^{-\lambda\tau}(e^{-\lambda\delta}-1)},\tag{11.4}$$

where $A(\lambda)=A_1(\lambda)A_2(\lambda)+B_1(\lambda)B_2(\lambda)$ and $B(\lambda)=B_1(\lambda)B_2(\lambda)$. Without any loss of generality, we can assume that A and B are coprime, that is, they do not have common zeros.

We are interested in investigating the robustness of stability of the Smith predictor w.r.t. an inaccurate modeling of the delay. This corresponds to the following:

Problem 11.1. *Find conditions on Δ such that the equation*

$$A(\lambda)-B(\lambda)e^{-\lambda\tau}+B(\lambda)e^{-\lambda(\tau+\delta)}=0,\tag{11.5}$$

where $A(\lambda),B(\lambda)$ are polynomials with $\deg(A(\lambda))\geq\deg(B(\lambda))$ and $A(\lambda)$ Hurwitz, has all its solutions in $\mathbb{C}^-$ when $|\delta|\leq\Delta$.

Remark 11.2. *The transfer function from a plant input disturbance $w(\lambda)$ to the plant output $y(\lambda)$ is given by*

$$\begin{aligned}H^{\delta}_{y,w}&=\frac{H_0(\lambda)\left(1+C(\lambda)H_0(\lambda)-C(\lambda)H_0(\lambda)e^{-\lambda\tau}\right)e^{-\lambda(\tau+\delta)}}{1+C(\lambda)H_0(\lambda)+C(\lambda)H_0(\lambda)e^{-\lambda\tau}(e^{-\lambda\delta}-1)},\\&=\frac{B_1(\lambda)\left(A(\lambda)-B(\lambda)e^{-\lambda\tau}\right)}{A_1(\lambda)(A(\lambda)+B(\lambda)e^{-\lambda\tau}(e^{-\lambda\delta}-1))},\end{aligned}\tag{11.6}$$

and hence its poles consist of the zeros of (11.5) *and the open-loop system poles, i.e., the zeros of $A_1(\lambda)$. Therefore, the Smith predictor can only be applied to stable open-loop systems. For modifications of the scheme, which are applicable to unstable open-loop systems, we refer the reader to Section* 12.5.2 *and the paper* [261] *and the references therein.*

Without being exhaustive, we cite some of the works in the control literature related to this problem. Some robust stability conditions (plant and delay uncertainty) using frequency-sweeping tests have been proposed for scalar systems in [410]. The corresponding criteria use the growth rate of the plant uncertainty towards high frequencies and do not include any information on the delay uncertainty upper bound. This result has been extended to handle more general multivariable feedback systems in [106]. A different criterion, including information on the delay uncertainty upper bound, was proposed in [340], where the tuning procedure is based on a proper selection of the closed-loop bandwidth, but the procedure explicitly requires no uncertainty on the (delay-free) plant's parameters.

Notice that the Smith predictor takes the delay out of the control loop because it involves a dynamic prediction of the output variable. As explained in [261, 322] and the next chapter, it is also possible to use a static prediction, where a predicted output is directly calculated from the current output and an integral over past inputs. It will be shown in the next chapter that the analysis of the delay sensitivity problem of such controllers reduces to a robustness problem of the form (11.5). Therefore, we will analyze the more general Problem 11.1 and interpret the results only in terms of Smith predictors.

In the previous work [272], the robustness of Smith predictors w.r.t. delay inaccuracy was studied in detail and various stability/instability characterizations were provided, under the condition that the transfer function $C(\lambda)H_0(\lambda)$ is strictly proper or, equivalently, that $\deg(A(\lambda)) > \deg(B(\lambda))$ in equation (11.5).

Here we *also* consider the case where $\deg(A(\lambda)) = \deg(B(\lambda))$, which includes Smith predictors, where the transfer function $C(\lambda)H_0(\lambda)$ is proper but not strictly proper. As we will see, this can lead to a sensitivity of stability w.r.t. *infinitesimal* delay mismatches; i.e., infinitesimal delay mismatches δ may destroy stability, and hence equation (11.5) may not be practically stable (the latter is defined as in [321, Section 3]):

Definition 11.3. *Equation* (11.5) *is practically stable w.r.t. delay mismatches* δ *when there exists a* $\bar{\delta} > 0$ *such that the solutions of* (11.5) *are in the open left half plane for all* $|\delta| < \bar{\delta}$.

Although the possible lack of practical stability has been observed by other authors in the context of Smith predictors [322, 321], a complete characterization and interpretation in terms of the system parameters will be given in this chapter.

The structure of the chapter is as follows. First, we consider the sensitivity of the stability of (11.5) w.r.t. infinitesimal delay mismatches δ. Thereby, we explain the instability mechanism and derive necessary and sufficient conditions for practical stability, based on two different approaches. The first approach, inspired by [13, 242], consists of first interpreting (11.5) as the characteristic equation of a neutral differential equation and then studying the sensitivity of the spectrum of the associated difference equation; the other approach consists of relating practical (in)stability of (11.5) with the analysis of infinitesimal delays in feedback loops, studied in [212]. In Section 11.3 of the chapter, we assume practical stability and—according to Problem 11.1—derive bounds on the maximal delay mismatch Δ. Then we comment on the geometry of the stability regions, using the results of Chapter 6, and illustrate the obtained results with a numerical example. Finally, we generalize the main results to the multivariable case.

11.2 ▪ Sensitivity of stability w.r.t. infinitesimal delay mismatches

We take a closer look at the sensitivity w.r.t. small delay mismatches. First, we explain the instability mechanism, and next we present necessary and sufficient conditions for practical stability.

11.2.1 ▪ Instability mechanism

Assume that $\deg(A(\lambda) = n$ and define

$$S = \lim_{|\lambda|\to\infty} \frac{B(\lambda)}{A(\lambda)}. \tag{11.7}$$

Then equation (11.5) can be written as

$$\lambda^n(1+Se^{-\lambda(\tau+\delta)}-Se^{-\lambda\tau})+q(\lambda)=0, \tag{11.8}$$

where $\lim_{|\lambda|\to\infty,\ \Re(\lambda)\geq 0} q(\lambda)/\lambda^n = 0$. This is also the characteristic equation of the delay-differential equation

$$\frac{d^n}{dt^n}(x(t)+Sx(t-(\tau+\delta))-Sx(t-\tau))+q\left(\frac{d}{dt}\right)x=0. \tag{11.9}$$

Note that for any $\delta \neq 0$, equation (11.9) is of *neutral* type when $S \neq 0$, while it is of *retarded* type for $\delta = 0$. In fact, a small perturbation $\delta \neq 0$ forms a noncompact perturbation of the time-integration operator (solution semigroup) associated with equation (11.9), which introduces an essential spectrum.

As shown in Chapter 1 and in [149, 13], it is precisely this essential spectrum which may cause a sensitivity of the stability w.r.t. infinitesimal parameter changes. It corresponds to sequences of roots of (11.5), whose moduli tend to infinity yet whose real parts have a finite limit. Since the essential spectrum of the solution semigroup of (11.9) coincides with the essential spectrum of the solution semigroup of the delay-difference equation

$$x(t) = -Sx(t-(\tau+\delta)) + Sx(t-\tau), \tag{11.10}$$

(see, e.g., [151]), we are led to the study of the stability of (11.10) or, equivalently, of the behavior of the roots of its characteristic equation

$$H(\delta,\lambda) := 1 + Se^{-\lambda(\tau+\delta)} - Se^{-\lambda\tau} = 0. \tag{11.11}$$

We now illustrate that equation (11.10) and, as a consequence, equation (11.5), may be unstable for arbitrarily small values of δ and explain why this is not in conflict with the continuous dependence of the characteristic roots w.r.t. the system parameter δ. Therefore, we plot in Figure 11.1 the characteristic roots of equation (11.10) with $S = 0.6$ and $\delta = \tau/n$ for different integer values of n. Note the spectrum is periodic w.r.t. shifts in the imaginary parts, following from the fact that (11.11) can be written as a polynomial in $e^{-\lambda\tau/n}$. Since the spectrum is empty for the limit case $\delta = 0$, the continuous dependence of the individual characteristic roots w.r.t. δ implies that these have to move off to infinity as $\delta \to 0$. However, since the rightmost characteristic roots grow unbounded without leaving the right half plane, instability is preserved for any $\delta \neq 0$.

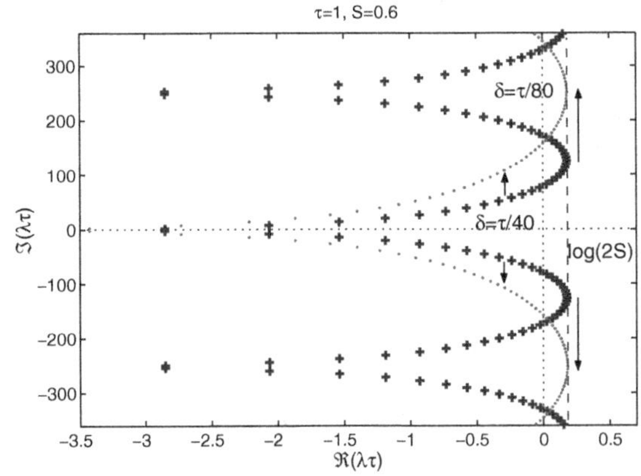

Figure 11.1. *Spectrum of the difference equation* (11.10) *for* $S = 0.6$ *and* $\delta = \tau/n$ *with* $n = 40$, *indicated with* (+), *and* $n = 80$, *indicated with* (o) [253].

In Figure 11.2 we show the corresponding (essential) spectrum of the operator $\mathcal{T}(1)$, which performs a time-integration of equation (11.10) over one time-unit. It consists of eigenvalues μ satisfying $\mu = \exp(\lambda)$, where λ is a characteristic root of (11.10). Note that a normalization in the delay is no longer possible. The fact that the essential spectra of the time-integration operators of equations (11.9) and (11.10) coincide implies that the characteristic roots of (11.10) with a large imaginary part are approximate characteristic roots of (11.9).

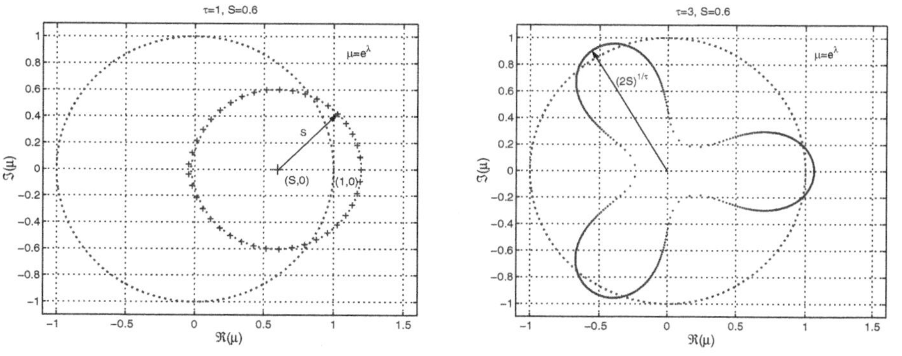

Figure 11.2. *Eigenvalues of the time-integration operator* $\mathcal{T}(1)$ *of the difference equation* (11.10) *for* $\delta = \tau/n$ *with* $S = -0.6$, $n = 40$ (x), *and* $n = 80$ (o). *One can show that for* $\tau = 1$, *the essential spectrum approaches a circle as* $\delta \to 0$ *(left), while for* $\tau = m \in \mathbb{N}$ *one recovers an "m-folded leaf" (right). For all values of* τ *the spectral radius of the essential spectrum is given by* $(2|S|)^{1/\tau}$ [253].

11.2.2 ▪ Conditions for practical stability

In order to check the stability of (11.10) for small δ, we have to analyze the set

$$Z(\delta) = \{\Re(\lambda) : H(\delta, \lambda) = 0\},$$

whose properties are described in [13, 242] and now here briefly rehearsed. In the difference equation (11.10) there are in fact two delays $\tau_1 = \tau + \delta$ and $\tau_2 = \tau$. When τ_1 and τ_2 are rationally independent (noncommensurate), the spectrum is quasi-periodic and $\bar{Z}(\delta)$ consists of an *interval*, characterized by $\bar{Z}(\delta) = \mathcal{I}(\delta)$, where

$$\begin{aligned} \mathcal{I}(\delta) = \{\alpha \in \mathbb{R} :\ & \exists \theta_1, \theta_2 \in [0,\ 2\pi) \text{ such that} \\ & 1 + Se^{-\alpha(\tau+\delta)}e^{-i\theta_1} - Se^{-\alpha\tau}e^{-i\theta_2} = 0\}. \end{aligned} \tag{11.12}$$

This result follows from substituting $\lambda = \alpha + i\beta$ in the characteristic equation and from the observation that with a suitable choice of β, $(\beta\tau_1, \beta\tau_2)\text{mod}2\pi$ is arbitrarily close to any given (θ_1, θ_2) by Kronecker's Theorem [155, Theorem 444].

When the delays τ_1 and τ_2 are rationally dependent (commensurate), the spectrum of (11.10) is periodic. The set $Z(\delta)$ consists of a finite number of *points*, as illustrated with Figure 11.1, and satisfies $Z(\delta) \subset \mathcal{I}(\delta)$. However, when considering a sequence of rationally dependent delays $(\tau + \delta_n, \tau)$ converging to rationally independent delays $(\tau + \delta, \tau)$, the interval $\mathcal{I}(\delta)$, described by (11.12), is arbitrarily well approximated by this set of points as $n \to \infty$, due to the fact that the map $\delta \in \mathbb{R} \to Z(\delta)$ is lower semicontinuous in the Hausdorff metric.

From the above analysis it follows that (practical)[25]) is stability of the difference equation (11.10) is determined by the maximal value $\alpha_m(\delta)$ of (11.12), defined by

$$1 - |S|e^{-\alpha_m(\delta)(\tau+\delta)} - |S|e^{-\alpha_m(\delta)\tau} = 0, \tag{11.13}$$

which converges[26] to

$$\alpha_M = \frac{\log 2|S|}{\tau}, \tag{11.14}$$

[25]Here we mean that stability should be robust for small deviations around the nominal value δ. Note that rationally dependent delays $(\tau+\delta, \tau)$ can always be perturbed to rationally independent delays by an arbitrarily small perturbation of δ.

[26]Actually α_m is a continuous function of $\delta \geq -\tau$.

as $\delta \to 0$. Taking into account the relation between the spectrum of (11.10) and the roots of (11.5), this leads to the following result:

Proposition 11.4. *Consider equation* (11.5) *and let S be defined by* (11.7). *If* $|S| < 1/2$, *then the asymptotic stability is preserved for small values of* δ. *If* $|S| > 1/2$, *then the equation is not practically stable.*

Proof. Instability of the difference equation (11.10), i.e., $\alpha_M > 0$ or $|S| > 1/2$, implies instability of (11.9)-(11.5), as follows from the arguments spelled out before. Hence, we only have to prove that the same holds for asymptotic stability (i.e., $\alpha_M < 0$ or $|S| < 1/2$) when δ is sufficiently small. Here, note that the difference equation only provides information on the essential spectrum of the solution semigroup associated with (11.9), while the latter also has a point spectrum.

Take a number ϵ satisfying $\alpha_M < \epsilon < 0$ and such that $A(\lambda)$ has no roots in the half plane $\mathbb{C}_\epsilon := \{\lambda \in \mathbb{C} : \Re(\lambda) > \epsilon\}$. Equation (11.8) can be rewritten as

$$1 + Se^{-\lambda(\tau+\delta)} - Se^{-\lambda\tau} = -q(\lambda)/\lambda^n. \tag{11.15}$$

For small values of δ, the modulus of the left-hand side is uniformly bounded below over $\mathbb{C}_\epsilon$ by a strictly positive constant. Since the right-hand size of (11.15) tends to zero for large $|\lambda|$, this implies the existence of numbers $M, \bar{\delta} > 0$ such that all the solutions of (11.15)-(11.5) in $\mathbb{C}_\epsilon$ satisfy $|\lambda| \leq M$ when $|\delta| \leq \bar{\delta}$. Define the compact set $\mathcal{S} := \{\lambda \in \bar{\mathbb{C}}_\epsilon : |\lambda| \leq M\}$. Since on $\mathcal{S}$ the function $A(\lambda) - B(\lambda)e^{-\lambda\tau} + B(\lambda)e^{-\lambda(\tau+\delta)}$ uniformly converges to the function $A(\lambda)$ as $\delta \to 0$, we can apply Lemma A.2, which states that the two functions have the same number of zeros in $\mathcal{S}$ when δ is sufficiently small. Therefore, equation (11.5) has no solutions in $\mathcal{S}$ for small δ and, as a consequence, in $\mathbb{C}_\epsilon$, and is asymptotically stable. □

Remark 11.5. *The condition* $|S| < 1/2$ *is a necessary and sufficient condition for the strong stability of the difference equation* (11.10); *see* Chapter 1.

Remark 11.6. *For* $|S| = 1/2$ *we can only conclude the existence of sequences of roots of* (11.5) *which approach the imaginary axis. However, the way of approaching the imaginary axis (from the left/from the right/oscillatory) may also depend on system parameters other than those of the difference equation* (11.10), *which only describes the limit case.*

Remark 11.7. *Stability or instability for small* δ *does not depend on the value of the nominal delay* τ. *However, expression* (11.14) *reveals the softening effect of increasing the delay on the (in)stability of difference equations, observed in* [149]. *When* $\tau \to \infty$, *we have* $\alpha_M \to 0$, *and hence unstable equations become less unstable and vice versa.*

Remark 11.8. *In the case where* (11.5) *represents the (closed-loop) characteristic equation of the Smith predictor* (11.3), *we have*

$$S = \lim_{|\lambda| \to \infty} \frac{B(\lambda)}{A(\lambda)} = \lim_{|\lambda| \to \infty} \frac{C(\lambda)H_0(\lambda)}{1 + C(\lambda)H_0(\lambda)}.$$

The corresponding instability result of Proposition 11.4 *has also been derived in* [321], *based on a Nyquist stability criterion.*

We now outline an alternative frequency domain approach for the analysis of our practical stability problem, based on the results of Logemann et al. [212]. Note that the zeros of (11.5) are the poles of the transfer function

$$\frac{B(\lambda)e^{-\lambda\tau}}{A(\lambda)-B(\lambda)e^{-\lambda\tau}+B(\lambda)e^{-\lambda(\tau+\delta)}}=\frac{M(\lambda)}{1+M(\lambda)e^{-\lambda\delta}}, \tag{11.16}$$

where the transfer function

$$M(\lambda)=\frac{B(\lambda)e^{-\lambda\tau}}{A(\lambda)-B(\lambda)e^{-\lambda\tau}}$$

is regular. Equation (11.16) can be interpreted as a system represented by $M(\lambda)$ and stabilized with unity feedback, in the presence of a small feedback delay δ. Precisely, the robustness of stability of such feedback systems w.r.t. the small feedback delay has been studied in [212]. Denote by $\mathscr{B}_M$ the set of poles of $M(\lambda)$ in $\mathbb{C}_+$. Then a direct application of [212, Theorem 1.1] yields:

Proposition 11.9. *Consider the system* (11.16) *and define*

$$\gamma=\limsup_{\substack{|\lambda|\to\infty\\ \lambda\in\mathbb{C}_+\setminus\mathscr{B}_M}}|M(\lambda)|.$$

If $\gamma<1$, then the asymptotic stability is preserved for small feedback delays $\delta>0$. If $\gamma>1$, then the system is not practically stable.

Note that only positive perturbations $\delta>0$ are considered in the paper [212] (in the framework of feedback delays), while in our case δ may be negative (recall that δ is a delay *mismatch*). However, based on the continuity properties of characteristic roots, one can easily show that the condition $\gamma<1$ also implies robustness of stability for small $\delta<0$. Obviously the conditions of Propositions 11.9 and 11.4 are equivalent, since

$$\gamma=\begin{cases}\frac{|S|}{1-|S|}, & |S|<1,\\ +\infty, & |S|\geq 1.\end{cases}$$

11.3 ▪ Stability analysis and critical delay mismatches

Consider equation (11.5) and assume that $|S|<1/2$, with S defined by (11.7). By Proposition 11.4, the asymptotic stability is then preserved for a small delay mismatch δ. We now characterize its maximal deviation Δ, that is, Problem 11.1.

From equation (11.13) it follows that the difference equation (11.10) remains asymptotically stable for all values of $\delta\geq-\tau$ (although α_m depends on δ, its sign cannot change). Furthermore, from [151, Theorem XII.10.4] it can be deduced that for any $\epsilon>0$, there are only a finite number of solutions of (11.8) with $\Re(\lambda)>\alpha_m+\epsilon$. Therefore, when increasing the delay mismatch, the essential spectrum of the solution semigroup of (11.9) remains stable and the transition from stability to instability is always of a finite-dimension nature, as in the case of equations of retarded type.

In order to calculate stability regions in δ, we look for solutions of (11.5) on the imaginary axis. Substituting $\lambda=j\omega$ yields

$$A(j\omega)-B(j\omega)e^{-j\omega\tau}=-B(j\omega)e^{-j\omega\tau}e^{-j\omega\delta}. \tag{11.17}$$

Then the critical delay mismatches and the corresponding characteristic roots on the imaginary axis can be computed as follows: the "crossing" frequencies ω_k are the positive solutions of the equation

$$| A(j\omega) - B(j\omega)e^{-j\omega\tau} | = | B(j\omega) |, \tag{11.18}$$

and for each ω_k, one can calculate the corresponding delay mismatches $\delta_{k,l}$ using the phase information of equation (11.17):

$$\delta_{k,l} = \frac{\angle\left(\frac{-B(j\omega_k)e^{-j\omega_k\tau}}{A(j\omega_k)-B(j\omega_k)e^{-j\omega_k\tau}}\right) + 2\pi l}{\omega_k}, \quad l \in \mathbb{Z}. \tag{11.19}$$

Remark 11.10. *When rewriting equation* (11.18) *as*

$$\left|1 - \frac{B(j\omega)}{A(j\omega)}e^{-j\omega\tau}\right| = \left|\frac{B(j\omega)}{A(j\omega)}\right|,$$

it becomes clear that under the assumed practical stability condition, i.e.,

$$|S| = |\lim_{|\lambda|\to\infty} B(\lambda)/A(\lambda)| < 1/2,$$

it only has a finite *number of solutions, which expresses the finite-dimensional nature of the transitions to instability, explained in Chapter* 1. *On the contrary, when the practical stability condition is not fulfilled (i.e., $|S| > 1/2$), equation* (11.18) *has an infinite sequence of solutions $\{\omega_k\}_{k\geq 1}$ with $|\omega_k| \to \infty$ and corresponding $|\delta_{k,l}| \to 0$ as $n \to \infty$, which precisely corresponds to the (destabilizing) behavior of the essential spectrum of the associated solution semigroup for small δ, illustrated in section* 11.2.1.

Define the sets $\Lambda_{+,0}$ and $\Lambda_{-,0}$ as follows [295]: if there does not exist any *positive* $\delta_{k,l}$, then $\Lambda_{+,0} = \{+\infty\}$, and

$$\Lambda_{+,0} = \{\delta_{k,l} > 0 \quad : \quad \delta_{k,l} \text{ given by (11.19)}\}, \tag{11.20}$$

elsewhere. Similarly, if there does not exist any *negative* $\delta_{k,l}$, then $\Lambda_{-,0} = \{-\tau\}$, and

$$\Lambda_{-,0} = \{\delta_{k,l} < 0 \quad : \quad \delta_{k,l} \text{ given by (11.19)}\}, \tag{11.21}$$

elsewhere. The smallest destabilizing delay mismatches, $\Delta_1 < 0$ and $\Delta_2 > 0$, are given by

$$\Delta_1 = \max \Lambda_{-,0}, \qquad \Delta_2 = \min \Lambda_{+,0}, \tag{11.22}$$

and the derived condition is *necessary and sufficient*. The results can be summarized as follows [295]:

Proposition 11.11. *Define the real function*

$$\mathcal{F}(\omega) = | A(j\omega) + B(j\omega)e^{-j\omega\tau} |^2 - | B(j\omega) |^2 .$$

The stability of the closed-loop system is guaranteed for any inaccurate modeling delay δ, $| \delta | \leq \Delta$, if

(i) *$\mathcal{F}(\omega)$ has no zeros. In such a case, the stability property is of* delay-independent *type, i.e., it may hold for any $\delta > -\tau$;*

(ii) *$\mathscr{F}(\omega)$ has at least one zero. In such a case, the stability property is of* delay-dependent *type and it holds for any $\delta \in (\Delta_1, \Delta_2)$ ($\Delta_1 < 0 < \Delta_2$), where Δ_1, Δ_2 are given by*

$$\Delta_1 = \max \Lambda_{-,0}, \qquad \Delta_2 = \min \Lambda_{+,0}.$$

where the sets $\Lambda_{\pm,0}$ are defined as in (11.20) *and* (11.21).

In this case, $\Delta < \min\{\Delta_1, \Delta_2\}$ and is always a finite value.

Remark 11.12. *If the following inequality holds:*

$$2|B(j\omega)| < |A(j\omega)| \quad \forall \omega \in \mathbb{R},$$

then the closed-loop system stability is guaranteed for all *inaccurate modelings of the delay. The same result can be found in* [340], *but the stability argument is different.*

11.4 ▪ Geometry of stability regions

In Chapter 6 we presented to some extent the geometry of the stability regions in the delay-parameter space of a linear system including two delays. In this section we adapt the corresponding approach and ideas to the (delay-, delay uncertainty-)parameter space under consideration.

11.4.1 ▪ Identification of crossing points

Define the following auxiliary characteristic function:

$$p(\lambda;\ \tau_1, \tau_2) := A(\lambda) + B(\lambda)e^{-\lambda\tau_1} - B(\lambda)e^{-\lambda\tau_2},$$

which is nothing else than the characteristic function of the closed-loop system of the Smith predictor when taking $\tau_1 = \tau$ and $\tau_2 = \tau + \delta$.

Let $\mathscr{T}$ denote the set of all points of $(\tau_1, \tau_2) \in \mathbb{R}_+^2$ such that $p(\lambda;\ \tau_1, \tau_2)$ has at least one zero on the imaginary axis. Any $(\tau_1, \tau_2) \in \mathscr{T}$ is known as a *crossing point*. The set $\mathscr{T}$, which is the collection of all crossing points, is called the set of *stability crossing curves*. Let $\mathscr{T}_\omega$ denote the set of all $(\tau_1, \tau_2) \in \mathbb{R}_+^2$ such that the auxiliary characteristic function $p(\lambda;\ \tau_1, \tau_2)$ has at least one zero $\lambda = j\omega$ on the imaginary axis. Let Ω be the set of all $\omega > 0$ for which there exists a pair (τ_1, τ_2) such that $p(j\omega;\ \tau_1, \tau_2) = 0$. We will refer to Ω as the *frequency crossing set*. Obviously,

$$\mathscr{T} = \{\mathscr{T}_\omega :\ \omega \in \Omega\}. \tag{11.23}$$

In what follows we consider the *nondegenerate* case satisfying the following assumption:

$$A(j\omega)B(j\omega) \neq 0 \qquad \forall \omega \in \Omega. \tag{11.24}$$

Introduce now

$$h(\lambda) := \frac{B(\lambda)}{A(\lambda)}, \tag{11.25}$$

which corresponds to the transfer function from the external reference signal r to the output y for the system free of delay: $C(\lambda)H_0(\lambda)/(1 + C(\lambda)H_0(\lambda))$. Next, introduce

$$D(\lambda;\ \tau_1, \tau_2) := 1 + h(\lambda)e^{-\lambda\tau_1} - h(\lambda)e^{-\lambda\tau_2}. \tag{11.26}$$

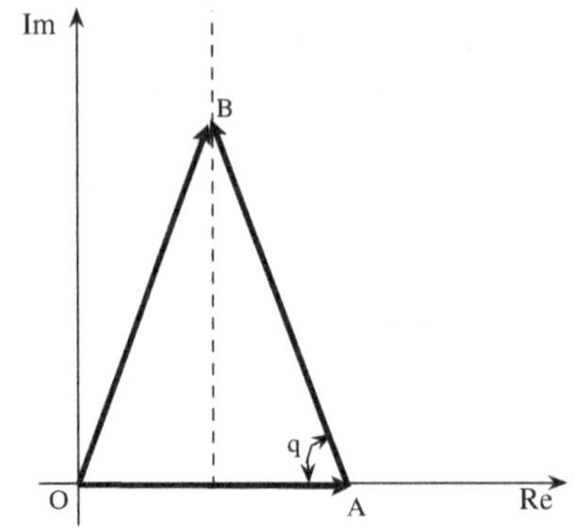

Figure 11.3. *Triangle formed by* 1, $h(\lambda)e^{-j\omega\tau_1}$, *and* $h(\lambda)e^{-j\omega\tau_2}$.

For given τ_1 and τ_2, and as long as (11.24) is satisfied, $p(\lambda;\ \tau_1,\tau_2)$ and $D(\lambda;\ \tau_1,\tau_2)$ share all the zeros in a neighborhood of the imaginary axis. As in Chapter 6, we can interpret the three terms in $D(\lambda;\ \tau_1,\tau_2)$ as three vectors in the complex plane, with the magnitudes 1, $|h(\lambda)|$, and $|h(\lambda)|$, respectively. So when we adjust the values of τ_1 and τ_2 we in fact adjust the directions of the vectors represented by the second and third terms. Equation (11.26) simply means that if we put the first two vectors head to tail, then we get the third vector. In other words, these vectors form an *isosceles triangle*; see Figure 11.3. These remarks allow us to conclude with the following proposition, whose proof is omitted:

Proposition 11.13. *For some* $(\tau_1,\tau_2)\in\mathbb{R}_+^2$, $D(\lambda;\ \tau_1,\tau_2)$ *has a zero* $\lambda = j\omega$, $\omega\neq 0$ *if and only if*

$$|h(j\omega)|\geq\frac{1}{2}. \tag{11.27}$$

As in the case of general quasi-polynomials including two distinct delays, discussed in Chapter 6, the crossing set Ω consists of a *finite number of intervals of finite length*. In what follows we denote these intervals as $\Omega_1,\Omega_2,\ldots,\Omega_N$, and without loss of generality we may assume that the intervals are ordered such that $\omega_1\in\Omega_{k_1}$, $\omega_2\in\Omega_{k_2}$, $k_1<k_2$ imply $\omega_1<\omega_2$. Next, if (11.27) is satisfied for $\omega=0$ and sufficiently small positive value of ω, then we take 0 as the left end of Ω_1: $\Omega_1=(0,\ \omega_1^r]$. In general, an end point of an interval Ω_k must be in one of the following situation:

Type 1. It satisfies the equation $|h(j\omega)|=\frac{1}{2}$.

Type 2. It equals 0.

We say that an interval is of Type 11 if both ends are of Type 1, and Ω_1 is of Type 21 if its left end is 0. Therefore, the crossing set Ω consists of a finite number of intervals of Type 11, with the possibility of the first interval Ω_1 being of Type 21. An application of the results of Chapter 6 then leads to:

Proposition 11.14. *The set* $\mathcal{T}^k$ *corresponding to the crossing set* Ω_k *consists of a series of curves belonging to one of the following categories:*

(i) *A series of closed curves (*Ω_k *is of Type* 11*).*

(ii) *A series of open-ended curves with both ends approaching* ∞ *(*Ω_k *is of Type* 21*).*

For a discussion on the degenerate cases, where assumption (11.24) is not satisfied, we refer the reader to [281].

11.4.2 ▪ Stability crossing curves: Smoothness and crossing directions

We briefly discuss some qualitative aspects regarding the crossing curves and the corresponding characteristic roots. The results on smoothness and the direction of crossing are derived straightforwardly from the general theory presented in Chapter 6.

Tangent and smoothness

A direct application of the implicit function theorem allows us to conclude with the following result:

Proposition 11.15. *The curve $\mathcal{T}^k$ is smooth everywhere except possibly at the points corresponding to a multiple root of the auxiliary characteristic function $p(\lambda;\ \tau_1, \tau_2)$.*

Direction of crossing

We discuss the direction in which the characteristic roots of p cross the imaginary axis as (τ_1, τ_2) deviates from the curve $\mathcal{T}^k$. We call the direction of the curve that corresponds to increasing ω the *positive direction*. The region on the left-hand side as we head in the positive direction of the curve is denoted as the *region on the left*. For a given $\omega \in \Omega_k$, let

$$R_l = \Re\left(\frac{1}{\lambda}\frac{\partial D(\lambda;\ \tau_1, \tau_2)}{\partial \tau_l}\right)_{\lambda=j\omega} = (-1)^{l-1}\Re\left(h(j\omega)e^{-j\omega\tau_l}\right),$$

$$I_l = \Im\left(\frac{1}{\lambda}\frac{\partial D(\lambda;\ \tau_1, \tau_2)}{\partial \tau_l}\right)_{\lambda=j\omega} = (-1)^{l-1}\Im\left(h(j\omega)e^{-j\omega\tau_l}\right), \qquad l = 1,2.$$

We have the following result:

Proposition 11.16. *Let $\omega \in \Omega_k$ and $(\tau_1, \tau_2) \in \mathcal{T}^k$ such that $j\omega$ is a zero of p with multiplicity one. As (τ_1, τ_2) crosses $\mathcal{T}$ from the region on the right to the region on the left, a pair of zeros of p crosses the imaginary axis to the right, through $\lambda = \pm j\omega$ if $R_2 I_1 - R_1 I_2 > 0$. The crossing is to the left if the inequality is reversed.*

11.4.3 ▪ Illustrative example

We investigate the delay sensitivity of the Smith predictor (11.2), applied to the system (11.1), where

$$H_0(\lambda) = \frac{1}{\lambda + a}, \qquad C(\lambda) = k_1(k_2\lambda + 1). \tag{11.28}$$

This corresponds to the analysis of Problem 11.1 with

$$A(\lambda) = (k_1 k_2 + 1)\lambda + (a + k_1), \qquad B(\lambda) = k_1(k_2\lambda + 1).$$

We assume $a > 0$ and $(a + k_1)/(k_1 k_2 + 1) > 0$, which guarantees internal stability of the closed-loop system. In case of a delay mismatch δ in the controller design, the solution semigroup associated with the closed-loop system has an essential spectrum when

$k_1k_2 \neq 0$, and thus stability may be sensitive to an infinitesimal delay mismatch. According to Proposition 11.4, the practical stability condition is given by

$$|S| = \left|\frac{k_1k_2}{1+k_1k_2}\right| < \frac{1}{2},$$

which implies $-1/3 < k_1k_2 < 1$.

Now we compute the stability region in δ as a function of the nominal delay τ when

$$a = 1, \quad k_1 = 2, \quad k_2 = 1/4. \tag{11.29}$$

Then the system is practically stable ($S = 1/3$), and for $\delta = 0$, there is one closed-loop characteristic root at $\lambda = -2$. In Figure 11.4 (left) we plot the (positive) solutions of equation (11.18) as a function of τ. As $\tau \to \infty$, the number of solutions tends to infinity.[27] Further, note that when a tuple (ω, τ) satisfies (11.18), this also hold for $(\omega, \tau + m2\pi/\omega)$, $m \in \mathbb{Z}$. In Figure 11.4 (right), we plot for each solution (i.e., possible crossing frequency) ω_k the corresponding smallest (positive and negative) values of the delay mismatch δ such that the characteristic equation (11.5) has roots at $\pm j\omega_k$. This way the stability region in δ is completely characterized. Notice that the quantities Δ_1 and Δ_2, defined in (11.22), are *no* continuous functions of the nominal delay τ. Furthermore, the maximal allowable delay mismatch Δ is a uniformly bounded function of τ, which implies that the maximal *relative* delay mismatch tends to zero as $\tau \to \infty$.

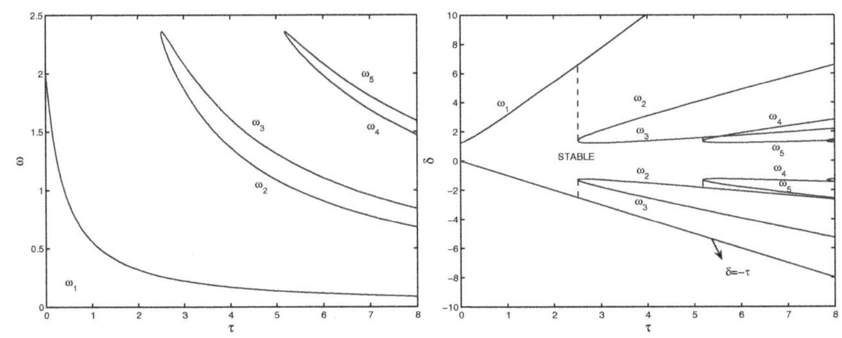

Figure 11.4. *Solutions of equation* (11.18) *when it represents the characteristic equation of the Smith predictor scheme, applied to the system* (11.28)–(11.29) *(left), and corresponding values of the critical delay mismatches δ (right)* [253].

The geometry of the stability crossing curves presented in the previous sections allows deriving the same condition as above. More precisely, by using the isosceles triangle, we obtain that the frequency crossing set Ω is defined by *only one* interval, $\Omega = (0, 2.37]$, which is of Type 21. Correspondingly, the set $\mathcal{T}$ in the delay-parameter set (τ_1, τ_2) consists of a series of open-ended curves with both ends approaching infinity, a fact confirmed in Figure 11.4 due to the linear (simple) transformation between (τ, δ) and (τ_1, τ_2). For the sake of brevity, the corresponding figure in $O\tau_1\tau_2$ is omitted.

[27] This fact is of importance since it proves that a finite-dimensional characterization of the characteristic roots on the imaginary axis by, e.g., matrix pencils [295], is not possible.

11.5 ▪ Multivariable case

In the multivariable case we have to analyze a characteristic equation of the form

$$\det\{D(\lambda)-N(\lambda)e^{-\tau\lambda}(1-e^{-\delta\lambda})\}=0, \tag{11.30}$$

where $D(\lambda)$, $N(\lambda)\in\mathbb{C}^{n\times n}$ are coprime polynomial matrices [183] such that the Hurwitz polynomial matrix $D(\lambda)$ is column reduced and such that the column degrees of $N(\lambda)$ are less than or equal to the corresponding column degrees of the polynomial matrix $D(\lambda)$, see [272].

11.5.1 ▪ Practical stability condition

We provide conditions on the matrix polynomials $D(\lambda)$ and $N(\lambda)$ for the (non-)robustness of stability of equation (11.30) w.r.t. infinitesimal delay mismatches δ. Our approach is again based on the interpretation of (11.30) as the characteristic equation of a neutral equation.

Define $d=\deg(D(\lambda))$ and denote by $\{\gamma_i\}_{i=1}^n$ the column degrees of $D(\lambda)$. Let

$$P(\lambda)=\begin{bmatrix}\lambda^{d-\gamma_1} & & & \\ & \lambda^{d-\gamma_2} & & \\ & & \ddots & \\ & & & \lambda^{d-\gamma_n}\end{bmatrix}, \tag{11.31}$$

and define the matrices D_0 and N_0 as follows:

$$D_0=\lim_{|\lambda|\to\infty}\frac{D(\lambda)P(\lambda)}{\lambda^d},\qquad N_0=\lim_{|\lambda|\to\infty}\frac{N(\lambda)P(\lambda)}{\lambda^d}. \tag{11.32}$$

With $r_\sigma(\cdot)$ denoting the spectral radius, Proposition 11.4 can be generalized to:

Proposition 11.17. *Consider the characteristic equation* (11.30). *Let* $S=r_\sigma(N_0D_0^{-1})$, *where* N_0 *and* D_0 *are defined in* (11.32).

If $|S|<1/2$, *then the asymptotic stability of equation* (11.30) *for* $\delta=0$ *is preserved for small values of* δ.

If $|S|>1/2$, *then equation* (11.30) *is not practically stable.*

Proof. Instead of equation (11.30) we may analyze the spectrum of

$$\det\{\bar{D}(\lambda)-\bar{N}(\lambda)e^{-\tau\lambda}(1-e^{-\delta\lambda})\}=0, \tag{11.33}$$

where $\bar{D}(\lambda)=D(\lambda)P(\lambda)$ and $\bar{D}(\lambda)=D(\lambda)P(\lambda)$ (notice that this transformation only introduces additional characteristic roots at zero). Since $D(\lambda)$ is column reduced, $\bar{D}(\lambda)$ is monic. Therefore, (11.33) can be written as

$$\det\{\lambda^d[I-N_0D_0^{-1}e^{-\tau\lambda}(1-e^{-\delta\lambda})]+Q(\lambda)\}=0, \tag{11.34}$$

where $\lim_{|\lambda|\to\infty,\,\Re(\lambda)\geq 0}|Q(\lambda)|/\lambda^d=0$, D_0 is the leading coefficient matrix of $\bar{D}(\lambda)$ (equivalently the leading column coefficient matrix of $D(\lambda)$) and N_0 the coefficient matrix of $\bar{D}(\lambda)$ corresponding to the degree d, i.e., these quantities satisfy (11.32).

Since (11.34) can be interpreted as the characteristic equation of a neutral functional differential equation, the sensitivity of its spectrum w.r.t. infinitesimal parameter changes is determined by the strong stability of the corresponding difference equation

$$x(t)+N_0D_0^{-1}x(t-(\tau+\delta))-N_0D_0^{-1}x(t-\tau)=0. \tag{11.35}$$

Recall from Chapter 1 that (11.35) is strongly stable if and only if

$$\sup\left\{r_\sigma(N_0D_0^{-1}e^{i\theta_1}-N_0D_0^{-1}e^{i\theta_2}):\ \theta_i\in[0,\ 2\pi],\ i=1,2\right\}<1,$$

which is equivalent to

$$r_\sigma(N_0D_0^{-1})<1/2.$$

The statement of the theorem follows. □

Remark 11.18. *When the column degrees of $N(\lambda)$ are strictly smaller than the corresponding column degrees of $D(\lambda)$ or, equivalently, when the transfer matrix $N(\lambda)D(\lambda)^{-1}$ is strictly proper, we have $S=0$ and thus practical stability, as expected from the high-frequency instability mechanism, outlined in Subsection* 11.2.1.

11.5.2 ▪ Stability domain

Under the assumption of *practical stability*, we characterize the stability region in the parameter δ. Substituting $\lambda=j\omega$ in (11.30) and some algebraic manipulations yield

$$\det\left(I-N(j\omega)D(j\omega)^{-1}e^{-j\omega\tau}(1-e^{-j\omega\delta})\right)=0. \tag{11.36}$$

Following from the relation

$$(1-e^{-j\theta})^{-1}=\frac{1}{2}-\frac{j}{2\tan(\theta/2)}\qquad\forall\theta\in(-\pi,\pi),$$

the critical delay mismatches and the corresponding characteristic roots on the imaginary axis can be computed in the following way: the "crossing frequencies" ω_k are the strictly positive solutions of one of the equations

$$\Re\left(\lambda_i\left(N(j\omega)D(j\omega)^{-1}e^{-j\omega\tau}\right)\right)=\frac{1}{2},\quad i=1,\dots,n, \tag{11.37}$$

with $\lambda_i(\cdot)$ denoting the ith eigenvalue. Next, for each ω_k, one can calculate the corresponding delay mismatches $\delta_{k,l}$ as follows:

$$\delta_{k,l}=\frac{-2\,\mathrm{atan}\left(\left(2\Im\left(\lambda_{\bar{k}}(N(j\omega_k)D(j\omega_k)^{-1}e^{-j\omega_k\tau})\right)\right)^{-1}\right)+2\pi l}{\omega_k},\quad l\in\mathbb{Z}, \tag{11.38}$$

where $\bar{k}\in\{1,\dots,n\}$ is such that

$$\Re\left(\lambda_{\bar{k}}\left(N(j\omega_k)D(j\omega_k)^{-1}e^{-j\omega_k\tau}\right)\right)=\frac{1}{2}.$$

Analogously to the scalar case we define the sets $\Lambda_{+,0}$ and $\Lambda_{-,0}$ as follows: if there does not exist any *positive* $\delta_{k,l}$, then $\Lambda_{+,0}=\{+\infty\}$, and

$$\Lambda_{+,0}=\left\{\delta_{k,l}>0\ \ :\ \ \delta_{k,l}\text{ given by (11.38)}\right\}, \tag{11.39}$$

elsewhere. Similarly, if there does not exist any *negative* $\delta_{k,l}$, then $\Lambda_{-,0} = \{-\tau\}$, and

$$\Lambda_{-,0} = \left\{\delta_{k,l} < 0 \quad : \quad \delta_{k,l} \text{ given by (11.38)}\right\}, \tag{11.40}$$

elsewhere. The smallest destabilizing delay mismatches, $\Delta_1 < 0$ and $\Delta_2 > 0$, are given by

$$\Delta_1 = \max \Lambda_{-,0}, \qquad \Delta_2 = \min \Lambda_{+,0}. \tag{11.41}$$

The results can be summarized as follows:

Proposition 11.19. *Define the real functions*

$$\mathcal{F}_i(\omega) = \Re(\lambda_i(N(j\omega)D(j\omega)^{-1}e^{-j\omega\tau})) - 1/2, \quad i = 1, \ldots, n.$$

The stability of the closed-loop system is guaranteed for any inaccurate modeling delay δ, $|\delta| \leq \Delta$, if

(i) *all functions $\mathcal{F}_i(\omega)$, $i = 1, \ldots, n$, have no zeros. In such a case, the stability property is of* delay-independent *type, i.e., it holds for any $\delta > -\tau$;*

(ii) *there exists a number $i \in \{1, \ldots, n\}$ such that $\mathcal{F}_i(\omega)$ has at least one zero. In such a case, the stability property is of* delay-dependent *type and it holds for any $\delta \in (\Delta_1, \Delta_2)$ ($\Delta_1 < 0 < \Delta_2$), where Δ_1, Δ_2 are given by*

$$\Delta_1 = \max \Lambda_{-,0}, \qquad \Delta_2 = \min \Lambda_{+,0},$$

where the sets $\Lambda_{\pm,0}$ are defined as in (11.39) *and* (11.40).

In this case, $\Delta < \min\{\Delta_1, \Delta_2\}$ and is always a finite value.

Remark 11.20. *The above analysis is based on isolating the factor $(1 - e^{-\lambda\delta})$ in the characteristic equation, whereas the analysis in Section* 11.3 *is based on isolating $e^{-\lambda\delta}$, which is not beneficial in the multivariable case. In the scalar case the two procedures are identical, since* (11.37), *respectively* (11.38), *become equivalent to* (11.18), *respectively,* (11.19), *when taking $D(\lambda) = A(\lambda)$ and $N(\lambda) = B(\lambda)$.*

To conclude, we provide some easy-to-check sufficient stability conditions:

Proposition 11.21. *If for all $\omega \geq 0$,*

$$\left\|N(j\omega)D(j\omega)^{-1}\right\|_2 < \frac{1}{2},$$

then the stability of (11.30) *is guaranteed for all $\delta \geq -\tau$.*

Proof. Since $|e^{-j\omega\tau}(e^{-j\omega\delta} - 1)| \leq 2$ for all ω and δ, the assumption of the proposition implies that

$$\|N(j\omega)D(j\omega)^{-1}e^{-j\omega\tau}(1 - e^{-j\omega\delta})\|_2 < 1 \tag{11.42}$$

for all ω and δ. It follows that (11.30) cannot have roots on the imaginary axis. □

Proposition 11.22. *If there exists a frequency $\omega > 0$ such that*

$$\left\|N(j\omega)D(j\omega)^{-1}\right\|_2 \geq \frac{1}{2}, \tag{11.43}$$

then the stability of (11.30) *is guaranteed for*

$$|\delta| < \min\left(\inf_{\omega \in S_\omega \setminus \{0\}} \frac{2}{\omega} \arcsin \frac{1}{2\|N(j\omega)D(j\omega)^{-1}\|_2}, \tau \right), \tag{11.44}$$

where

$$S_\omega = \left\{ \omega \geq 0 : \left\| N(j\omega)D(j\omega)^{-1} \right\|_2 \geq \frac{1}{2} \right\}.$$

Proof. Notice that

$$\|N(j\omega)D(j\omega)^{-1}e^{-j\omega\tau}(e^{-j\omega\delta} - 1)\|_2 = 2\|N(j\omega)D(j\omega)^{-1}\|_2 \left| \sin \frac{\omega\delta}{2} \right|.$$

When δ satisfies (11.44) we have that (11.42) holds for all $\omega \geq 0$ and asymptotic stability follows. □

Remark 11.23. *The above propositions also hold when replacing* $\|N(j\omega)D(j\omega)^{-1}\|_2$ *with* $r_\sigma(N(j\omega)D(j\omega)^{-1})$. *Although this significantly reduces conservatism, the obtained conditions may still be far from necessary because no phase information of equation* (11.30) *is exploited (unlike the conditions of Proposition* 11.19).

11.6 ▪ Notes and references

We have analyzed the robustness of Smith predictors w.r.t. an inaccurate modeling of the delay, which corresponds to Problem 11.1 in the SISO case and to the analysis of equation (11.30) in the MIMO case. When $\deg(A(\lambda)) = \deg(B(\lambda))$ or when not all column degrees of $N(\lambda)$ are strictly smaller than the corresponding column degrees of $D(\lambda)$, the closed-loop system may not be practically stable. We have derived necessary and sufficient conditions for practical stability using two different approaches, one based on an interpretation in terms of neutral equations and the other based on the analysis of small feedback delays in infinite-dimensional control systems.

We have focused on the first approach, since it also provides insight into the instability mechanism. Under the assumption of practical stability, we have characterized the maximal delay mismatch such that stability is maintained.

For the scalar cases the geometry of the stability regions in the space (nominal delay, delay mismatch) was briefly discussed by using the methodology presented in Chapter 6. Such an approach was further exploited in motion synchronization control of distributed multisubsystems with invariant local natural dynamics [73], physically-based collaborative simulations under ring-like network configurations [74], haptics [208], and congestion control (see, for instance, [91] and Chapter 13).

This chapter is based on the papers [253, 280, 281] and Chapter 3 of [278]. To the best of our knowledge the results of Section 11.5, summarized in Proposition 11.19, have not been published so far in the literature.

Chapter 12

Controlling unstable systems using finite spectrum assignment

12.1 ▪ Introduction

We consider the linear finite-dimensional system with input delay

$$\dot{x}(t) = Ax(t) + Bu(t-\tau), \quad x \in \mathbb{R}^d, \quad u \in \mathbb{R}, \tag{12.1}$$

where we assume that the matrix A is *not* Hurwitz and that the pair (A,B) is stabilizable. An approach for the stabilization and control of (12.1), called *finite spectrum assignment* [227, 403], can be interpreted as follows: a prediction of the state variable over one delay interval is generated first and, next, a feedback of the predicted state is applied, thereby compensating the effect of the time delay. This results in a closed-loop system with a finite number of characteristic roots, which can be freely assigned. Mathematically, with the feedback law

$$\begin{aligned} u(t) &= K^T x_p(t, t+\tau) \\ &= K^T \left(e^{A\tau} x(t) + \int_0^\tau e^{A\theta} B u(t-\theta) d\theta \right), \end{aligned} \tag{12.2}$$

where $x_p(t_1, t_2)$ is the prediction of $x(t)$ at $t = t_2$, based on values of x and u for $t \le t_1$, the characteristic equation of the closed-loop system is given by

$$\det\left(\lambda I - A - BK^T\right) = 0. \tag{12.3}$$

This elimination of the delay is employed in the so-called process model control techniques [405] such as, for example, the celebrated Smith predictor discussed in the previous chapter. When applied to (12.1) and (12.2) it can also be interpreted as the effect of a model transformation [8, 204]. For generalizations of the finite spectrum assignment approach to a broader class of time-delay systems than (12.1), we refer the reader to [227, 405, 8]. More generally, the state of the art in prediction-based feedback is described in [201, 20].

Note that the application of the control law (12.2) at time-instant t requires the availability of the full state x at time t, unlike the Smith predictor discussed in Chapter 11. Extensions to the output feedback case and relations with the Smith predictor are discussed at the end of this chapter.

There are two difficulties related to the implementation of the control law (12.2). First, an exact knowledge of the delay is required for having a finite number of closed-loop characteristic roots, and hence an investigation of the effect of a delay mismatch is critical. Second, the application of the control law requires the computation of an integral over past inputs. Obtaining this term as the solution of a differential equation must be discarded because it involves an unstable pole-zero cancellation when the matrix A is unstable; see [227]. It is then suggested to realize the control law by means of a numerical computation of the integral term *on-line*, which involves some approximation. In this chapter we assume that the distributed delay is approximated with a *sum of pointwise delays* by applying a numerical quadrature rule. To state this more precisely, define $\mathscr{C}([0,\ \tau],\mathbb{C}^d)$ as the space of continuous functions from $[0,\ \tau]\subset\mathbb{R}$ to $\mathbb{C}^d$, equipped with the supremum norm. A quadrature rule on $[0,\ \tau]$ is a sequence of maps $\{\mathscr{I}_n\}_{n\geq 1}$ from $\mathscr{C}([0,\ \tau],\mathbb{C}^d)\rightarrow\mathbb{C}$, defined as

$$\mathscr{I}_n(f)=\sum_{j=1}^{n} h_{j,n} f(\theta_{j,n}), \quad h_{j,n}>0, \quad \theta_{j,n}\in[0,\tau], \tag{12.4}$$

where we assume that the following convergence property is satisfied:

$$\forall f\in\mathscr{C}([0,\ \tau],\mathbb{C}^d) \quad \forall\epsilon>0\,\exists\bar{n}\in\mathbb{N}:\ |I_n(f)-\int_0^{\tau} f(\theta)d\theta|<\epsilon \quad \forall n\geq\bar{n}. \tag{12.5}$$

When the quadrature formulae (12.4) are used to approximate the integral term in (12.2), we end up with a sequence of control laws

$$u(t)=K^T\left(e^{A\tau}x(t)+\sum_{j=1}^{n} h_{j,n}e^{A\theta_{j,n}}Bu(t-\theta_{j,n})\right). \tag{12.6}$$

The effect of this semidiscretization of the control law on the closed-loop stability will be analyzed in detail.

The structure of the chapter reflects our main goal, that is, giving an overview of the existing stability results on the implementation of distributed delay control laws and on the effects of a delay mismatch. After some preliminaries we discuss the implementation of the integral term in (12.2). We address a possible instability mechanism when using the control law (12.6), as reported in [381, 380, 99, 267]. Then we discuss conditions for a safe implementation [270, 250] and outline modifications of the control law to remove the resulting restrictions [250, 271, 380, 104, 337]. Subsequently, we study the effect of a delay mismatch between plant and controller. Finally, we comment on the output feedback case and make connections with the previous chapter. Throughout the chapters, plots of characteristic roots are extensively used in order to make the main ideas and results apparent.

12.2 ▪ Preliminaries

The initial data for both the system (12.1)–(12.2) and the system (12.1) and (12.6) are $x(0)\in\mathbb{R}^d$, $u_0\in\mathscr{C}([-\tau,\ 0],\mathbb{R})$. For $t\in[0\ \tau]$, the closed-loop system becomes

$$\dot{x}(t)=Ax(t)+B\ u_0(t-\tau).$$

For $t\geq\tau$, we have

$$\begin{aligned}Bu(t-\tau)&=BK^T\left(e^{A\tau}x(t-\tau)+\int_0^{\tau} e^{A\theta}Bu(t-\theta-\tau)d\theta\right)\\&=BK^T\left(e^{A\tau}x(t-\tau)+\int_0^{\tau} e^{A\theta}(\dot{x}(t-\theta)-Ax(t-\theta))d\theta\right),\end{aligned}$$

where the right-hand derivative of x should be taken at time zero, and we can write the system (12.1) and (12.2) in the form

$$\frac{d}{dt}\mathcal{L}(x_t) = Ax(t) + BK^T e^{A\tau} x(t-\tau) - BK^T A \int_0^\tau e^{A\theta} x(t-\theta) d\theta,$$

where the map $\mathcal{L} : \mathcal{C}([-\tau\ 0], \mathbb{R}^d) \to \mathbb{R}^d$ is defined as

$$\mathcal{L}(\phi) = \phi(0) - BK^T \int_0^\tau e^{A\theta} \phi(-\theta) d\theta.$$

Because (12.1)–(12.2) is a Volterra equation of the second kind, the growth of its solutions is determined by the roots of its characteristic equation (see [227])

$$\det\left\{\lambda\left(I - BK^T \int_0^\tau e^{(A-\lambda I)\theta} d\theta\right) - A - BK^T e^{(A-\lambda I)\tau} + BK^T A \int_0^\tau e^{(A-\lambda I)\theta} d\theta\right\} = 0, \tag{12.7}$$

which can be simplified to (12.3). This makes the finite spectrum assignment property apparent.

Analogously, with the approximated control law (12.6) and for $t \geq \tau$, the closed-loop system can be written as

$$\frac{d}{dt}\mathcal{N}_n(x_t) = Ax(t) + BK^T e^{A\tau} x(t-\tau) - BK^T A \sum_{j=1}^{n} h_{j,n} e^{A\theta_{j,n}} x(t-\theta_{j,n}), \tag{12.8}$$

where the map $\mathcal{N}_n : \mathcal{C}([-\tau_n,\ 0], \mathbb{R}^d) \to \mathbb{R}^d$ is defined as

$$\mathcal{N}_n(\phi) = \phi(0) - BK^T \sum_{j=1}^{n} h_{j,n} e^{A\theta_{j,n}} \phi(-\theta_{j,n}),$$

with $\tau_n = \max_j \theta_{j,n}$.

Note that equation (12.8) is a delay-differential equation of neutral type. Under mild assumptions on the integration rule (12.4), the map $\mathcal{N}_n$ is atomic at zero,[28] guaranteeing existence and uniqueness of solutions for initial conditions $\phi \in \mathcal{C}([-\tau_n,\ 0], \mathbb{R}^d)$. The associated delay-difference equation of (12.8) is given by $\mathcal{N}_n(x_t) = 0$, that is,

$$x(t) = BK^T \sum_{j=1}^{n} h_{j,n} e^{A\theta_{j,n}} x(t-\theta_{j,n}). \tag{12.9}$$

Recall from Chapter 1 that the asymptotic behavior of the solutions, and thus stability of the neutral equation (12.8), is determined by the spectral radius $r(\mathcal{T}_N^n(t))$, satisfying

$$r(\mathcal{T}_N^n(1)) = e^\alpha, \qquad \alpha = \sup\{\Re(\lambda) : \det(\Delta_N^n(\lambda)) = 0\},$$

where the characteristic matrix[29] Δ_N^n is given by

$$\Delta_N^n(\lambda) := \left(\lambda \Delta_D^n(\lambda) - A - BK^T e^{(A-\lambda I)\tau} + BK^T A \sum_{j=1}^{n} h_{j,n} e^{(A-\lambda I)\theta_{j,n}}\right),$$

[28]This property makes the system (12.1) and (12.6) causal and allows us to write the control input u at the present time as a function of the present state and *past* inputs.

[29]The superscript n refers to the number of abscissa of the quadrature rule.

and $\Delta_D^n(\lambda)$ is defined as follows:

$$\Delta_D^n(\lambda) := \left(I - BK^T \sum_{j=1}^{n} h_{j,n} e^{(A-\lambda I)\theta_{j,n}}\right).$$

In a similar way, stability of the difference equation (12.9) is determined by the spectral radius

$$r_\sigma(\mathcal{T}_D^n(1)) = e^{\beta}, \qquad \beta = \sup\{\Re(\lambda) : \det(\Delta_D^n(\lambda)) = 0\}.$$

An important property w.r.t. the stability analysis in this chapter is the relation

$$r_e(\mathcal{T}_N^n(1)) = r_\sigma(\mathcal{T}_D^n(1)), \tag{12.10}$$

where $r_e(.)$ denotes the radius of the essential spectrum.

12.3 ▪ Implementation of the integral

We study the effect on stability of the implementation of the control law (12.2) as (12.6).

12.3.1 ▪ Instability mechanism

A starting point of the research on the implementation of distributed delay control laws was the paper [381], which illustrated that the closed-loop system (12.1) and (12.6) may be unstable for *arbitrarily large* values of n, even when the ideal closed-loop system (12.1) and (12.2) is exponentially stable—the latter expressed by the Hurwitz stability of matrix $A + BK^T$. This paradox can intuitively be explained with the occurrence of unstable characteristic roots with a large modulus for the approximated closed-loop system. When the approximation becomes better, some characteristic roots tend to the characteristic roots of the limit case, while the others move off to infinity. When some characteristic roots do so without leaving the right half plane, instability persists. This is now illustrated with an example:

Example 12.1. *Consider the scalar system*

$$\dot{x}(t) = x(t) + u(t-1) \tag{12.11}$$

and the control law

$$u(t) = -2\, x_p(t, t+1) = -2\left(e\, x(t) + \int_0^1 e^{\theta} u(t-\theta)d\theta\right), \tag{12.12}$$

which assigns one closed-loop characteristic root $\lambda = -1$. When the integral term in (12.12) *is discretized using the forward rectangular rule, i.e., using* (12.4) *with*

$$\theta_{j,n} = \frac{j-1}{n}, \quad h_{j,n} = \frac{1}{n}, \quad j = 1, \ldots, n,$$

the control law becomes

$$u(t) = -2\left(ex(t) + \frac{1}{n}\sum_{j=1}^{n} e^{\frac{j-1}{n}} u\left(t - \frac{j-1}{n}\right)\right). \tag{12.13}$$

In Figure 12.1 *(left) the characteristic roots of the closed-loop system* (12.11) *and* (12.13) *are shown for $n = 40$ and $n = 60$. As $n \to \infty$, one characteristic root converges to the assigned*

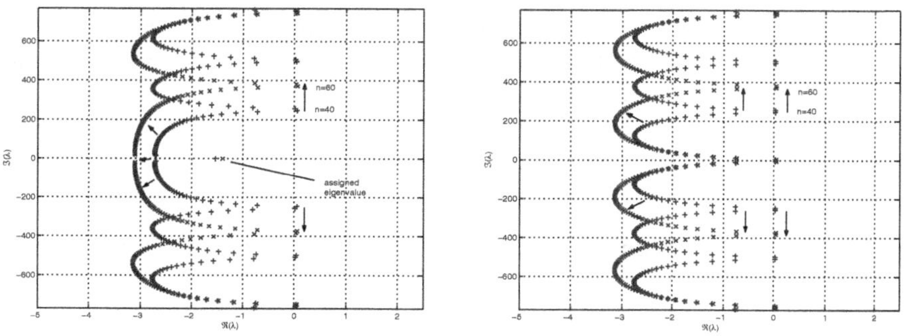

Figure 12.1. *(Left) Closed-loop characteristic roots of the system* (12.11) *and* (12.13) *for* $n = 40$ (+) *and* $n = 60$ (x). *(Right) Characteristic roots of the difference equation* (12.14) [251].

characteristic root $\lambda = -1$*, while all the characteristic roots, introduced by the approximation, move off to infinity. However, stability is not obtained. The sequences of characteristic roots, whose imaginary parts tend to infinity, yet whose real parts have a finite limit, are explained by the neutral type of the closed-loop system. In Figure* 12.1 *(right) we show the characteristic roots of the associated difference equation*

$$u(t) = -2\left(\frac{1}{n}\sum_{j=1}^{n} e^{\frac{j-1}{n}} u\left(t - \frac{j-1}{n}\right)\right). \tag{12.14}$$

As expected from theoretical considerations (related to property (12.10)*), the closed-loop characteristic roots with a large modulus, but small real part, are well approximated by characteristic roots of the difference equation* [151].

The instability mechanism is due to the *neutral type* of the approximated closed-loop system (12.1) and (12.6), in contrast with the retarded type of the ideal closed-loop system (12.1)–(12.2). Hence for any n the approximation involves a noncompact perturbation of the solution semigroup $\mathcal{T}_N^n(t)$, associated with (12.1) and (12.2), which introduces an essential spectrum; see [250]. Having the radius of this essential spectrum larger than one results in the sensitivity of stability. Alternatively, a frequency domain interpretation, including links with (lack of) w-stability [119], is presented in [267]: the integral in (12.2) has a smoothing effect on its input, unlike any finite sum approximation. This is reflected in the property that the sequence

$$\left\{ \sup_{\omega \geq 0} \left| \int_0^{\tau} e^{(A - j\omega I)\theta} d\theta - \sum_{j=1}^{n} h_{j,n} e^{(A - j\omega I)\theta_{j,n}} \right| \right\}_{n \geq 1}$$

does not converge to zero, as $n \to \infty$. The high-frequency error is related to the unstable characteristic roots of the closed-loop system with large imaginary parts.

12.3.2 ▪ Stability conditions

Because the Hurwitz stability of $A + BK^T$ does not imply the stability of (12.1) and (12.6) for large values of n, additional conditions are needed to guarantee that also the characteristic roots, due to the approximation, are in the open left half plane. A comparison of the spectral plots in Figure 12.1 suggests that hyperbolic stability properties of the closed-loop system (12.1) and (12.6) are tightly related to stability properties of the difference equation (12.9). This is indeed the case, leading to very simple and easy-to-check stability criteria, which we now review.

Necessary condition

The well-known result which states that a necessary condition for the stability of the neutral equation (12.8) is given by the stability of the difference equation $\mathcal{N}_n(x_t) = 0$, with

$$\mathcal{N}_n(x_t) = x(t) - \sum_{j=1}^{n} BK^T e^{A\theta_{j,n}} x(t-\theta_{j,n}), \tag{12.15}$$

(a corollary of 12.10), and the relation of (12.15) with the equation $\mathcal{L}(x_t) = 0$, with

$$\mathcal{L}(x_t) = x(t) - \int_0^{\tau} BK^T e^{A\theta} x(t-\theta)d\theta, \tag{12.16}$$

lead to the following necessary condition for a safe implementation, which slightly generalizes [270, Theorem 1]:

Theorem 12.2. *Consider the system* (12.1)–(12.2) *and a quadrature rule satisfying* (12.5). *Assume that the closed-loop system* (12.1) *and* (12.6) *is asymptotically stable for large values of* n. *Then the characteristic equation of* (12.16),

$$\det\left\{I - BK^T(\lambda I - A)^{-1}\left(I - e^{-(\lambda I - A)\tau}\right)\right\} = 0, \tag{12.17}$$

has all its roots in the closed left half plane.

As an illustration we compare in Figure 12.2 the rightmost roots of equation (12.17), applied to the example (12.11)–(12.12), with characteristic roots of the difference equation (12.14). Notice that the real parts of the roots of (12.17) correspond to the position of chains of characteristic roots of the closed-loop system (12.11) and (12.13) for large n.

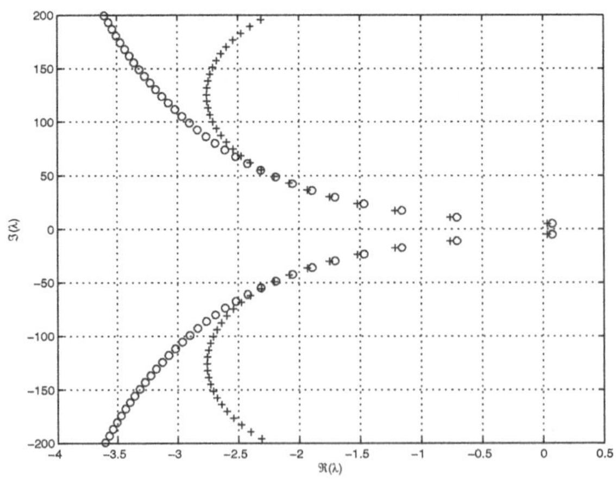

Figure 12.2. *Rightmost roots of equation* (12.17) *(indicated with* o*) and characteristic roots of the difference equation* (12.14), *where* $n = 40$ (+). *The latter are shown on a different scale in Figure* 12.1 [251].

Necessary and sufficient condition

The numerical experiments with the example could suggest that the necessary condition, following from Theorem 12.2, is close to sufficient. However, in [250] it is proven that this is generally *not* the case, because Theorem 12.2 does not take into account the fact that the radius of the essential spectrum of the solution semigroup, associated with the neutral equation (12.8), i.e., $r_e(\mathcal{T}_N^n(1))$, is *not* continuous in the delays $\theta_{j,n}$. See [13, 149, 152, 242] and the references therein for more information on delay robustness of neutral equations and related questions. A consequence for the system (12.1) and (12.6) is that closed-loop stability may depend on the type of integration rule used and be sensitive to infinitesimal perturbations of the abscissa $\theta_{j,n}$, as we now illustrate.

Example 12.3. *We revisit Example* 12.1 *and discretize the control law* (12.12) *as*

$$u(t)=-2\left(e\,x(t)+\sum_{j=1}^{n}h_{j,n}e^{\theta_{j,n}}u\left(t-\theta_{j,n}\right)\right), \tag{12.18}$$

where

$$\theta_{j,n}=\begin{cases}\frac{j-1}{n}, & j \text{ even},\\ \frac{j-4/5}{n}, & j \text{ odd},\end{cases}\qquad h_{j,n}=\tfrac{1}{n},\quad j=1\ldots n. \tag{12.19}$$

Notice that the modified integration rule with parameters (12.19) *also satisfies the convergence property* (12.5)*. In Figure* 12.3 *we show the characteristic roots of the closed-loop system* (12.11) *and* (12.18)*, as well as the characteristic roots of the associated difference equation*

$$u(t)=-2\left(\sum_{j=1}^{n}h_{j,n}e^{\theta_{j,n}}u(t-\theta_{j,n})\right),\qquad h_{j,n},\theta_{j,n}\text{ given by (12.19)}. \tag{12.20}$$

and roots of equation (12.17)*. Although the rightmost roots of* (12.17) *are also well approximated by characteristic roots of* (12.20)*, they no longer determine the stability of* (12.20) *and of the closed-loop system* (12.11) *and* (12.18) *for large n.*

By making infinitesimal perturbations of the abscissa $\theta_{j,n}$ in (12.19)*, the rightmost characteristic roots of the difference equation can have their real part arbitrarily close to the value α_n, indicated on the figure, but no larger than $\alpha_n+\epsilon$, for any $\epsilon>0$. In fact, such a value α_n*

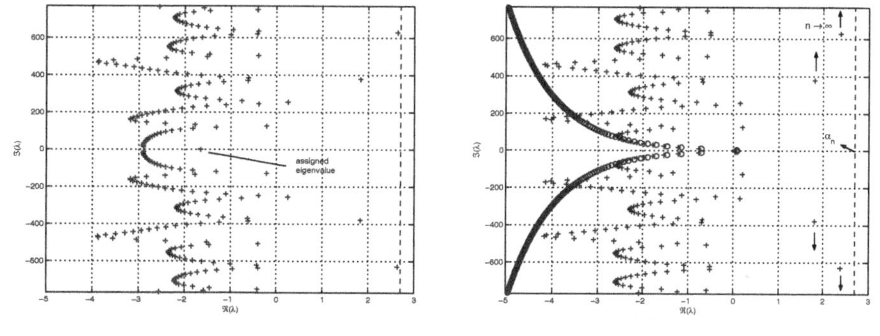

Figure 12.3. *(Left) Closed-loop characteristic roots of the system* (12.11) *and* (12.18)–(12.19) *for $n = 40$. (Right) Characteristic roots of the associated difference equation* (12.20) (+) *and roots of equation* (12.17) (o)*. Unlike the situation displayed in Figure* 12.1*, stability of the difference equation for large n is no longer determined by the rightmost roots of equation* (12.17) [251].

determines stability of the difference equation when subject to small variations in the delays, called strong stability *in* [152]. *In* [250, *Section* 4] *it is described how* α_n *can be computed analytically. Furthermore, it is shown that* $\lim_{n\to\infty} \alpha_n$ *exists and is* independent *of the type of quadrature rule.*

Remark 12.4. *The sensitivity of stability w.r.t. infinitesimal perturbations, as well as the high-frequency instability mechanism, described in Section* 12.3.1, *are phenomena which are related to the sensitivity of stability of some boundary controlled hyperbolic PDEs, feedback controlled descriptor systems, and neutral-type systems against small delays in the control loop, as reported in, e.g.,* [87, 88, 89, 154, 211, 275, 212, 213, 214]. *A sensitivity of stability w.r.t. infinitesimal modeling errors may also occur with the Smith predictor control scheme* [353]*; see Chapter* 11 *and* [322, 253] *and the references therein.*

Because arbitrarily small perturbations of the abscissa $\theta_{j,n}$ may destroy stability of the closed-loop system (12.1) and (12.6), yet are inevitable in any practical application, they should be taken into account in a definition of a safe implementation. This leads to the following definition:

Definition 12.5. *Consider the system* (12.1) *and* (12.2), *where* $A+BK^T$ *is Hurwitz, and the quadrature rule* (12.4). *The implementation* (12.6) *of the control law* (12.2) *is safe if the following two conditions are satisfied:*

1. *There exists a number* $\bar{n} \in \mathbb{N}$ *such that the closed-loop system* (12.1) *and* (12.6) *is asymptotically stable for all* $n \geq \bar{n}$.
2. *For each* $n \geq \bar{n}$, *there exist constants* $\Delta\theta_{j,n} > 0$ *such that the control law*

$$u(t)=K^T\left(e^{A\tau}x(t)+\sum_{j=1}^{n} h_{j,n}e^{A(\theta_{j,n}+\delta\theta_{j,n})}Bu(t-(\theta_{j,n}+\delta\theta_{j,n}))\right)$$

achieves asymptotic stability for all $|\delta\theta_{j,n}| < \Delta\theta_{j,n}$.

In the sense of this definition, an almost necessary and sufficient stability condition is given in [250, Theorem 1], which does *not* depend on the type of integration rule. Essentially it corresponds to a strong stability requirement for the difference equation (12.9) for large n:

Theorem 12.6. *Consider the system* (12.1) *and* (12.2), *and assume that* $A+BK^T$ *is Hurwitz. Let*

$$S=\int_0^\tau |K^T e^{A\theta}B|d\theta. \tag{12.21}$$

If $S<1$, *then the control law* (12.2) *can be safely implemented as* (12.6), *in the sense of Definition* 12.5.

If $S>1$, *then the control law* (12.2) *cannot be safely implemented.*

In the multiple input case, a *sufficient* condition for a safe implementation is given by

$$\int_0^\tau ||K^T e^{A\theta}B||d\theta < 1;$$

see [250, Section 6].

12.3.3 ▪ Removing restrictions

The main advantage of the control law (12.2) lies in the fact that all the closed-loop characteristic roots can be freely assigned. A disadvantage is the difficulty of computing the control law on-line, which involves the evaluation of the integral. In particular, for the implementation with a sum of pointwise delays the stability condition of Theorem 12.6, i.e., $S<1$, puts severe restrictions on stabilizability and performance, which are shown in [250] to be comparable to the case of a static, nonpredictive, state feedback controller, $u(t)=K^T x(t)$. In this section we briefly comment on possible modifications of the control law (12.6) to remove these restrictions.

Adding a low-pass filter

The instability mechanism, as explained in Section 12.3.1, is a *high-frequency* mechanism, related to the occurrence of unstable characteristic roots with *arbitrarily large* imaginary parts. A closer look at the problem reveals that the latter are caused by the throughput at infinity of past inputs in equation (12.6) and, therefore, can be avoided by including a *low-pass filter* in the control dynamics. This is now illustrated:

Example 12.7. *We reconsider the system* (12.11) *and* (12.13) *and modify the control law by adding a first order low-pass filter to*

$$\left\{\begin{array}{l} \dot z(t)=-f z(t)-2f\left\{e x(t)+\frac{1}{n}\sum_{j=1}^{n} e^{\frac{j-1}{n}} u\left(t-\frac{j-1}{n}\right)\right\}, \\ u=z(t). \end{array}\right. \tag{12.22}$$

Due to the filter, the closed-loop system is of retarded *type. Its characteristic roots are shown in Figure* 12.4 *for different values of n. For sufficiently large values the closed-loop system is asymptotically stable.*

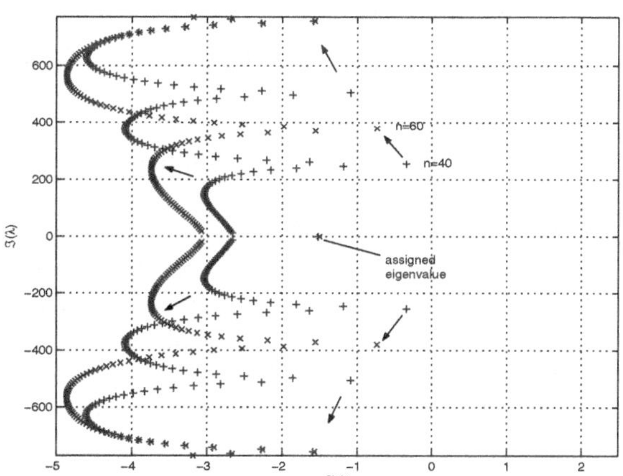

Figure 12.4. *Closed-loop characteristic roots of* (12.11) *and* (12.22) *for* $f=100$ *and* $n=40$ (+), *respectively* $n=60$ (x) [251].

This idea is generalized in [271]. Since *any strictly proper linear system* (A_f, B_f, C_f) has a *low-pass* filtering property, the dynamic control law

$$\left\{\begin{array}{ll} \dot z(t) & =A_f z(t)+B_f\, x_p(t,t+\tau), \\ u(t) & =C_f z(t) \end{array}\right. \tag{12.23}$$

is suggested. It is shown that the closed-loop characteristic roots are equal to the characteristic roots of the finite-dimensional system

$$\begin{cases} \dot{x}(t) = Ax(t) + BC_f z(t), \\ \dot{z}(t) = A_f z(t) + B_f x(t), \end{cases} \tag{12.24}$$

hence also assignable using standard design methods for systems without delays. Furthermore, a discretization using (12.4) preserves stability for large values of n.

Using piecewise constant inputs

In [380] and [337, Section 3] the input $u(t)$ is kept *piecewise constant* in time-intervals of length Δ, inspired by the implementation with a digital controller. Then, at the sampling times, the system (12.1) is completely equivalent with a discrete system. When $\Delta = \tau/p$, $p \in \mathbb{N}$, this discrete system takes the form

$$\bar{x}(k+1) = A_d \bar{x}(k) + B_d \bar{u}(k-p), \tag{12.25}$$

where $\bar{x}(k) = x(k\Delta)$, $\bar{u}(k) = u(k\Delta)$, and

$$A_d = e^{A\Delta}, \qquad B_d = \left(\int_0^{\Delta} e^{As}\, ds \right) B.$$

As in the continuous-time case, the delay can be compensated using a prediction: for the control law

$$\begin{aligned} \bar{u}(k) &= K_d^T \bar{x}_p(k, k+p) \\ &= K_d^T \left(A_d^p \bar{x}(k) + \sum_{n=1}^{p} A_d^{n-1} B_d \bar{u}(k-n) \right), \end{aligned} \tag{12.26}$$

the characteristic equation of the closed-loop system is given by

$$\det\left(zI - (A_d + B_d K_d^T)\right) = 0.$$

Because the system (12.25) and (12.26) is fully discrete, the maximal possible frequency is given by $1/(2\Delta)$ (the celebrated Nyquist–Shannon criterion) and, therefore, sensitivity of stability w.r.t. arbitrarily small perturbations of the parameters of (12.26) is not possible.

Note that the control law (12.26) can be considered as a full discretization of the continuous control law (12.2), using a particular quadrature rule. However, in [380] it is also illustrated that one cannot take any quadrature rule, satisfying (12.5), and obtain stability for sufficiently small values of Δ after a full discretization. For instance, applying Simpson's rule may lead to instability. The instability mechanism is related to the one in the continuous-time case, where approximations lead to characteristic roots with arbitrarily large frequencies. In the discrete-time case, unstable modes may occur with the maximal frequency[30] $1/(2\Delta)$, which tends to infinity as $\Delta \to 0$.

12.4 ▪ Delay mismatch

We investigate the (additional) effect of a delay mismatch between plant and controller. We assume that there is a constant uncertainty δ on the system delay such that (12.1) becomes

$$\dot{x}(t) = Ax(t) + Bu(t - \tau - \delta). \tag{12.27}$$

[30] Such a mode corresponds to a negative real characteristic root of the fully discretized system.

An application of the control law (12.2) to (12.27) leads to a closed-loop system whose stability properties are determined by the roots of the characteristic equation

$$\det\left\{\lambda\left(I-BK^T\int_0^{\tau}e^{(A-\lambda I)\theta}d\theta\right)-A-BK^Te^{(A-\lambda I)\tau}e^{-\lambda\delta}\right.$$
$$\left.+BK^TA\int_0^{\tau}e^{(A-\lambda I)\theta}d\theta\right\}=0, \tag{12.28}$$

which simplifies to

$$\det\left\{D_1(\lambda)-N_1(\lambda;\,\tau)e^{-\lambda\tau}(1-e^{-\lambda\delta})\right\}=0, \tag{12.29}$$

where

$$D_1(\lambda)=\lambda I-A-BK^T, \qquad N_1(\lambda;\,\tau)=-BK^Te^{A\tau}.$$

Similarly, an application of the control law (12.6) to (12.27) results in the characteristic equation

$$\det\left\{D_2(\lambda;\,\tau)-N_2(\lambda;\,\tau)e^{-\lambda\tau}(1-e^{-\lambda\delta})\right\}=0, \tag{12.30}$$

where

$$D_2(\lambda;\,\tau)=\lambda\left(I-BK^T\sum_{j=1}^{n}h_{j,n}e^{(A-\lambda I)\theta_{j,n}}\right)-A-BK^Te^{(A-\lambda I)\tau}$$
$$+BK^TA\sum_{j=1}^{n}h_{j,n}e^{(A-\lambda I)\theta_{j,n}},$$
$$N(\lambda;\,\tau)=-BK^Te^{A\tau}.$$

Equation (12.29) can be interpreted as the characteristic equation of a system of retarded type with pointwise delays. If the condition $S<1$ of Theorem 12.6 is satisfied and n is sufficiently large, then (12.30) corresponds to a neutral system with pointwise delays and a corresponding *strongly stable* difference equation. In both cases, a sensitivity of hyperbolic stability properties w.r.t. an infinitesimal perturbation of δ, in particular the transition from $\delta=0$ to $\delta=0+$, is not possible. So if $A+BK^T$ is Hurwitz, then the closed-loop system is practically stable in the sense of Chapter 11.

Stability regions with δ as free parameter can be computed using the approach of Section 11.5.2, since (12.29) and (12.30) are of the form (11.30). Note that D and N in (11.30) are assumed to be polynomial matrices that do not depend on τ, yet the analysis and results of Section 11.5.2 do not rely on these assumptions.

12.5 ▪ Output feedback

We generalize the results of the previous sections to the static output-feedback case. Next, we discuss dynamic output-feedback and make connections with the Smith predictor scheme discussed in Chapter 11.

12.5.1 ▪ Static output feedback

Assume in what follows that only a scalar output y is available for measurement:

$$y(t)=F^Tx(t), \quad F\in\mathbb{R}^{d\times 1}.$$

In this case the time delay cannot be direct compensated by using a prediction of the output in the feedback loop, since the computation of the prediction,

$$y_p(t,\, t+\tau) = F^T x_p(t,\, t+\tau) = F^T\left(e^{A\tau}x(t) + \int_0^{\tau} e^{A\theta}Bu(t-\theta)d\theta\right),$$

requires the availability of the full state $x(t)$. Instead, one may use the *fictitious* output

$$\bar{y}(t) = F^T e^{-A\tau}x(t), \tag{12.31}$$

whose prediction,

$$\begin{aligned}\bar{y}(t,\, t+\tau) &= F^T e^{-A\tau}\left(e^{A\tau}x(t) + \int_0^{\tau} e^{A\theta}Bu(t-\theta)d\theta\right)\\ &= y(t) + \int_0^{\tau} F^T\, e^{A(\theta-\tau)}Bu(t-\theta)d\theta\end{aligned} \tag{12.32}$$

depends only on the input and the output $y(t)$. The control law

$$\begin{aligned}u(t) &= k\, \bar{y}_p(t,\, t+\tau)\\ &= k\left(y(t) + \int_0^{\tau} F^T\, e^{A(\theta-\tau)}Bu(t-\theta)d\theta\right)\end{aligned} \tag{12.33}$$

results in a closed-loop system whose characteristic equation is given by

$$\det(\lambda I - A - kBF^T e^{-A\tau}) = 0,$$

and hence the stabilizablity of (12.1) using a control law of the form (12.33) depends on the output-feedback stabilizability of the corresponding undelayed system using the fictitious output $\bar{y}$.

To characterize the effects on stability of an implementation of the integral term in (12.33) and of a delay mismatch, the results of Sections 12.3–12.4 can be directly applied, provided that the following substitution is made:

$$K^T \leftarrow kF^T e^{-A\tau}.$$

For instance, the condition for a safe implementation of the integral from Theorem 12.6 becomes

$$|k| < \left(\int_0^{\tau} \left|F^T e^{A(\theta-\tau)}B\right| d\theta\right)^{-1}.$$

12.5.2 ▪ Dynamic output feedback and relations with Smith predictors

The approach of the previous section can be generalized to dynamic feedback of the prediction of the fictitious output (12.31). Let (A_c, B_c, F_c, G_c) be a minimal realization of the controller and apply the feedback

$$\begin{aligned}\dot{z}(t) &= A_c z(t) + B_c(\bar{y}_p(t,\, t+\tau) - r(t)),\\ u(t) &= F_c z(t) + G_c(\bar{y}_p(t,\, t+\tau) - r(t)),\end{aligned} \tag{12.34}$$

where r is an external reference signal. With

$$\begin{aligned} H_0(\lambda) &= F^T(\lambda I - A)^{-1}B, \\ \bar{H}_0(\lambda) &= F^T e^{-A\tau}(\lambda I - A)^{-1}B, \\ C(\lambda) &= F_c^T(\lambda I - A_c)^{-1}B_c + G_c, \end{aligned}$$

we have (with uppercase letters denoting the signals in the frequency domain)

$$\begin{aligned} U(\lambda) &= C(\lambda)(R(\lambda) - \bar{Y}_p(\lambda)) \\ &= C(\lambda)\left(R(\lambda) - Y(\lambda) - \int_0^\tau F^T e^{A(\theta-\tau)}Be^{-\lambda\theta}d\theta\, U(\lambda)\right), \end{aligned} \tag{12.35}$$

and we can formally derive

$$\begin{aligned} U(\lambda) &= \frac{C(\lambda)}{1 + C(\lambda)\int_0^\tau F^T e^{A(\theta-\tau)}Be^{-\lambda\theta}d\theta}(R(\lambda) - Y(\lambda)) \\ &= \frac{C(\lambda)}{1 + C(\lambda)(\bar{H}_0(\lambda) - H_0(\lambda)e^{-\lambda\tau})}(R(\lambda) - Y(\lambda)). \end{aligned}$$

A comparison with the control law (11.2) reveals that the control law (12.34) corresponds to a modification of the Smith predictor. The modification consists of replacing

$$H_0(\lambda) - H_0(\lambda)e^{-\lambda\tau}$$

with

$$\bar{H}_0(\lambda) - H_0(\lambda)e^{-\lambda\tau}, \tag{12.36}$$

where $\bar{H}_0$ is such that the transfer function (12.36) has a *finite impulse response* and is implemented statically by computing an integral online. This approach is well known in the context of extending the Smith predictor to open-loop unstable plants; see, for instance, the survey [322] and [413].

The transfer functions of the closed-loop system (12.1) and (12.34) from the reference signal r, as well as an input disturbance w, to the output y are given by

$$\begin{aligned} H_{y,r}(\lambda) &= \frac{C(\lambda)H_0(\lambda)}{1 + C(\lambda)\bar{H}_0(\lambda)}e^{-\lambda\tau}, \\ H_{y,w}(\lambda) &= \frac{H_0(\lambda)(1 + C(\lambda)(\bar{H}_0(\lambda) - H_0(\lambda)e^{-\lambda\tau}))}{1 + C(\lambda)\bar{H}_0(\lambda)}e^{-\lambda\tau}. \end{aligned}$$

Note that, due to a static implementation of (12.36), the open-loop poles no longer appear as poles of the transfer function $H_{y,w}$, in contrast to the classical Smith predictor for which we have expression (11.6).

The implementation of the control law (12.34) requires a discretization of the integral in (12.32). Applying the quadrature rule (12.4) results in

$$\begin{aligned} \dot{z}(t) &= A_c z(t) + B_c(\bar{y}_p^d(t, t+\tau) - r(t)), \\ u(t) &= F_c z(t) + G_c(\bar{y}_p^d(t, t+\tau) - r(t)), \\ \bar{y}_p^d(t, t+\tau) &= y(t) + \sum_{j=1}^n h_{j,n} F^T e^{A(\theta_{j,n}-\tau)}Bu(t - \theta_{j,n}). \end{aligned} \tag{12.37}$$

From (12.35) and (12.37) it is easy to see that Theorem 12.6 can be generalized to:

Theorem 12.8. *Consider the feedback system* (12.1) *and* (12.34), *which is assumed to be internally stable. If*

$$|G_c| < \left(\int_0^{\tau} |F^T e^{A(\theta-\tau)} B| d\theta \right)^{-1},$$

then the control law (12.34) *can be safely implemented as* (12.37), *in the sense of Definition* 12.5. *If*

$$|G_c| > \left(\int_0^{\tau} |F^T e^{A(\theta-\tau)} B| d\theta \right)^{-1},$$

then the control law (12.34) *cannot be safely implemented.*

Remark 12.9. *If* $G_c = 0$, *then the control law can be safely implemented because it behaves as a low-pass filter. The beneficial effect of a low-pass filter in the control loop was illustrated in Section* 12.3.3.

When there is a constant delay mismatch between plant and controller and the plant delay is given by $\tau + \delta$, the transfer function from the reference signal r to the plant output y becomes

$$H^{\delta}_{y,r}(\lambda) = \frac{C(\lambda)H_0(\lambda)e^{-\lambda(\tau+\delta)}}{1 + C(\lambda)\bar{H}_0(\lambda) + C(\lambda)H_0(\lambda)e^{-\lambda\tau}(e^{-\lambda\delta} - 1)},$$

and when, in addition, the integral in the control law is discretized, the transfer function becomes

$$H^{\delta}_{y,r}(\lambda) = \frac{C(\lambda)H_0(\lambda)e^{-\lambda(\tau+\delta)}}{1 + C(\lambda)\left(\sum_{j=1}^{n} h_{j,n} F^T e^{A(\theta_{j,n}-\tau)} B e^{-\lambda\theta_{j,n}}\right) + C(\lambda)H_0(\lambda)e^{-\lambda(\tau+\delta)}}.$$

In both cases a sensitivity of stability w.r.t. an infinitesimal delay mismatch δ is not possible, since the closed-loop system is of retarded type (due to the fact that H_0 is strictly proper and C proper). The maximal allowable delay mismatch can be computed by applying the techniques of Chapter 11.

12.6 ▪ Notes and references

We have studied the problem of the implementation of controllers arising in the context of finite spectrum assignment. Both the effect of an approximation of distributed delays by pointwise delays and the effect of a delay mismatch between plant and controller were studied. In the former case, a high-frequency instability mechanism was outlined, necessary and sufficient conditions for the preservation of stability for a sufficiently fine discretization were provided, and approaches were proposed to overcome the induced limitations. In the latter case, it was shown that a sensitivity of stability w.r.t. infinitesimal delay mismatches does not occur since the proposed controllers have a bounded high-frequency gain. Procedures to compute the maximal achievable delay mismatch were outlined. While the chapter focused on static state feedback, extensions to dynamic output feedback were also discussed, thereby making connections with the results of the previous chapter.

The results presented in this chapter are based on [251, 250, 261, 271, 322] and the references therein.

Chapter 13

Congestion control algorithms in networks

A communication network consists of a collection of "elements" (network users and/or sources) interconnected to transfer information/data from one *node* to another node through some (transmission or communication) *links*. A high-performance network represents a communication network able to support a large variety of applications, which can be transferred at high speed and with low (communication) delay. One of its dominant features is the *large scale* of the system [285]. There exist basically two ways to model communication networks systems using (discrete- or continuous-time) stochastic representations and/or deterministic (continuous-time) approximations.

In what follows, we shall consider only *continuous-time fluid approximations* due to their simplicity of representation as dynamical systems. We do not insist on one way to derive such models or on their explicit interest for congestion control analysis, but we point out the discussions proposed in [344] (see also [30, 65] for some connections between fluid-continuous and discrete-time representations). Further philosophical problems concerning the interactions between networks and feedback control theory can be found in [285].

Basically, the goal of a network congestion control algorithm is to adequately throttle the input rate at each source of a connection so that router buffer overflows are avoided. Roughly speaking, congestion can arise when packets arrive at the router at a rate $r(t)$ which is larger than the capacity of the output link so that the router queue builds up until it is full and it starts to drop packets. The congestion control proposed by Jacobson [175], and introduced in the TCP, proposes to address this problem implementing a distributed end-to-end algorithm that closes the loop at the end point of the connection and does not require any explicit feedback from the routers. The only (implicit) feedback signal considered by TCP is represented by packet loss events that are interpreted by the congestion control algorithm as synonymous to network congestion.

In what follows, we will consider the analysis of some congestion control algorithms from the literature by using some of the (algebraic, geometric) approaches and methods proposed in this monograph. More precisely, we shall analyze the asymptotic stability of three distinct fluid approximation models, and we will explicitly focus on *delay-induced instabilities*, that is, the way in which the stability is affected by the delay presence.

The first example is a simple model including an integrator (modeling the bottleneck queue) and two time delays (modeling forward and backward delays), which is a model considered in the control literature by Mascolo [230] (see also [229]), where the author proposed a congestion control algorithm based on a Smith predictor plus a proportional controller. In this setting, we will focus on estimating bounds for delay uncertainty such that the stability of the overall control scheme is still guaranteed. The proposed approach is based on some appropriate geometric arguments.

The second model was introduced by Izmailov [173, 174], and it represents a second-order linear system that describes the dynamics of a single connection between a source (controlled by an access regulator) and a distant node with a constant transmission capacity. The particularity of the model is the existence of two (independent) delays: round-trip time and the control-time interval, respectively. As in the previous case study, the stability analysis will be performed in the delay-parameter space, and connections with output-feedback control by using delay terms will be emphasized.

Finally, the third model was introduced by Misra, Gong, and Towsley in [268] and is of TCP congestion avoidance type. More precisely, this last model is a nonlinear second-order system including only one single delay (round-trip time) with the following features: the first differential equation describes the TCP window dynamics, and the second equation models the bottleneck queue behavior. Some appropriate transformations of the original systems will be considered for deriving the local asymptotic stability w.r.t. the system's parameters. Finally, we shall emphasize also the *chaotic behavior* of the system.

13.1 ▪ Smith predictor–based congestion control

13.1.1 ▪ Congestion control model

In [230], a model of Internet flow and congestion control as a time-delay system is provided. In particular, the model consists of a feedback loop in which two time delays are present as shown in Figure 13.1: T_1 models the propagation time of a packet from source to the bottleneck queue and T_2 models the propagation time from the bottleneck to the destination and then back to the sender. The round-trip time of the connection is $T = T_1 + T_2$. It has to be noted that considering a constant RTT is a modeling simplification that is often employed in the literature to make the model more tractable. The simple integrator $1/\lambda$ models the bottleneck queue that is filled (or drained) by the rate mismatch $r(t) - b(t)$, where $b(t)$ is the available bandwidth. The controller is a proportional Smith predictor with gain k that computes the rate $r(t)$ to match the available bandwidth $b(t)$ and to produce a stable output. The reason for using a simple proportional controller is that in this way the closed-loop dynamics can be made that of a first-order system with time constant $1/k$ delayed by T_1. Thus, the step response of the system can be made faster by increasing the proportional gain k providing an always stable system without oscilla-

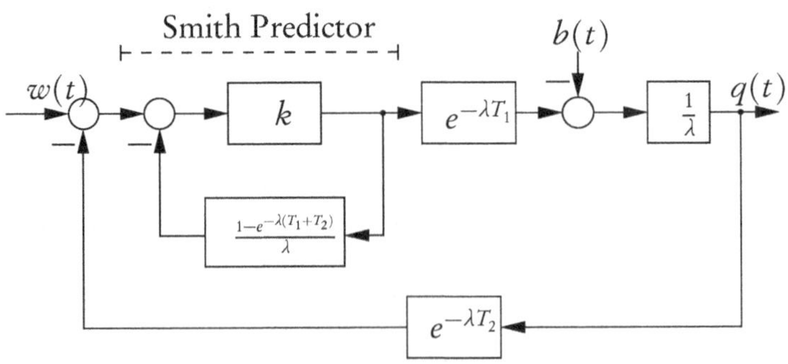

Figure 13.1. *Functional block of the congestion control model.*

tions or overshoots. This choice provides a controller in which only *one design parameter* (gain k) has to be tuned, which has a direct influence on the dynamics of the output. It is worth noting that in [320], by following optimal control arguments, a primary controller that inverts the plant model and that adds a pole in the origin is proposed, with the aim of making the closed-loop transfer function a first-order system plus dead-time that is stable for any positive value of the gain. In the case of the integral plant that is considered in this section, this means that the primary controller is a simple gain k. Finally, the input signal $w(t)$ models the congestion window or the advertised window, that is used by the congestion control algorithm to bound in-flight packets [230]. A remarkable feature of this model is that different variants of TCP congestion control algorithms can be modeled in a unified framework by proper input shaping of the proportional Smith predictor controller [232]. Moreover, in [230] it has been shown that the Smith predictor models the self-clocking property of TCP congestion control that is essential for the implementation of an effective congestion control algorithm [16].

13.1.2 ▪ Stability crossing curves of the congestion control model

In order to characterize the effect induced by the delay uncertainty on the stability of the considered feedback system, we apply the geometric approach introduced in Chapter 6. It is worth noticing that the delay-free model of the plant is $G_0(\lambda) = 1/\lambda$ and the controller transfer function is $C(s) = k$.

We suppose that the system described above is affected by a delay uncertainty Δ which is bounded by $\delta > 0$, i.e., $|\Delta| < \delta$. By considering the delay uncertainty, the characteristic equation of the system can be rewritten as follows:

$$1+\frac{k}{\lambda}-\frac{k}{\lambda}e^{-\tau_1\cdot\lambda}(1-e^{-\Delta\lambda})=0, \tag{13.1}$$

where $\tau_1 = T_1 + T_2$ represents the nominal RTT of the considered connection, which is used in the Smith predictor, and $\tau_2 = \tau_1 + \Delta$ is the actual plant time delay.

By multiplying both sides of (13.1) by $\lambda/(s\lambda+k)$ we obtain

$$1-\frac{k}{\lambda+k}e^{-\tau_1\lambda}+\frac{k}{\lambda+k}e^{-(\Delta+\tau_1)\lambda}=0. \tag{13.2}$$

We are interested in characterizing the stability of the system when τ_1, τ_2, and k vary in $\mathbb{R}_+$. By making the change of variable $z = \lambda/k$ we obtain

$$1-\frac{1}{z+1}e^{-h_1 z}+\frac{1}{z+1}e^{-h_2 z}=0, \tag{13.3}$$

where $h_1 = k\tau_1$ and $h_2 = k\tau_2$, which reduces the free parameters to two. It is worth noticing that the transformation from (13.2) to (13.3) simply involves a *scaling* of the closed-loop eigenvalues by $1/k$, thus indicating a natural trade-off between gain and delay since when k increases, the closed-loop poles approach to the imaginary axis (see, for instance, Chapter 10 or [300]).

We are now ready to study the stability of the original system in the h_1, h_2-plane regardless of the value of the proportional gain k. First, some simple computations prove that the crossing set is made by the single interval $\Omega = [0, \sqrt{3}]$, which means that the stability crossing curves in the h_1, h_2-plane are open curves which extend to infinity when $\omega \to 0$. Furthermore, the stability crossing curves of the considered system are parametrized as follows:

$$h_1^{u\pm}(\omega) = \frac{-\arctan\omega + 2u\pi \pm \arccos\left(\frac{\sqrt{1+\omega^2}}{2}\right)}{\omega}, \tag{13.4}$$

$$h_2^{v\pm}(\omega) = \frac{-\arctan\omega + (2v-1)\pi \mp \arccos\left(\frac{\sqrt{1+\omega^2}}{2}\right)}{\omega}. \tag{13.5}$$

13.1.3 ▪ Estimating bounds on delay uncertainty

Let's take a closer look at Figure 13.2. Consider first the h_2 axis ($h_1 = 0$), which means that we are using a simple proportional controller without the Smith predictor. By starting from the origin and increasing the value of h_2, the first curve is crossed at $h_2 = \pi/2$ which means that, as expected, the system becomes unstable for $h_2 > \pi/2$. On the other hand, the axis h_1 represents the system in which no delay affects the plant, but the Smith predictor is in the controller. Figure 13.2 shows that the system is stable for all the delays in the Smith predictor. Next, it is worth mentioning that points below (above) the bisector represent the case of negative (positive) delay mismatch, whereas points lying on the positive bisector represent the case of perfect matching of nominal delay τ_1 with the actual delay τ_2. Indeed, if we move on this line, no curves will be crossed since the Smith predictor provides a stable system regardless of the value of the proportional gain k.

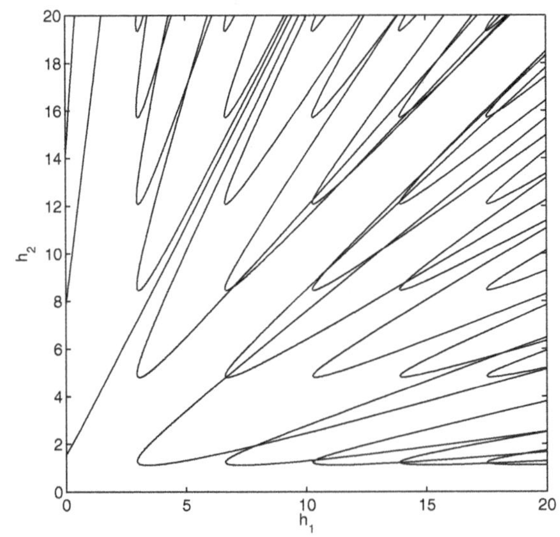

Figure 13.2. *Stability crossing curves for the considered system* [91].

In order to characterize the robustness of the system w.r.t. the delay uncertainties, we compute the maximum delay mismatch which still preserves stability. Thus, the problem here is to look for the maximum deviation δ with respect to a generic point (τ_1^*, τ_1^*) with $\tau_1^* \geq 0$, which lies on the positive bisector such that the system is stable for any (τ_1, τ_2) which satisfies $|\tau_2 - \tau_1^*| < \delta$. We remark that solving the maximum admissible delay uncertainty problem is equivalent to finding the minimum distance between the stability crossing curves and a generic point on the positive bisector of the h_1, h_2-plane.

Thus, for any $\tau_1^* > 0$ we have to solve

$$\delta(\tau_1^*) = \min_{u,v} \min_{\tau_2^{v\pm} \in \mathcal{T}} |\tau_2^{v\pm} - \tau_1^*| \tag{13.6}$$

so that the maximum delay to retain stability is

$$\delta = \min_{\tau_1^* \in \mathbb{R}_+} \delta(\tau_1^*). \tag{13.7}$$

We have the following result (see, for instance, [91] for further details):

Proposition 13.1. *A necessary and sufficient condition for the asymptotic stability of the system independent of the value of the nominal delay* τ_1 *is*

$$|\Delta| < \frac{\alpha}{k}, \tag{13.8}$$

where Δ *is the delay uncertainty,* $\alpha \cong 1.4775$*, and* k *is the proportional gain of the controller.*

Proof. Consider the stability crossing curves in the parameters space h_1, h_2. In order to find the minimum distance between such curves and a generic point of positive bisector of the h_1, h_2-plane, we evaluate the tangent to the crossing curves with direction parallel to the positive bisector:

$$\frac{dh_2}{dh_1} = 1 \Leftrightarrow \frac{dh_2}{d\omega}\frac{d\omega}{dh_1} = 1 \Leftrightarrow \frac{dh_1}{d\omega} = \frac{dh_2}{d\omega}. \tag{13.9}$$

Towards this purpose we look for a subset $\overline{\mathcal{T}}$ of the stability crossing curves $\mathcal{T}$ that are the "closest" curves to the positive bisector. By considering a generic curve $\mathcal{T}_{u,v}$ and by applying (13.4) and (13.5), it turns out that for all u, v and for all $\omega \in \Omega$ it holds that $h_2^{v+} - h_1^{u+} < h_2^{v-} - h_1^{u-}$ so that it is sufficient to consider only the curves $\mathcal{T}_{u,v}^+$ in the region $h_2 > h_1$ and the curves $\mathcal{T}_{u,v}^-$ in the region $h_2 < h_1$, since they are the closest ones to the positive bisector. Thus, without loss of generality, we can refer to the generic curves of $\mathcal{T}$ as $\mathcal{T}_{u,u+i}$ for all i and u in the integers. Straightforward computations on (13.4) and (13.5) give $h_2^{u+i} - h_1^u > h_2^{u+i-1} - h_1^u$, which means that when i decreases, the curves $\mathcal{T}_{u,u+i}$ will move downwards in the h_1, h_2-plane. Figure 13.3 shows the values of u and v for the curves $\mathcal{T}_{u,u}$ and $\mathcal{T}_{u,u+1}$. It is then easy to show that if we set $v = u$ we obtain the closest curves to the positive bisector in the region $h_2 < h_1$ whereas the curves with $v = u + 1$ are those which are closest to the positive bisector in the region $h_2 > h_1$. In conclusion we can restrict our search to the set $\overline{\mathcal{T}} = \mathcal{T}_{u,u}^- \cup \mathcal{T}_{u,u+1}^+$ for all $u \in \mathbb{Z}$. Let us consider the region $h_2 > h_1$ i.e., we consider the subset $\mathcal{T}_{u,u+1}^+$. By considering (13.9) after straightforward computations we get the following equation:

$$\arccos\left(\frac{\sqrt{\omega^2+1}}{2}\right) + \frac{\omega^2}{\sqrt{\omega^2+1}\sqrt{3-\omega^2}} + \pi\left(v - u - \frac{1}{2}\right) = 0, \tag{13.10}$$

with $\omega \in \Omega$. By letting $v = u + 1$ the equation (13.10) can be numerically solved to find the unique solution $\overline{\omega} \cong 1.3483$ rad/s in Ω which is independent of u. If we substitute this value in (13.4) and (13.5), we obtain

$$h_1(\overline{\omega}) = h_1 = 4.6601u - 0.2654, \qquad h_2(\overline{\omega}) = h_2 = 4.6601v - 3.4480.$$

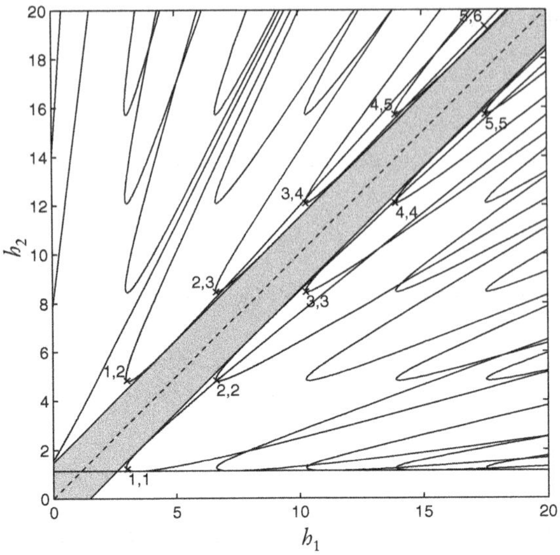

Figure 13.3. *Stability regions for the considered system. Stability region found in Proposition* 13.1 *is marked in grey* [91].

Thus, all the points belonging to the curves $\mathscr{T}^{+}_{u,u+1}$ having a tangent which is parallel to the positive bisector lie on the line:

$$h_2 = h_1 + 1.4775. \tag{13.11}$$

For this reason we can conclude that the maximum uncertainty in the h_1, h_2 coordinates is 1.4775. The proof is completed by recalling that $h_1 = k\tau_1$ and $h_2 = k\tau_2$ and that $\tau_2 = \tau_1 + \Delta$. Thus, we finally obtain

$$h_2 - h_1 < 1.4775 \Rightarrow k\Delta < 1.4775 \Rightarrow \Delta < \frac{1.4775}{k}. \tag{13.12}$$

The same procedure can be followed in the case $v = u$ leading to the inequality

$$h_1 - h_2 < 1.4775 \Rightarrow -k\Delta > 1.4775 \Rightarrow \Delta > -\frac{1.4775}{k}. \tag{13.13}$$

Thus, by considering both (13.12) and (13.13) we obtain (13.8). In order to prove the necessity of the condition (13.8) let us consider the curves $\mathscr{T}_{u,u+1}$. The points of the curve $\mathscr{T}_{u,u+1}$ that correspond to the frequency $\omega = 1.3483$ rad/s lie on the line described by (13.11) so that the maximum delay uncertainty admissible for those points is exactly α/k. If we select a larger value for δ, the system will become unstable at least on those points. This concludes the proof. □

Remark 13.2. *The fact that the maximum uncertainty allowed does not depend on the nominal delay τ_1 makes the controller effective even for large nominal delay values. Next, Proposition* 13.1 *is a* necessary and sufficient *condition for the robust stability of the system that is valid independent of the value of the nominal delay τ_1. This does not prevent the fact that, for a specific value of τ_1, the system is stable for a value of Δ that is greater than α/k. Nevertheless, Proposition* 13.1 *guarantees that if the maximum entity of the delay error Δ is known,*

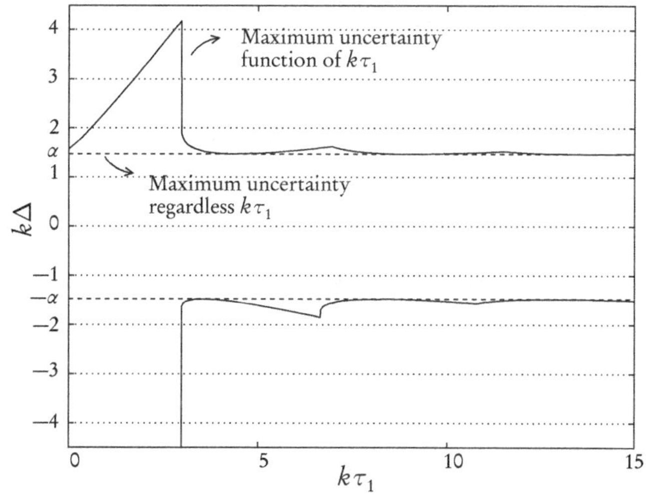

Figure 13.4. *Maximum delay uncertainty to retain asymptotic stability as function of $k\tau_1$ (continuous black curves) or independent of τ_1 (dashed black lines)* [91].

the controller gain k can be tuned so that for any value of the nominal delay τ_1 the stability of the system is guaranteed. Figure 13.4 *shows $k\Delta$ as function of $k\tau_1$ and can be obtained by employing equations* (13.4) *and* (13.5). *The two curves shown in the solid lines delimit the stability region in the case when $k\tau_1$ is variable, whereas the two dashed lines represent the stability boundary independent of τ_1, i.e., the robust stability condition* (13.8). *It is worth noticing that the stability bounds as a function of $k\tau_1$ shown in Figure* 13.4 *recover the results found in* [134], *where a Nyquist approach was employed.*

Finally, we observe that, although (13.8) *can be conservative for some values of the nominal delay (see Figure* 13.4), *it represents a simple and practical rule that does not require the designer to make difficult assumptions regarding the nominal delay. In fact, the nominal delay, i.e., the RTT of the connection, can vary in a range between* 0.01s *in the case of a wired connection up to* 1s *and more in the case of wireless connections such as in the case of satellite paths.*

Remark 13.3. *By using different approaches and related analytical arguments, this result improves the robust stability condition $|\Delta| < 1/k$ found in* [231] *and in* [273].

Proposition 13.4. *The system is stable, independent of the value of τ_1, if the delay uncertainty Δ satisfies the following inequality:*

$$-\tau_1 < \Delta < -\tau_1 + \frac{\beta}{k}, \tag{13.14}$$

with $\beta = 1.1188$.

Proof. The proof follows along the same lines as Proposition 13.1, and thus is omitted. □

Remark 13.5. *The condition* (13.14) *implicitly requires the delay uncertainty Δ to be negative, i.e., the nominal delay τ_1 always should be below the actual delay of the plant τ_2. Thus, condition* (13.14) *has no particular meaning for the characterization of controller robustness, since the sign of the uncertainty is not known a priori.*

13.2 ▪ Algorithms for single connection models with two delays

We present the stability analysis of classes of second-order linear systems including multiple delays in rational dependence, which cover some of the feedback control algorithms proposed by Izmailov [173, 174]. The stability analysis will make use of some algebraic and geometric arguments proposed in the previous chapters. Some connections with the output-feedback control problem will also be emphasized.

13.2.1 ▪ Model and related remarks

In [173, 174] Izmailov proposed the following deterministic models of a single connection between a source controlled by an access regulator and a distant node with a constant transmission capacity μ:

$$\begin{cases} \dot{x}_1(t) = x_2(t-\tau_1) - \mu, \\ \dot{x}_2(t) = -a(x_1(t-\tau_2) - \bar{X}) - b(x_1(t-\tau_2-r) - \bar{X}) \end{cases} \tag{13.15}$$

and

$$\begin{cases} \dot{x}_1(t) = x_2(t-\tau_1) - \mu, \\ \dot{x}_2(t) = -a(x_1(t-\tau_2) - \bar{X}) - b(x_1(t-\tau_2-r) - \bar{X}), \\ \qquad -c\left(x_1\left(t-\tau_2-\frac{r}{2}\right) - \bar{X}\right), \end{cases} \tag{13.16}$$

where x_1 represents the buffer contents, x_2 the current input rate, and $\bar{X}$ the target value, respectively. Using the new variable $y(t) = x_1(t) - \bar{X}$, systems (13.15) and (13.16) lead to the following second-order delay equations with two discrete and independent delays τ and r:

$$\ddot{y}(t) + ay(t-\tau) + by(t-\tau-r) = 0 \tag{13.17}$$

and, respectively,

$$\ddot{y}(t) + ay(t-\tau) + by(t-\tau-r) + cy\left(t-\tau-\frac{r}{2}\right) = 0, \tag{13.18}$$

where the "total" delay $\tau = \tau_1 + \tau_2$ represents the *round-trip time*, and r is the *control-time interval*.

As seen in Niculescu [291, 293], the delays τ and r have a *stabilizing* effect if the gains a, b, and c satisfy some appropriate assumptions, even if the system free of delays is not asymptotically stable. In this sense consider the case $a > |b|$, $b < 0$ in (13.17), which corresponds to a system that is an *oscillator* if it is free of delays, but which is stable for sufficiently *small* delays $\tau, r \neq 0$.

In what follows, we shall focus on the stability analysis of (13.17) and (13.18). More explicitly, we shall consider a more general model including both systems (13.17) and (13.18) as particular examples. This model takes the form

$$\ddot{y}(t) + \sum_{k=0}^{n} a_k y(t-\tau_k) = 0, \tag{13.19}$$

with appropriate initial condition, and with $\tau_k = (k\tau)/n + r$, $k = 1, 2, \ldots, n$. In other words, the delays τ_k, $k = 1, 2, \ldots, n$, are *rational dependent* and depend on two rationally independent delays τ and r (see Appendix A.4 for the interdependency of numbers)

$$\begin{bmatrix} \tau_1 \\ \tau_2 \\ \vdots \\ \tau_n \end{bmatrix} = \begin{bmatrix} \frac{1}{n} & 1 \\ \frac{2}{n} & 1 \\ \vdots & \vdots \\ 1 & 1 \end{bmatrix} \begin{bmatrix} \tau \\ r \end{bmatrix}.$$

The system (13.19) can be rewritten in a first-order form as follows:

$$\dot{x}(t) = Ax(t) + \sum_{k=0}^{n} A_k x\left(t - \frac{k}{n} r - \tau\right), \tag{13.20}$$

where $x = [y \;\; \dot{y}]^T$ and

$$A = \begin{bmatrix} 0 & 1 \\ 0 & 0 \end{bmatrix}, \qquad A_k = \begin{bmatrix} 0 & 0 \\ -a_k & 0 \end{bmatrix}, \qquad k = 0, \ldots, n. \tag{13.21}$$

The analysis in terms of delays can be reduced to the following two steps (one parameter-based analysis at each step):

(i) first, analyze the stability of (13.19) if $\tau = 0$, but $r \neq 0$, that is, the stability of a system including commensurate delays, multiple of r/n, and,

(ii) second, assume also $\tau \neq 0$, and perform the analysis w.r.t. to the second delay parameter τ while assuming the first delay r fixed.

In other words, we are searching first for the delay intervals in r for which the corresponding system with $\tau = 0$ is asymptotically stable or the number of unstable roots is as small as possible, and next, for any delay value r fixed inside such intervals we are looking for the delay intervals in τ for which the stability is guaranteed; that is, we are looking for the characteristic roots crossing the imaginary axis, and also for their crossing direction.

To summarize, the crossing curves in the parameter space $Or\tau$ are defined by using an explicit definition of the delay τ as a function of the parameter r, $\tau = \tau(r)$. Further remarks on a different way to compute the stability crossing curves and the corresponding stability regions in the delay-parameter space are presented in the next section. In particular, we will exploit an interesting property of this system, namely the fact that *any sufficiently small delay* r can stabilize (13.17) if $\tau = 0$ (oscillator subject to a delay output feedback). Note that the assumptions $a > 0$ and $b < 0$ were already encountered in the work of Izmailov (see, for instance, [174]), but the argument was completely different.

13.2.2 ▪ Linear stability analysis

Let us follow the procedure mentioned above. We start by assuming $\tau = 0$. In this case, the system (13.20) becomes a second-order system with commensurate delays, for which the methodology proposed in Chapter 6 for deriving delay intervals guaranteeing stability/instability works.

In this sense, we now introduce the following matrix pencil $\Lambda_1 \in \mathbb{C}^{2n \times 2n}$:

$$\Lambda_1(z) = z \begin{bmatrix} 1 & & 0 & 0 \\ & \ddots & & \\ 0 & & 1 & 0 \\ 0 & \ldots & 0 & a_n \end{bmatrix} + \begin{bmatrix} 0 & -1 & \ldots & 0 & 0 & 0 & \ldots & 0 \\ & & \ddots & & & & & \\ 0 & & & & & & & -1 \\ -a_n & -a_{n-1} & \ldots & -a_1 & 0 & a_1 & \ldots & a_{n-1} \end{bmatrix}. \tag{13.22}$$

As mentioned in the previous chapters, the characteristic roots crossing the imaginary axis and the corresponding critical delay values can be obtained from the *generalized eigenvalue distribution* of the matrix pencil Λ_1. Using the formalism and the notations in Chapter 6, the generalized eigenvalue $z_0 \in \sigma(\Lambda_1)$ such that

$$\sigma\left(A_0 + \sum_{k=1}^{n} A_k z_0^k\right) \cap j\mathbb{R}^* \neq \emptyset$$

corresponds to a crossing eigenvalue, and thus it belongs to the delay crossing generator set $\mathscr{T}_{1,g}$ that will define the corresponding (crossing) frequencies

$$\omega_0 \in \Omega_{1,z_0} := \left\{\omega \in \mathbb{R}_+^* : \quad j\omega \in \sigma\left(A_0 + \sum_{k=1}^{n} A_k z_0^k\right)\right\}$$

and critical delay values

$$r \in \mathscr{T}_{1,\omega_0} := \left\{\frac{\operatorname{Log}(\bar{z}_0)}{j\omega_0} + \frac{2\pi\ell}{\omega_0} > 0 : j\omega_0 \in \sigma\left(A + \sum_{k=1}^{n_d} A_k z_0^k\right) \setminus \{0\}, \quad \ell \in \mathbb{Z}\right\}.$$

As expected, the set of crossing frequencies Ω_1 includes a *finite* number of frequency values, and it is explicitly given by $\Omega_1 := \bigcup_{z \in \mathscr{T}_{1,g}} \Omega_{1,z}$. By similarity, we will have the set of critical delays $\mathscr{T}_1$ given by $\mathscr{T}_1 := \bigcup_{\omega \in \Omega_1} \mathscr{T}_{1,\omega}$. Under the assumption of *simple* crossings, the *crossing direction* w.r.t. any critical delay value $r \in \mathscr{T}_1$ corresponding to the crossing frequency ω_0 and to the delay crossing generator z_0 is given by the sign of the following quantity:

$$\Re\left[\sum_{k=1}^{n} k a_k z_0^k\right] > 0 \ (< 0). \tag{13.23}$$

If such a quantity is positive (negative), we will have a crossing towards instability (stability).

In what follows, we reduce our analysis only to the first delay intervals guaranteeing stability with respect to r and τ. In other words, we will compute the *stability region* which is the "closest" to the origin of the parameter space. In the case of *small delays*, we have the following stability result, which simply extends the remarks concerning the stabilization of oscillatory systems by using delays in the output feedback control laws (see also Chapter 10), and the remarks above concerning the crossing direction:

Proposition 13.6 (see [293]). *Assume that*

$$\sum_{k=0}^{n} a_k > 0, \qquad \sum_{k=1}^{n} k a_k < 0. \tag{13.24}$$

Then there exists a sufficiently small positive value $\epsilon > 0$ *such that* (13.20) *with* $r = \epsilon$, $\tau = 0$ *is asymptotically stable.*

Sketch of the proof. The first inequality simply says that the characteristic function of the second-order system free of delays has two critical roots on the imaginary axis, and the second inequality ensures that the crossing direction is towards stability for small delays. Indeed, in such a case, the crossing direction is given by the relation (13.23) with $z_0 = 1$. Thus, since the first crossing is towards stability, it follows that the asymptotic stability is guaranteed for small delays. □

The next result gives the complete characterization of the *first switch*, that is, the case when the roots of the characteristic equation associated to the original system cross the imaginary axis towards instability when the delay parameter r is varying from 0 to $+\infty$.

Proposition 13.7 (see [293]). *The system* (13.20) *with* $\tau = 0$, *satisfying the inequalities* (13.24), *is asymptotically stable for all delay values* r *satisfying*

$$0 < r < r_1(a_0, a_1, \dots, a_n) := \min\{\xi > 0: \quad \xi \in \mathcal{T}_1\}. \tag{13.25}$$

Furthermore, if $r = 0$ *or* $r = r_1(a_0, \dots, a_n)$, *the corresponding associated characteristic equation has at least one pair of complex conjugate roots on the imaginary axis.*

Proposition 13.8 (see [293]). *The system* (13.20) *satisfying the constraints* (13.24) *is asymptotically stable for all delays* r *and* τ *satisfying the following conditions:*

$$\begin{cases} 0 < r < r_1(a_0, \dots, a_n), \\ 0 \le \tau < \tau_{1,r} := \min\limits_{\omega_s} \left\{ \frac{1}{\omega_s} \cdot \tan^{-1}\left(\frac{-\sum_{k=1}^{n} a_k \sin \frac{\omega_s k r}{n}}{\sum_{k=0}^{n} a_k \cos \frac{\omega_s k r}{n}} \right) \right\}, \end{cases} \tag{13.26}$$

where ω_s *belongs to the set of positive solutions of the equation*

$$\omega^4 = \sum_{k=0}^{n} a_k^2 + 2 \sum_{k=1}^{n} \sum_{h=0}^{k-1} a_k a_h \cos\left(\frac{(k-h)\omega r}{n} \right). \tag{13.27}$$

Furthermore, if the chosen delay r *and the solution* $\tilde{\omega}_s$ *defining the corresponding upper bound* $\tau_{1,r}$ *in* (13.26) *satisfy the condition*

$$\tilde{\omega}_s^3 > \frac{1}{2} \sum_{k=1}^{n} \sum_{h=0}^{k-1} \frac{a_k a_h (k-h) r}{n} \sin\left(\frac{(k-h)\tilde{\omega}_s r}{n} \right), \tag{13.28}$$

then

(i) *the system* (13.19) *is unstable for* $\tau = \tau_{1,r} + \epsilon$, *with* $\epsilon > 0$ *sufficiently small.*

(ii) *there does not exist any* $\tau > \tau_{1,r}$ *such that the system* (13.19) *is asymptotically stable provided equation* (13.27) *has only one positive solution.*

Condition (i) simply says that the first crossing w.r.t. the delay parameter τ is towards instability at $\tau = \tau_{1,r}$ and at the frequency ω_s. Furthermore, an explicit characterization of the corresponding set of crossing frequencies Ω_2 (with respect to the parameter τ for a fixed r) is given in the form of the nonlinear equation (13.27). It is important to point out that, for a fixed r, this equation has a *finite number of roots*. Next, condition (ii) characterizes the situation where, for a fixed delay parameter r value, $\text{card}(\Omega_2) = 1$.

Remark 13.9. *Equation* (13.27) *always has at least one positive solution for* ω *in the interval*

$$\left[\min_{i \in \mathcal{I}_n} \sqrt{\sum_{k=1}^{n} (-1)^{i_k} a_k^2}, \max_{i \in \mathcal{I}_n} \sqrt{\sum_{k=1}^{n} (-1)^{i_k} a_k^2} \right], \tag{13.29}$$

where

$$\mathcal{I}_n = \{ i = (i_1, \dots, i_n) : i_k \in \{1, 2\}, \forall k = 1, \dots, n \}$$

is an appropriate index family. In [290] *Niculescu therefore concludes that the* upper *bound on* τ *will* always be finite, *that is, one may expect a sequence of stability/instability delay intervals in terms of* τ_0, *with* instability persistence *for sufficiently large delays. In other words, there exists a finite value* $\bar{\tau}_r$ *such that the system is unstable for* $\tau > \bar{\tau}_r$.

Remark 13.10. *Based on the results above, it is easy to see that* reducing *the* control-time interval *will increase the* sensitivity *of the control algorithm with respect to the* RTT. *In conclusion, a "wait-and-act" strategy in terms of control-time interval will substantially improve the* robustness *of the control algorithm. Such a result was largely discussed in* [291, 293].

If necessary or desirable, more precise stability information can be obtained by directly computing and monitoring the rightmost characteristic roots of the closed-loop system. This is illustrated in Figure 13.5. The characteristic roots of the system (13.18) with $a = 2.25$, $b = c = -1$, $\tau = 0$ are shown as a function of the control interval r, computed with the software package DDE-BIFTOOL [100]. By automatic continuation in the two-parameter space (r, τ) of solutions, for which characteristic roots lie on the imaginary axis, the stability region in this parameter space is obtained; see Figure 13.6 (left). For the optimal value of the control interval (w.r.t. robustness in the RTT, the rightmost characteristic roots are shown in Figure 13.6 (right).

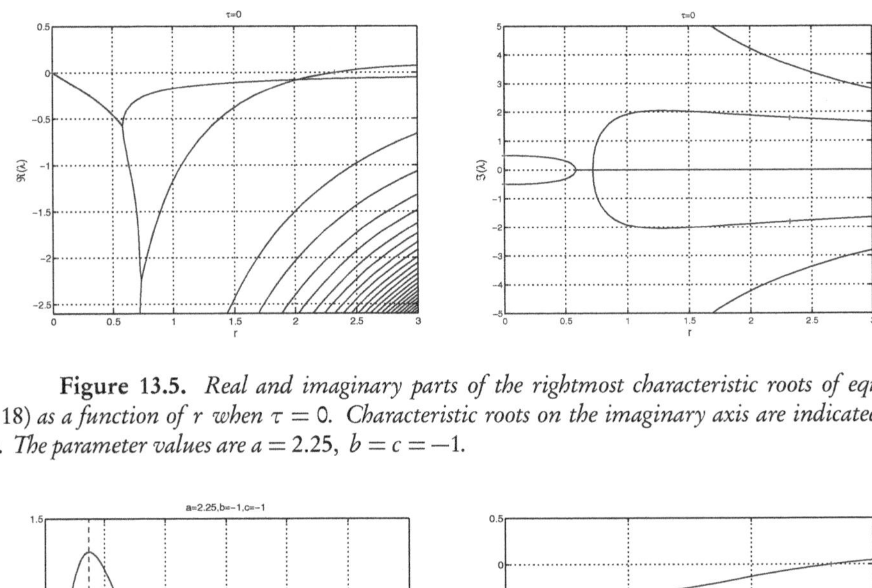

Figure 13.5. *Real and imaginary parts of the rightmost characteristic roots of equation* (13.18) *as a function of* r *when* $\tau = 0$. *Characteristic roots on the imaginary axis are indicated with* (+). *The parameter values are* $a = 2.25$, $b = c = -1$.

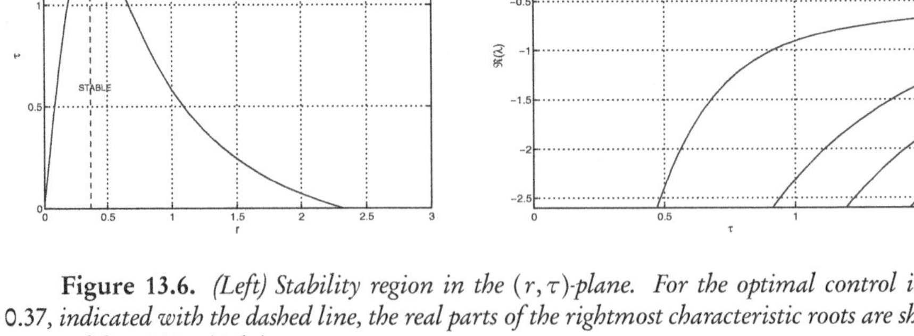

Figure 13.6. *(Left) Stability region in the* (r, τ)*-plane. For the optimal control interval* $r \approx 0.37$, *indicated with the dashed line, the real parts of the rightmost characteristic roots are shown as a function of the RTT* τ *(right).*

13.2.3 ▪ Interpretations and discussions

The analysis above was based on analytically computing the bound of τ as a function of r and taking into account the delay intervals on r, for which (asymptotic) stability exists. In other words, we tried to use r as a *stabilizing* delay parameter, and then τ as a *destabilizing* delay parameter, that is, we explicitly computed the delay margin in terms of τ as a function of r. The problem considered in the previous sections concerned the analysis of the effects induced by the delays for some $(n+1)$-tuple of parameters $(a_0, a_1, \ldots, a_n)$.

In what follows, we shall consider two particular ways to re-interpret the problem in light of the results proposed in the previous chapters.

Output feedback stabilization

It is easy to see that the stability of the initial second-order system including two or three (rationally-dependent) delays

$$\ddot{y}(t) + ay(t-\tau) + by(t-\tau-r) = 0$$

and, respectively,

$$\ddot{y}(t) + ay(t-\tau) + by(t-\tau-r) + cy\left(t-\tau-\frac{r}{2}\right) = 0,$$

can be interpreted as an output feedback control problem: *controlling a double integrator by using two or three delay blocks*, that is, the system $H_{yu}(\lambda) = 1/\lambda^2$ with the control laws

$$u(t) = -ay(t-\tau) - by(t-\tau-r)$$

in the first case, and

$$u(t) = -ay(t-\tau) - by(t-\tau-r) - cy(t-\tau-r/2)$$

in the second case. It is quite clear that the (two or three) delay blocks above are distinct since $\tau < \tau + r/2 < \tau + r$ for all $(\tau, r) \in \mathbb{R}_+^2$. As discussed in Chapter 10, *two delay blocks* are sufficient for getting asymptotic stability of the corresponding closed-loop scheme.

Let us consider this particular case of two delay blocks controlling the chain including two integrators. A possible use for the algorithm above is for tuning the gains a and b by using the delay information available. Indeed, the RTT τ is always measured, and the control-time interval r can be selected as discussed by Izmailov in his papers [173, 174]. In this sense, we can apply, for instance, the pole placement idea mentioned in Chapter 10. The characteristic function of the closed-loop scheme is

$$p(\lambda;\ \tau, r) := \lambda^2 + ae^{-\lambda\tau} + be^{-\lambda(\tau+r)}.$$

Consider now that the parameters (a, b) are such that the double root at the origin is "moved" in $\mathbb{C}_-$ at $\lambda = -\varepsilon$ as a double root for sufficiently small $\varepsilon > 0$. In such a case, we will need the following conditions to hold simultaneously:

$$\begin{cases} p(-\varepsilon;\ \tau, r) = 0, \\ p'(-\varepsilon;\ \tau, r) = 0. \end{cases}$$

This results in the following system of linear equations in (a, b):

$$\begin{cases} a + be^{\varepsilon r} = -\varepsilon^2 e^{-\varepsilon\tau}, \\ a\tau + b(\tau + r)e^{\varepsilon r} = 2\varepsilon e^{-\varepsilon\tau}, \end{cases}$$

with solution

$$\begin{cases} a = \varepsilon \frac{2+\tau\varepsilon}{re^{\varepsilon\tau}}, \\ b = -\varepsilon \frac{2-(\tau+r)\varepsilon}{re^{\varepsilon(\tau+r)}}. \end{cases}$$

It is not difficult to prove that for sufficiently small values of $\varepsilon > 0$, the characteristic roots of $p(\lambda;\ \tau, r)$ are located in $\mathbb{C}_-$, and we can use the argument above to "adapt" the gains as a function of the parameters (τ, r). Larger values for ε can also be considered, but then we need to be sure that there are no other characteristic roots coming from $\mathbb{C}_-$ across the imaginary axis. Reciprocally, the idea above can be used also to find a pair of stabilizing delay values (τ, r) for a given pair of gain values (a, b). Next, some *robustness region* around the corresponding pair $(\tau(a, b), r(a, b))$, for which the closed-loop stability is preserved, can be determined.

On the geometry of the crossing curves in the delay-parameter space

Reconsider the initial problem including only four parameters (a, b, τ, r). It is easy to see that for fixed parameters (a, b), the problem of characterizing the stability regions in the delay-parameter space is reduced to finding all the stability crossing curves, and, next, defining some partition of the delay-parameter space in stability (instability) domains as outlined in Chapter 6. In the particular case under consideration, the corresponding characteristic function is given by $\lambda^2 + ae^{-\lambda\tau} + be^{-\lambda(\tau+r)}$, which is of the form

$$p(\lambda;\ \tau_1, \tau_2) := \lambda^2 + ae^{-\lambda\tau_1} + be^{-\lambda\tau_2}, \tag{13.30}$$

where $\tau_2 \geq \tau_1$. Such a case study enters into the framework of the analysis suggested in Chapter 6 by choosing $p_0(\lambda) = \lambda^2$, $p_1(\lambda) = a$, $p_2(\lambda) = b$. The corresponding rational transfer functions $a_1(j\omega) := p_1(j\omega)/p_0(j\omega) = a/(j\omega)^2$ and $a_2(j\omega) = p_2(j\omega)/p_0(j\omega) := b/(j\omega)^2$ are properly defined for all positive ω except $\omega = 0$. On the other hand, if $a + b \neq 0$, $\lambda = 0$ is not a characteristic root of the quasi-polynomial p. As discussed in the previous chapters, such an assumption is standard and serves to avoid a characteristic root at the origin that is *invariant* w.r.t. delay changes.

Simple computations prove that the corresponding frequency crossing set Ω is reduced to the interval

$$\Omega = \left[\sqrt{\|a| - |b\|}, \sqrt{|a| + |b|}\right], \tag{13.31}$$

which is of Type 13 (or 23), if we assume that $|a| > |b|$ (or $|a| < |b|$). The explicit computation of the stability regions in the delay-parameter space is straightforward. We refer the reader to Chapter 6 for the details.

13.3 ▪ TCP/AQM congestion avoidance models with one delay

We study the stability of some nonlinear models for the behavior of congested routers. First, we propose and discuss the models. Next, we sketch some important scaling properties of these models. Finally, we study the attractors and their stability properties.

13.3.1 ▪ Model and related remarks

Recently, some models describing accurately the behavior of congested routers in TCP/AQM (Transmission Control Protocol/Active Queue Management) networks were presented in [268, 188, 166]. As expected, such models are described by nonlinear differential equations with time delay, where the delay represents the corresponding RTT in the network.

The model of [268, 166, 165] consists of the following coupled nonlinear differential equations with time-varying delay:

$$\dot{W}(t) = \frac{1}{R(t)} - \frac{1}{2}\frac{W(t)W(t-R(t))}{R(t-R(t))}p(t-R(t)), \tag{13.32}$$

$$\dot{Q}(t) = \begin{cases} N(t)\frac{W(t)}{R(t)} - C, & q > 0, \\ \max\left(N(t)\frac{W(t)}{R(t)} - C, 0\right), & q = 0, \end{cases} \tag{13.33}$$

where $W(t)$ denotes the average of TCP window-size (packets), $Q(t)$ is the average of queue length (packets), $R(t)$ is the RTT (secs) , C is the queue capacity (packets/sec), $N(t)$ is the number of TCP sessions, and $p(\cdot)$ is the probability function of a packet mark. The queue length $Q(t)$ and window-size $W(t)$ are positive. The probability function of a packet mark $p(\cdot)$ takes values only in $[0,1]$. The RTT can be decomposed as

$$R(t) = \frac{Q(t)}{C} + \tau_p, \tag{13.34}$$

where τ_p is the propagation delay (secs).

The first differential equation describes the *TCP window control dynamic*. Indeed, the first term $\frac{1}{R(t)}$ describes the window's additive increase phase, and the second term $\frac{W(t)}{2}$ describes the multiplicative decreasing phase (including the packet marking probability). Different AIMD (additive-increasing multiplicative-decreasing) continuous-time models can be found in [188, 217]. Note also the excellent overview of existing fluid approximation–based approaches proposed in [217]. Equation (13.33) describes the bottleneck queue length as the difference between the packet arrival rate $\frac{NW}{R}$ and the link capacity C, assuming that there are no internal dynamics in the bottleneck (roughly speaking, a simple integrator).

Using fluid flow models like (13.32)–(13.34), AQM can be interpreted as a *feedback control problem*, where the control action consists of marking packets (with probability p) as a function of the measured queue length Q; see [166].

As in [165] we shall assume that the TCP load $N(t)$ and the RTT $R(t)$ are time-invariant, i.e., $N(t) \equiv N$ and $R(t) \equiv R$. The latter may be a good approximation when the RTT is dominated by the propagation delay. This occurs when the capacity C of the link is large [165]. Furthermore, as presented in [165, 249], considering that the probability marking function $p(\cdot)$ is *proportional* to the queue length, i.e., $p(t) = K\,q(t)$, the system under consideration becomes

$$\dot{W}(t) = \frac{1}{R} - \frac{W(t)W(t-R)}{2R}K\,q(t-R), \tag{13.35}$$

$$\dot{Q}(t) = \begin{cases} N(t)\frac{W(t)}{R} - C, & Q > 0, \\ \max\left(N(t)\frac{W(t)}{R} - C, 0\right), & Q = 0. \end{cases} \tag{13.36}$$

The unique equilibrium point of (13.35)–(13.36) is given by

$$W^* = \frac{RC}{N}, \quad Q^* = \frac{2N^2}{R^2C^2K}.$$

In Section 5 of [166] a linearized stability analysis of the equilibrium point was performed in the frequency domain, where the variation of RTT was taken into account, yet some of the delay effects were treated as high-frequency uncertainty. The references [165, 249] contain a Lyapunov-based (non-)local stability analysis of the equilibrium when making the additional simplification of (13.35) to

$$\dot{W}(t) = \frac{1}{R} - \frac{W(t)^2}{2R} K\, q(t-R). \tag{13.37}$$

In [165] the authors proved that when the delay is equal to zero the equilibrium point of the system (13.37) and (13.36) is asymptotically stable for all $K > 0$. When the delay is different from zero a Lyapunov–Razumikhin approach was used to show the asymptotic stability of the equilibrium point of (13.37) for sufficiently small $\frac{K}{N} > 0$. In [249] this result was refined and sufficient conditions on the parameters for local stability and estimates of the attraction domain were derived using a less conservative Lyapunov–Krasovskii approach.

The structure of this section is as follows: after some brief comments on a transformation of state and time, we completely characterize the linear stability region of the steady state solution of (13.35)–(13.36) as a function of the model parameters. More explicitly, only one delay interval guarantees the asymptotic stability of the linearized model. Then we take the nonlinearities into account and study the global behavior of the solutions.

13.3.2 ▪ Transformation

When the RTT R is assumed to be constant one can apply a transformation of state and time to (13.35)–(13.36), yielding

$$\begin{aligned} \dot{w}(t) &= 1 - \frac{w(t)w(t-1)}{2} k q(t-1), \\ \dot{q}(t) &= \begin{cases} w(t) - c, & q > 0, \\ \max(w(t) - c, 0), & q = 0, \end{cases} \end{aligned} \tag{13.38}$$

where $w = W$, $q = Q/N$, $t^{(\text{new})} = t^{(\text{old})}/R$, and $c = (RC)/N$, $k = KN$. The importance of this transformation lies in the fact that the four model parameters (K, N, C, R) are reduced to only two parameters (k, c). This facilitates the study of the dependence of the attractors and their stability properties on the system's parameters. It also allows us to display stability regions w.r.t. *all* parameters in only one figure.

Note that only one of the "new" parameters depends explicitly on the RTT R. Furthermore, such a dependence is *linear*, a fact which simplifies the analysis of delay effects on stability.

13.3.3 ▪ Stability analysis

Equilibrium

In the normalized coordinates (w, q) the unique equilibrium point is given by

$$(w^*, q^*) = \left(c, \frac{2}{kc^2}\right). \tag{13.39}$$

Linearization around it results in the second-order differential equation in $\tilde{q} := q - q^*$,

$$\ddot{\tilde{q}}(t) + \frac{1}{c}\dot{\tilde{q}}(t) + \frac{1}{c}\dot{\tilde{q}}(t-1) + \frac{kc^2}{2}\tilde{q}(t-1) = 0,$$

whose characteristic equation is given by

$$H(\lambda) := \lambda^2 + \frac{1}{c}\lambda + \frac{1}{c}\lambda e^{-\lambda} + \frac{kc^2}{2} e^{-\lambda} = 0. \tag{13.40}$$

To characterize the stability region in the (k,c)-plane we first fix $c>0$ and consider the stability region as a function of k. We have the following result:

Proposition 13.11. *For each value of* c*, there exists exactly* one *stability interval as a function of* k*, i.e.,* $k\in(0,\bar{k}(c))$ *with* $\bar{k}(c)\in\bar{\mathbb{R}}_+$.

Proof. For $k=0$ the characteristic equation reduces to $H(\lambda)=\lambda(\lambda+\frac{1}{c}+\frac{1}{c}e^{-\lambda})$. Following from [79], the rightmost eigenvalue is equal to zero and isolated. The continuity of this eigenvalue w.r.t. k implies the existence of a root function $r(k)$, satisfying $r(0)=0$ and $H(r(k))=0$. Differentiating this identity w.r.t. k at the point $k=0$, we arrive at

$$r'(0)=-\frac{c^3}{4}<0. \tag{13.41}$$

Therefore, the linearized system has one unstable real eigenvalue for small $k<0$. Instability follows for *all* $k<0$, because a zero eigenvalue cannot occur for $k\neq 0$ and, as a consequence, eigenvalues can only cross the imaginary axis in complex conjugate *pairs* as k is varied.

Equation (13.41) also implies asymptotic stability for small $k>0$. The stability can only be lost when eigenvalues cross the imaginary axis. When an imaginary eigenvalue $\lambda=j\omega$ occurs for, say $k=k^*$, one can compute

$$\left.\frac{d\Re(\lambda)}{dk}\right|_{\lambda=j\omega,k=k^*}=\frac{\omega^2c^2}{2}\frac{\left(\frac{1}{c}\omega^2+\frac{k^*c^2}{2}\omega^2+\frac{k^*}{2}+\frac{k^*c}{2}\right)}{\left(\frac{1}{c}\omega^2-\frac{k^*c}{2}+\frac{\omega^2}{c^2}+\frac{k^*c^2\omega^2}{2}\right)^2+\left(\frac{\omega^3}{c}-\omega k^*c^2-\frac{k^*c\omega}{2}\right)^2}>0. \tag{13.42}$$

Therefore, eigenvalues can only cross the imaginary axis towards instability as k is increased, and thus only one stability interval is possible. Notice that (13.42) holds under the implicit assumption that the imaginary eigenvalues are simple. An easy calculation, which is omitted, excludes the nongeneric case of having imaginary eigenvalues with a multiplicity larger than one. □

The critical value $\bar{k}(c)$ in Proposition 13.11 corresponds to a subcritical *Hopf bifurcation* of the original nonlinear system. By numerical continuation of such Hopf bifurcation in the two-parameter space (k,c), the stability region can be computed, and the result is shown in Figure 13.7. See [342] for theory on continuation and bifurcation analysis and [100, 101] for the numerical tool DDE-BIFTOOL.

While the technique of numerical continuation of Hopf bifurcations to separate stability/instability regions of a steady state solution in a two-parameter space is applicable in general, the method of D-subdivision [196] (largely discussed in Chapter 6) only applies to specific problems, such as the example above, but allows us to obtain *analytical* expressions for the boundary: when substituting $\lambda=j\omega$ in (13.40), some simple computations yield an implicit expression of the relation $\bar{k}(c)$:

$$\begin{cases} c=\frac{1+\cos\omega}{\omega\sin\omega},\\ \bar{k}=\frac{2\omega^4(\sin\omega)^2}{(1+\cos\omega)^2}, \quad \omega\in(0,\ \pi). \end{cases} \tag{13.43}$$

The above analysis of the linearized fluid model illustrates how analytical and numerical tools can *complement* each other to obtain a complete solution of an analysis problem. Numerically, we computed a curve separating stable and unstable parameter pairs in

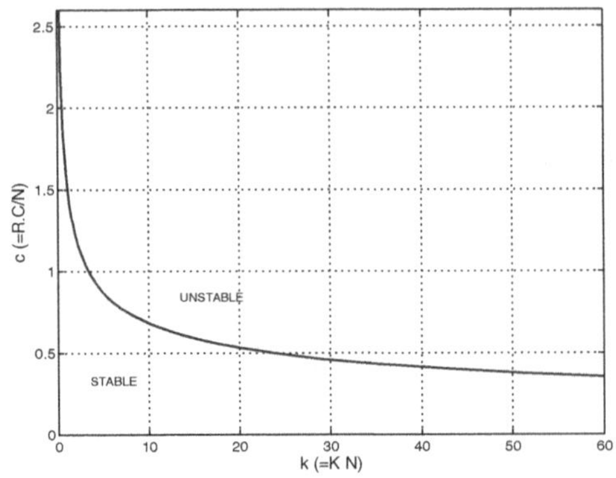

Figure 13.7. *Linear stability region of the steady state solution in the (k,c)-plane. The curve is obtained by computing a branch of Hopf bifurcations* [254].

the (k,c)-plane. Analytically, we obtained qualitative information which proves that this curve bounds the *whole* stability region.

Other attractors

A first observation is that solutions of (13.38) cannot grown unbounded, even when the steady state solution is (locally) exponentially unstable:

Proposition 13.12. *All the solutions of the system* (13.38) *are bounded.*

Proof. First, we show by contradiction that the function $t \to w(t)$ is bounded along a solution. Therefore, assume that w is unbounded and denote by $t = t_m$ the smallest time such that $w(t) = w_m$, where $w_m := |w_0| + c + 3 + 2/\sqrt{k}$. Since $\dot{w}(t) < 1$ for $t \geq 0$, we have $t_m \geq 2$ and $w(t) \geq w_m - 2$ for all $t \in [t_m - 2,\ t_m - 1]$. Consequently,

$$q(t_m - 1) = q(t_m - 2) + \int_{t_m-2}^{t_m-1} \dot{q}(t)dt \geq \min_{t\in[t_m-2,\ t_m-1]} \dot{q}(t) \geq |w_0| + 1 + 2/\sqrt{k} \geq 1.$$

This implies $\dot{w}(t_m) \leq 1 - (w_m(w_m - 2)k)/2 < 0$ and we have a contradiction, and thus w is bounded. Similarly, assume that q is unbounded and denote by $t = t_m$ the first time such that $q(t) = q_m$, where q_m is a sufficiently large number. The boundedness of w implies the boundedness $\dot{q}$, and thus t_m must grow unbounded as $q_m \to \infty$. Furthermore, for large t_m there exists a number $M > 0$ such that in the time-interval $[t_m/2,\ t_m]$, the first equation of (13.38) can be written as

$$\dot{w}(t) = 1 - w(t)w(t-1)\alpha(t), \quad \alpha(t) > M. \tag{13.44}$$

By taking q_m sufficiently large, M also can be chosen arbitrarily large. For sufficiently large values, (13.44) and the boundedness of w imply that $w(t_m) < c$, as follows from the method of steps. Thus $\dot{q}(t_m) < 0$, and we have a contradiction. □

As a consequence there exist other attractors in addition to the equilibrium point. Now we provide some qualitative and quantitative information on these attractors. Since a complete bifurcation analysis is beyond the scope of this book, we focus on the particularities, due to the delayed damping term in (13.38) (period doubling route to chaos) and the discontinuity in the right-hand side of (13.38) (superstable limit cycles).

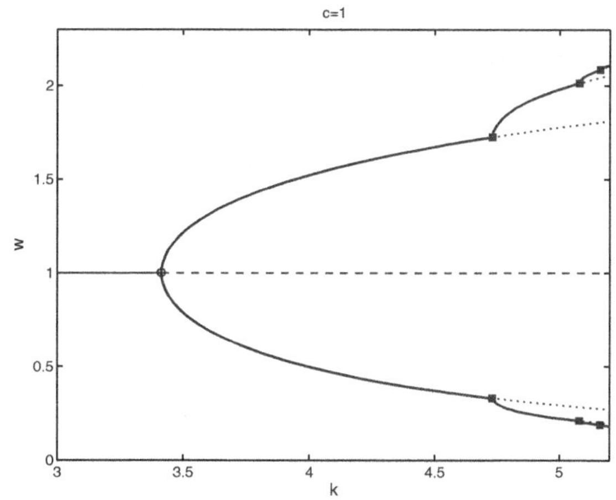

Figure 13.8. *Bifurcation diagram of the system* (13.38) *with* $c = 1$ *and free parameter* k. *Stable (thin line) and unstable* (---) *steady state solution. Stable (thick lines) and unstable* $(\cdots)$ *branches of periodic solutions (maximum and minimum values of* $w(t)$ *are shown). Connections are formed by a Hopf bifurcation* (o) *and by period doubling bifurcations* (□). *As* k *is increased a periodic doubling route to chaos occurs. The chaotic attractor is shown in Figure* 13.9 [254].

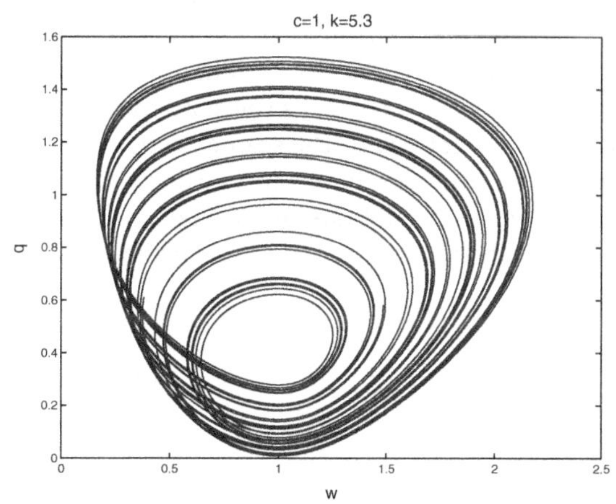

Figure 13.9. *Chaotic attractor of the system* (13.38) *for* $c = 1$ *and* $k = 5.3$ [254].

Chaotic behavior

In Figure 13.8 we show a bifurcation diagram of (13.38), when k is the free parameter and $c = 1$ is fixed, computed with the help of DDE-BIFTOOL [100]. Recall that for small $k > 0$ the unique steady state solution is locally asymptotically stable and that stability is lost in a subcritical Hopf bifurcation as k is increased. In the Hopf bifurcation a branch of stable *periodic solutions* emanates. The latter become unstable after a period doubling bifurcation, where a new branch of stable, *period doubled* periodic solutions emanates. A sequence of period doubling bifurcations ultimately leads to chaos. In Figure 13.9 we plot the chaotic attractor for $k = 5.3$. Chaotic behavior is inherent to TCP/IP traffic. The work of Veres and Boda [389], where chaotic behavior was detected and analyzed in simulations with the ns-2 simulator [103], showed that TCP itself can cause or contribute to chaotic behavior as a *deterministic* system (in previous works a large number of

ON-OFF sources with *random* periods were rather seen as a source of chaos in TCP). The analysis above shows that already the simple second-order deterministic model (13.35)–(13.36) exhibits chaos, which thus supports this proposition.

The chaotic behavior in (13.38) is clearly caused by the nonlinear delayed damping term in the first equation, which is proportional to $-w(t)w(t-1)$ (notice that q is strictly larger than zero along the attractor shown in Figure 13.9, and hence the discontinuity in the right-hand side of (13.38) does not contribute). Therefore, it is expected that also the model proposed in [188],

$$\dot{x}(t) = k(w - x(t-\tau)p(x(t-\tau))),$$

which describes the dynamics of a collection of flows all using a single resource and sharing the same gain parameter k, may exhibit chaotic behavior for particular choices of the function $p(.)$, which can again be interpreted as the fraction of packets indicating congestion. Finally, an analogous instability mechanism leading to chaos occurs in the *delayed logistic equation* [372].

Superstable limit-cycles

When one lets the size of the attractor grow, by changing the system's parameters, it may ultimately simplify to a nonsmooth limit-cycle, which contains a segment where $q \equiv 0$. When $q = 0$ for a sufficiently large time, this limit-cycle is *superstable*, meaning that the effect of perturbations around it disappears in a *finite* time. In Figure 13.10 we plot such a superstable limit-cycle. The discontinuity in the right-hand side of (13.38) creates a mechanism which resets the state to the same value each time before starting a new loop (more precisely, to the segment $(w,q) = (c+\theta, 0)$, $\theta \in [-1, 0]$). Notice that such nonsmooth solutions have a physical explanation: when a buffer becomes empty, it remains so until the arrival rate of packages, which increases because the window size is increased, exceeds the maximal capacity of the link.

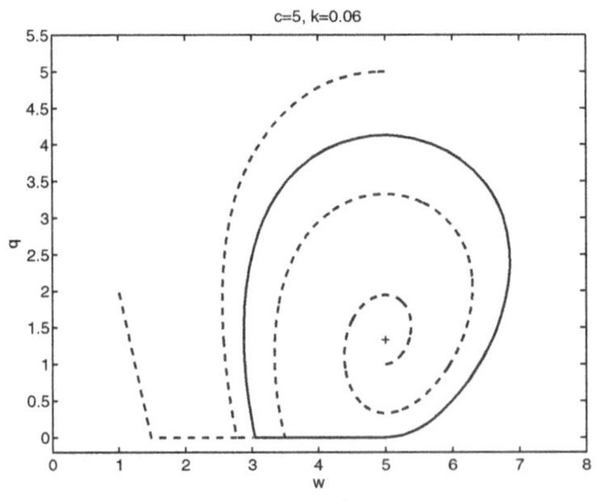

Figure 13.10. *Attractor of* (13.38) *for* $c = 5$ *and* $k = 0.06$ *(solid line), as well as a few trajectories (dashed lines) and the unstable steady state solution* (+)*. Around the attractor, perturbations disappear in a finite time* [254].

13.4 ▪ Notes and references

In this chapter, we presented particular fluid approximation models encountered in congestion control analysis of high-performance networks. Without discussing the importance and particularities of such models, our interest was to point out some specific behaviors induced by the delay presence. Further discussions and a more complete list of references can be found in Srikant's overview on existing models and methods for analyzing Internet congestion control algorithms [356]. Other models and some remarks and an application of Lyapunov techniques for the analysis of some nonlinear fluid approximation models with delays can be found in [249, 302].

The application of the Smith predictor to congestion control mechanisms was proposed by Mascolo in [229, 230]. The model was reconsidered in this chapter but our focus was on computing the maximal allowable uncertainty on the delay parameter (round-trip time) such that the stability of the corresponding closed-loop system is still preserved. The presentation made use of the geometric approach proposed in Chapter 6 and followed closely the presentation proposed by [91].

The second model under consideration is due to Izmailov [173, 174] and describes the behavior of a single connection between a source controlled by an access regulator and a distant node. One of the main features of this model is the presence of two delays: the RTT and some control-time interval. The proposed analysis follows closely the approaches considered in [291, 293]. A different analysis can be found in [412], where the characterization of stability regions in the delay-parameter space was derived by using a dual-locus diagram (an extension of the well-known Nyquist diagram). Some connections with the output feedback stabilization problem in the presence of multiple delays, and some remarks concerning the geometry of the stability regions, have been added and complete the presentation. Further remarks on the case when the round-trip time is time-varying can be found in [302] (using the integral quadratic constraints (IQC)-based approach).

Next, we focused on some nonlinear time-delay system describing the behavior of congested routers in TCP/AQM networks. A stability and bifurcation analysis of a fluid flow–like model was performed. The stability region in the parameter space of the unique equilibrium point was completely characterized by combining analytical and numerical tools. A direct computation of periodic solutions and a continuation procedure revealed a period doubling route to chaos. The presence of a chaotic attractor in the low-order, deterministic model supports the assumption that chaotic behavior is inherent in the TCP mechanism. The presentation of the results follows closely [254].

Chapter 14

Consensus problems with distributed delays, with traffic flow applications

14.1 ▪ Introduction

It is well known that the traffic dynamics are inherently time delayed because of the limited sensing and acting capabilities of drivers against velocity and position variations. The undesirable effects of a mismanaged traffic flow in social and economic life have made this interdisciplinary problem interesting and challenging over the years, and it becomes critical in the context of increasing highway traffic [126, 157].

Without any deep discussions on the modeling of the traffic dynamics, this book concerns the stability analysis of some (microscopic) linear system including distributed delays. The idea of using delays in traffic flow dynamics is not new, and, to the best of the authors' knowledge, was pointed out in the 1960s (see, for instance, [64]). According to its origin (see, e.g., [126]), we can classify the delays in the traffic flow dynamics as follows: physiological delays (mainly induced by human operators), mechanical time delays (time needed for the vehicle's response after some driver's action), and delays in the vehicle's action. Such a classification is far from complete. For instance, the physiological delay can be further classified as follows: sensing, perception, response, selection, and programming delay. Finally, it is important to point out that such delays are in order of seconds, although different ranges are stated in the literature, or a function of the correlation of the human reactions with acceleration/deceleration (see, e.g., [349, 126]). Since the presence of delays may drastically change the dynamics of the model free of delays, the analysis of delay effects on the stability properties of the corresponding system is necessary.

One of the simplest models often discussed in the literature is the (microscopic) car following model, describing the behavior of multiple vehicles under the influence of a single constant time delay [64, 157, 333]. In general, two spatial configurations are dealt with: the linear and the ring configuration. For the sake of brevity, in what follows we consider only the *ring configuration* when discussing the traffic flow application, but the obtained results can also be applied to the linear configuration. The linear model of [64] can be written conceptually as follows (inspired by some delay-free models of Reuschel from the 1950s):

$$\dot{v}_k(t) = \alpha_k(v_{k-1}(t-\tau) - v_k(t-\tau)), \quad k = 1, \ldots, p, \tag{14.1}$$

where p is the number of considered vehicles and $v_0 = v_p$. The left-hand side represents the *acceleration* of the kth vehicle, and the right-hand side expresses the *velocity difference* of consecutive vehicles. A natural extension of the model above is by taking into account multiple cars,[31] as discussed in [333]. Such a model is attributed to Kuhne in the survey [350].

However, various issues are not considered in the model (14.1) or its multiple car following version. For instance, humans retain a short-term memory of past events and this may affect their control decision strategy. Such behavior cannot be described by using pointwise (or discrete) delays in the model. Furthermore, the driver's perception and interpretations of the stimuli depend on various parameters and are different from one driver to another. As pointed out in [350], a more realistic model should include a *delay distribution* over the time that depicts the human behavior in average. Conceptually, defining the delay distribution represents a challenging problem itself and is far from being solved. In [350], the authors proposed three types of delay distributions: a uniform distribution, a γ-distribution, and a γ-distribution with a gap, where the gap corresponds to the minimum reaction time of the humans with respect to some external signals and/or stimuli. In this chapter we shall assume the third type of distribution. Remarks and discussions on its applications to other problems from engineering and biology can be found in [278].

The above discussions lead us to the stability analysis of the model

$$\dot{v}_k(t) = \sum_{i=1}^{p-1} \alpha_{k,\lfloor k-i \rfloor} \int_0^\infty f(\theta)(v_{\lfloor k-i \rfloor}(t-\theta) - v_k(t-\theta))\mathrm{d}\theta, \quad k = 1, \ldots, p, \tag{14.2}$$

where $f(\cdot)$ denotes the delay kernel, and where the notation $\lfloor \cdot \rfloor$ stands for

$$\lfloor l \rfloor = \begin{cases} l, & l = 1, \ldots, p, \\ \lfloor l+p \rfloor, & l < 1. \end{cases}$$

We assume that

$$\alpha_{k,l} \geq 0, \quad k = 1, \ldots, p, \ l = 1, \ldots, p, \ k \neq l, \tag{14.3}$$

$$\alpha_{k,\lfloor k-1 \rfloor} > 0, \quad k, \ldots, p. \tag{14.4}$$

From the application point of view (14.4) is natural, as it expresses that a driver always takes into account the preceding car in his or her reaction. Since the delay distribution corresponds to a gamma-distribution with a gap, the kernel f is given by

$$f(\xi) = \begin{cases} 0, & \xi < \tau, \\ \frac{(\xi-\tau)^{n-1} e^{-\frac{\xi-\tau}{T}}}{T^n (n-1)!}, & \xi \geq \tau, \end{cases} \tag{14.5}$$

where $n \in \mathbb{N}$, $T > 0$, and $\tau \geq 0$. Note that $f(\xi) \geq 0$ for all $\xi \geq 0$ and $\int_0^\infty f(\xi) d\xi = 1$. The gap is defined by τ, and the corresponding *average delay* of (14.5) satisfies

$$\tau_m = \int_0^\infty \xi f(\xi)\, d\xi = \tau + nT. \tag{14.6}$$

As $T \to 0+$, the kernel (14.5) tends to a Dirac impulse centered at $\xi = \tau$, and (14.2) therefore reduces to a system with a pointwise delay τ. As we shall see, the transition to $T = 0$ is smooth from a stability analysis point of view, as the stability determining eigenvalues are continuous w.r.t. $T \geq 0$.

[31]The drivers observes not only the vehicle in front of them, but also other vehicles.

The aim of this chapter is to perform a stability analysis of (14.2)–(14.5) with respect to the parameters (T,τ) and n, which determine the shape of the delay distribution. More precisely, for a given value of n we will determine regions in the (T,τ) space such that for all initial conditions a consensus is reached, that is, the cars eventually reach the same speed.

The stability analysis of (14.2)–(14.5) can be interpreted as the study of a consensus protocol of a multiagent system with a fixed, directed network topology and a distributed delay in the communication channels. More precisely, let the directed graph

$$\mathcal{G}(V,\mathbb{A}) \tag{14.7}$$

be characterized by the node set $V=\{1,\dots,p\}$ and a weighted adjacency matrix $\mathbb{A}$ with zero diagonal and nondiagonal entries $\alpha_{k,l}$. Let each node correspond to an agent whose dynamics are described by

$$\dot{v}_k(t)=u_k(t), \quad k=1,\dots,p. \tag{14.8}$$

Then (14.2) is obtained by applying the following protocol to (14.8):

$$u_k(t)=\sum_{(k,l)\in E}\alpha_{k,l}\int_0^{\infty} f(\theta)(v_l(t-\theta)-v_k(t-\theta))\mathrm{d}\theta, \quad k=1,\dots,p.$$

Note that the corresponding problem for an undirected graph (where $\mathbb{A}$ is symmetric) and a constant time delay was investigated in [312]. Note also that (14.4) implies that the graph $\mathcal{G}$ is strongly connected. It is important to mention that the results derived in this chapter remain valid when this condition is replaced by the weaker condition that the graph is strongly connected.

The structure of the chapter is as follows. Section 14.2 contains the necessary extensions of the stability theory of Chapter 1. Section 14.3 is devoted to the characterization of the region in the delay-parameter space for which the system (14.2)–(14.5) reaches a consensus. Some illustrative examples are presented in Section 14.4. In Section 14.5 the stability analysis of other types of models encountered in the literature is commented on. Some concluding notes and references complete the chapter.

Throughout the chapter, the following functions will be used:

Definition 14.1. *For $n\in\mathbb{N}$, let $g_n:\mathbb{R}_+\to\mathbb{R}_+$ be such that $y=g_n(x)$ is the positive solution of $|y(1+jy)^n|=x$.*

14.2 ▪ Extension of stability theory to systems with distributed delays

Motivated by the structure of (14.2)–(14.5) we extend the stability theory for systems with pointwise delays, presented in Chapter 1, to systems with distributed delays of the form

$$\dot{x}(t)=A\int_0^{\infty} f(\theta)x(t-\theta)d\theta, \tag{14.9}$$

where $x(t)\in\mathbb{C}^{p\times 1}$, $A\in\mathbb{C}^{p\times p}$ and f is given by (14.5). A second extension consists of characterizing consensus problems in the frequency domain for both systems with constant and distributed delays.

A solution of (14.9) is uniquely determined for an initial condition ϕ, which belongs to the set $\mathscr{F}((-\infty,0],\mathbb{C}^{p\times 1})$, defined as

$$\mathscr{F}((-\infty,0],\mathbb{C}^{p\times 1}) := \left\{\phi\in\mathscr{C}((-\infty,0],\mathbb{C}^{p\times 1}) \colon \|\phi\|_f := \int_{-\infty}^{0} \|f(-\theta)\phi(\theta)\|_2 d\theta < \infty\right\}$$

and equipped with $\|\cdot\|_f$. Denote by $t\in(-\infty,\infty)\to x(\phi)(t)$ the forward solution of (14.9) with initial condition ϕ. In this way, stability definitions can be formulated in a similar way as for systems with constant delays; see, e.g., [151] for the latter. We say, for instance, that the zero solution of (14.9) is asymptotically stable if and only if

$$\forall\epsilon>0\,\exists\delta>0\,\forall\phi\in\mathscr{F}((-\infty,0],\mathbb{C}^{p\times 1})\ \|\phi\|_f<\delta\Rightarrow\forall t\ge 0\ \|x(\phi)(t)\|_2<\epsilon,$$
$$\forall\phi\in\mathscr{F}((-\infty,0],\mathbb{C}^{p\times 1})\ \lim_{t\to\infty} x(\phi)(t)=0.$$

The substitution of a sample solution of the form $x(t)=e^{\lambda t}X$, with $X\in\mathbb{C}^{p\times 1}$, in (14.9) leads us to the *characteristic equation*

$$\det\left(\lambda I - A\frac{e^{-\lambda\tau}}{(1+\lambda T)^n}\right)=0, \tag{14.10}$$

which can be factored as

$$\Pi_{k=1}^{p}\left(\lambda-\frac{\mu_k e^{-\lambda\tau}}{(1+\lambda T)^n}\right)=0, \tag{14.11}$$

with μ_k, $k=1,\dots,p$, the eigenvalues of A. As we shall see, the roots distribution of (14.10)–(14.11) determines the stability properties of (14.9). However, the commonly used arguments, which are based on a spectral decomposition of the solutions (see, for instance, [93, 151]) cannot be directly applied to a system of the form (14.9). A major obstacle is the fact that functions of the form $e^{\lambda t}X$, $t\le 0$, do not belong to the space $\mathscr{F}((-\infty,0],\mathbb{C}^{p\times 1})$ if $\Re(\lambda)<1/T$. We shall therefore develop arguments based on a *comparison system*.

Formally, with

$$y(t)=\int_0^{\infty} f(\theta+\tau)x(t-\theta)d\theta=\int_{-\infty}^{t} f(t+\tau-\theta)x(\theta)d\theta,$$

we get

$$\left\{\begin{array}{lcl} y'(t) & = & \int_{-\infty}^{t} f'(t+\tau-\theta)x(\theta)d\theta \\ & \vdots & \\ y^{(n-1)} & = & \int_{-\infty}^{t} f^{(n-1)}(t+\tau-\theta)x(\theta)d\theta \\ y^{(n)}(t) & = & f^{(n-1)}(\tau)x(t)+\int_{-\infty}^{t} f^{(n)}(t+\tau-\theta)x(\theta)d\theta \end{array}\right. ,$$

which leads to

$$\left(T\frac{d}{dt}+I\right)^n y(t)=T^n f^{(n-1)}(\tau)x(t)+\int_{-\infty}^{t}\left(T\frac{d}{dt}+I\right)^n f(t+\tau-\theta)d\theta$$
$$=x(t).$$

We conclude that a solution $x(\phi)(t)$ of (14.9) satisfies

$$\begin{cases} \dot{x}(t) = Ay(t-\tau), \\ \left(T\frac{d}{dt}+I\right)^n y(t) = x(t) \end{cases} \tag{14.12}$$

for $t \geq 0$ if the initial value problem (14.12) is accordingly initialized with

$$\begin{cases} x(0) = \phi(0), \\ y(\theta) = \int_{-\infty}^{\theta} f(\tau+\theta-\xi)\phi(\xi)d\xi, \quad \theta \in [-\tau,\, 0], \\ y^{(i)}(0) = \int_{-\infty}^{0} f^{(i)}(\tau-\xi)\phi(\xi)d\xi, \quad i = 1,\dots,n-1. \end{cases} \tag{14.13}$$

Note that the integrals in the right-hand side of (14.13) are defined and bounded because $f^{(i)}(\xi)$, $i = 1,\dots,n-1$, has the same asymptotic behavior as $f(\xi)$ as $\xi \to \infty$. When letting $z = [x^T\; y^T\; y'^T \cdots (y^{(n-1)})^T]^T$, the *comparison system* (14.12) can be written in a first-order form as

$$\dot{z}(t) = \bar{A}z(t) + \bar{B}(t-\tau), \tag{14.14}$$

where

$$\bar{A} = \begin{bmatrix} 0 & 0 & 0 & \cdots & 0 \\ 0 & 0 & I & & \\ & & & \ddots & \\ & & & & I \\ \frac{I}{T^n} & -\binom{n}{n}\frac{1}{T^n} & & \cdots & -\binom{n}{1}\frac{1}{T^1} \end{bmatrix} \quad \text{and} \quad \bar{B} = \begin{bmatrix} 0 & A & 0 & \cdots & 0 \\ 0 & 0 & 0 & & 0 \\ \vdots & & & & \vdots \\ 0 & & \cdots & & 0 \end{bmatrix}.$$

The initial conditions for (14.14) are assumed to belong to the space $\mathscr{C}([-\tau,\, 0], \mathbb{C}^{(n+1)p\times 1})$. The next lemma summarizes the established relation between the solutions of (14.9) and (14.14) and addresses a partial converse:

Lemma 14.2. *If $x(t)$, $t \in \mathbb{R}$, is a solution of (14.9), then there exists a solution $z(t), t \geq -\tau$, of (14.14) such that $[I\; 0\cdots 0]z(t) = x(t)$ for all $t \geq -\tau$. If (14.14) has a solution of the form $Ze^{\lambda t}$, $t \geq -\tau$, where $\Re(\lambda) \geq 0$ and $Z \in \mathbb{C}^{(n+1)p\times 1} \setminus \{0\}$, then $[I\; 0\cdots 0]Ze^{\lambda t}, t \in \mathbb{R}$, is a nontrivial solution of (14.9).*

Proof. The first assertion follows from the above construction, and from an extension of (14.13) on the interval $[-\tau,\, 0]$.

To prove the second assertion, we partition Z according to the structure of $\bar{A}$ and $\bar{B}$ as $Z = [X^T\; Y_0^T \cdots Y_{n-1}^T]^T$. Substituting $Ze^{\lambda t}$ in (14.14) yields

$$\left(\lambda I - A\frac{e^{-\lambda\tau}}{(1+\lambda T)^n}\right)X = 0, \tag{14.15}$$

$$Y_i = \frac{\lambda^i}{(1+\lambda T)^n}X, \quad i = 0,\dots,n-1. \tag{14.16}$$

It follows that $Z \neq 0$ if and only if $X \neq 0$. Furthermore, (14.15) implies that $Xe^{\lambda t}$ satisfies (14.9) for all $t \in \mathbb{R}$. The function $Xe^{\lambda t}$, $t \in \mathbb{R}$, is a solution since $\Re(\lambda) \geq 0$, and thus $Xe^{\lambda t}$, $t \leq 0$, belongs to $\mathscr{F}((-\infty,\, 0], \mathbb{C}^{p\times 1})$. □

For the system (14.14) a well-established stability theory exists [151, 93]. For instance, its zero solution is asymptotically stable if and only if all the roots of its characteristic equation,

$$\det(\lambda I - \bar{A} - \bar{B}e^{-\lambda\tau}) = 0 \tag{14.17}$$

or, equivalently,

$$\det\left(\begin{bmatrix} \lambda I & -Ae^{-\lambda\tau} \\ -I & (1+\lambda T)^n \end{bmatrix}\right) = 0, \tag{14.18}$$

are in $\mathbb{C}_-$. Note that (14.18) reduces to (14.10) if $\lambda \neq -1/T$. Combining this result with Proposition 14.2 results in:

Proposition 14.3. *The zero solution of* (14.9) *is asymptotically stable if and only if all roots of* (14.10) *are in* $\mathbb{C}_-$.

Next, we derive conditions on the roots of (14.18) for which the system (14.9) solves a consensus problem. This stability property is defined in the following way:

Definition 14.4. *The system* (14.9) *solves a consensus problem if and only if*

$$\forall \phi \in \mathscr{F}((-\infty, 0], \mathbb{C}^{p\times 1}) \ \lim_{t\to\infty} x(\phi)(t) = \chi(\phi)\, E_0,$$

where $\chi(\phi) \in \mathbb{C}$ *and* $E_0 = [1 \cdots 1]^T$. *The function* $\chi : \mathscr{F}((-\infty, 0], \mathbb{C}^{p\times 1}) \to \mathbb{C}$ *is called the consensus functional. The system* (14.9) *solves a nontrivial consensus problem if and only if it solves a consensus problem and the consensus functional is not identically zero.*

We follow the same methodology as for the asymptotic stability condition: we first address a consensus problem for a system with a constant delay, and next we treat (14.9) using Lemma 14.2:

Lemma 14.5. *The system*

$$\dot{x}(t) = A_0 x(t) + A_1 x(t-\tau), \tag{14.19}$$

with initial condition $\phi \in \mathscr{C}([-\tau, 0], \mathbb{C}^{p\times 1})$, *solves a nontrivial consensus problem if and only if all its characteristic roots are in the open left half plane, except for a zero root with multiplicity one, and* $(A_0 + A_1)E_0 = 0$, *with* $E_0 = [1 \cdots 1]^T$. *The consensus functional* $\chi : \mathscr{C}([-\tau, 0], \mathbb{C}^{p\times 1}) \to \mathbb{C}$ *can be expressed as*

$$\chi(\phi) = \frac{V_0^T \left(x(\phi)(\hat{t}) + A_1 \int_{\hat{t}-\tau}^{\hat{t}} x(\phi)(\theta)\, d\theta \right)}{V_0^T (I + \tau A_1) E_0}, \tag{14.20}$$

where V_0 *is the left null vector of* $(A_0 + A_1)$ *and*

$$\hat{t} \geq p\tau - \limsup_{r\to\infty} \frac{\log F(r)}{r}, \qquad F(r) = \max_{|\lambda|=r} \det\left(\lambda I - A - Be^{-\lambda\tau}\right).$$

If, in addition, $A_0 = 0$, *then the consensus functional is given by*

$$\chi(\phi) = \frac{V_0^T \phi(0)}{V_0^T E_0}. \tag{14.21}$$

Proof. The first assertion is a trivial corollary of the spectrum determined growth property of the solutions of (14.19); see, e.g., [151, 93]. The assertions regarding the form of the consensus functional follow from a spectral decomposition of the solutions, using Theorem 8.4 of [93]. In such a decomposition the stationary term corresponds to the consensus functional and can be isolated by employing orthogonality properties of left and right eigenfunctions. See [252] for details. □

Proposition 14.6. *The system* (14.9) *solves a nontrivial consensus problem if and only if all roots of* (14.10) *are in the open left half plane, except for a root at zero with multiplicity one, and* $AE_0 = 0$, *with* $E_0 = [1 \cdots 1]^T$. *The corresponding consensus functional* $V : \mathscr{F}((-\infty, 0], \mathbb{C}^{p\times 1}) \to \mathbb{C}$ *satisfies*

$$\chi(\phi) = \frac{V_0^T \phi(0)}{V_0^T E_0}, \tag{14.22}$$

where V_0 *is the left eigenvector of* A *corresponding to the zero eigenvalue.*

Proof. The first assertion is based on Lemma 14.2. Expression (14.22) is obtained by simplifying the consensus functional for the corresponding comparison system (14.12), to which Proposition 14.6 is applicable. See [252] for details. □

Remark 14.7. *Expressions* (14.21) *and* (14.22) *also follow from a geometric argument. As in both cases* $V_0^T \dot{x}(t) = 0$ *the solutions* $x(\phi)(t)$ *are constrained to the plane* $V_0^T x = V_0^T \phi(0)$ *for all* $t \geq 0$. *Furthermore, a constant stationary solution must be a multiple of* E_0. *Thus* $x^*(\phi) = \lim_{t\to\infty} x(\phi)(t)$ *satisfies the equations*

$$\begin{cases} V_0^T x^*(\phi) = V_0^T \phi(0), \\ x^*(\phi) = \chi(\phi) E_0, \end{cases}$$

which can be interpreted as the intersection of the plane through $\phi(0)$ *and perpendicular to* V_0 *with a line with slope* E_0. *A similar argument was used in Section* X *of* [312].

14.3 ▪ Conditions for the realization of a consensus

We perform a stability analysis of the system (14.2)–(14.5) in the (T, τ) parameter space. In particular, we give necessary and sufficient conditions such that a consensus is reached for all initial conditions. We present only the main results. For detailed proofs we refer the reader to [252].

The system (14.2)–(14.5) can be written in the form (14.9), yet has some special properties due to the induced structure of A, which we outline first. Next, we present the main results.

14.3.1 ▪ Prerequisites

The system (14.2)–(14.5) is of the form (14.9), where $A = [a_{k,l}]$ is defined as

$$a_{k,l} = \begin{cases} \alpha_{k,l}, & k \neq l, \\ -\sum_{i=1,\, i\neq k}^{p} \alpha_{k,i}, & k = l. \end{cases} \tag{14.23}$$

Note that in the context of multiagent systems $-A$ is typically called the graph Laplacian of (14.7). By construction A has the following property:

Property 14.8. *All eigenvalues of A, defined by* (14.23), *are in* $\mathbb{C}_-$, *except for a zero eigenvalue with multiplicity one.*

Note that zero also appears as a root of (14.10) and (14.23), whatever the values of T, τ, and n. If all other roots are in $\mathbb{C}_-$, we have from Proposition 14.6 that the system (14.2)–(14.5) solves a (nontrivial) consensus problem with delay. In the car following application the consensus variables are the speed of the vehicles. This means that, whatever the initial values, the speed of the vehicles will eventually converge to a common value (which depends on the initial values).

In what follows we shall use the following terminology to characterize parameter values in the (T,τ) space for which a consensus is reached:

Definition 14.9. *The consensus region of* (14.2)–(14.5) *in the* (T,τ) *parameter space is the set of parameters* (T,τ) *for which the system* (14.2)–(14.5) *solves a consensus problem.*

Property 14.8 and the factorization of the characteristic equation (14.10) as (14.11) lead us to the study of the zeros location of the function

$$\xi(\lambda;\, T,\tau) := \lambda(1+\lambda T)^n e^{\lambda\tau} - \mu, \quad \mu \in \mathbb{C}_-, \tag{14.24}$$

as a function of the parameters T and τ. We have the following result, which can be proven using the two-step approach of Chapter 6 consisting of first characterizing the crossing frequencies, where zeros can cross the imaginary axis as the parameters are changed, and, second, characterizing the corresponding crossing direction:

Proposition 14.10. *If* μ *is real and* $n=1$, *then the zeros of* (14.24) *are in* $\mathbb{C}_-$ *if and only if* $T\in[0,\ \infty)$ *and* $\tau\in[0,\ \tau_\mu(T))$. *Otherwise, the zeros are in* $\mathbb{C}_-$ *if and only if* $T\in[0,\ T_\mu)$ *and* $\tau\in[0,\ \tau_\mu(T))$.

14.3.2 ▪ Computation of stability regions

Taking into account the factorization of (14.10) as (14.11), Property 14.8, and Proposition 14.10, we obtain the following characterization of the consensus region (cf. Definition 14.9) of the system (14.2)–(14.5) in the (T,τ) space:

Theorem 14.11. *If* $n=1$ *and all eigenvalues of A, defined by* (14.23), *are real, then the consensus region of* (14.2)–(14.5) *in the* (T,τ) *plane is unbounded and characterized by*

$$T\in[0,\infty), \quad \tau\in[0,\tau^*(T)),$$

where

$$\tau^*(T) = \min_{k=1,\dots,p,\ \mu_k\neq 0} \frac{|\angle(\mu_k)| - \angle(j\omega_k(T)(1+j\omega_k(T))^n)}{\frac{\omega_k(T)}{T}} \tag{14.25}$$

and $\omega_k(T) = g_n(T|\mu_k|)$.

Otherwise, the consensus region is bounded and characterized by

$$T\in[0,T^*), \quad \tau\in[0,\tau^*(T)),$$

where

$$T^* = \min_{k=1,\ldots,p,\ \Im(\mu_k)>0} \frac{\tan\left(\dfrac{\angle(\mu_k)-\frac{\pi}{2}}{n}\right)}{|\mu_k|\left[\cos\left(\dfrac{\angle(\mu_k)-\frac{\pi}{2}}{n}\right)\right]^n} \tag{14.26}$$

and $\tau^*(T)$ *is given by* (14.25).

Based on this result the consensus region of (14.2)–(14.5) can be computed fully *automatically*. For large p the overall computational complexity is determined by the computation of the eigenvalues of the p-by-p matrix A.

Theorem 14.11 does not make assumptions on the multiplicity of the eigenvalues of A and is generally applicable. If A has eigenvalues with multiplicity larger than one, then the stability study of (14.2)–(14.5) is even facilitated, as not all factors in (14.11) are different. The following proposition clarifies the connection between multiple eigenvalues of A and multiple eigenvalues of the comparison system (14.14) of (14.2)–(14.5):

Proposition 14.12. *Let* $\hat{\mu}$ *be a nonzero eigenvalue of* A *with multiplicity* m_1 *and a corresponding eigenspace of dimension* m_2*. Then the roots of*

$$\lambda(1+\lambda T)^n e^{\lambda\tau} - \hat{\mu} = 0 \tag{14.27}$$

with multiplicity m_3 *are eigenvalues of the comparison system* (14.14) *with multiplicity* $m_1 m_3$ *and an eigenspace of dimension* m_2*. Furthermore, if* $m_3 = 1$*, then these roots smoothly depend on the parameters* T *and* τ*.*

Remark 14.13. *If* $m_1 > m_2$*, then the roots of* (14.27) *with multiplicity one (this is, for instance, always the case for roots on the imaginary axis) are multiple, nonsemisimple eigenvalues of* (14.14)*, yet they smoothly depend on the parameters* T *and* τ*. Small changes of* T *and* τ *do not lead to a splitting of these multiple eigenvalues.*

The next proposition reveals a scaling property of the consensus region:

Proposition 14.14. *If the matrix* A *is scaled with a factor* $\epsilon > 0$*, then the consensus region of* (14.2)–(14.5) *in the* (T,τ) *plane is scaled with a factor* ϵ^{-1} *in both directions.*

Remark 14.15. *Proposition* 14.14 *implies an inherent trade-off between the rate with which the undelayed system* ($\tau = T = 0$) *reaches a consensus (determined by the rightmost nonzero eigenvalue of* A*) and the robustness of this stability property w.r.t. delays. Such an observation was already made in* [312]*, where the case of a symmetric matrix* A *and a pointwise delay was dealt with.*

In the remainder of this section, we refine Theorem 14.11 to two special cases where exploiting the additional structure leads to a *simpler* characterization of the consensus region, and also allows an *analytical* expression for the solutions corresponding to an onset of instability. The latter can be obtained from the eigenfunctions corresponding to the characteristic roots on the imaginary axis (which are the eigenvalues of the infinitesimal generator of the evolution operator associated with the system; see Chapter 1). The following result corresponds to the situation where all cars/drivers have identical behavior and the reaction of a driver is determined only by the preceding car:

Proposition 14.16. *Consider the system* (14.2)–(14.5), *where*

$$\alpha_{k,l} = \begin{cases} \alpha, & \lfloor k-l \rfloor = 1, \\ 0 & \textit{otherwise}. \end{cases} \tag{14.28}$$

If $n=1$ and $p=2$, then the consensus region in the (T,τ)-plane is unbounded and characterized by

$$T \in [0,\infty), \quad \tau \in [0,\tau^*(T)),$$

where

$$\tau^*(T) = \frac{\frac{\pi}{p} - n\arctan(\omega(T))}{\frac{\omega(T)}{T}}, \qquad \omega(T) = g_n\left(2\alpha T \sin\left(\frac{\pi}{p}\right)\right). \tag{14.29}$$

Otherwise, the consensus region is bounded and characterized by

$$T \in [0,T^*), \quad \tau \in [0,\tau^*(T)),$$

where

$$T^* = \frac{\tan\left(\frac{\pi}{pn}\right)}{2\alpha \sin\left(\frac{\pi}{p}\right)\left(\cos\left(\frac{\pi}{pn}\right)\right)^n} \tag{14.30}$$

and $\tau^*(T)$ is given by (14.29).

For $\tau = \tau^*(T)$ the stationary solutions are backwards traveling waves:

$$\begin{bmatrix} v_1^s(t) \\ v_2^s(t) \\ \vdots \\ v_p^s(t) \end{bmatrix} = C_1 \begin{bmatrix} \cos\left(\frac{\omega(T)}{T}t + \varphi\right) \\ \cos\left(\frac{\omega(T)}{T}t + \varphi - \frac{2\pi}{p}\right) \\ \vdots \\ \cos\left(\frac{\omega(T)}{T}t + \varphi - \frac{2\pi(p-1)}{p}\right) \end{bmatrix} + C_2 \begin{bmatrix} 1 \\ 1 \\ \vdots \\ 1 \end{bmatrix}, \tag{14.31}$$

where $\omega(T)$ is defined as in (14.29) *and the constants C_1, C_2, and ϕ depend on the initial conditions.*

Second, we consider the case where (14.2)–(14.5) is of the form (14.9), with the matrix A symmetric. Although this is not a realistic assumption from point of view of the car *following* application, it makes sense in the context of consensus algorithms for multiagent systems. The symmetry of A there corresponds to an *undirected* network topology.

Proposition 14.17. *Consider the system* (14.2)–(14.5) *with A symmetric. If $n=1$, then the consensus region of* (14.2)–(14.5) *in the (T,τ)-plane is unbounded and characterized by*

$$T \in [0,\infty), \quad \tau \in [0,\tau^*(T)),$$

where

$$\tau^*(T) = \frac{\frac{\pi}{2} - n\arctan(\omega(T))}{\frac{\omega(T)}{T}}, \qquad \omega(T) = g_n(T|\lambda_{\max}(A)|). \tag{14.32}$$

Otherwise, the consensus region is bounded and characterized by

$$T \in [0,T^*), \quad \tau \in [0,\tau^*(T)),$$

where

$$T^* = \frac{\tan\left(\frac{\pi}{2n}\right)}{|\lambda_{\max}(A)|\left[\cos\left(\frac{\pi}{2n}\right)\right]^n} \tag{14.33}$$

and $\tau^*(T)$ *is given by* (14.32).

If, in addition,

$$\alpha_{k,l} = \begin{cases} \alpha, & \lfloor k-l \rfloor = 1 \text{ or } \lfloor l-k \rfloor = 1, \\ 0 & \textit{otherwise}, \end{cases} \tag{14.34}$$

then

$$\lambda_{\max}(A) = \begin{cases} -4\alpha \ (\textit{multiplicity } 1), & p \textit{ even}, \\ -2\alpha\left(1+\cos\left(\frac{\pi}{p}\right)\right) \ (\textit{multiplicity } 2), & p \textit{ odd}. \end{cases}$$

The stationary solutions for $\tau = \tau^*(T)$ *take the form*

$$\begin{bmatrix} v_1^s(t) \\ \vdots \\ v_{p-1}^s(t) \\ v_p^s(t) \end{bmatrix} = C_1 \begin{bmatrix} (-1)^{p-1} \\ \vdots \\ (-1) \\ 1 \end{bmatrix} \cos\left(\frac{\omega(T)}{T} t + \varphi_1\right) + C_2 \begin{bmatrix} 1 \\ \vdots \\ 1 \\ 1 \end{bmatrix} \tag{14.35}$$

if p is even, and

$$\begin{aligned} \begin{bmatrix} v_1^s(t) \\ \vdots \\ v_{p-1}^s(t) \\ v_p^s(t) \end{bmatrix} &= C_3 \begin{bmatrix} (-1)^{p-1}\cos\left(\frac{\pi(p-1)}{p}\right) \\ \vdots \\ (-1)\cos\left(\frac{\pi.1}{p}\right) \\ 1 \end{bmatrix} \cos\left(\frac{\omega(T)}{T} t + \varphi_2\right) \\ &+ C_4 \begin{bmatrix} (-1)^{p-1}\sin\left(\frac{\pi(p-1)}{p}\right) \\ \vdots \\ (-1)\sin\left(\frac{\pi.1}{p}\right) \\ 0 \end{bmatrix} \cos\left(\frac{\omega(T)}{T} t + \varphi_3\right) + C_5 \begin{bmatrix} 1 \\ \vdots \\ 1 \\ 1 \end{bmatrix} \end{aligned} \tag{14.36}$$

if p is odd. The constants $C_1, \ldots, C_5$ *and* $\varphi_1, \ldots, \varphi_3$ *depend on the initial conditions.*

Remark 14.18. *The consensus functional satisfies*

$$V(\phi) = \frac{1}{p}\,[1 \cdots 1]\phi(0).$$

This follows from (14.6), *taking into account that* $V_0 = E_0$ *if A is symmetric. Hence, under the conditions of the above proposition an* average *consensus problem is solved, in the sense that all components of a solution* $x(\phi)(t)$ *converge to the average of these components at the starting time, i.e.,* $\phi(0)$. *Note that* $\phi(\theta)$, $\theta < 0$, *has* no *influence on the limit reached.*

Remark 14.19. *Expression* (14.33) *reduces to the statement of Theorem* 10 *in* [312] *if* $T \to 0+$.

Let us briefly compare the stationary solutions (14.31) with (14.35)–(14.36). In the former case, the directed "network topology" (a driver only reacts—with some delay—on its predecessor and not the other way around) naturally leads to a backwards traveling wave. In the latter case one would, from the symmetry of the coupling, intuitively expect a stationary wave, where subsequent agents oscillate in antiphase. This is indeed the case for (14.35), which holds if p is even. However, if p is odd, such a solution is *incompatible* with the ring configuration, and (14.36) holds. If p is large, (14.36) can be seen as an approximation of a stationary wave with subsequent agents oscillating in antiphase that is compatible with the ring configuration.

14.4 ▪ Examples

As a first example we compute the consensus regions in the (T, τ)-plane of system (14.2)–(14.5) with $n = 1$ and

$$A = \begin{bmatrix} -5 & 0 & 0 & 5 \\ 1 & -1 & 0 & 0 \\ 0 & 1 & -1 & 0 \\ 0 & 0 & 5 & -5 \end{bmatrix}. \qquad (14.37)$$

The eigenvalues of this matrix are given by

$$\mu_1 = -6, \quad \mu_2 = \bar{\mu}_3 = -3 + j, \quad \mu_4 = 0.$$

An application of Theorem 14.11 yields the consensus region

$$T \in [0, 3), \quad \tau \in [0, \tau^*(T));$$

the function $T \to \tau^*(T)$ is displayed in Figure 14.1 as a solid curve. The dotted curves bound the "stability" regions of the auxiliary equations

$$\lambda(1 + \lambda T)e^{\lambda\tau} - \mu_{1,2} = 0, \qquad (14.38)$$

which are described by Proposition 14.10. The stability region corresponding μ_1 is unbounded as μ_1 is real and $n = 1$.

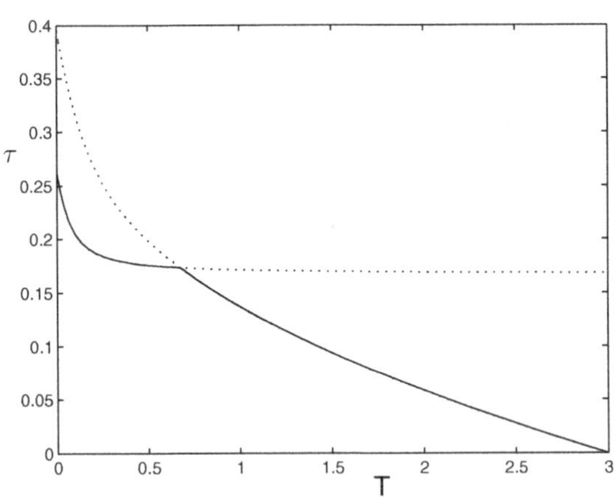

Figure 14.1. *Boundary of the consensus regions of* (14.2)–(14.5) *with parameters* (14.37) *(solid curve). Boundaries of stability regions of* (14.38) *(dotted curves)* [252].

To illustrate the asymptotic behavior when the number of cars is large, we take a system satisfying condition (14.28) of Proposition 14.16. Figure 14.2 shows the consensus region in the (T, τ)-plane for $\alpha = 2$, $n = 1$, and $p = 2^k$, $k = 1, \ldots, 4$. It it follows from

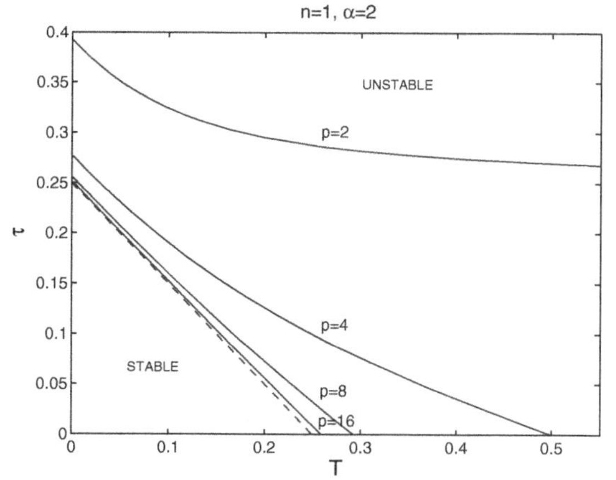

Figure 14.2. *Boundary of the consensus region of a system satisfying* (14.28), *with parameters* $\alpha = 2$ *and* $n = 1$ [252].

(14.29) that as $p \to \infty$, the boundary of the consensus region uniformly converges to the function

$$\tau_l^*(T) = \frac{1}{2\alpha} - nT,$$

indicated in Figure 14.2 with a dashed line.

Finally, we consider the system

$$\dot{v}_k(t) = \sum_{l=1}^{3} \alpha_{k,\lfloor k-l \rfloor} \int_0^\infty f(\theta)(v_{\lfloor k-l \rfloor}(t-\theta) - v_k(t-\theta))\mathrm{d}\theta, \quad k = 1,\ldots,1000, \quad (14.39)$$

where f is given by (14.5), with $n = 2$. The parameters

$$\begin{aligned} &\alpha_{k,\lfloor k-1 \rfloor} \in [1,\ 5], \\ &\alpha_{k,\lfloor k-2 \rfloor} \in \left[0,\ \tfrac{3}{4}\alpha_{k,\lfloor k-1 \rfloor}\right], \\ &\alpha_{k,\lfloor k-3 \rfloor} \in \left[0,\ \tfrac{3}{4}\alpha_{k,\lfloor k-2 \rfloor}\right], \quad k = 1,\ldots,1000, \end{aligned} \quad (14.40)$$

are randomly generated according to a *uniform distribution* over the above intervals. For 30 sets of parameters obtained in this way, the consensus region in the (T, τ)-plane was computed. The results are displayed in Figure 14.3.

14.5 ▪ Other models

For general time-delay systems of retarded type with multiple constant delays and distributed delays with gamma-distribution kernels, stability and/or consensus regions of equilibria in a two-parameter spaces can be computed semiautomatically by numerical continuation; see, for instance, [218] and the package DDE-BIFTOOL [100]. Such an approach involves the discretization of an infinite-dimensional evolutionary operator, associated with the time-delay system, to compute the rightmost eigenvalues. Roughly speaking, the computation of the boundary of a stability or consensus region involves solving r eigenvalue problems of dimension $pq \times pq$, where p is the dimension of the system, q denotes the number of discretization points, and r is the number of points on the stability crossing curves where stability information is checked. When using Theorem 14.11

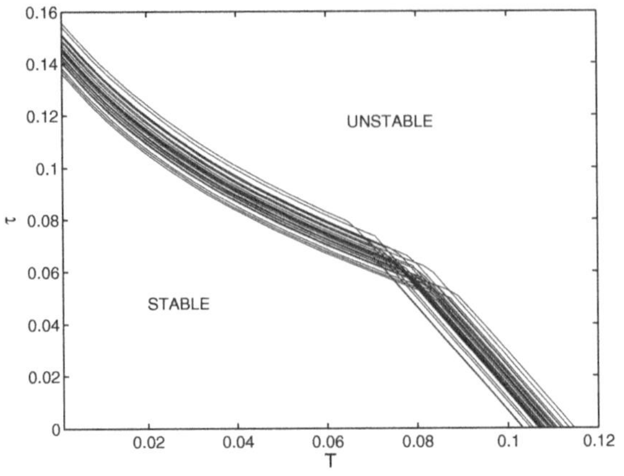

Figure 14.3. *Consensus region of* (14.39)–(14.40) *for* 30 *different data sets* [252].

only, one eigenvalue problem of dimension $p \times p$ needs to be solved to determine the complete stability region of (14.2)–(14.5) in the (T, τ)-space. The underlying reason is that the structure of the system allowed a decomposition into small subproblems, which is apparent from the form of the characteristic equation (14.11).

Let us now take a brief look at the so-called optimal velocity models, also frequently encountered in the literature. The linearization around the equilibrium of the models studied in [15], respectively [90, 315, 316] and the references therein, takes the form

$$\tau_k \ddot{x}_k(t) + \dot{x}_k(t-\tau) = \alpha(x_{\lfloor k-1 \rfloor}(t-\tau) - x_k(t-\tau)), \quad k = 1, \ldots, p, \tag{14.41}$$

respectively,

$$\tau_k \ddot{x}_k(t) + \dot{x}_k(t) = \alpha(x_{\lfloor k-1 \rfloor}(t-\tau) - x_k(t-\tau)), \quad k = 1, \ldots, p. \tag{14.42}$$

In both cases, x_k denotes the position of the kth vehicle. The left-hand side models the dynamics of the vehicle and the right-hand side is the reference velocity, which is a function of the distance to the preceding vehicle and models the behavior of the driver. Note that (14.41) and (14.42) can be generalized to

$$\begin{aligned} \ddot{x}_k(t) = \int_0^\infty f(\theta) \Bigg(\Bigg(\sum_{i=1}^{p-1} \alpha_{k,\lfloor k-i \rfloor} (x_{\lfloor k-i \rfloor}(t-\theta) - x_k(t-\theta)) \Bigg) \\ - \beta_k \dot{x}_k(t-\theta) \Bigg) \mathrm{d}\theta - \gamma_k \dot{x}_k(t), \; k = 1, \ldots, p, \end{aligned} \tag{14.43}$$

with f given by (14.5). General purpose tools like DDE-BIFTOOL for the stability and bifurcation analysis of time-delay systems can be applied directly to (14.43). However, if all vehicles (but not necessarily the drivers) have similar characteristics, that is, $\alpha_k \equiv \alpha, \beta_k \equiv \beta$, then the characteristic equation can again be factorized as

$$0 = \det\left(\lambda(\lambda+\gamma)I - (A - \beta\lambda I)\frac{e^{-\lambda\tau}}{(1+\lambda T)^n} \right) = \Pi_{k=1}^p \left(\lambda(\lambda+\gamma) - \frac{(\mu_k - \beta\lambda)e^{-\lambda\tau}}{(1+\lambda T)^n} \right),$$

where A is given by (14.23) and $\mu_1, \ldots \mu_p$ denote its eigenvalues. Also here, it is beneficial to exploit this decomposition into small subproblems, in particular if the number of vehicles is large.

14.6 ▪ Notes and references

The stability analysis of a linear system including a γ-distributed delay with a gap for modeling traffic flow dynamics was considered. A complete characterization of the regions in the delay-parameter space, where a consensus is reached for all initial conditions, was obtained. In particular, by exploiting the structure of the system, analytical expressions were derived for the bounds on the parameters of the delay distribution. These expressions give rise to a fully automatic computation of the consensus region, whose complexity is determined by the computation of the eigenvalues of one matrix with dimensions equal to the number of vehicles. Some illustrative examples were also presented.

From a theoretical point of view, some stability theory for linear systems with γ-distributed delays was developed. As this type of distributed delays is characterized by kernels with an infinite support, which prohibits a full spectral decomposition of the solutions, the relation between the growth properties of the solutions and the roots of an appropriate characteristic equations was established via a comparison system with constant delays. Necessary and sufficient conditions for the realization of a consensus problem and an explicit construction of the consensus functional were provided for both systems with constant and distributed delays.

The results presented in this chapter are based on [252, 278] and the references therein.

Chapter 15

Synchronization of delay-coupled oscillators

15.1 ▪ Introduction

Synchronization is an important problem in the study of coupled dynamical systems, motivated by applications in science and engineering. In this chapter we consider the synchronization problem in networks of identical nonlinear oscillators, where the signal exchanges are affected by time delays. The main interest lies in gaining insight into the occurrence of synchronized behavior and in studying its dependence on parameters of the coupling (coupling strength, delay) and on the topology of the network. Because synchronization is a notion of relative stability, the existing methods for the stability analysis of nonlinear time-delay systems can be directly applied or adapted to the synchronization problem, possibly combined with a boundedness argument of the solutions. Such methods are almost exclusively time-domain–based and heavily rely on an appropriately chosen Lyapunov function(al). Although they are applicable to the study of complex synchronized behavior (e.g., synchronized chaotic behavior), the resulting conditions for synchronization are typically in the form of sufficient, yet not necessary, conditions, from which it may be hard to extract qualitative insight into the synchronization mechanisms. Inspired by this observation, a distinct approach is taken which builds on the stability and bifurcation analysis of a special type of synchronized solutions for which necessary and sufficient stability conditions can be obtained, namely synchronized equilibria.

Specifically, we consider p identical nonlinear oscillators described by

$$\begin{aligned} \dot{x}_i(t) &= f(x_i) + Bu_i(t), \\ y_i(t) &= Cx_i(t), \qquad i = 1,\dots,p, \end{aligned} \tag{15.1}$$

where $x_i \in \mathbb{R}^n$, $i = 1,\dots,p$, $B, C \in \mathbb{R}^{n\times 1}$ and $f : \mathbb{R}^n \mapsto \mathbb{R}^n$ is twice continuously differentiable. We further assume that for $u_i = 0$ the system (15.1) has at least one unstable equilibrium of focus type, which we denote by x^* in what follows.

In order to describe the coupling between the oscillators, we define a directed graph

$$\mathcal{G}(\mathcal{V}, G), \tag{15.2}$$

characterized by the node set $\mathcal{V} = \{1,\dots,p\}$ and a weighted adjacency matrix G with zero diagonal entries and nondiagonal entries equal to $\alpha_{i,l} \geq 0$. The corresponding edge set $\mathcal{E}$ satisfies $(i,l) \in \mathcal{E}$ if and only if $\alpha_{i,l} \neq 0$. Next, we couple the systems (15.1) by means of the "control" law

$$u_i(t) = k\left(\sum_{(i,l)\in\mathcal{E}} \alpha_{i,l}\,(y_l(t-\tau) - y_i(t))\right), \quad i = 1,\dots,p, \tag{15.3}$$

where $k > 0$ represents the gain parameter and τ the transmission delay. We assume the transmission delay to be fixed and independent of the nodes. It is important to point out that we do not assume that G is symmetric.

The aim of this chapter is to study the effect of the coupling (15.3), with k and τ as parameters, on the synchronization of the systems (15.1), and to reveal synchronization mechanisms and conditions. Hereto a complete characterization will be made of the stability-instability regions of the synchronized equilibrium $(x^*,\dots,x^*)$ of (15.1) and (15.3) in the (k,τ) parameter space. Note that achieving stability can be seen as an extension of the use of Pyragas-type feedback [329] to stabilize an unstable equilibrium as in [169], though, in contrast to the original idea of the Pyragas feedback, the time-delay τ is not linked to any possible periodicity in the system. It can also be interpreted as a situation where so-called oscillator death is achieved [11]. The proposed approach to compute stability regions in the (k,τ) parameter space is inspired by Chapter 5, where an overview of methods for computing stability regions in parameter spaces of linear control systems with delays is presented, yet not in the context of networks of interconnected systems. Beyond the stability analysis of (synchronized) equilibria, the goal is to gain insight into and reveal explanations for the occurrence of more complex synchronized behavior.

A motivation and overview of synchronization problems and results can be found in [363, 407]. An overview of the available results on synchronization problems with coupling delays is presented in Section II of [311]. As is apparent from this overview, most existing results assume a coupling similar to (15.3), but where also the output $y_i(t)$ is delayed over an interval of length τ. This situation is technically more tractable; however, it is less realistic from an application point of view. A control law of the form (15.3) is employed in [172, 310, 311]. In [172] the anticipative synchronization problem of two systems is considered. In [310] a symmetric network of four systems is analyzed. In [311] the synchronization problem in general networks with delays is considered. Necessary conditions on the network topology and delays for the existence of synchronized solutions are presented, and sufficient conditions for asymptotic synchronization are stated. However, no qualitative analysis is performed, and a different methodology is taken, relying on time-domain methods and, in particular, Lyapunov functionals. Synchronization in general networks is also addressed in [24, 325] and the references therein, but without taking into account delay effects. The latter effects are, however, crucial. Finally, in [51] a complete bifurcation analysis of a model for three interacting neurons with delay in the interconnections is presented, under the assumption of a ring configuration and bidirectional coupling.

The structure of the chapter is as follows. In Section 15.2 some preliminary results are presented, which include necessary conditions for the occurrence of synchronized solutions of (15.1) and (15.3). In Section 15.3 a linearized stability analysis of synchronized equilibria is performed. In Section 15.4 the results are applied to networks of coupled Lorenz systems and complemented with a nonlinear stability analysis. Particular attention will be paid to the asymptotic behavior of stability properties for large values of the gain parameter and on the derivation of generic results that do not depend on the network topology and the number of oscillators. Finally, some extensions are discussed in Section 15.5. The conclusions are formulated in Section 15.6.

15.2 ▪ Preliminaries

15.2.1 ▪ Assumptions on the graph

The following assumptions are made throughout the chapter:

Assumption 15.1. *The graph $\mathcal{G}$ is strongly connected.*

Assumption 15.2. *The adjacency matrix G satisfies*

$$\sum_{l=1}^{p} \alpha_{i,l} = 1, \quad i = 1, \ldots, p.$$

The first assumption is natural in the context of synchronization with delayed coupling; the second assumption will be motivated in Section 15.2.2. The following results are direct corollaries.

Corollary 15.3. *G has a simple eigenvalue equal to one, and $[1 \ \cdots \ 1]^T$ is the corresponding eigenvector. Furthermore, there exists a vector $\gamma = [\gamma_1 \ \cdots \gamma_p]^T$ such that*

$$\gamma_l > 0, \ 1 \leq l \leq p, \quad \sum_{l=1}^{p} \gamma_l = 1, \quad \gamma^T(G - I) = 0.$$

Corollary 15.4. *All eigenvalues of G have modulus less than or equal to one.*

In what follow we denote the eigenvalues of G as $\lambda_i(G)$, $1 \leq i \leq p$, where we take the convention

$$\lambda_1(G) = 1.$$

15.2.2 ▪ A coordinate transformation

Define the matrix $\tilde{G}$:

$$\tilde{G} = \begin{bmatrix} 0 & \alpha_{2,3} & \alpha_{2,4} & \cdots & \alpha_{2,p} \\ \alpha_{3,2} & 0 & \alpha_{3,4} & & \alpha_{3,p} \\ & & & \ddots & \\ \alpha_{p,2} & \alpha_{p,3} & \cdots & \alpha_{p,p-1} & 0 \end{bmatrix} - \begin{bmatrix} 1 \\ 1 \\ \vdots \\ 1 \end{bmatrix} \begin{bmatrix} \alpha_{1,2} & \alpha_{1,3} & \cdots & \alpha_{1,p} \end{bmatrix}, \quad (15.4)$$

which satisfies the following property:

Property 15.5. $\sigma(\tilde{G}) = \sigma(G) \setminus \{1\}$.

By means of the new variables

$$\begin{cases} e_2(t) &= x_2(t) - x_1(t) \\ & \vdots \\ e_p(t) &= x_p(t) - x_1(t) \end{cases},$$

we can bring (15.1) and (15.3) into the form

$$\left\{\begin{array}{l} \dot{x}_1(t)=f(x_1(t))+kBC\left(\sum_l \alpha_{1,l}\right)(x_1(t-\tau)-x_1(t))+kBC\sum_l \alpha_{1,l}\, e_l(t-\tau),\\ \begin{bmatrix} \dot{e}_2(t)\\ \vdots \\ \dot{e}_p(t)\end{bmatrix}=\begin{bmatrix} f(x_1+e_2)-f(x_1)\\ \vdots \\ f(x_1+e_p)-f(x_1)\end{bmatrix}-k\left(\begin{bmatrix} \sum_l \alpha_{2,l} & & \\ & \ddots & \\ & & \sum_l \alpha_{p,l}\end{bmatrix}\otimes BC\right)\begin{bmatrix} e_2(t)\\ \vdots \\ e_p(t)\end{bmatrix}\\ \qquad +k\,\tilde{G}\otimes BC^T\begin{bmatrix} e_2(t-\tau)\\ \vdots \\ e_p(t-\tau)\end{bmatrix}+\left(k\begin{bmatrix} \sum_l \alpha_{1,l}-\sum_l \alpha_{2,l}\\ \sum_l \alpha_{1,l}-\sum_l \alpha_{3,l}\\ \vdots \\ \sum_l \alpha_{1,l}-\sum_l \alpha_{p,l}\end{bmatrix}\otimes BC\right)(x_1(t)-x_1(t-\tau)). \end{array}\right. \tag{15.5}$$

From this equation it can be seen that a synchronized solution, characterized by

$$e_2 \equiv 0,\ldots,e_p \equiv 0,$$

can exist only in three cases:

1. the delay is equal to zero;
2. the overall motion is τ-periodic;
3. Assumption 15.2 holds.

Because we are primarily interested in explaining complex synchronized behavior in the presence of delays in the coupling, we can take Assumption 15.2 without losing generality, and equations (15.5) simplify to

$$\dot{x}_1 = f(x_1(t))+kBC(x_1(t-\tau)-x_1(t))+kBC\sum_{l=1}^{p}\alpha_{1,l}\, e_l(t-\tau), \tag{15.6}$$

$$\begin{bmatrix} \dot{e}_2\\ \vdots \\ \dot{e}_p\end{bmatrix}=\begin{bmatrix} f(x_1+e_2)-f(x_1)-kBCe_2\\ \vdots \\ f(x_1+e_p)-f(x_1)-kBCe_p\end{bmatrix}+k\,\tilde{G}\otimes BC\begin{bmatrix} e_2(t-\tau)\\ \vdots \\ e_p(t-\tau)\end{bmatrix}. \tag{15.7}$$

The solutions on the synchronization manifold are characterized by

$$\dot{x}_1(t)=f(x_1(t))+kBC(x_1(t-\tau)-x_1(t)). \tag{15.8}$$

If all the solutions of the coupled system converge to a bounded forward invariant set, then the synchronization between the agents is achieved locally if the linearization of (15.7),

$$\begin{bmatrix} \dot{e}_2(t)\\ \vdots \\ \dot{e}_p(t)\end{bmatrix}=\begin{bmatrix} \left(\frac{\partial f}{\partial x}(x_1(t))-kBC\right)e_2(t)\\ \vdots \\ \left(\frac{\partial f}{\partial x}(x_1(t))-kBC\right)e_p(t)\end{bmatrix}+k\,\tilde{G}\otimes BC\begin{bmatrix} e_2(t-\tau)\\ \vdots \\ e_p(t-\tau)\end{bmatrix}, \tag{15.9}$$

is uniformly asymptotically stable. In order to simplify the analysis, we let R and I be defined as

$$R=\{i\in\{2,\ldots,p\}:\ \Im(\lambda_i(G))=0\},$$
$$I=\{i\in\{2,\ldots,p\}:\ \Im(\lambda_i(G))>0\},$$

and we let T_r be a matrix satisfying

$$T_r^{-1}\tilde{G}T_r = D,$$

where D is a block triangular matrix whose diagonal blocks are given by

$$\{\lambda_i(G):\ i\in R\}\cup\left\{\begin{bmatrix} \Re(\lambda_i(G)) & \Im(\lambda_i(G)) \\ -\Im(\lambda_i(G)) & \Re(\lambda_i(G)) \end{bmatrix}:\ i\in I\right\}.$$

The matrices T_r and D always exist by the identity $\sigma(D)=\sigma(\tilde{G})$ and Property 15.5.

From the state transformation induced by the matrix $(T_r\otimes I)$ it follows that its zero solution of (15.9) is uniformly asymptotically stable if the following equations are uniformly asymptotically stable:

$$\dot{\xi}_i(t)=\left(\frac{\partial f}{\partial x}(x_1(t))-kBC\right)\xi_i(t)+k\lambda_i(G)BC\xi_i(t-\tau)\ \ \forall i\in I, \tag{15.10}$$

$$\begin{aligned}\begin{bmatrix} \dot{\xi}_i \\ \dot{\eta}_i \end{bmatrix} &= I\otimes\left(\frac{\partial f}{\partial x}(x_1(t))-kBC\right)\begin{bmatrix} \xi_i \\ \eta_i \end{bmatrix} \\ &\quad + k\begin{bmatrix} \Re(\lambda_i(G)) & \Im(\lambda_i(G)) \\ -\Im(\lambda_i(G)) & \Re(\lambda_i G)) \end{bmatrix}\otimes BC\begin{bmatrix} \xi_i(t-\tau) \\ \eta_i(t-\tau) \end{bmatrix}\quad \forall i\in J.\end{aligned} \tag{15.11}$$

Equivalently, a full triangularization of $\tilde{G}$ results in the analysis of

$$\dot{\xi}_i(t)=\left(\frac{\partial f}{\partial x}(x_1(t))-kBC\right)\xi_i(t)+k\lambda_i(G)BC\xi_i(t-\tau)\quad \forall i=2,\dots,p, \tag{15.12}$$

at the price of some of the equations (15.12) being complex valued if $\tilde{G}$ has complex eigenvalues.

Remark 15.6. *The analysis of networks using the master stability function* [325, 66] *is based on a similar decomposition of the error dynamics. For* $\tau = 0$*, this function maps* $z\in\mathbb{C}$, $\Re(z)\le 0$ *to the largest Lyapunov exponent of*

$$\dot{\xi}_i = \frac{\partial f}{\partial x}(x_1(t))\xi_i(t)+zBC\xi_i(t). \tag{15.13}$$

Equation (15.13) *is related to* (15.12) *with* $\tau=0$ *via*

$$z=k\lambda_i(G-I).$$

Unlike the undelayed case considered in the literature, the dynamics on the synchronization manifold, governed by (15.8)*, depend on both* k *and* τ*, since the coupling is invasive if* $\tau\neq 0$*. Furthermore, in the presence of delay both parameters* k *and* $\lambda_i(G)$ *can no longer be simultaneously absorbed in the variable* z.

15.3 ▪ Stability analysis of synchronized equilibria

When we linearize the system (15.1) and (15.3) around the synchronized equilibrium $(x^*,\cdots x^*)$, we obtain

$$\begin{bmatrix} \dot{v}_1(t) \\ \vdots \\ \dot{v}_p(t) \end{bmatrix} = I\otimes(A-kBC)\begin{bmatrix} v_1(t) \\ \vdots \\ v_p(t) \end{bmatrix} + kG\otimes BC\begin{bmatrix} v_1(t-\tau) \\ \vdots \\ v_p(t-\tau) \end{bmatrix}, \tag{15.14}$$

where

$$A = \frac{\partial f}{\partial x}(x^*).$$

15.3.1 ▪ The characteristic equation

Factorization

The characteristic function of (15.14) is given by

$$f(\lambda;\, k,\tau) := \det F(\lambda;\, k,\tau),$$

where the characteristic matrix F is defined as

$$F(\lambda;\, k,\tau) = I \otimes (\lambda I - A + kBC) - G \otimes kBCe^{-\lambda\tau}. \tag{15.15}$$

If we factorize $G = T\Lambda T_c^{-1}$, where $\Lambda \in \mathbb{C}^{p\times p}$ is triangular and $T_c \in \mathbb{C}^{p\times p}$, the characteristic function becomes

$$\begin{aligned} f(\lambda;\, k,\tau) &= \det\left(I \otimes (\lambda I - A + kBC_c) - T_c\Lambda T_c^{-1} \otimes kBC_c e^{-\lambda\tau}\right) \\ &= \det\left(T_c^{-1} \otimes I\right)\det\Big(I \otimes (\lambda I - A + kBC) \\ &\qquad - T_c\Lambda T_c^{-1} \otimes kBCe^{-\lambda\tau}\Big)\, \det(T_c \otimes I) \\ &= \det\left(I \otimes \left(\lambda I - A + kBC - kBC\lambda_i(G)e^{-\lambda\tau}\right)\right) \\ &= \Pi_{i=1}^{p} f_i(\lambda;\, k,\tau), \end{aligned} \tag{15.16}$$

where

$$\begin{aligned} &f_i(\lambda;\, k,\tau) := \det F_i(\lambda;\, k,\tau), \\ &F_i(\lambda;\, k,\tau) := \lambda I - A + kBC - kBC\lambda_i(G)e^{-\lambda\tau}, \quad i = 1,\dots,p. \end{aligned}$$

Remark 15.7. *This factorization of the characteristic function can also be obtained from the factorization of* (15.6) *and* (15.7) *into* (15.6) *and* (15.12) *when taking into account that* $x_1(t) \equiv x^*$. *It follows from this observation that the zeros of*

$$f_1(\lambda;\, k,\tau) = \det(\lambda I - A + kBC - kBCe^{-\lambda\tau})$$

describe the dynamics of the linearization of the "nominal" system (15.8), *while the zeros of*

$$f_2(\lambda;\, k,\tau),\dots,f_p(\lambda;\, k,\tau)$$

describe the behavior of the synchronization error dynamics.

Eigenspaces and behavior on the onset of instability

We investigate the eigenspace of the characteristic matrix (15.15), corresponding to a characteristic root. For reasons of simplicity we restrict ourselves to the generic case where all the eigenvalues of the adjacency matrix G are simple. Let E_i be the eigenvector of G corresponding to the eigenvalue $\lambda_i(G)$, $i = 1,\dots,p$. By Corollary 15.3, we have $E_1 = [1\, 1\, \cdots\, 1]^T$.

If for some $l \in \{1,\dots,p\}$, the equation

$$f_l(\lambda;\, k,\tau) = 0$$

has a simple root at $\lambda = \hat{\lambda}$ such that

$$F_l(\hat{\lambda};\, k, \tau)\, V = 0, \quad V \in \mathbb{C}^{m\times 1}, \tag{15.17}$$

then it can be verified that

$$F(\hat{\lambda};\, k, \tau)\,(E_l \otimes V) = 0. \tag{15.18}$$

This implies that the linearized system (15.14) has an exponential solution

$$\begin{bmatrix} v_1(t) \\ \vdots \\ v_p(t) \end{bmatrix} = c(E_l \otimes V)e^{\hat{\lambda}t} = c \begin{bmatrix} e_{l,1}V \\ \vdots \\ e_{l,p}V \end{bmatrix} e^{\hat{\lambda}t}, \tag{15.19}$$

with the constant c depending on the initial conditions.

The above analysis can be generalized to the case where the zero $\hat{\lambda}$ of f_l has multiplicity larger than one: if the vectors $(V_r, \ldots, V_1)$ form a Jordan chain of length r of F_l corresponding to the eigenvalue $\hat{\lambda}$ (see [170] for the definition of a Jordan chain of a nonlinear eigenvalue problem), then the vectors

$$(E_l \otimes V_r, \ldots, E_l \otimes V_1)$$

form a Jordan chain of F. Moreover, the corresponding solution of (15.14) takes the form

$$\begin{bmatrix} v_1(t) \\ \vdots \\ v_p(t) \end{bmatrix} = \sum_{i=1}^{r} c_i \left(\sum_{k=1}^{i} \frac{t^{i-k}}{(i-k)!}(E_l \otimes V_k) \right) e^{\hat{\lambda}t}, \tag{15.20}$$

where the constants $(c_1, \ldots, c_r)$ depend on the initial condition. The following can be concluded:

- In an exponential solution of (15.14) corresponding to a zero of $f_l(\lambda;\, k, \tau)$, the relation between the state variables of an individual subsystem is determined by the Jordan system of F_l, while the relation between the corresponding state variables of the different subsystems is *solely* determined by the eigenvector E_l, corresponding to the lth eigenvalue of the adjacency matrix G. This implies that all modes can be classified in at most p types, based on the relations between the behavior of the different subsystems. The bifurcations of the synchronized equilibria of the original nonlinear system can be classified in the same way.

- The modes induced by the zeros of $f_1(\lambda;\, k, \tau)$ all correspond to synchronized behavior of the different subsystems because $E_1 = [1 \cdots 1]^T$. This is expected because they describe the dynamics on the synchronization manifold. By Corollary 15.3, the occurrence of these modes is independent of the topology of the network. The modes induced by $E_2, \ldots, E_p$ correspond to the synchronization error dynamics.

15.3.2 ▪ Computation of stability regions in the delay parameter

For a fixed value of k the following propositions allow us to characterize delay values corresponding to characteristic roots of (15.14) on the imaginary axis:

Proposition 15.8. *For every* $i \in \{1, \ldots, p\}$ *we have*

$$f_i(0;\, k, 0) = 0 \Leftrightarrow f_i(0;\, k, \tau) = 0 \quad \forall \tau \geq 0. \tag{15.21}$$

Proof. The relation (15.21) is implied by the identity $f_i(0;\, k,\tau) = f_i(0;\, k,0)$. □

The following two proposition are direct adaptations of Proposition 6.6:

Proposition 15.9. *The following assertions hold:*

1. *The equation*
$$f_i(\lambda;\, k,\tau) = 0, \quad i \in \{1,\ldots,p\}, \tag{15.22}$$
has a root $j\omega$, $\omega > 0$, for some value of $\tau \geq 0$ if and only if there exists a complex number z such that
$$|z| = 1, \qquad j\omega \in \sigma(A - kBC + kBC|\lambda_i(G)|z). \tag{15.23}$$

2. *The corresponding delay values are given by*
$$\mathcal{T}_\omega := \bigcup \{\mathcal{T}_\omega^z : z \text{ satisfies } (15.23)\}, \tag{15.24}$$
with
$$\mathcal{T}_\omega^z := \left\{ \tfrac{\angle(\lambda_i(G)\bar{z}) + 2\pi\ell}{\omega} : \ell \in \mathbb{N} \right\}. \tag{15.25}$$

3. *If $j\omega$ is a simple root of* (15.22) *for the delay value $\tau = \tau_0$ and $\tau_0 \in \mathcal{T}_\omega^z$, then by sweeping the delay through τ_0 the root crosses the imaginary axis towards instability (stability) if*
$$\Re\left(\frac{u^* BC v j\omega z}{u^* v} \right) < 0\,(>0), \tag{15.26}$$
where u and v are left and right null vectors of
$$j\omega I - A + kBC - kBC|\lambda_i(G)|z.$$

Proposition 15.10. *The conditions* (15.23) *imply*

$$\left\{ \begin{array}{l} |z| = 1, \\ \det\left(\begin{bmatrix} 0 & -I \\ I \otimes k|\lambda_i(G)|(BC)^T & (A - kBC) \oplus (A - kBC)^T \end{bmatrix} \right. \\ \qquad \left. + z \begin{bmatrix} I & 0 \\ 0 & k|\lambda_i(G)|(BC) \otimes I \end{bmatrix} \right) = 0. \end{array} \right. \tag{15.27}$$

By combining the above results with a continuation argument we obtain the following algorithm for computing stability/instability regions of (15.14) in the delay parameter space:

ALGORITHM 15.1. Stability regions of (15.14) in the delay parameter space.

1. Repeat for $i = 1,\ldots,p$:

   Compute the zeros of the polynomial $f_i(\lambda;\, k,0)$ in the closed right half plane.
   Free the delay parameter. Compute all delay values for which $f_i(\lambda;\, k,\tau)$ has a zero on the positive imaginary axis and the corresponding crossing direction from Propositions 15.9 and 15.10, in the following way:

(a) Compute all solutions of the eigenvalue problem (15.27).

(b) For every z satisfying (15.27), compute the eigenvalues on the positive imaginary axis of the matrix

$$A-kBC+kBC|\lambda_i(G)|z.$$

(c) Use the results of steps (a) and (b) to determine all pairs (ω, z) satisfying (15.23).

(d) Determine all critical delay values as well as the corresponding crossing direction of the corresponding zeros of f_i on the imaginary axis using (15.24)–(15.26) (in the generic case of simple zeros).

2. Combine the obtained results for all $i \in \{1,\dots,p\}$. This yields a full characterization of the stability regions of (15.14) in the delay-parameter, because all critical delay values and stability switches are covered.

Remark 15.11. *Step* 1(a) *is facilitated by the following symmetry property of the generalized eigenvalue problem: a number* $z \in \mathbb{C}\setminus\{0\}$ *satisfies the second condition of* (15.27) *if and only if* $\bar{z}^{-1}$ *satisfies this condition.*

Remark 15.12. *The argument* $\angle\lambda_i(G)$ *does not affect the solutions of* (15.23) *and* (15.26). *From* (15.24) *a change of the argument leads only to a shift of the critical delay values. This property will be apparent in the examples presented in Section* 15.4.

As an alternative to Algorithm 15.1 the curves separating stability-instability regions in the (k,τ) parameter space can be computed as Hopf bifurcation curves by numerical continuation (see Chapter 5). On the one hand, the advantage of numerical continuation is that curves in the two-parameter space (k,τ) are directly computed in a computationally efficient way (whereas Algorithm 15.1 only sweeps the delay parameter for a fixed value of the gain parameter and needs to be repeated for a set of gain values chosen on a grid). On the other hand, isolated curves may not be automatically detected since starting values are required in a continuation procedure. The latter problem does not occur with Algorithm 15.1 as it is based on a complete description of critical delay values.

The computations for the numerical examples presented in Section 15.4.2 are based on numerical continuation using the package DDE-BIFTOOL [100], where Algorithm 15.1 is used to generate starting values for the curves. The asymptotic analysis of coupled Lorenz systems presented in Section 15.4.1 is based on Propositions 15.8–15.10, on which Algorithm 15.1 relies.

15.4 ▪ Application to coupled Lorenz systems

In this section the nonlinear oscillators (15.1) are specified as Lorenz systems:

$$\left\{\begin{array}{l} \dot{x}_{i,1}(t)=\sigma(x_{i,2}(t)-x_{i,1}(t)),\\ \dot{x}_{i,2}(t)=rx_{i,1}(t)-x_{i,2}(t)-x_{i,1}(t)x_{i,3}(t)+u_{i,1}(t),\\ \dot{x}_{i,3}(t)=-bx_{i,3}(t)+x_{i,1}(t)x_{i,2}(t)+u_{i,2}(t), \end{array}\right. \tag{15.28}$$
$$\left\{\begin{array}{l} y_{i,1}(t)=x_{i,2}(t),\\ y_{i,2}(t)=x_{i,3}(t)-r, \quad i=1,\dots,p, \end{array}\right.$$

where

$$u_i=[u_{i,1}\ u_{i,2}]^T, \qquad y_i=[y_{i,1}\ y_{i,2}]^T.$$

The parameter values are given by

$$\sigma = 10, \quad r = 28, \quad b = 8/3. \tag{15.29}$$

Note that for $u_i \equiv 0$ each Lorenz system has three equilibria given by

$$(0,0,0), \quad \left(\pm\sqrt{b(r-1)}, \pm\sqrt{b(r-1)}, r-1\right), \tag{15.30}$$

the latter two corresponding to unstable foci. Furthermore, with the parameter values (15.29) each system exhibits a chaotic attractor [362].

If we linearize the coupled system (15.28) and (15.3) around the synchronized equilibrium

$$(x^*, \dots, x^*), \quad x^* = (\pm\sqrt{b(r-1)}, \pm\sqrt{b(r-1)}, r-1), \tag{15.31}$$

then we obtain the linear system (15.14), where the matrices are specified as

$$A = \begin{bmatrix} -\sigma & \sigma & 0 \\ 1 & -1 & \mp\sqrt{b(r-1)} \\ \pm\sqrt{b(r-1)} & \pm\sqrt{b(r-1)} & -b \end{bmatrix}, \quad B = C^T = \begin{bmatrix} 0 & 0 \\ 1 & 0 \\ 0 & 1 \end{bmatrix}. \tag{15.32}$$

It is easy to show that the stability of the linearized system does not depend on which equilibrium x^* in (15.31) is considered. Therefore, we will restrict ourselves to the one in the positive octant.

In what follows we analyze the stability properties of the synchronized equilibria (15.31) in the (k, τ) parameter space. First, we study the asymptotic behavior for large values of the gain parameters in Section 15.4.1. For the standard parameters (15.29) this will allow us to make assertions about stability regions, stability switches, and emerging behavior, which do *not* depend on the network topology. Next we present several numerical examples in Section 15.4.2. Finally, we perform a nonlinear stability analysis of (15.28) and (15.3) in Section 15.4.3.

15.4.1 ▪ Asymptotic behavior for large gain values

We first state some technical lemmas, whose proofs can be found in [258]:

Lemma 15.13. *For large values of k the zeros of the functions* $f_i(\lambda;\, k, 0)$, $i = 2, \dots, p$, *are in the open left half plane. Furthermore, the system* (15.14) *and* (15.32) *has exactly two characteristic roots in the closed right half plane for* $\tau = 0$, *which are equal to the unstable eigenvalues of A.*

Lemma 15.14. *Assume that* $|\lambda_i| < 1$. *Then for large values of k the zeros of the function*

$$f_i(\lambda;\, k, \tau)$$

are in the open left half plane for all values of the delay parameter.

Lemma 15.15. *Assume that* $|\lambda_i| = 1$. *If* $(z(k), \omega(k))$ *satisfies* (15.23) *for all* $k > 0$, *then*

$$\lim_{k\to\infty} z(k) = 1.$$

A combination of the above results leads us to the following main result:

Theorem 15.16. *Consider a network of coupled Lorenz systems* (15.28) *with parameters* (15.29) *and coupling* (15.3). *Assume that the network satisfies Assumptions* 15.1 *and* 15.2.

Then there exists a number $\hat{k} > 0$ *and a function*

$$\tau^* : [\hat{k}, \infty] \to \mathbb{R}_+, \quad k \mapsto \tau^*(k), \tag{15.33}$$

satisfying the following properties:

1. *There is a constant* $\tilde{k} > \hat{k}$ *such that for every* $k > \tilde{k}$*, the synchronized equilibrium has two characteristic roots in the open right half plane for all* $\tau \in [0, \tau^*]$*, while it is asymptotically stable for* $\tau \in (\tau^*, \tau^* + \epsilon)$*, with* ϵ *sufficiently small.*
2. *At* $\tau = \tau^*$ *a synchronization preserving Hopf bifurcation occurs.*
3. *For all* $k \in [\hat{k}, \infty]$ *we can factor*

$$\tau^*(k) = \frac{\upsilon(k)}{k}, \tag{15.34}$$

where

$$\lim_{k \to \infty} \upsilon(k) = 0.586004. \tag{15.35}$$

Furthermore, the number $\hat{k}$ *and the function* (15.33) *are independent of the number of subsystems and of the network topology.*

Proof. By Lemmas 15.13 and 15.14 the functions $f_i(\lambda, k, \tau)$, where $|\lambda_i(G)| < 1$, have their zeros in the left half plane for all values of τ if k is sufficiently large. So for large k all stability switches in the delay parameter space are due to the functions $f_i(\lambda; k, \tau)$ such that $|\lambda_i| = 1$, and (15.23) simplifies to

$$j\omega \in \sigma(A - kBC + kBCz), \quad |z| = 1. \tag{15.36}$$

If we set $z = 1 + \frac{j\rho}{k}$, then this expression becomes

$$j\omega \in \sigma(A + j\rho BC), \tag{15.37}$$

under the constraint

$$\left|1 + j\frac{\rho}{k}\right| = 1. \tag{15.38}$$

We analyze the solutions of (15.37)–(15.38) as $k \to \infty$. From Lemma 15.15 and the constraint (15.38) it follows that ρ/k must converge to zero along the *real axis* as $k \to \infty$. Hence the asymptotic behavior for $k \to \infty$ is determined by the solutions (ω, ρ) of (15.37), where ρ is restricted to be real. To find these solutions, matrix pencil techniques can be used, similarly to Proposition 15.10. Equation (15.37) namely implies

$$-j\omega \in \sigma\left(A^T - j\rho(BC)^T\right), \tag{15.39}$$

and under the condition $\rho \in \mathbb{R}$, (15.37) and (15.39) imply

$$\det\left(A \oplus A^T + \rho j[(BC) \otimes I - I \otimes (BC)^T]\right) = 0. \tag{15.40}$$

Thus all solutions of (15.37) under the constraint $\rho \in \mathbb{R}$ can be directly computed by calculating the real solutions of the eigenvalue problem (15.40) in the first step, in order to obtain a finite number of candidate values for ρ, and then solving (15.37) for ω. With the parameter values (15.29) and with matrices (15.32) these solutions are given by $(\hat{\rho}_1, \hat{\omega}_1)$ and $(\hat{\rho}_2, \hat{\omega}_2)$, where

$$\begin{aligned} \hat{\rho}_1 &= -3.9959064, & \hat{\omega}_1 &= 6.8189034, \\ \hat{\rho}_2 &= 5.2236045, & \hat{\omega}_1 &= 14.811554. \end{aligned} \tag{15.41}$$

As a consequence, for large values of k the solutions of (15.36) are

$$(\omega, z) = (\omega_l(k), z_l(k)), \quad l = 1, 2,$$

where

$$z_l(k) = 1 + j\frac{\rho_l(k)}{k}, \quad l = 1, 2,$$

and

$$\lim_{k\to\infty} \rho_l(k) = \hat{\rho}_l, \quad \lim_{k\to\infty} \omega_l(k) = \hat{\omega}_l, \quad l = 1, 2.$$

Next, from (15.24) and (15.41) it follows that for sufficiently large k, the *first* critical delay value, as the delay is increased from zero, is given by

$$\tau^*(k) := \frac{\angle(\bar{z}_1(k))}{\omega_1(k)} = \frac{\upsilon(k)}{k},$$

where

$$\begin{aligned} \lim_{k\to\infty} \upsilon(k) &= \frac{1}{\hat{\omega}_1} \lim_{k\to\infty} k \; \angle(\bar{z}_1(k)) \\ &= -\frac{1}{\hat{\omega}_1} \lim_{k\to\infty} k \; \mathrm{atan}(\rho_1(k)/k) \\ &= -\frac{\hat{\rho}_1}{\hat{\omega}_1} = 0.586004. \end{aligned}$$

As this switch is due to a zero of $f_1(\lambda;\ k, \tau)$, it is independent of the network topology, and the emanating solutions have the form

$$\begin{bmatrix} y_1(t) \\ \vdots \\ y_p(t) \end{bmatrix} = \begin{bmatrix} V \\ \vdots \\ V \end{bmatrix} e^{j\omega t},$$

where $F_1(j\omega;\ k, \tau)V = 0$, i.e., synchronization is preserved in the emanating solutions.

Finally, we consider the crossing direction of the characteristic roots on the imaginary axis for

$$(k, \tau) = (k, \tau^*(k))$$

when the delay is varied. According to (15.26) the crossing direction is determined by the sign of

$$s(k) := \Re\left(\frac{u(k)^* BC v(k) j\omega_1(k) z_1(k)}{u(k)^* v(k)}\right),$$

where $u(k)$ and $v(k)$ are left and right null vectors of

$$j\omega_1(k)I - A + kBC - kBCz_1(k) = j\omega_1(k)I - A - BCj\rho_1(k).$$

It follows that

$$\lim_{k\to\infty} s(k) = \Re\left(\frac{\hat{u}^* BC \hat{v} j \hat{\omega}_1}{\hat{u}^* \hat{v}}\right),$$

with $\hat{u}$ and $\hat{v}$ left and right null vectors of $j\hat{\omega}_1 I - A - BCj\hat{\rho}_1$. For the parameters (15.29) and with matrices (15.32) we arrive at

$$\lim_{k\to\infty} s(k) = 3.4980241 > 0.$$

Thus for large k the first stability switch, which occurs at $\tau = \tau^*(k)$, is towards stability, and it results in asymptotic stability by Lemma 15.13.

When putting together the above results the statements of the theorem follow. □

15.4.2 ▪ Examples

We illustrate the obtained results with several examples with different network topologies. The computations of stability regions are done as described in Subsection 15.3.2.

Ring topology, unidirectional coupling

We consider a ring topology with unidirectional coupling, described by the adjacency matrix

$$G = \begin{bmatrix} 0 & \cdots & & 0 & 1 \\ 1 & 0 & & & \\ & \ddots & \ddots & & \\ & & & 1 & 0 \end{bmatrix} \in \mathbb{R}^{p\times p}, \tag{15.42}$$

which has the following properties:

$$\lambda_l(G) = e^{-j\frac{2\pi(l-1)}{p}}, \quad E_l = \begin{bmatrix} 1 \\ e^{-j\frac{2\pi(l-1)}{p}} \\ \vdots \\ e^{-j\frac{2\pi(p-1)(l-1)}{p}} \end{bmatrix}$$

for $l = 1, \ldots, p$.

If (15.17) is satisfied for $\hat{\lambda} = j\omega$, $\omega > 0$, then the emanating solution (15.19) becomes

$$\begin{bmatrix} v_1(t) \\ \vdots \\ v_p(t) \end{bmatrix} = c \begin{bmatrix} V\, e^{j\omega t} \\ V\, e^{j\omega t - \frac{2\pi(l-1)}{p}} \\ V\, e^{j\omega t - \frac{4\pi(l-1)}{p}} \\ \vdots \\ V\, e^{j\omega t - \frac{2(p-1)\pi(l-1)}{p}} \end{bmatrix}. \tag{15.43}$$

It can be interpreted as a traveling wave solution, where the agents follow each other with a phase shift of $360(l-1)/p$ degrees. Therefore, if the characteristic root $\hat{\lambda}$ on the imaginary axis corresponds to a Hopf bifurcation of the original nonlinear system (15.1) and (15.3) for a critical value of some free parameter, we refer to this bifurcation as a "Hopf $360(l-1)/p$" bifurcation. In a sense, this type of traveling wave solution strongly resembles the gait of an animal even though the underlying oscillator is different from a Lorenz oscillator.

With the individual agents taken as Lorenz systems (15.28) with parameters (15.29) and with $p = 4$ and $p = 12$, we display in Figure 15.1 the stability regions in the delay-parameter space of the synchronized equilibria (15.31). The Hopf 0 bifurcation curves

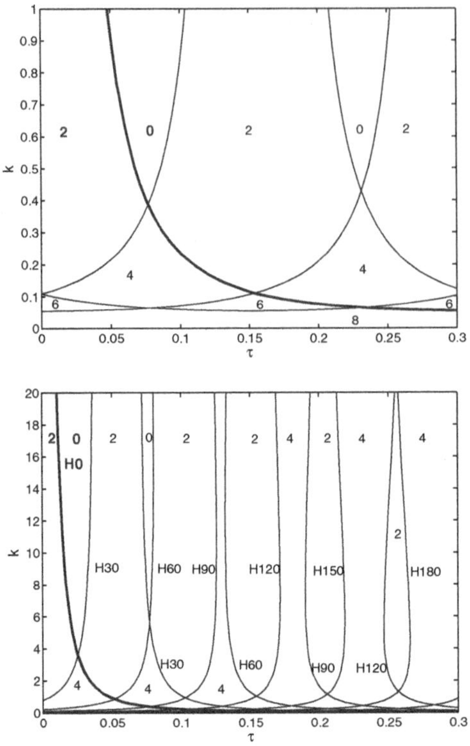

Figure 15.1. *Stability regions of the synchronized equilibrium* (15.31) *of Lorenz systems* (15.28)–(15.29) *coupled in a ring configuration described by* (15.42)*, for* $p = 4$ *(above) and* $p = 12$ *(below). The numbers refer to the characteristic roots in the closed right half plane. The quantities indicated in bold are independent of the network topology and of the number of subsystems* [258].

are independent of the number of subsystems, because they are induced by the zeros of $f_1(\lambda; k, \tau)$. The first corresponds to the function (15.33). By Theorem 15.16 the quantities indicated in bold on the figure are independent of the number of agents and of the network topology.

Ring topology, bidirectional coupling

A ring topology with bidirectional coupling between the agents is described by the matrix

$$G = \frac{1}{2}\begin{bmatrix} 0 & 1 & & & 1 \\ 1 & 0 & 1 & & \\ & \ddots & \ddots & \ddots & \\ & & 1 & 0 & 1 \\ 1 & & & 1 & 0 \end{bmatrix} \in \mathbb{R}^{p\times p}, \tag{15.44}$$

satisfying

$$\lambda_l(G) = \cos\left(\frac{2\pi}{p}(l-1)\right), \quad l = 1, \ldots, q,$$

where $q = (p+2)/2$ if p is even and $q = (p+1)/2$ if p is odd. All eigenvalues have multiplicity two, except for $\lambda_1(G) = 1$ and, if p is even, $\lambda_{\frac{p+2}{2}}(G)$. The corresponding eigenvectors are

$$\left[\begin{array}{cccc} \cos\left(\frac{2\pi(l-1).(p-1)}{p}\right) & \cdots & \cos\left(\frac{2\pi(l-1).1}{p}\right) & 1 \end{array}\right]^T$$

and

$$\left[\begin{array}{cccc} \sin\left(\frac{2\pi(l-1).(p-1)}{p}\right) & \cdots & \sin\left(\frac{2\pi(l-1).1}{p}\right) & 0 \end{array}\right]^T.$$

Note that if all subsystems are Lorenz systems described by (15.28)–(15.29), then for large values of k the stability switches are associated only with the eigenvalues ± 1 and corresponding eigenvectors $[1\ 1\ \cdots 1]^T$ and $[1\ (-1)\ (-1)^2 \cdots\ (-1)^{p-1}]^T$ (see Lemma 15.14). They result in either synchronized motion or standing waves. This is due to the bidirectional coupling and is in contrast to the case of unidirectional coupling addressed above, where traveling wave solutions naturally appear.

For $p = 4$ the stability regions in the (k, τ) parameter space are shown in Figure 15.2. Note from the right pane that the number of characteristic roots in the right half plane changes from two to six when crossing the horizontal curve. This is due to the double eigenvalue of G at zero.

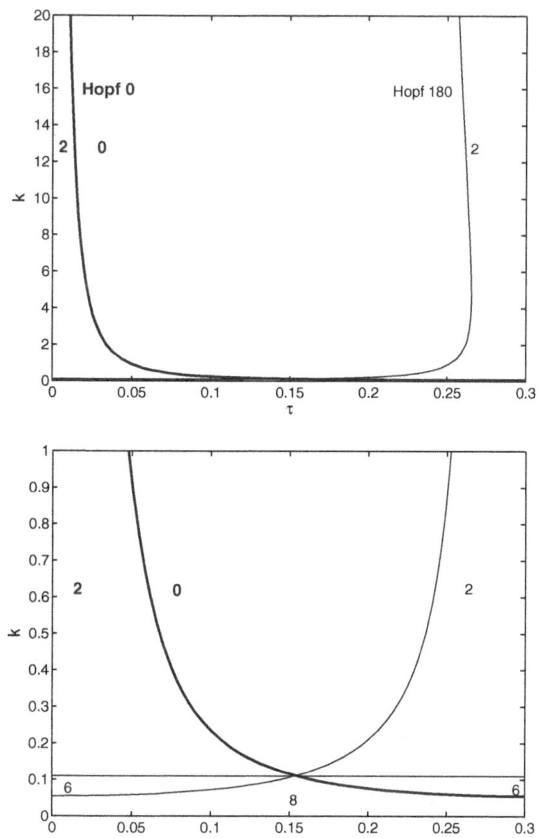

Figure 15.2. *Stability regions of the synchronized equilibrium* (15.30) *of Lorenz systems* (15.28)–(15.29) *coupled in a ring configuration described by* (15.44), *on two different scales* [258].

Cross topology

The topology induced by the matrix

$$G=\begin{bmatrix} 0 & 1 & 0 & 0 & 0 & 0 & 0 \\ \frac{1}{2} & 0 & \frac{1}{2} & 0 & 0 & 0 & 0 \\ 0 & \frac{1}{4} & 0 & \frac{1}{4} & 0 & \frac{1}{4} & \frac{1}{4} \\ 0 & 0 & \frac{1}{2} & 0 & \frac{1}{2} & 0 & 0 \\ 0 & 0 & 0 & 1 & 0 & 0 & 0 \\ 0 & 0 & 1 & 0 & 0 & 0 & 0 \\ 0 & 0 & 1 & 0 & 0 & 0 & 0 \end{bmatrix} \tag{15.45}$$

is displayed in Figure 15.3. The eigenvalues and corresponding eigenvectors of G are

$$\begin{bmatrix} \lambda_1(G) & \cdots & \lambda_7(G) \end{bmatrix}=\begin{bmatrix} 1 & \frac{1}{2} & -\frac{1}{2} & -1 & \frac{\sqrt{2}}{2} & -\frac{\sqrt{2}}{2} & 0 \end{bmatrix}$$

and

$$\begin{bmatrix} E_1 \,|\, \cdots \,|\, E_7 \end{bmatrix} = \left[\begin{array}{c|c|c|c|c|c|c} 1 & 2 & 2 & 1 & \sqrt{2} & \sqrt{2} & 0 \\ 1 & 1 & -1 & -1 & 1 & -1 & 0 \\ 1 & -1 & -1 & 1 & 0 & 0 & 0 \\ 1 & 1 & -1 & -1 & -1 & 1 & 0 \\ 1 & 2 & 2 & 1 & \sqrt{2} & -\sqrt{2} & 0 \\ 1 & -2 & 2 & -1 & 0 & 0 & -1 \\ 1 & -2 & 2 & -1 & 0 & 0 & 1 \end{array}\right].$$

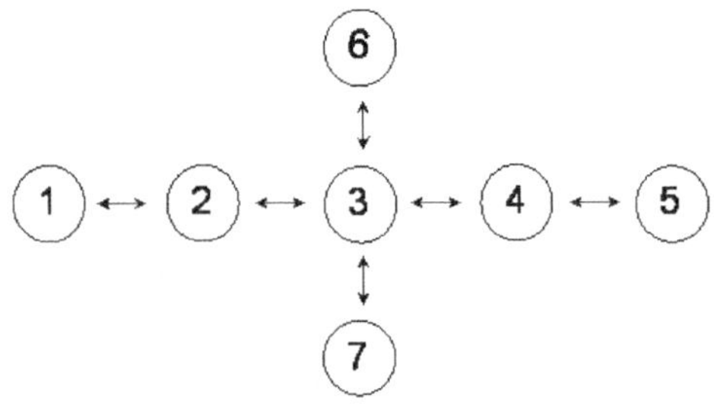

Figure 15.3. *Network topology with adjacency matrix* (15.45) [258].

With all agents described by (15.28)–(15.29) we display in Figure 15.4 the stability regions in the (k, τ) parameter space of the synchronized equilibria (15.31). The type of Hopf bifurcation is displayed by indicating the corresponding eigenvector of G. Following from Lemma 15.14, only the Hopf curves corresponding to eigenvalues on the unit circle of G persist for large k, namely $\lambda_1(G) = 1$ and $\lambda_4(G) = -1$. The bold curve once again corresponds to the function (15.33).

15.4.3 ▪ Beyond the linearized stability analysis

The exact linearized stability analysis in the previous sections illustrates that for large values of the coupling gain and small values of the delay, coupled Lorenz systems exhibit some generic behavior, independent of the network topology. In what follows, the results are strengthened by a nonlinear stability analysis.

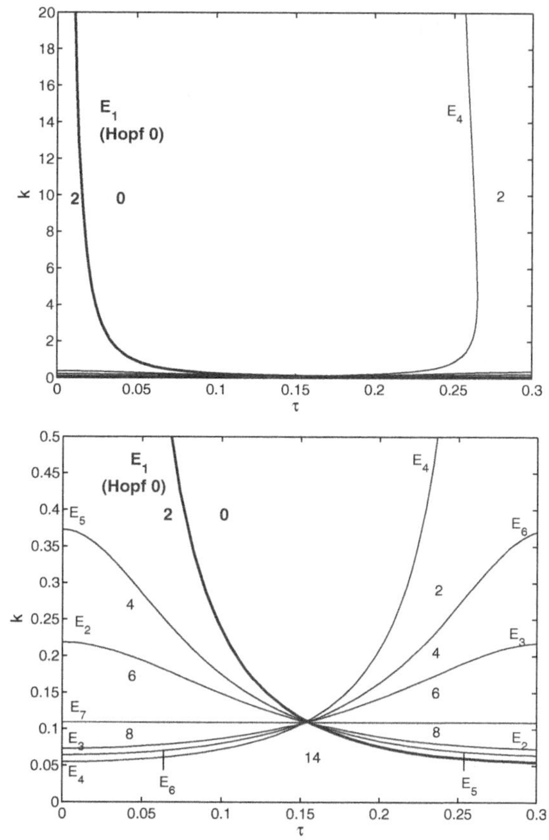

Figure 15.4. *Stability regions in the* (k,τ) *parameter space of the synchronized equilibrium of coupled Lorenz systems on two different scales. The network topology is described by* (15.45) [258].

We reconsider Theorem 15.16 and make some observations. The presence of the synchronization preserving Hopf bifurcation at $\tau = \tau^*(k)$, the fact that for large values of k the functions $f_2(\lambda;\ k,\tau),\dots,f_p(\lambda;\ k,\tau)$, that describe the synchronization error around the synchronized equilibrium, have all zeros in the open left half plane for all $\tau \in [0,\ \tau^*(k)]$, and the observed synchronized behavior in our experiments for all $\tau \in [0,\ \tau^*(k)]$ suggest that asymptotic synchronization can also be achieved for all $\tau \in [0,\ \tau^*(k))$, though the dynamics on the synchronization manifold are no longer characterized by the presence of stable equilibria. Furthermore, the asymptotic behavior of the curve $k \mapsto \tau^*(k)$, described by (15.34)–(15.35), suggests that the natural parameters in the analysis are rather the gain parameter k and the normalized delay parameter $k\tau$. These observations hold and are apparent from the following theorem:

Theorem 15.17. *Consider a network of coupled Lorenz systems* (15.28) *with parameters* (15.29) *and coupling* (15.3). *Assume that the network satisfies Assumptions* 15.1 *and* 15.2. *Let*

$$y = (y_1,\dots,y_p)$$

and define the functions

$$V_s(y) := \sum_{i=1}^{p} \gamma_i V(y_i), \qquad H_s(y) := \sum_{i=1}^{p} \gamma_i H(y_i),$$

where γ *is defined as in Corollary* 15.3,

$$V(y_i) := \frac{1}{2} y_i^T y_i,$$

and

$$H(y_i) := y_{i,1}^2 + b y_{i,2}^2 + b r y_{i,2}. \tag{15.46}$$

The following results hold:

1. *All solutions of* (15.28) *and* (15.3) *are bounded and converge to the set* Ω*, defined as*

$$\Omega := \left\{ x \in \mathbb{R}^{3p} : \; V_s(y) < v_m \text{ and } |x_{i,1}| \le \sqrt{\frac{2v_m}{\gamma_i}}, \; i = 1, \dots, p \right\}, \tag{15.47}$$

   where the constant $v_m > 0$ *is such that*

$$V_s(y) \ge v_m \Rightarrow H_s(y) \ge 0. \tag{15.48}$$

2. *The set* Ω *is a forward invariant set of* (15.28) *and* (15.3).

3. *For all* $C > 0$*, there exists a number* $\hat{k} > 0$ *such that all synchronized solutions in* Ω *exhibit asymptotically stable error dynamics whenever* $k > \hat{k}$ *and* $k\tau < C$.

Sketch of proof. Because the coupling affects the dynamics on the synchronization manifold for $\tau \neq 0$, proving boundedness properties of the solutions is a necessary step in the analysis. The first and second statements are due to a semipassivity-like property of the individual oscillators, more precisely the fact that the derivative of the function $V(y_i)$ along the solutions of (15.28) satisfies

$$\dot{V} = -H(y_i) + y_i^T u_i,$$

with $H(y_i) > 0$ for large values of $\|y_i\|$. The proofs rely on a composed Lyapunov–Krasovskii functional and a Lyapunov–Razumikhin function for the output y, where, inspired by [77], the components are weighted by the left eigenvector γ of the adjacency matrix G. The third statement of the theorem follows from the uniform stability of the null solution of (15.12) for large k when x_1 is confined to a compact set. This is proved using techniques from $\mathcal{L}_2$ gain analysis [382], where $x_1(t)$ in (15.12) is interpreted as a time-varying perturbation.

For a detailed proof we refer the reader to [258]. □

The main results of this section, Theorems 15.16 and 15.17, are graphically displayed in Figure 15.5. Whereas Theorem 15.17 only makes assertions about preservation of synchronized behavior in the emanating solutions in the Hopf 0 bifurcation of the synchronized equilibrium, Theorem 15.17 states that for an arbitrary value of C, asymptotic synchronization can be achieved for all $\tau \in [0, C/k)$. If k is chosen such that asymptotic synchronization is guaranteed for $\tau \in [0, \tau^*(k) + \epsilon]$ with $\epsilon > 0$ some small number, and if the system is initialized close to the synchronized equilibrium and the delay parameter slowly swept from $\tau^*(k) + \epsilon$ to zero, then the attractor of the solution evolves from the stable synchronized equilibrium to synchronized chaotic behavior for $\tau = 0$, because the synchronization between the agents is maintained throughout *every* bifurcation. Recall that the dynamics on the synchronization manifold are described by (15.8), which reduces to $\dot{x}_1 = f(x_1)$ for $\tau = 0$.

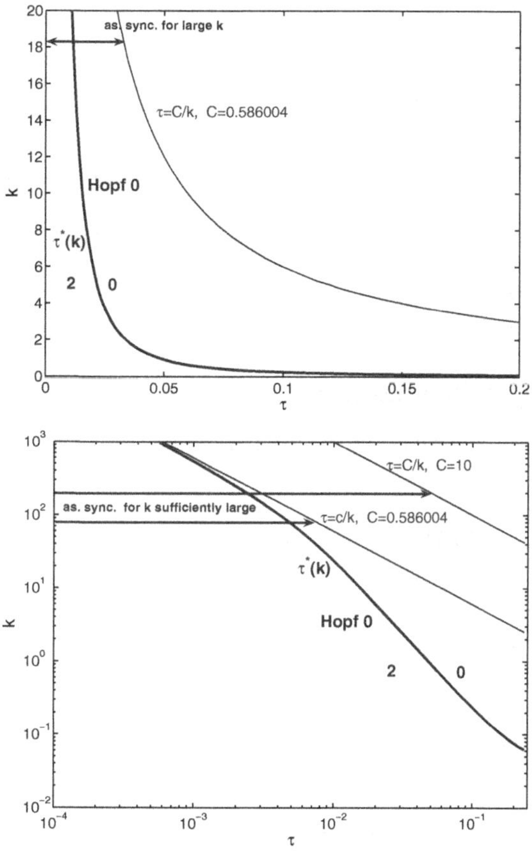

Figure 15.5. *Graphical illustration of the results of Theorems* 15.16 *and* 15.17 *for large values of the coupling gain k on a linear scale (above) and on a logarithmic scale (below). The quantities indicated in bold are independent of the number of agents and of the network topology* [258].

15.5 ▪ Generalizations

We outline how the approach can be generalized to another frequently encountered type of coupling and to the analysis of partial synchronization.

15.5.1 ▪ Other types of coupling

Assume now that the coupling is described by

$$u_i(t) = k\left(\sum_{(i,l)\in\mathscr{E}} \alpha_{i,l}\,(y_l(t-\tau)-y_i(t-\tau))\right), \quad i=1,\ldots,p. \tag{15.49}$$

The difference with (15.3) is that also the self-feedback is delayed. Although this might sound unnatural, as the delay is expected only in the transmission of signals from one agent to another, there are several applications where the form (15.49) appears, ranging from neuronal networks to mechanical systems (see [360] and references therein). An example concerns the traffic flow problems addressed in Chapter 14, where (15.49) appears in models for humans adjusting the distance between consecutive cars. In this case, k represents the sensitivity of a driver to the velocity difference, y_i represents the ith vehicle's velocity, and τ is the driver's reaction delay. The implications of changing (15.3) to (15.49) are the following.

1. The control law has become *noninvasive*, in the sense that it vanishes on the synchronization manifold, characterized by $x_1 = \cdots = x_p$. Therefore, the dynamics on the synchronization manifold are solely determined by the dynamics of one uncontrolled system. Consequently, if it has an unstable equilibrium, the corresponding synchronized equilibrium of the coupled systems can never be stable.

2. Assumption 15.2 is not longer needed to guarantee the existence of a synchronization manifold.

15.5.2 ▪ Partial synchronization

Partial synchronization, also called cluster synchronization, is defined in the following way:

Definition 15.18. *Denote the initial conditions of* (15.1) *and* (15.3) *or* (15.49) *by* $\phi := (\phi_1, \ldots, \phi_p)$. *The coupled systems are said to (locally) partially synchronize if for every* $R > 0$ *there is a* $\delta > 0$ *such that for* $||\phi||_2 < R$, *the inequality* $||\phi_i - \phi_j||_s < \delta$ *implies*

$$\lim_{t\to\infty} ||x_i(\phi_i)(t) - x_j(\phi_j)(t)||_2 = 0$$

for at least two but not all $i, j \in \{1, 2, \ldots, p\}$.

Under Assumption 15.1, the occurrence of partial synchronization requires the presence of a symmetry property in the network, which can be translated into the presence of a so-called partial synchronization manifold [361].

Definition 15.19. *The linear subspace*

$$M := \{x \in \mathbb{R}^{np} \mid x_i = x_j \text{ for some pairs } \{i,j\} \subset \{1,2,\ldots,p\}\} \tag{15.50}$$

is called a partial synchronization manifold for the systems (15.1) *and* (15.3) *or* (15.49) *if it is a forward invariant set, i.e., a set invariant with respect to the coupled systems' dynamics.*

Letting $x(t) = [x_1(t)^T \ \cdots \ x_p(t)^T]^T$, solutions lying in the manifold (15.50) satisfy a relation of the form

$$x(t) = (\Phi \otimes I_n)x(t) \quad \forall t \geq 0, \tag{15.51}$$

where Φ is a p-by-p permutation matrix. As an illustration, in the case of partial synchronization in a network of 4 agents, where agents 1 and 2 synchronize as well as agents 3 and 4, the partial synchronization manifold takes the form

$$M = \{x \in \mathbb{R}^{4n} \mid x_1 = x_2,\ x_3 = x_4\}, \tag{15.52}$$

and the corresponding permutation matrix is given by

$$\Phi = \begin{bmatrix} 0 & 1 & 0 & 0 \\ 1 & 0 & 0 & 0 \\ 0 & 0 & 0 & 1 \\ 0 & 0 & 1 & 0 \end{bmatrix}. \tag{15.53}$$

It is important to note that condition (15.51) can also be written as

$$x(t) \in \ker\left(I_{pn} - (\Phi \otimes I_n)\right) \quad \forall t \geq 0. \tag{15.54}$$

In the light of these observations, an important step in analyzing the occurrence of partially synchronous regimes lies in determining the set of permutation matrices Φ for which $\ker\left(I_{pn} - (\Phi \otimes I_n)\right)$ is a forward invariant set for (15.1) and (15.3) or (15.49), i.e., a set for which the implication

$$(I_{pn} - (\Phi \otimes I_n))\phi(s) = 0, \quad s \in [-\tau, 0] \Rightarrow (I_{pn} - (\Phi \otimes I_n))x(\phi)(t) = 0 \quad \forall t \geq 0,$$

holds, where $x(\phi)(t)$ is the solution at time t with initial condition $x(s) = \phi(s)$, $s \in [-\tau, 0]$. In [379] necessary and sufficient conditions for the occurrence of partial synchronization manifolds are derived. These conditions, and the construction of the permutation matrices, simply follow from the eigenvalues and eigenvectors of the graph Laplacian. It should be noted that in the special case of full synchronization the conditions simplify to Assumption 15.2.

The fact that for some perturbation matrix Φ the subspace $\ker\left(I_{pn} - (\Phi \otimes I_n)\right)$ is a forward invariant set guarantees the presence of partially synchronous solutions. However, whether such solutions are robust against small perturbations and observed in experiments depends on whether the invariant set attracts neighboring solutions. For the coupling (15.49) this relative stability problem has been addressed in [361] using a Lyapunov-based approach. In [379] it is shown how qualitative correct predictions about the occurrence and type of partially synchronous regimes can be made based on the structure of the solutions in the unstable manifold of the synchronized equilibrium, similarly to the approach of Section 15.3.

15.6 ▪ Notes and references

We studied the synchronization of coupled nonlinear oscillators with delay in the coupling, (15.1) and (15.3), with the emphasis on coupled Lorenz systems. First, the state transformation to (15.5) led us to necessary conditions on the network topology for the existence of synchronized solutions. Next, we performed a stability analysis of synchronized equilibria in a (gain, delay) parameter space. Instrumental to this study, we employed a factorization of the characteristic equation, which separates the nominal behavior and the synchronization error dynamics, and we revealed the precise role of the eigenvalues and the eigenvectors of the adjacency matrix of the graph in the behavior of the solutions. The latter allowed us to classify the modes of the system, as well as the Hopf bifurcation curves and the emerging behavior on the onset of instability. As a result of this analysis for the case of coupled Lorenz systems we proved that for sufficiently large gain values, there always exists a stability interval in the delay parameter space that does not contain the zero delay value. Furthermore, this behavior is generic because both the critical delay value, $\tau^*(k)$, and the type of corresponding bifurcation (a synchronization preserving Hopf bifurcation, in the sense that if the delay is reduced beyond the critical value, the equilibrium becomes unstable without losing the synchronization between the agents) do not depend on the network topology and the number of agents. Finally, these results were complemented with a nonlinear stability analysis which, among others, showed that by choosing the gain parameter sufficiently large, asymptotic synchronization can actually be achieved over any finite interval in the normalized delay $k\tau$, again independently of the network. As we have pointed out, the methodology extends to a coupling of the form (15.49) and to the study of partial synchronization.

Instead of directly deriving synchronization conditions for the nonlinear system (15.1) and (15.3), the methodology was based on considering first the linearized stability problem around a synchronized equilibrium, which can be exactly solved. Such an approach directly leads to insight into the problem, because not only the stability regions in the (k,τ) parameter space can be characterized but also the solutions on the onset of instability, by considering the structure of the eigenspaces in the bifurcations. In addition, the gained qualitative insight and observations may lead to a better targeted and less conservative nonlinear stability analysis. This was illustrated in this chapter with coupled Lorenz systems. The formulation and proof of Theorem 15.17 were namely based on the following properties suggested by the linear stability analysis: (i) behavior independent of the network for large coupling gains and small delays, (ii) natural parameters $(k,k\tau)$ rather than (k,τ), and (iii) instead of analyzing stability of the full error dynamics (15.9) directly, it is preferred to analyze the decoupled systems (15.12), where the magnitudes of the eigenvalues of the adjacency matrix suggest the natural type of criterion to be used.

If the agents are not completely identical, then in general (perfectly) synchronized solutions do not exist. This can be seen from (15.5) where terms related to the deviations appear in the right-hand side. Though the analysis in the chapter has been performed step by step using a particular decomposition or factorization, holding for identical agents and uniform delays only, the final results for the coupled system (presence of a synchronized steady state solution, its stability regions and Hopf bifurcation curves in the (k,τ) plane, the structure of the eigenfunctions corresponding to the Hopf bifurcations) will be slightly perturbed only if the differences between the agents and the delay parameters are sufficiently small. This means that Theorem 15.16 remains approximately valid in the sense that for large k and particular values of τ there exists an almost synchronized equilibrium, which is stable but loses stability beyond $\tau \approx \tau^*$, while maintaining the solutions close to being synchronized. This indicates that for τ sufficiently small, the synchronization error dynamics exhibits an attractor whose size can be made arbitrarily small by reducing the difference between the agents. The effect of a small time-variation of delays and other system parameters around a nominal value can be taken into account using the ideas of [244], where the time-varying parameters are essentially treated as perturbations of time-invariant parameters. However, to analyze the effect of large variations or the effects of a time-varying network topology, time-domain methods become necessary, at the cost of introducing conservatism in the analysis.

The presented results are based on [258, 379] and the references therein.

Chapter 16

Stability analysis of delay models in biosciences

16.1 ▪ Introduction

Most of the models encountered in biosciences that represent competition between populations, epidemics, or respiration control mechanisms include a particular common element used to describe a reaction chain (distributed character [223]), a transport process (breathing process [18, 19] in the physiological circuit controlling the carbon dioxide (CO_2) level in the blood [394]), storing nutrients or cell cycles (in the case of controlling the supply of nutrients to a growing population of microorganisms in some chemostat [352, 14]), and latency and short intercellular phases (in epidemics as, for example, cell-to-cell spread models in a particular compartment, such as the bloodstream [83]). Such an element is nothing else than the *transport* and/or *propagation delay*, which can be discrete or distributed, constant or time-varying.

The list of applications of delay models in biosciences is extensive. For instance, one can cite a density-dependent feedback mechanism for responding to changes in population density (that never takes place *instantaneously*; see, e.g., [123, 202]), the spread (*propagation*) of infections within a family, and epidemics with intermediate classes (that is, the presence of "individuals" for a given period such that they are "exposed, but not (*necessarily*) yet infected"; see, e.g., [167]), or *recurrent diseases*, as suggested in various relapse-recovery models (after a given period, an infective returns to being fully susceptible again [167]). Further remarks, and discussions on delay models in biosciences, together with a large list of references, can be found in [144, 14, 284].

Without discussing the assumptions needed to derive the biological models considered in what follows, and without considering the problems related to the model representation, this chapter focuses on the stability analysis of the linearization of some of the delay models mentioned above. The chapter is organized as follows: Section 16.2 is devoted to the stability analysis of some genetic regulatory networks. Section 16.3 is devoted to the stability analysis of the linearization of a two-compartment representation including one transport delay. Next, Section 16.4 is devoted to the stability analysis of some immune dynamics model encountered in chronic leukemia and including four delays in a large range. The geometry of the stability regions is derived in the corresponding delay-parameter space. Various interpretations complete the presentation. Finally, some notes and references end the chapter.

16.2 ▪ Delay effects on the stability of some biochemical network models

In the 1960s, Goodwin [122] introduced a low-order dynamical system that became the quintessential example of a biochemical oscillator based on negative feedback. Such a model is at the origin of most feedback-based approaches for describing cell regulatory mechanisms (see, e.g., [121] and the references therein).

16.2.1 ▪ Delay model and its linearization

Consider a simple two-compartment model for a protein (end product)-mRNA self-repression including delays during the translation and transcription processes and represented by a second-order system of the form

$$\begin{cases} \dot{x}_1(t) = -\lambda_1 x_1(t) + c_1 x_2(t-\tau_1), \\ \dot{x}_2(t) = -\lambda_2 x_2(t) + g(x_1(t-\tau_2)), \end{cases} \tag{16.1}$$

where $c_1 \in \mathbb{R}_+$ and $\lambda_i \in \mathbb{R}_+$ $(i = 1,2)$ describe the translation and the degradation effects, respectively. The feedback mechanism (g) here appears as a "propagation" of effects among the components of the system such that one component predicts a change (for example, increase/decrease, inhibition/activation) in its own production rate and is defined by an appropriate monotone function. The variables x_1 and x_2 denote the concentrations of the mRNA (messenger RNA) and of the protein (end product), and the rate dx/dt is defined by the balance between mRNA synthesis and the end product consumption as shown by equations (16.1). Note that the delay τ_1 (τ_2) defines the time-lag between the initiation of the translation (transcription) and the appearance of the mature protein (mRNA). For the use of delays in describing chemical or biochemical kinetics, we refer the reader to [367, 334]; by kinetics, we understand the assignment of rate function to each reaction [105].

A simple change of variables $x_2(t-\tau_1) \mapsto x_2(t)$ allows us to rewrite the system (16.1) as an equivalent system with only one delay $\tau = \tau_1 + \tau_2$:

$$\begin{cases} \dot{x}_1(t) = -\lambda_1 x_1(t) + c_1 x_2(t), \\ \dot{x}_2(t) = -\lambda_2 x_2(t) + g(x_1(t-\tau)). \end{cases} \tag{16.2}$$

To the best of the authors' knowledge, [219] (see also [223]) was the first study that extended the classical Goodwin's model by incorporating a time-delay representing the dependency of one chemical component on other chemical component. There exists a long list of references concerning Goodwin-like oscillators and their dynamics, and its renewed interest in the control literature appeared in the context of monotone systems [6]. In [98], the author considered a generalization of (16.1) consisting in some cyclic system with delay described by a system of delay-differential equations of the form

$$\begin{cases} \dot{x}_i(t) = -\lambda_i x_i(t) + g_i(x_{i+1}(t)), \quad i = 1,\ldots,m-1, \\ \dot{x}_m(t) = -\lambda_m x_m(t) + g_m(x_1(t-\tau)), \end{cases} \tag{16.3}$$

where g_i are monotone functions and the system is subject to negative feedback. Delay systems with this structure can be also found in the modeling of neural networks using $g_i(x) = \alpha_i \tanh(\beta_i x)$ (see, for instance, [127] or the models describing associative memory in Hopfield networks as mentioned in [21, 290]). Finally, another important special case in biochemical models is that in which the functions $g_i(x)$ have the Hill function form

$$f(x) = \frac{a}{b + x^p} + c, \quad a, b > 0, c \geq 0, \quad \text{with} \quad p \in \mathbb{N}^*. \tag{16.4}$$

Similarly to [98], we consider in what follows the linearization

$$\begin{cases} \dot{x}_i(t) = -\lambda_i x_i(t) + k_i x_{i+1}(t), & i = 1, \ldots, m-1, \\ \dot{x}_m(t) = -\lambda_m x_m(t) + k_m x_1(t - \tau) \end{cases} \tag{16.5}$$

of (16.3) around its unique equilibrium point defined by $(\overline{x}_1, \ldots, \overline{x}_m)$, where $k_i = g_i'(\overline{x}_{i+1})$, $i = 1, \ldots, m-1$, and $k_m = g_m'(\overline{x}_1)$. Therefore, the characteristic equation of the linearized system of (16.1) and (16.2) around the equilibrium $\overline{x} = (\overline{x}_1\ \overline{x}_2)^T$ has the form

$$(\lambda + \lambda_1)(\lambda + \lambda_2) + k e^{-\lambda \tau} = 0, \tag{16.6}$$

where $k = -c_1 g'(\overline{x}_1)$, and $\tau = \tau_1 + \tau_2$ represents the total delay (translation, transcription). Equation (16.6) can be interpreted as the closed-loop of a second-order system subject to a delayed output feedback (see, for instance, Chapter 10). On the other hand, the linearized version of the biochemical system in [98] becomes

$$H(\lambda; k, \tau) \triangleq (\lambda + \lambda_1) \ldots (\lambda + \lambda_m) + k e^{-\lambda \tau} = 0 \tag{16.7}$$

for some $m \in \mathbb{N}$, $m \geq 2$, and $k = -\Pi_{i=1}^m k_i$.

Remark 16.1. *For instance, if, instead of a selective memory in* (16.1), *the delay dependency follows a γ-distribution law (as suggested and discussed by* [223]), *the corresponding model is similar to the original Goodwin oscillatory system* [122] *free of delay, but with an increased number of variables.*

Finally, it is worth mentioning that combined positive and negative feedback, or *time delays*, can generate oscillations and complex behaviors in genetic systems [354].

16.2.2 ▪ Stability analysis: Computing the delay margin

Let us focus on computing the *frequency crossing set* Ω defined as the set of all real "frequencies" ω for which there exist at least a set of parameters $(\lambda_1, \ldots, \lambda_m, k, \tau)$ such that $H(j\omega; k, \tau) = 0$.

Remark 16.2. *Obviously, $H(j\omega, k, \tau) = 0 \Leftrightarrow \overline{H(j\omega, k, \tau)} = 0$. In other words, in what follows we only need to consider positive frequencies ω, that is, $\Omega \subset (0, \infty)$.*

Next, it is easy to see that $H(j\omega; k, \tau) = 0$ can be expressed as $-kH_{yu}(j\omega) = e^{-j\omega\tau}$, where

$$H_{yu}(j\omega) \triangleq \frac{1}{\Pi_{i=1}^m (j\omega + \lambda_i)}.$$

The geometric interpretation of the previous relation is quite simple: critical characteristic roots on the imaginary axis are obtained at the intersection of the plot $-kH_{yu}(j\omega)$ with the unit circle of the complex plane.

It is noteworthy that $\lim_{\omega \to \infty} |kH_{yu}(j\omega)| = 0$, which simply means that the curve $-kH_{yu}(j\omega)$ will always end at the origin of the complex plane. Since $kH_{yu}(j\omega)$ is a smooth curve, the intersection with the unit circle always exists (in other words, crossing always exists) if

$$|kH_{yu}(0)| = \left| \frac{k}{\Pi_{i=1}^{m} \lambda_i} \right| > 1.$$

We recall that the system (16.7) is called *hyperbolic* if it has no characteristic roots crossing the imaginary axis when the delay parameter is increased from 0 to ∞. In such a case, the stability of the system free of delay will be preserved for all positive delay. In other words, the system is *delay-independent* asymptotically stable if it is stable in the case free of delays. Under the assumption of stability of the strictly proper transfer function kH_{yu}, the frequency-sweeping test $|kH_{yu}(j\omega)| < 1$ for all $\omega \in \mathbb{R}_+$ is nothing else than the Tsypkin criterion encountered in control engineering area (see, for instance, [290]). Geometrically speaking, we have delay-independent asymptotic stability if and only if the graph of $-kH_{yu}(j\omega), \omega \in \mathbb{R}_+$, stays inside the unit circle of the complex plane (with some eventual tangency at the point $(-1,0)$ at the "frequency" $\omega = 0$) under the assumption of a delay-free stable system.

In what follows, let us consider the system described by (16.7) with

$$|\, k \,| \leq \Pi_{i=1}^{m} |\lambda_i|. \tag{16.8}$$

Proposition 16.3. *Under the assumption* $|\, k \,| \leq \Pi_{i=1}^{m} |\lambda_i|$, *the delay system given by the characteristic equation* (16.7) *is delay-independent stable (unstable) if the system free of delay is stable (unstable).*

Proof. The result is straightforward since

$$|kH_{yu}(j\omega)| = \frac{|k|}{\sqrt{\prod_{i=1}^{m}(\omega^2 + \lambda_i^2)}} \leq \frac{|k|}{\prod_{i=1}^{m} |\lambda_i|} = |kH_{yu}(0)| \leq 1. \tag{16.9}$$

□

Consider now the system described by (16.7) under the assumption

$$|\, k \,| > \Pi_{i=1}^{m} |\lambda_i|. \tag{16.10}$$

In this case it is easy to see that increasing the delay induces the presence of characteristic roots crossing $j\mathbb{R}$. Let us first consider the delay-free stable case.

Proposition 16.4. *Under the assumption that the condition*

$$|\, k \,| > \Pi_{i=1}^{m} |\lambda_i|$$

holds, the system (16.7) *has only one crossing frequency* $\omega^* > 0$. *Furthermore, if the corresponding delay-free system is stable, then* (16.7) *is asymptotically stable for all* $\tau \in [0, \tau_m)$, *where the delay margin* τ_m *is the smallest positive value given by*

$$\tau_m = \frac{1}{\omega^*} \left(2h\pi + \angle(-k) - \sum_{i=1}^{m} \arctan \frac{\omega^*}{\lambda_i} \right), \quad h \in \mathbb{Z}. \tag{16.11}$$

Proof. In fact, we will prove the following statement: *if condition* (16.10) *holds, the equation*

$$(j\omega + \lambda_1)\ldots(j\omega + \lambda_m) + ke^{-j\omega\tau} = 0 \tag{16.12}$$

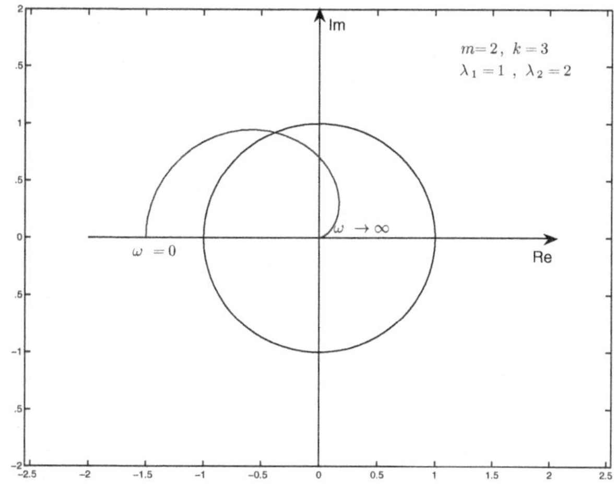

Figure 16.1. *The intersection between the unit circle of the complex plane and the graph of* $-kH_{yu}(j\omega)$ *when* $|kH_{yu}(j\omega)| > 1$ [283].

has only one positive solution ω^* (see Figure 16.1). It is clear that the roots of a delay-free stable system with only one crossing frequency can cross the imaginary axis only towards instability as the delay increases. Thus, the system will be asymptotically stable for all $\tau \in [0, \tau_m)$, where τ_m is the delay corresponding to the first crossing.

Using the modulus condition, equation (16.12) becomes

$$\left|\prod_{i=1}^{m}(j\omega + \lambda_i)\right| = |k| \tag{16.13}$$

or, equivalently,

$$\prod_{i=1}^{m}(\omega^2 + \lambda_i^2) = k^2. \tag{16.14}$$

Since condition (16.10) holds, we simply deduce that there exists at least one "frequency" ω satisfying the previous equation. On the other hand, the application $\omega \in \mathbb{R}_+ \mapsto \prod_{i=1}^{m}(\omega^2 + \lambda_i^2) \in \mathbb{R}_+$ is an increasing function of ω. Therefore, equation (16.14) has exactly one solution ω^*, which is also the *unique* solution of (16.12). Some simple computations prove that the delay values corresponding to the crossing frequency ω^* are given by the right-hand side of formula (16.11). Obviously, τ_m is the smallest positive value given by (16.11) since we are interested in finding the largest interval guaranteeing the asymptotic stability of the system (16.7). □

Next, we suppose that condition (16.10) still holds, but the delay-free system associated to (16.7) is unstable. In the proof of Proposition 16.4 we have seen that the frequency crossing set of (16.7) contains only one element. Then, we can conclude that (16.7) cannot be stabilized by increasing the delay value if the system free of delay has more than one simple unstable root. A more general result is stated in [191]:

Proposition 16.5. *Consider the class of quasi-polynomials* $q(s) + ke^{-s\tau}$ *such that* $q(s)$ *is unstable and there is no* $\tau > 0$ *such that* $\tau q(s) + q'(s)$ *is a Hurwitz stable polynomial. Then* $q(s) + ke^{-s\tau}$ *is unstable for all pairs* $(k, \tau) \in \mathbb{R} \times \mathbb{R}_+$.

However, the previous result provides only a *necessary condition* for the existence of some stabilizing delay τ when the delay-free system is unstable.

In what follows, we shall prove that the particular class of systems described by (16.7) allows a complete answer for this particular delay stabilization problem. Precisely, we have already derived the delay interval preserving the system's stability when the delay-free system is stable, and now we shall see that the delay-free unstable system cannot be stabilized by increasing the delay value.

In order to do that, let us consider the case that is not already treated by Proposition 16.5. In other words, we consider the case where $q(s)$ *has only one simple unstable root.* Therefore, equation (16.7) is simply rewritten as

$$(\lambda-\lambda_1)(\lambda+\lambda_2)\dots(\lambda+\lambda_m)+ke^{-\lambda\tau}=0, \qquad \lambda_1,\dots,\lambda_m>0. \tag{16.15}$$

It is worth mentioning that we analyze the behavior of (16.15) for the sake of completeness, even if the meaningful λ_i, $i=1,2$, in (16.1) must be positive.

Proposition 16.6. *When the condition* (16.10) *holds, the delay-free unstable system defined by the characteristic equation* (16.7) *remains unstable for all* $\tau>0$.

Proof. In order to prove the result, we derive the crossing direction associated to the unique crossing frequency ω^* of the system (16.10). Straightforward computations lead to

$$\Re\left(\frac{d\lambda}{d\tau}\right)_{\lambda=j\omega}=\sum_{k=1}^{m}\frac{1}{\omega^2+\lambda_i^2}.$$

Therefore $\Re(\frac{d\lambda}{d\tau})_{\lambda=j\omega}>0$, which means that all the crossings are towards instability and the system cannot be stabilized by increasing the delay value. □

Remark 16.7. *It is noteworthy that* $\Re(\frac{d\lambda}{d\tau})_{\lambda=j\omega}$ *is strictly positive, which means that all the characteristic eigenvalues (roots) that arrive on the imaginary axis will cross it. In other words, there exists no tangentially stable point.*

16.2.3 ▪ Illustrative example

First, we consider the biological system (16.2) with the translation and degradation effects given by $c_1=4/3$, $\lambda_1=1$, $\lambda_2=2$, and the feedback mechanism described by the Hill function $g(x)=1+10/(x^2+1)$. Next, we derive the overall delay τ that guarantees the system's stability. The equilibrium point of the system is $(\overline{x}_1\,\overline{x}_2)=(2,\frac{3}{2})$ and $g'(\overline{x}_1)=-\frac{8}{5}$. Thus, $k=\frac{32}{15}$ and the corresponding characteristic equation is

$$(\lambda+1)(\lambda+2)+\frac{32}{15}e^{-\lambda\tau}=0. \tag{16.16}$$

We can directly verify that the delay-free system is stable. On the other hand, since $k>\lambda_1\lambda_2$, the system is not delay-independent stable. Therefore, there exists some delay margin τ_m such that for $\tau\in[0,\tau_m)$ the system is stable and for $\tau\geq\tau_m$ the system is unstable. Following the proof of Proposition 16.4, we derive the crossing frequency $\omega^*\simeq 0.3285$. Explicitly, we have derived the solution of $|kH_{yu}(j\omega)|=1$ or, equivalently,

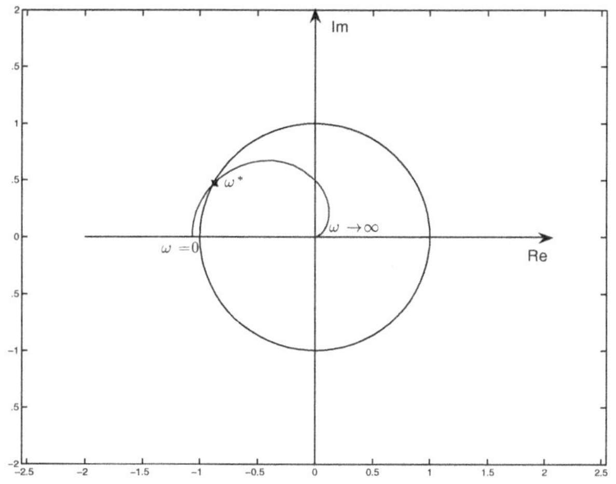

Figure 16.2. *The intersection between the unit circle of the complex plane and the graph of* $-kH_{yu}(j\omega)$ *for the system* (16.16) [283].

the intersection of $-kH_{yu}(j\omega)$ and the unit circle (see Figure 16.2). The formula (16.11) leads to the delay margin

$$\tau_m = \frac{1}{\omega^*}\left(\pi - \arctan\omega^* - \arctan\frac{\omega^*}{2}\right) \simeq 8.1017.$$

16.3 ▪ Delay effects on stability in some human respiration models

The human respiratory system is an extremely complicated mechanism, and a large variety of dynamical models describing its behavior exists. In general, the delays represent the *(circulatory) transport time* between the lungs and the peripheral and central chemoreceptors.

16.3.1 ▪ Delay model and its linearization

In what follows, we consider a two-compartment representation (lungs and tissues) as an interconnection between some "plant" (in which CO_2 exchange takes place) and some "controller" (which regulates the CO_2 partial pressures in the body), as discussed by Vielle and Chauvet [394].

The model is written as follows:

$$\begin{cases} \dot{P}_T(t) = \dfrac{\bar{Q}}{V_T}(P_L(t) - P_T(t)) + \dfrac{\bar{M}}{\alpha V_T}, \\ \dot{P}_L(t) = \dfrac{\alpha\bar{Q}B}{V_L}(P_T(t) - P_L(t)) - \dfrac{1}{V_L}(P_L(t) - P_1)F(P_L(t-\tau)), \end{cases} \tag{16.17}$$

where the variables P_L, P_T denote the CO_2 partial pressure and $F(\cdot)$ is the controller function. Here, the subscript "L" ("T") denotes "lungs"("tissues"). Next, the parameters V_T (volume, tissues), V_L (volume, lungs), B (barometric pressure minus the water vapor pressure), $\bar{Q}$ (blood flow), $\bar{M}$ (CO_2 metabolic production rate), α (CO_2 dissociation curve slope), and P_1 are positive.

The transport delay appears in the equation by the controller action F, which is an appropriate nonlinear function of the CO_2 partial pressure P_L. Further discussions on the way such a controller is defined can be found in [394] and the references therein.

It is easy to see that (16.17) has a *unique equilibrium* point $(\bar{P}_T, \bar{P}_L)$. By introducing the new variables $x_1(t) = P_T(t) - \bar{P}_T$, $x_2(t) = P_L(t) - \bar{P}_L$ and neglecting the nonlinear second-order terms, system (16.17) can be linearized as follows:

$$\begin{cases} \dot{x}_1(t) = -ax_1(t) + ax_2(t), \\ \dot{x}_2(t) = bx_1(t) - (b+c)x_2(t) - dx_2(t-\tau), \end{cases} \tag{16.18}$$

where the constants (a,b,c,d) are all *positive* and given by

$$a = \frac{\bar{Q}}{V_T}, \quad b = \frac{\alpha \bar{Q} B}{V_L}, \quad c = \frac{F(\bar{P}_L)}{V_L}, \quad d = \frac{F'(\bar{P}_L)(\bar{P}_L - P_I)}{V_L}. \tag{16.19}$$

Such a delay system has a particular structure. More precisely, it can be written in the form

$$\dot{x}(t) = A_0 x(t) + b_0 c_0^T x(t-\tau), \tag{16.20}$$

where $x \in \mathbb{R}^2$ is given by $x^T = [x_1 \; x_2]^T$ and b_0 and c_0 are column matrices. As seen in Chapter 1, the characteristic function of such a system is

$$p(\lambda;\, \tau) := Q(\lambda) + P(\lambda)e^{-\lambda\tau} = Q(\lambda)\left(1 + h(\lambda)e^{-\lambda\tau}\right), \tag{16.21}$$

where $h(\lambda) = P(\lambda)/Q(\lambda)$, and the polynomials P and Q are given by

$$Q(\lambda) = \lambda^2 + \lambda(a+b+c) + ac, \qquad P(\lambda) = \lambda d + ad. \tag{16.22}$$

Since a, b, c, and d are strictly positive, it is clear that $p(\lambda;\, 0)$ is Hurwitz stable, that is, all its roots have strictly negative real parts. Furthermore, the polynomial Q is also Hurwitz, and thus the quasi-polynomial $p(\lambda;\, \tau)$ shares the same characteristic roots on the imaginary axis with the analytic function $q(\lambda;\, \tau) := 1 + h(\lambda)e^{-\lambda\tau}$.

In what follows we will analyze the stability of the model (16.20) by taking into account the particular form and properties of its characteristic function $p(\lambda;\, \tau)$.

16.3.2 ▪ Stability analysis and delay intervals

In Chapter 6, we proposed an algorithm for computing delay intervals guaranteeing asymptotic stability by using the function $q(\lambda;\, \tau)$ instead of the original quasi-polynomial $p(\lambda;\, \tau)$. In what follows, we will explicitly apply such an algorithm to our stability analysis problem. Since the strictly proper transfer h is stable (the denominator Q is Hurwitz), it follows that it is bounded on the imaginary axis. In this context, it is easy to see that the crossing set Ω, that is, the set of all crossing roots $\omega \in \mathbb{R}_+$ w.r.t. the imaginary axis, is given by the solution of the polynomial equation $|P(j\omega)| = |Q(j\omega)|$. As expected, only two situations can occur: (i) $|P(j\omega)| < |Q(j\omega)|$ for all $\omega \in \mathbb{R}_+^*$, or (ii) there exists at least one frequency $\omega^* > 0$ such that $|P(j\omega^*)| = |Q(j\omega^*)|$. Let us analyze each case separately.

Delay-independent stability

Condition (i) simply says that system (16.20) is hyperbolic, that is, there are no roots crossing w.r.t. the imaginary axis if the delay parameter is increased from 0 to ∞. In such a case, the stability of the system free of delays is preserved for all positive delays. In other words, the system is *delay-independent* asymptotically stable. Under the assumption of stability of the strictly proper transfer function h, the *frequency-sweeping* test $|\, h(j\omega)\,| < 1$ for all $\omega \in \mathbb{R}_+$ is nothing else than the Tsypkin criterion (see, e.g., [290]). Geometrically speaking, we have delay-independent asymptotic stability if and only if the graph of $-h(j\omega)$ for all $\omega \in \mathbb{R}_+^*$ stays inside the unit circle of the complex plane, with some eventual tangency at the point $(-1,0)$ for the frequency $\omega = 0$, under the assumption of a delay-free stable system. A *necessary condition* for no crossing w.r.t. the imaginary axis is $|\, Q(0)\,| \geq |\, P(0)\,|$, that is, $c \geq d$. It is important to point out that we can have the equality above since 0 is not a root of the characteristic function $p(\lambda;\, \tau)$ for any positive delay $\tau \in \mathbb{R}_+$. Geometrically speaking, the equality simply describes the tangency property mentioned above, since $P(0) = ad$ will be equal to $Q(0) = ac$ if $c = d$.

Now let us check whether the condition $c \geq d$ is also *sufficient* for getting delay-independent stability. Simple computations prove that there do not exist crossing roots if and only if the second-order equation

$$x^2 + \left[(a+b)^2 + 2bc + c^2 - d^2\right]x + a^2(c^2 - d^2) = 0 \tag{16.23}$$

has *no strictly positive* roots. This last condition holds if and only if the inequalities

$$\begin{cases} a^2(c^2 - d^2) \geq 0, \\ (a+b)^2 + 2bc + c^2 - d^2 \geq 0 \end{cases}$$

are satisfied simultaneously: under the constraint of positive parameters a, b, c, and d. In conclusion, we obtain the following simple *delay-independent stability* condition:

$$c \geq d. \tag{16.24}$$

Figure 16.3 depicts such a delay-independent stability condition for some positive c, and d such that $c > d$. As explained in Chapter 6, the plot of $-P(j\omega)/Q(j\omega)$ and its intersection with the unit circle are not sufficient to conclude about asymptotic stability of the corresponding system. We can only expect to detect the occurrence of crossing roots for some delay values. Finally, notice that the same condition was obtained in [394], but using a different argument.

Delay-dependent stability

As expected, the characteristic function $p(\lambda;\, \tau)$ has roots crossing the imaginary axis if and only if the second-order equation (16.23) has *at least one strictly positive root*. Based on the particular form of this equation, it follows that such a situation appears if and only if

$$d > c. \tag{16.25}$$

More precisely, if $d > c$, equation (16.23) will always have two real roots of opposite sign. In other words, *only one root* x_+ will be *positive*, that is, the *frequency crossing set* Ω is given by: $\Omega = \{\omega_+\}$, where $\omega_+ = \sqrt{x_+}$, with x_+ the only positive root of (16.23). This corresponds to the situation shown in Figure 16.4.

In conclusion, since only one crossing root exists, the corresponding crossing direction is always *towards instability*. Thus, the system will be asymptotically stable for *all*

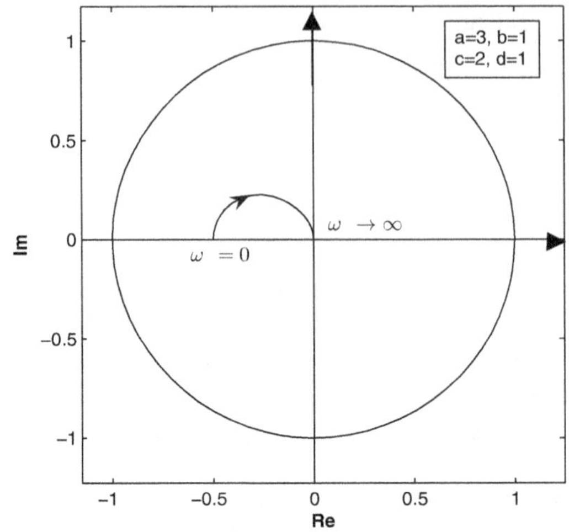

Figure 16.3. *Delay-independent stability due to the intersection between the ratio curve $-h(j\omega)=-P(j\omega)/Q(j\omega)$ and the unit circle being empty and the system free of delay being asymptotically stable.*

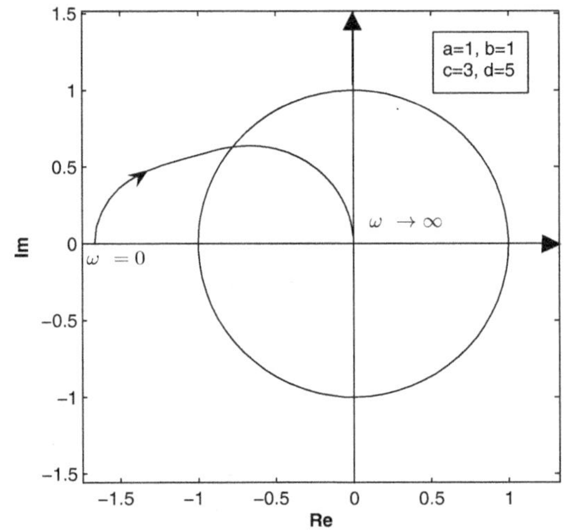

Figure 16.4. *Delay-dependent stability due to the intersection between the ratio curve $-h(j\omega)=-P(j\omega)/Q(j\omega)$ and the unit circle consisting of only one point, and containing existing points outside and inside the unit circle.*

delays $\tau \in [0,\tau_m)$, where the delay margin τ_m is given by the formula

$$\tau_m = \min\left\{\tau>0:\ \tau = \frac{\angle\left(-\frac{P(j\omega_+)}{Q(j\omega_+)}\right)+2\pi l,\ l\in\mathbb{Z}}{\omega_+}\right\}, \tag{16.26}$$

with ω_+ the unique element of the set Ω.

The corresponding crossing direction is towards instability, and it is independent of the parameter values. Therefore, we can conclude that the linearized second-order system is asymptotically stable for all $\tau \in [0,\tau_m)$. Furthermore, it is *unstable* for all $\tau \geq \tau_m$.

Finally, notice that the same stability conditions were obtained in [394], but using a slightly different argument.

16.3.3 ▪ Discussions and interpretations

As pointed out by Vielle and Chauvet [394], the respiration model above includes a two-compartment representation (lungs and tissues) subject to a general class of controllers. The approach considered takes into consideration the particular structure of the linearized two-compartment model. Indeed, the corresponding delay matrix is of rank one, which simplifies the analysis. We believe that such a property was not sufficiently exploited in the literature. The next section, devoted to the analysis of some models for the human immune system's dynamics in leukemia, provides a further argument for this thesis.

Finally, our sensitivity analysis with respect to the delay-parameter proves that the essential quantities for delay-induced instabilities are the parameters c and d given by

$$c = \frac{F(\bar{P}_L)}{V_L}, \qquad d = \frac{F'(\bar{P}_L)(\bar{P}_L - P_I)}{V_L},$$

that is, the evaluation of the continuous controller function F and of its derivative F' relating air flow in the lungs with delayed partial pressure in arterial blood, under the "standard" assumption that the partial pressure in arterial blood is greater than "outside." Please note that the partial pressures in arterial blood and in the lungs are identical (equilibrium) [394]. In conclusion, the delay-induced stability is completely characterized by the value F'/F evaluated at $\bar{P}_L$ and compared to $1/(\bar{P}_L - P_I)$ (which are both directly defined by the corresponding equilibrium), and by the controller law. Some discussions on the linear and Hill controllers and appropriate (physiological) assumptions on F can be found in [394].

16.4 ▪ Delays in immune dynamics models in leukemia

In what follows, we consider a second-order nonlinear model that describes immune dynamics in chronic myelogenous leukemia after a bone marrow transplantation. The particularity of the system is the presence of *four* distinct *delays* in some large range.

16.4.1 ▪ Delay model and its linearization

In what follows, we shall consider a simplification (and modification) of the nonlinear model proposed by [92] to describe the post-transplantation dynamics of the immune response to chronic myelogenous leukemia (CML). The original model from [92] tracks the time evolution of six cell populations (cancer cells, antidonor T cells, general patient blood cells, antihost T cells, anticancer T cells, and general donor blood cells). Hence, to simplify the stability analysis, we consider the reduced system (16.27). This reduced system considers only the anticancer T cell population, $T(t)$, and the active and dying cancer populations, $C_A(t)$ and $C_D(t)$. The total population of cancer cells is denoted by $C(t)$, i.e., $C(t) = C_A(t) + C_D(t)$. All other variables are constant and nonnegative.

To be more precise, the T cells are moving through different compartments depending on their "decisions" or probabilistic events that occur when interacting with a cancer cell (rate kCT). Now, if the T cell ignores the cancer cells with some probability p_2, it passes through an "Ignore" compartment and returns to the base after some delay σ. Now, if the T cell reacts with probability p_1, it passes to the "React" compartment. Then it also

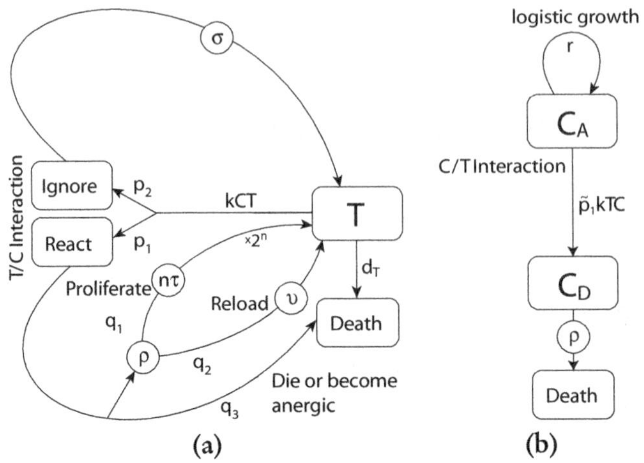

Figure 16.5. (a) *Evolution of anticancer T cells;* (b) *Evolution of cancer cells, C, in the modified model* [299].

has possibilities q_1, q_2, and q_3 of proliferating, reloading (after killing the cancer cell), or dying. Each of these subcompartments (proliferate, reload, and die) has its own time delay:

$$\left\{\begin{aligned} \frac{dT(t)}{dt} &= -d_T T(t) - kC(t)T(t) + p_2 kC(t-\sigma)T(t-\sigma) \\ &\quad + 2^N p_1 q_1 kC(t-\rho-N\tau)T(t-\rho-N\tau) \\ &\quad + p_1 q_2 kC(t-\rho-\upsilon)T(t-\rho-\upsilon), \\ \frac{dC_A(t)}{dt} &= rC_A(t)(1-C_A(t)/K) - \tilde{p}_1 kC_A(t)T(t), \\ \frac{dC_D(t)}{dt} &= \tilde{p}_1 kC_A(t)T(t) - \tilde{p}_1 kC_A(t-\rho)T(t-\rho). \end{aligned}\right. \tag{16.27}$$

The stages of the evolution of anticancer T cells are demonstrated in Figure 16.5(a). These cells interact with the cancer population, C. In the equation for the T cell population, we define the total cancer population C to be the sum, $C_A + C_D$, of the two cancer subpopulations. In addition, active cancer cells, C_A, multiply at a logistic growth rate indicated by the closed loop at the top of Figure 16.5(b). The logistic parameter r represents the net growth rate of cancer.

Furthermore, we assume the past history is constant before time 0. In other words, we set $T(t) = T_0$, $C_A(t) = C_{A,0}$, $C_D(t) = C_{D,0}$ for $t \leq 0$, where the following compatibility condition holds:

$$C_{D,0} = \tilde{p}_1 kC_{A,0}T_0. \tag{16.28}$$

The compatibility condition follows by noticing that when T and C_A remain constant, the equation for $C_D(t)$ in (16.27) implies (16.28). Note that condition (16.28), although biologically reasonable, is technically stronger than necessary. In fact, the more general condition

$$C_D(0) = \tilde{p}_1 k \int_{-\rho}^{0} C_A(u)T(u)du \tag{16.29}$$

applies to any biologically reasonable (e.g., continuous and nonnegative) choice of past history for C_A and T. This condition also guarantees the nonnegativity of $C_D(t)$. Indeed, from (16.27), we obtain

$$C_D(t) = \left[C_D(0) - \tilde{p}_1 k \int_{-\rho}^{0} C_A(u)T(u) \right] + \tilde{p}_1 k \int_{t-\rho}^{t} C_A(u)T(u)du,$$

and (16.29) implies that the constant term above equals 0. The nonnegativity of $T(t)$ and $C_A(t)$ is guaranteed by the form of (16.27). In the system (16.27) there are *four distinct delays*, namely σ, $\rho + N\tau$, $\rho + \upsilon$, and ρ. These constants respectively represent the time for nonreactive interactions between T cells and cancer cells (σ), the time for reactive interactions (ρ), the time for one round of cell division (τ), the T cell recovery time after killing a cancer cell (υ), and the average number of T cell divisions after stimulation (N). The relevant values, taken from [92], are approximately $\sigma = 0.0007$ days $= 1$ min; $\rho = 0.0035$ day $= 5$ min; $\tau = 1$ day; $\upsilon = 1$ day. Here, N is between 1 and 8 and probably close to 3. The model includes multiple delays in a large range starting from one minute up to several days. Due to the scale difference, without any loss of generality we can define ρ and σ as *small delays* and the remaining delays as *large*.

16.4.2 ▪ Delay model and its linearization

We start by deriving the linearized model of the system (16.27) and study its stability in the delay-free case. The particular structure of the linearized model leads to a nice form of the corresponding characteristic equation in the presence of delays. Its structure will allow us to define an appropriate *auxiliary system* with only the *small delays* needed in the next section to complete the analysis. For convenience, let

$$\begin{array}{lll} b_1 = d_T, & b_5 = p_1 q_2 k, & \tilde{\tau} = \rho + N\tau, \\ b_2 = k, & c_1 = r, & \tilde{\upsilon} = \rho + \upsilon, \\ b_3 = p_2 k, & c_2 = r/K, & \\ b_4 = 2^N p_1 q_1 k, & c_3 = \tilde{p}_1 k, & \end{array} \tag{16.30}$$

and rewrite (16.27) as

$$\begin{cases} \dfrac{dT(t)}{dt} = -b_1 T(t) - b_2(C_A(t) + C_D(t))T(t) \\ \qquad + b_3(C_A(t-\sigma) + C_D(t-\sigma))T(t-\sigma) \\ \qquad + b_4(C_A(t-\tilde{\tau}) + C_D(t-\tilde{\tau}))T(t-\tilde{\tau}) \\ \qquad + b_5(C_A(t-\tilde{\upsilon}) + C_D(t-\tilde{\upsilon}))T(t-\tilde{\upsilon}), \\ \dfrac{dC_A(t)}{dt} = c_1 C_A(t) - c_2 C_A(t)^2 - c_3 C_A(t)T(t), \\ \dfrac{dC_D}{dt} = c_3 C_A(t)T(t) - c_3 C_A(t-\rho)T(t-\rho). \end{cases} \tag{16.31}$$

We note that all parameters in (16.30) are positive. Let $b = -b_2 + b_3 + b_4 + b_5$. Then the fixed points, $(T_0, C_{A,0}, C_{D,0})$, of (16.31) are solutions to

$$\begin{cases} 0 = -b_1 T_0 + b(C_{A,0} + C_{D,0})T_0 = (-b_1 + b(C_{A,0} + C_{D,0}))T_0, \\ 0 = (c_1 - c_2 C_{A,0} - c_3 T_0)C_{A,0}. \end{cases}$$

Note from (16.31) that since $C_{A,0}$ and T_0 are constant, we have $C_{D,0} = \rho c_3 C_{A,0} T_0$ (see (16.28)), so

$$\begin{cases} 0 = (-b_1 + b(1 + \rho c_3 T_0)C_{A,0})T_0, \\ 0 = (c_1 - c_2 C_{A,0} - c_3 T_0)C_{A,0}. \end{cases}$$

From these equations, we immediately calculate two fixed points $(T_0, C_{A,0}, C_{D,0})$ given by $(0,0,0)$, $(0,K,0)$. In the case where $T_0 \neq 0$, we obtain the quadratic equation

$$\rho b c_3^2 T_0^2 + b c_3 (1 - \rho c_1) T_0 + b_1 c_2 - b c_1 = 0. \tag{16.32}$$

This means that we can determine $C_{A,0}$ and $C_{D,0}$ by the following equations:

$$\begin{cases} C_{A,0} = (c_1 - c_3 T_0)/c_2, \\ C_{D,0} = \rho c_3 C_{A,0} T_0. \end{cases} \tag{16.33}$$

Hence, the system may contain up to two additional fixed points, for a total of between 2 and 4 fixed points, depending on the coefficients of (16.32).

The *fixed point* $(0,0,0)$ represents the ideal outcome, where the cancer population is entirely eliminated and the cancer-reactive T cells become unnecessary and disappear. We will later show that, unfortunately, this fixed point is a *saddle* regardless of the values of the parameters, which means that this fixed point is unattainable. Next, the *fixed point* $(0,K,0)$ represents the case where cancer expands to its carrying capacity, K, and the cancer-reactive T cells die off completely. This is the most undesirable state, and we will later show that the fixed point is *unstable* for biologically reasonable parameter choices. Finally, the *fixed points* determined by (16.32) represent the scenarios where the cancer and T cell populations coexist. This means that cancer is not entirely eliminated but is controlled in part by the immune response. For biologically reasonable parameters, we will show that this state can be *stable* for appropriate values of the delays.

To study the stability of (16.31), we linearize the nonlinear system around the fixed point $(T_0, C_{A,0}, C_{D,0})$ and obtain

$$\begin{cases} \dfrac{dT(t)}{dt} = -\tilde{b}_1 T(t) - b_2 T_0 (C_A(t) + C_D(t)) \\ \qquad + b_3 (C_{A,0} + C_{D,0}) T(t-\sigma) + b_3 T_0 (C_A(t-\sigma) + C_D(t-\sigma)) \\ \qquad + b_4 (C_{A,0} + C_{D,0}) T(t-\tilde{\tau}) + b_4 T_0 (C_A(t-\tilde{\tau}) + C_D(t-\tilde{\tau})) \\ \qquad + b_5 (C_{A,0} + C_{D,0}) T(t-\tilde{\upsilon}) + b_5 T_0 (C_A(t-\tilde{\upsilon}) + C_D(t-\tilde{\upsilon})), \\ \dfrac{dC_A(t)}{dt} = \tilde{c}_1 C_A(t) - c_3 C_{A,0} T(t) - c_3 T_0 C_A(t), \\ \dfrac{dC_D(t)}{dt} = c_3 C_{A,0} T(t) + c_3 T_0 C_A(t) - c_3 C_{A,0} T(t-\rho) - c_3 T_0 C_A(t-\rho), \end{cases} \tag{16.34}$$

where $\tilde{b}_1 = b_1 + b_2(C_{A,0} + C_{D,0})$ and $\tilde{c}_1 = c_1 - 2c_2 C_{A,0}$.

16.4.3 ▪ Stability analysis in the delay-parameter space

The characteristic equation of (16.34) is $\det(A(e^{-\lambda}) - \lambda I) = 0$, where

$$\begin{aligned} A_{11}(e^{-\lambda}) &= -\tilde{b}_1 + b_3(C_{A,0} + C_{D,0})e^{-\sigma\lambda} + b_4(C_{A,0} + C_{D,0})e^{-\tilde{\tau}\lambda} \\ &\quad + b_5(C_{A,0} + C_{D,0})e^{-\tilde{v}\lambda}, \\ A_{12}(e^{-\lambda}) &= -b_2 T_0 + b_3 T_0 e^{-\sigma\lambda} + b_4 T_0 e^{-\tilde{\tau}\lambda} + b_5 T_0 e^{-\tilde{v}\lambda}, \\ A_{13}(e^{-\lambda}) &= -b_2 T_0 + b_3 T_0 e^{-\sigma\lambda} + b_4 T_0 e^{-\tilde{\tau}\lambda} + b_5 T_0 e^{-\tilde{v}\lambda}, \\ A_{21}(e^{-\lambda}) &= -c_3 C_{A,0}, \\ A_{22}(e^{-\lambda}) &= \tilde{c}_1 - c_3 T_0, \\ A_{23}(e^{-\lambda}) &= 0, \\ A_{31}(e^{-\lambda}) &= c_3 C_{A,0} - c_3 C_{A,0} e^{-\rho\lambda} = c_3 C_{A,0}\eta(\rho,\lambda), \\ A_{32}(e^{-\lambda}) &= c_3 T_0 - c_3 T_0 e^{-\rho\lambda} = c_3 T_0 \eta(\rho,\lambda), \\ A_{33}(e^{-\lambda}) &= 0, \end{aligned}$$

where we let $\eta(\rho,\lambda) = 1 - e^{-\rho\lambda}$. Expanding this determinant, we obtain the following characteristic equation:

$$p_0(\rho,\lambda) + p_1(\rho,\lambda)e^{-\sigma\lambda} + p_2(\rho,\lambda)e^{-\tilde{\tau}\lambda} + p_3(\rho,\lambda)e^{-\tilde{v}\lambda} = 0, \tag{16.35}$$

where

$$\begin{aligned} p_0(\rho,\lambda) &= -\lambda^3 + (-\tilde{b}_1 + \tilde{c}_1 - c_3 T_0)\lambda^2 + (\tilde{b}_1\tilde{c}_1 - b_1 c_3 T_0 - b_2 c_3 C_{D,0} T_0)\lambda \\ &\quad - (b_2 c_3 C_{A,0} T_0)\lambda\eta(\rho,\lambda) + (b_2 \tilde{c}_1 c_3 C_{A,0} T_0)\eta(\rho,\lambda), \\ p_1(\rho,\lambda) &= b_3 C_{A,0} p_{\rho,\text{aux}}(\lambda), \\ p_2(\rho,\lambda) &= b_4 C_{A,0} p_{\rho,\text{aux}}(\lambda), \\ p_3(\rho,\lambda) &= b_5 C_{A,0} p_{\rho,\text{aux}}(\lambda) \end{aligned}$$

for

$$p_{\rho,\text{aux}}(\lambda) = (1+\rho c_3)\lambda^2 + (\rho c_3^2 T_0 - \tilde{c}_1(1+\rho c_3))\lambda + (c_3 T_0)s\eta(\rho,\lambda) - (\tilde{c}_1 c_3 T_0)\eta(\rho,\lambda).$$

We now fix ρ, and introduce an *auxiliary system* that is associated with the small delay σ and is given by the following characteristic function:

$$p_{\rho,\sigma}(\lambda) = p_0(\rho,\lambda) + p_1(\rho,\lambda)e^{-\sigma\lambda}. \tag{16.36}$$

After finding the stable interval for ρ, we can easily characterize the *stability crossing curves* of $p_{\rho,\sigma}(s)$ given by (16.36) in the delay-parameter space defined by the small delays ρ and σ. In this sense, we make the following assumption:

Assumption 16.8. *Let $\mathscr{I}_\rho \subset \mathbb{R}_+$, and $\mathscr{I}_\sigma \subset \mathbb{R}_+$ be some real intervals such that there exists some $\delta > 0$ such that $p_{\rho,\sigma}(\lambda) \neq 0$ for all the pairs $(\sigma,\rho) \in \mathscr{I}_\sigma \times \mathscr{I}_\rho$, and for all $s \in \mathscr{V}_\delta$, where $\mathscr{V}_\delta$ is defined by*

$$\mathscr{V}_\delta = \{\lambda \in \mathbb{C} : \quad -\delta < \Re(\lambda) < \delta\}. \tag{16.37}$$

The assumption above can be seen as a *regularity* condition for the original linearized model. It means that there exists some delay intervals such that $p_{\rho,\sigma}$ is *invertible* in some neighborhood $\mathcal{V}_\delta$ of the imaginary axis for all the pairs $(\sigma,\rho) \in \mathcal{I}_\sigma \times \mathcal{I}_\rho$. It is important to note that the assumption above is not restrictive.

Identification of the crossing points, and crossing set characterization

Since we assume that ρ is fixed, to simplify notation we will write $p_2(\lambda)$ and $p_3(\lambda)$ instead of $p_2(\rho,\lambda)$ and $p_3(\rho,\lambda)$. We have the following result:

Proposition 16.9. *Assume that the auxiliary system given by* (16.36) *satisfies Assumption* 16.8. *Define*

$$a_{\tilde{\tau},\tilde{\upsilon}}(\lambda) = 1 + a_{\tilde{\tau}}(\lambda)e^{-\lambda\tilde{\tau}} + a_{\tilde{\upsilon}}(\lambda)e^{-\lambda\tilde{\upsilon}}, \tag{16.38}$$

where $a_{\tilde{\tau}}(\lambda) = \frac{p_3(s)}{p_{\rho,\sigma}(\lambda)}$ *and* $a_{\tilde{\upsilon}}(\lambda) = \frac{p_4(s)}{p_{\rho,\sigma}(\lambda)}$, *for all* $(\sigma,\rho) \in \mathcal{I}_\sigma \times \mathcal{I}_\rho$. *Then for any* $(\sigma,\rho) \in \mathcal{I}_\sigma \times \mathcal{I}_\rho$, *the characteristic equation associated with* (16.35) *and* $a_{\tilde{\tau},\tilde{\upsilon}}(s)$ *have the same solutions in a neighborhood* $\mathcal{V}_\delta$ *of the imaginary axis, where*

$$\mathcal{V}_\delta := \{\lambda \in \mathbb{C} : \quad \delta \geq \Re(\lambda) > -\delta\}$$

for some $\delta > 0$.

Proof. The proof follows from the continuity argument with respect to the delay parameters, and from the equivalence between the characteristic equations (16.38) and (16.35) if $a_{\rho,\sigma}(\lambda) \neq 0$ in some vertical strip including the imaginary axis (Assumption 16.8). □

Weak T/C interactions and delay-independence type results

Consider first the case of *weak T/C cell interactions.* Without any loss of generality, a weak T/C interaction simply means a reduced probability of interactions of anticancer and cancer cells. In other words, the weak T/C interaction describes the situations when the anticancer cells will tend to "ignore" the cancer cells.

Roughly speaking, such a T/C interaction will be translated in "small" values for the coefficients $b_4 = 2^N p_1 q_1 k$ and $b_5 = p_1 q_2 k$, which may correspond to the case when *no crossing* in the delay-parameter space defined by large delays exists. In conclusion, a weak T/C interaction may correspond to some *delay-independent* type of property with respect to the delay parameters under consideration, and the last argument will give us a way to define a *measure* for characterizing the interaction character.

With the notations and the results above, we have the following:

Proposition 16.10 (delay-independence in large delays). *Assume that the auxiliary system given by the characteristic equation* (16.36) *satisfies Assumption* 16.8, *and that* $a_{\tilde{\tau},\tilde{\upsilon}}(0) \neq 0$, *where* $a_{\tilde{\tau},\tilde{\upsilon}}$ *is defined by* (16.38). *Then the following statements are equivalent:*

(a) *If the auxiliary system* (16.36) *is stable for some pair* $(\rho,\sigma) \in \mathcal{I}_\rho \times \mathcal{I}_\sigma$, *and if the system free of delays* ($\sigma = \rho = \tilde{\tau} = \tilde{\upsilon} \equiv 0$) *given by* (16.34) *when all delays are set to* 0 *is stable, then the system* (16.31) *is stable for all pairs* $(\tilde{\tau},\tilde{\upsilon}) \in \mathbb{R}_+ \times \mathbb{R}_+$, *and there does not exist any root crossing the imaginary axis when the delays* $\tilde{\tau}$, *and* $\tilde{\upsilon}$ *are increased in* $\mathbb{R}_+$.

(b) *The following frequency-sweeping test holds:*

$$\frac{|C_{A,0}||p_{\rho,\text{aux}}(j\omega)|}{|p_{\rho,\sigma}(j\omega)|} < \frac{1}{(2^N q_1 + q_2)p_1 k} \quad \forall \omega > 0. \tag{16.39}$$

The same equivalence holds if the stability property is replaced by the instability of the system with a prescribed number of unstable roots.

Proof. It is easy to see that the condition (16.39) is equivalent to the condition $a_{\tilde{\tau},\tilde{\upsilon}}(j\omega) \neq 0$ for all $\omega > 0$. Then the equivalence between conditions (a) and (b) above follows straightforwardly since $a_{\tilde{\tau},\tilde{\upsilon}} \neq 0$ whenever $\omega = 0$, and thus $a_{\tilde{\tau},\tilde{\upsilon}} \neq 0$ on the whole imaginary axis. Next, no crossing with respect to the imaginary axis means that the stability property valid for $\tilde{\tau} = \tilde{\upsilon} = 0$ is valid for all $(\tilde{\tau},\tilde{\upsilon}) \in \mathbb{R}_+ \times \mathbb{R}_+$. The instability property follows similarly. $\square$

Remark 16.11 (weak T/C interaction measure). *The frequency-sweeping test* (16.39) *can be used to define a measure for characterizing the T/C interaction type in the following sense: the T/C interaction will be called* weak *if the probabilities (q_1, q_2) and the average number of cell division N verify the condition*

$$2^N q_1 + q_2 < \frac{1}{p_1 k} \cdot \frac{1}{\sup_{\omega \in \mathbb{R}} \frac{|C_{A,0}||p_{\rho,\mathrm{aux}}(j\omega)|}{|p_{\rho,\sigma}(j\omega)|}}. \tag{16.40}$$

This condition gives the corresponding T/C interaction measure. *It becomes clear that the average number N of cell division plays a central role in defining the T/C interaction character, since the function $\mathbb{N} \ni N \mapsto 2^N q_1 + q_2 \in \mathbb{R}$ is an increasing function.*

Remark 16.12 (delay-independence in three delays). *For all three delays σ, $\tilde{\tau}$, and $\tilde{\upsilon}$, the frequency-sweeping test is*

$$\frac{|C_{A,0}||p_{\rho,\mathrm{aux}}(j\omega)|}{|p_{\rho,\sigma}(j\omega)|} < \frac{1}{|b_3|+|b_4|+|b_5|} = \frac{1}{k(p_2 + p_1(2^N q_1 + q_2))} \quad \forall \omega > 0. \tag{16.41}$$

This relation can be derived in the same manner as (16.39).

Strong T/C interactions and identification of the crossing points

Let us characterize now the strong T/C interactions. Inspired by the geometric approach proposed in Chapter 6, the condition that $a_{\tilde{\tau},\tilde{\upsilon}}$ defined by (16.38) has at least one root $j\omega_0$ on the imaginary axis is reduced geometrically to the condition that the "lengths" 1, $|a_\upsilon(j\omega_0)|$, and $|a_\mu(j\omega_0)|$ define a triangle. Thus, some simple computations lead to the following criterion for the *identification of the crossing points*:

Proposition 16.13. *Assume that the auxiliary system given by the characteristic equation* (16.36) *satisfies Assumption* 16.8. *Then each $\omega \in \mathbb{R}_+$ can be a solution of the characteristic equation associated to Σ for some $(\tilde{\tau}, \tilde{\upsilon}) \in \mathbb{R}_+^2$ if and only if*

$$\frac{1}{(2^N q_1 + q_2) p_1 k} \leq \frac{|C_{A,0}||p_{\rho,\mathrm{aux}}(j\omega)|}{|p_{\rho,\sigma}(j\omega)|} \leq \frac{1}{|2^N q_1 - q_2| p_1 k}. \tag{16.42}$$

Then, the *crossing set* Ω will be defined by all $\omega \in \mathbb{R}_+$, for which the frequency condition (16.42) holds. In conclusion, the algorithm for identifying the crossing points can be summarized as follows:

- first, we represent graphically

$$\frac{|C_{A,0}||p_{\rho,\mathrm{aux}}(j\omega)|}{|p_{\rho,\sigma}(j\omega)|}$$

against ω, and

- next we analyze the intersection of this graph with two parallel lines to the ω-axis: $1/((2^N q_1+q_2)p_1 k)$ and $1/(|2^N q_1-q_2|p_1 k)$, respectively.

Remark 16.14. *It is easy to see that*

$$\frac{|C_0|\sqrt{\tilde{c}_1^2+\omega^2}}{|p_{\rho,\sigma}(j\omega)|}\to 0,$$

when $\omega\to+\infty$, and in conclusion $\infty\notin\Omega$. In other words, Ω is bounded.

As discussed in Chapter 6, the characterization of the stability crossing curves for a general system including *two delays* was based on an important property of the corresponding *stability crossing set*. More precisely, such a set consisted of a finite number of intervals of finite length. In our case, in order to completely characterize the crossing curves in the parameter space defined by the large delays, it is useful to have a similar property.

Assumption 16.15. *The following condition holds:*

$$\frac{d}{d\omega}\left(\frac{|C_{A,0}||p_{\rho,\mathrm{aux}}(j\omega)|}{|p_{\rho,\sigma}(j\omega)|}\right)\neq 0 \text{ whenever } h(\omega)=\frac{1}{|2^N q_1\pm q_2|p_1 k}. \tag{16.43}$$

This assumption simply requires that the corresponding differentiable function satisfy some nondegenerate property at the corresponding upper and lower bounds given by (16.42). With the remarks and the assumptions above, we have:

Proposition 16.16. *Under Assumptions* 16.8 *and* 16.15, *the crossing set Ω consists of a finite number of intervals of finite length.*

Proof. Define the function $h:\mathbb{R}_+\mapsto\mathbb{R}_+$ by

$$h(\omega):=\frac{|C_{A,0}||p_{\rho,\mathrm{aux}}(j\omega)|}{|p_{\rho,\sigma}(j\omega)|}. \tag{16.44}$$

Thus, the condition (16.42) defining the crossing set Ω is rewritten as

$$\frac{1}{(2^N q_1+q_2)p_1 k}\le h(\omega)\le\frac{1}{|2^N q_1-q_2|p_1 k}.$$

Since Ω is bounded (see Remark 16.14), it is sufficient to prove that $h(\omega)=\frac{1}{(2^N q_1\pm q_2)p_1 k}$ has a finite number of solutions in $[0,\omega_{\max}]$, where $\omega_{\max}=\sup\{\omega|\omega\in\Omega\}$. This will be shown by contradiction. Assume, for example, that $h(\omega)=\frac{1}{(2^N q_1+q_2)p_1 k}$ has an *infinite* number of solutions ω_i, $i = 1,2,\ldots$, with $[0, \omega_{\max}]$. Then

$$h(\omega_i) = \frac{1}{(2^N q_1 + q_2) p_1 k}, \quad \omega_i \in [0, \omega_{\max}], \quad i = 1,2,\ldots.$$

In such a case, there exists a subseries i_m such that $\lim_{i_m \to \infty} \omega_{i_m} = \omega_0$. But this last condition simply means that

$$\frac{d}{d\omega} h(\omega) = \lim_{i_m \to \infty} \frac{h(\omega_{i_m}) - h(\omega_0)}{\omega_{i_m} - \omega_0} = 0,$$

contradicting Assumption 16.15. □

Characterization of the crossing curves

The next step is to characterize the crossing curves of the system (16.31) or, equivalently, all the crossing curves satisfying $a_{\tilde{\tau},\tilde{\upsilon}}(s) = 0$ for $s = j\omega$, $\omega \in \Omega$. By using the same arguments as the ones proposed in Chapter 6, we have:

Proposition 16.17. *Under Assumptions* 16.8 *and* 16.15, *the stability crossing curves* $\mathcal{T}^k$ *corresponding to* Ω_k *must be an intersection of* $\mathbb{R}^2_+$ *with a series of curves belonging to one of the following categories:*

A. *A series of closed curves.*

B. *A series of spiral-like curves with axes oriented either horizontally, vertically, or diagonally.*

C. *A series of open ended curves with both ends approaching* ∞.

The smoothness of curves and the associated crossing direction can be characterized using the general methodology of Chapter 6.

16.4.4 ▪ Illustrative example and discussions

Stability without delays

For our application in [92], we estimated the parameters to be

$$\begin{array}{lll} d_T = 0.2, & p_1 = 0.5, & \rho = 0.0035, \\ r = 0.2, & p_2 = 0.5, & \sigma = 0.0007, \\ k = 1, & q_1 = 0.5, & \tilde{\tau} = 2.0035, \\ N = 2, & q_2 = 0.5, & \tilde{\upsilon} = 1.0035. \\ K = 200, & \tilde{p}_1 = 0.5, & \end{array} \tag{16.45}$$

Hence, b_1, c_1, and b are of order 1 or 0.1. A stability analysis of the fixed points of (16.31) leads to the following results.

- **Fixed point FP-I:** $(T_0, C_{A,0}, C_{D,0}) = (0,0,0)$ For the fixed point $(T_0, C_{A,0}, C_{D,0}) = (0,0,0)$, the characteristic equation given by (16.35) is

$$0 = \lambda^3 + (\tilde{b}_1 - \tilde{c}_1)\lambda^2 - \tilde{b}_1 \tilde{c}_1 \lambda = \lambda(\lambda^2 + (b_1 - c_1)\lambda - b_1 c_1),$$

regardless of the delays. Substituting the parameters from (16.45), we obtain $\lambda(\lambda^2 - 0.04) = 0$, which has roots -0.2, 0, and 0.2. Note that due to the format of the equation and the assumption that all parameters are positive. This characteristic equation is always unstable, even with other biologically relevant parameters.

- **Fixed point FP-II:** $(T_0, C_{A,0}, C_{D,0}) = (0, K, 0)$ In the case where all delays are set to 0 and the parameters from (16.45) are used, we obtain the fixed point $(T_0, C_{A,0}, C_{D,0}) = (0, 200, 0)$. Then, the characteristic equation given by (16.35) is

$$\lambda^3 - 149.6\lambda^2 - 29.96s = \lambda(\lambda^2 - 149.6\lambda - 29.96) = 0,$$

  which has roots -149.8, 0, and 0.2. Hence, this fixed point is unstable when all delays are set to 0.

- **Fixed point FP-III:** The fixed point $(T_0, C_{A,0}, C_{D,0})$ is determined by (16.32) and (16.33). When nondelay parameter values from (16.45) are used, equation (16.32) for T_0 is

$$0.1875\rho T_0^2 + (0.375 - 0.075\rho)T_0 - 0.1498 = 0. \qquad (16.46)$$

  Since the constant term is negative, the equation yields only one positive solution for T_0, which means only one biologically feasible fixed point exists.

In the case where all delays are set to 0, we obtain $T_0 = 0.3995$, and from (16.33), we obtain $C_{A,0} = 0.2667$ and $C_{D,0} = 0$. Thus, from (16.35), we obtain the characteristic equation

$$\lambda(\lambda^2 + 2.4333 \times 10^{-4}\lambda + 3.9947 \times 10^{-2}) = 0,$$

which has roots 0 and $-0.0001 \pm 0.1999j$. Hence, the linearized system is marginally stable when all delays are set to 0.

In this way we arrive at:

Proposition 16.18. *The fixed points of the linearized system without delay have the following properties:*

(a) *The fixed point FP-I* $(T_0, C_0) = (0, 0, 0)$ *is a saddle;*

(b) *the fixed point FP-II* $(T_0, C_0) = (0, K, 0)$ *is a saddle;*

(c) *the fixed point FP-III, determined by* (16.32) *and* (16.33), *has two eigenvalues in the open left half plane and one eigenvalue at* 0.

It is clear that FP-III is of *most interest* to us, because it represents the case where the immune response controls the cancer population.

Stability of FP-III with delays

In what follows, we will perform a more thorough analysis of the roots of the characteristic equation (16.35) for FP-III with respect to the delay values. As stated in Proposition 16.18, the characteristic equation for FP-III has two conjugate eigenvalues with real part less than 0 and one eigenvalue at 0.

In fact, the system has eigenvalue zero no matter what the four delays are. The zero eigenvalue is due to the description of the dying cancer cell population. Indeed, a close inspection of the equation for C_D in (16.27) indicates that an initial condition with positive C_D and zero C_A and T results in a permanently positive population of dying cells. However, in practice, this type of disturbance is not present due to the fact that all dying cells have a finite life span, ρ. Therefore, the mode corresponding to the zero eigenvalue can be easily separated from the rest of the system. Indeed, we may replace the third differential equation of (16.27) by the equivalent integral equation (modulo a constant), i.e.,

$$C_D(t) = \tilde{p}_1 k \int_{-\rho}^{0} C_A(t+w)T(t+w)dw,$$

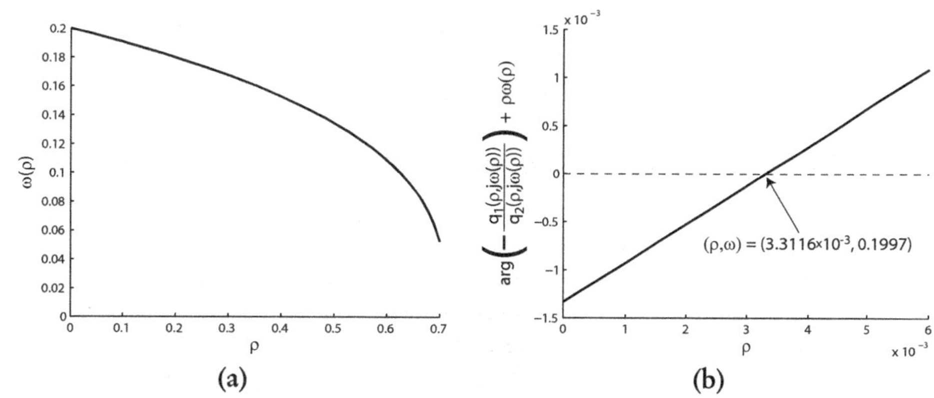

Figure 16.6. (a) *Graph of the stability crossing point $\omega(\rho)$ as a function of the time delay ρ based on the implicit relation* (16.48). (b) *Graph of* $\arg\left(-\frac{q_0(\rho,j\omega(\rho))}{q_1(\rho,j\omega(\rho))}\right)+\rho\omega(\rho)$ *as a function of ρ. We are interested in when the expression equals* 0, *thus satisfying* (16.47) [299].

which will eliminate the mode with the zero eigenvalue. Hence, we study the behavior of the other eigenvalues excluding the fixed eigenvalue at 0. If we set all delays to 0 except ρ, we obtain the characteristic equation

$$\begin{aligned}
0 &= p_0(\rho,\lambda)+p_1(\rho,\lambda)+p_2(\rho,\lambda)+p_3(\rho,\lambda)\\
&= -\lambda^3+(-\tilde{b}_1+\tilde{c}_1-c_3T_0+(b_3+b_4+b_5)C_{A,0}(1+\rho c_3))\lambda^2\\
&\quad+\left[\tilde{b}_1\tilde{c}_1-b_1c_3T_0-b_2c_3C_{D,0}T_0+(b_3+b_4+b_5)C_{A,0}(\rho c_3^2T_0-\tilde{c}_1(1+\rho c_3))\right]\lambda\\
&\quad+bc_3C_{A,0}T_0(\lambda-\tilde{c}_1)-bc_3C_{A,0}T_0(\lambda-\tilde{c}_1)e^{-\rho\lambda},
\end{aligned}$$

which is of the form

$$q_0(\rho,\lambda)+q_1(\rho,\lambda)e^{-\rho\lambda}=0, \tag{16.47}$$

where the coefficients of $q_0(\rho,\lambda)$ and $q_1(\rho,\lambda)$ depend on ρ. To locate stability crossings, we are interested in purely imaginary roots $\lambda=j\omega$, and from the characteristic equation it follows that

$$\frac{q_0(\rho,j\omega)}{q_1(\rho,j\omega)}=1. \tag{16.48}$$

Hence, we can implicitly express ω as a function of ρ. Figure 16.6(a) shows a plot of ω as a function of ρ for $\rho\in[0,1]$. Once we have ω as a function of ρ, we can use the equation

$$\arg\left(-\frac{q_0(\rho,j\omega(\rho))}{q_1(\rho,j\omega(\rho))}\right)+\rho\omega(\rho)=0$$

to determine which values of ρ solve (16.47). Next, Figure 16.6(b) shows a plot of

$$\arg\left(-\frac{q_0(\rho,j\omega(\rho))}{q_1(\rho,j\omega(\rho))}\right)+\rho\omega(\rho)$$

as a function of ρ. The value of ρ at which this expression equals 0 is $\rho=3.3116\times10^{-3}$. Hence, other than the permanent root at 0, the system has stable eigenvalues for $\rho\in[0,0.0033)$ when all other delays are set to 0. The boundary for instability is close to (and slightly less than) the estimated value of ρ, which is 0.0035. To preserve stability, we re-estimate the value of ρ to be 0.003 days (4.3 minutes). From the above reasoning, we know that the system is stable (i.e., there are no roots with real part strictly greater than 0) when $\rho = 0.003$ and all the other delays are equal to 0.

16.4.5 ▪ Crossing curves of the large delay $\tilde{\tau}$ versus the small delay σ

One way of visualizing the crossing surface of (16.34) is to fix two delays, in this case ρ and $\tilde{\upsilon}$, and determine the crossing curves for the other two delays, in this case σ and $\tilde{\tau}$. In this section, we assume that $\rho = 0.003$ (and later 0.001). We then fix $\tilde{\upsilon} = 0$ and look at the crossing curves with respect to the small delay σ and the large delay $\tilde{\tau}$. Figure 16.7(a) shows the resulting crossing curves. As we see from Figure 16.7(a), the system is not stable for large σ and $\tilde{\tau}$. There is a small stable region around (0,0). However, for $\rho = 0.003$, it nearly vanishes, and hence it is hard to determine numerically. As a result, in Figure 16.7(b), we plot a close-up view of the small stable region around the origin for $\rho = 0.001$. This small stable region quickly diminishes as ρ increases and completely vanishes when $\rho = 3.3116 \times 10^{-3}$ as discussed at the end of Section 16.4.4. Since σ is a small delay (~ 0.0007), while $\tilde{\tau}$ is large (~ 2.0035), this small stable region may be relevant for σ, but it is too small to be relevant for $\tilde{\tau}$.

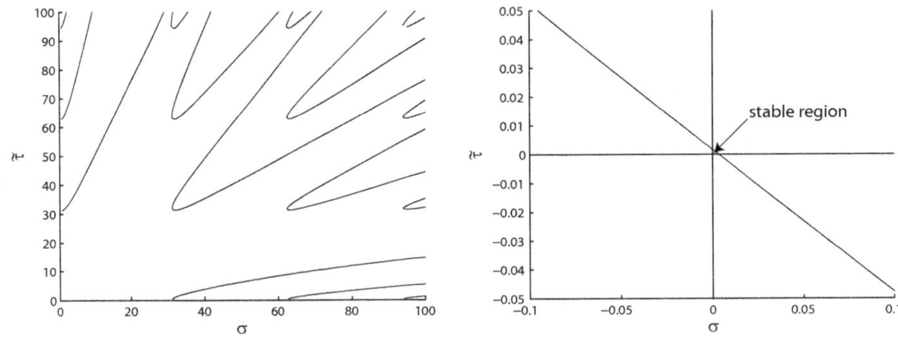

(a) (b)

Figure 16.7. *Crossing curves with respect to σ and $\tilde{\tau}$.* (a) *The zoomed-out plot shows the overall structure of the crossing curves for when $\rho = 0.003$ and $\tilde{\upsilon} = 0$.* (b) *The zoomed-in plot shows the small region of stability around the origin for when $\rho = 0.001$ and $\tilde{\upsilon} = 0$. Note that only nonnegative delays make sense* [299].

For comparison, when $\rho = 0.001$ and $\tilde{\upsilon} = 0.005$, the stable region around the origin disappears, and the system is not stable for any (positive) choice of σ and $\tilde{\tau}$. Figure 16.8 shows that the stable region around (0,0) disappears when $\tilde{\upsilon} = 0.005$. From this figure, it is apparent that the stability of the system is sensitive to small changes in the delay $\tilde{\upsilon}$.

However, it is interesting to note that around $\tilde{\upsilon} = 25$, the stable region around 0 reappears. The new stable region is also about two orders of magnitude larger, making it of relevant size to include reasonable values for $\tilde{\tau}$. Figure 16.9 shows crossing curves of σ versus $\tilde{\tau}$ when $\tilde{\upsilon} = 25$. We will inspect this stable region again in the next section.

16.4.6 ▪ Crossing curves of the large delays $\tilde{\tau}$ and $\tilde{\upsilon}$ versus the small delay σ

We take $N = 1$ (i.e., we assume that T cells divide once upon stimulation by cancer cells) and assume that $\tilde{\tau} = \tilde{\upsilon}$. When $N = 1$, this assumption makes sense, since $\tilde{\tau} = \rho + N\tau$ and $\tilde{\upsilon} = \rho + \upsilon$, and both τ and υ are approximately equal to 1 (see also (16.30)). In this case,

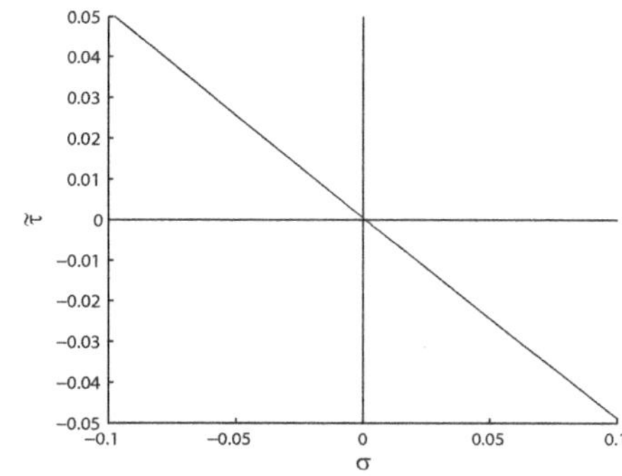

Figure 16.8. *Crossing curves for* σ *and* $\tilde{\tau}$ *when* $\rho = 0.001$ *and* $\tilde{\upsilon} = 0.005$. *The stable region around the origin disappears for this value of* $\tilde{\upsilon}$ [299].

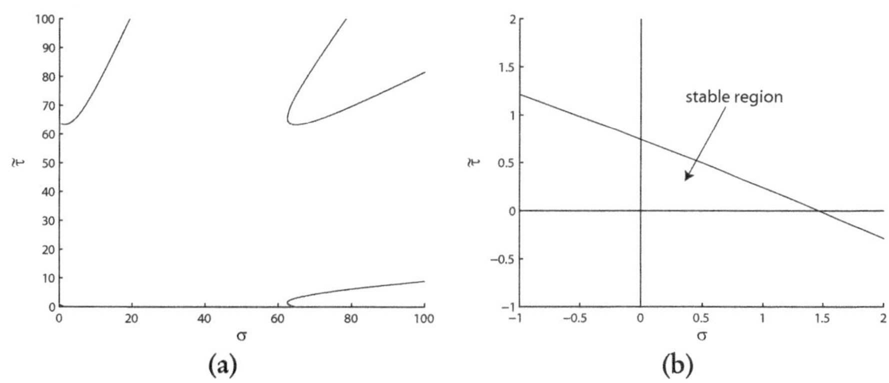

Figure 16.9. *Crossing curves for* σ *and* $\tilde{\tau}$ *when* $\rho = 0.003$ *and* $\tilde{\upsilon} = 25$. (a) *The zoomed-out plot shows the overall structure of the crossing curves.* (b) *The zoomed-in plot shows the region of stability around the origin. The stability region around the origin is about two to three orders of magnitude larger than the one in Figure* 16.7(b) [299].

the characteristic equation (16.35) can be rewritten as

$$p_0(\rho,s)+p_1(\rho,s)e^{-\sigma s}+(p_2(\rho,s)+p_3(\rho,s))e^{-\tilde{\tau}s}=0, \tag{16.49}$$

and we can perform the stability analysis with respect to the two delays σ and $\tilde{\tau}$ $(=\tilde{\upsilon})$.

Figure 16.10 shows crossing curves for σ and the large delays ($\tilde{\tau}$ and $\tilde{\upsilon}$) when $\rho = 0.003$. As before, there is a small stable region around the origin.

Crossing curves of the two large delays: $\tilde{\upsilon}$ versus $\tilde{\tau}$

We take $\rho = 0.001$ as done previously. We then fix σ and determine the crossing curves of $\tilde{\upsilon}$ versus $\tilde{\tau}$. Figure 16.11 shows crossing curves when $\sigma = 0$ and $\sigma = 0.0007$. As before, there is a small stable region around the origin.

As remarked in Section 16.4.5 (Figure 16.9), there is another region of stability for large values of $\tilde{\upsilon}$. In Figure 16.12, we set $\rho = 0.001$ and $\sigma = 0.0007$ and plot the crossing curves with respect to $\tilde{\tau}$ and $\tilde{\upsilon}$. As shown in Figure 16.12(b), there is a relatively large stable region away from the origin. In fact, the system can still be stable for $\tilde{\tau} = 1$ if $\tilde{\upsilon}$ is between 27 and 30.8. In fact, if $\tilde{\upsilon} = 30.5$, then $\tilde{\tau}$ can be as large as 1.5.

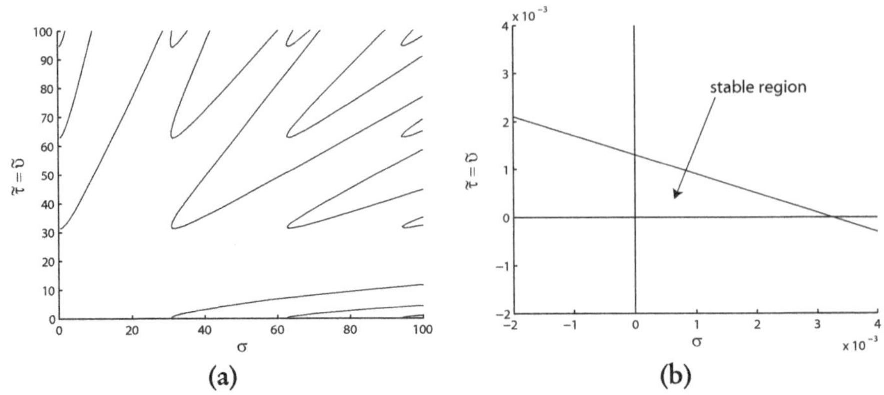

Figure 16.10. *Stability region for $N = 1$, $\tilde{\tau} = \tilde{\upsilon}$, and $p = 0.03$. Also, take $\rho = 0.003$. (a) The zoomed-out plot shows the overall structure of the crossing curves. (b) The zoomed-in plot shows the small region of stability around the origin. The stability region around the origin is comparable in size to the one in Figure* 16.7(b) [299].

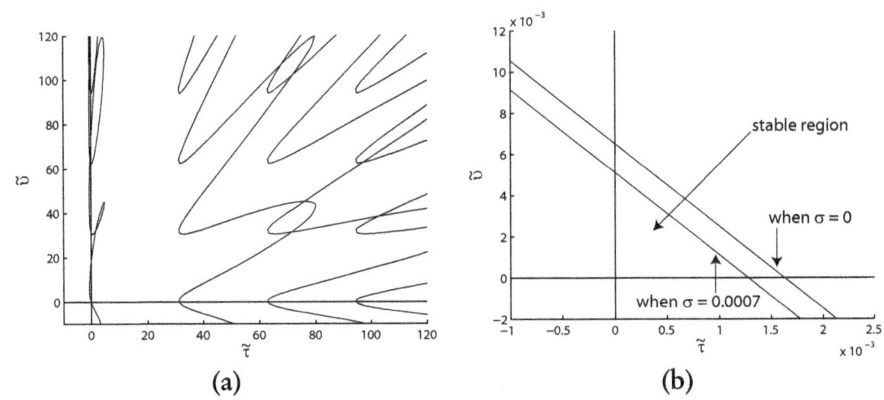

Figure 16.11. *Crossing curves for $\tilde{\tau}$ and $\tilde{\upsilon}$ when $\rho = 0.001$. (a) The zoomed-out plot shows the crossing curves when $\sigma = 0$. The figure hardly changes when $\sigma = 0.0007$. (b) The zoomed-in plot shows the small stable regions near the origin when $\sigma = 0$ and $\sigma = 0.0007$. As shown in the figure, the stable region shrinks slightly as σ increases from 0 to 0.0007* [299].

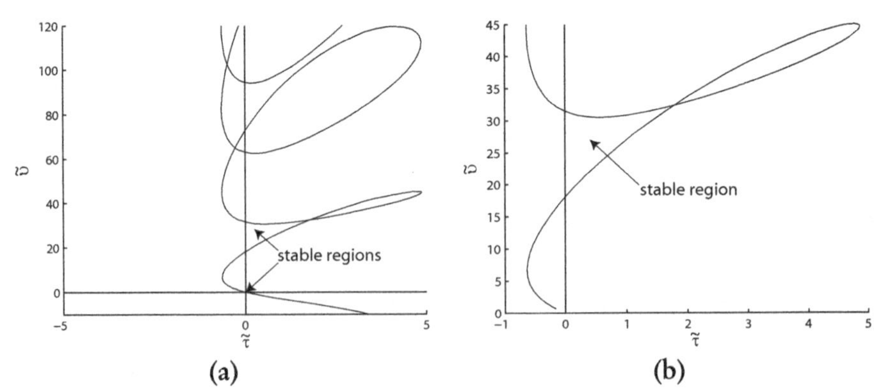

Figure 16.12. *Crossing curves with respect to $\tilde{\tau}$ and $\tilde{\upsilon}$ when $\rho = 0.001$ and $\sigma = 0.0007$. (a) The zoomed-out plot shows the overall structure of the crossing curves. (b) There is a relatively large region of stability that is isolated from the origin* [299].

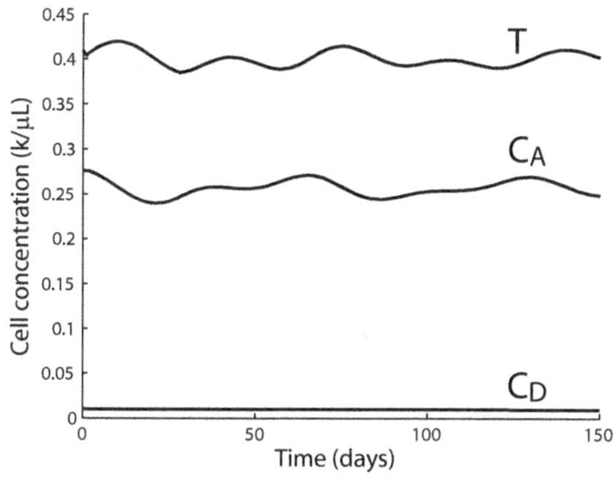

Figure 16.13. *Time evolution of the linearized DDE system for* $(\rho, \sigma, \tilde{\tau}, \tilde{\upsilon}) = (0.001, 0.0007, 1, 28)$. *The plot shows the recentered cell concentrations* $C_A(t) + C_{A,0}$, $C_D(t) + C_{D,0}$, $T(t) + T_0$, *where* $C_A(t)$, $C_D(t)$, *and* $T(t)$ *are as in* (16.34) *and* T_0, $C_{A,0}$, $C_{D,0}$ *correspond to* FP-III. *The system is stable but converges very slowly* [299].

Solutions of the linearized system (16.34) for delay values $(\rho, \sigma, \tilde{\tau}, \tilde{\upsilon})$ equal to $(0.001, 0.0007, 1, 28)$ are shown in Figure 16.13. As shown by our analysis, the linearized system is stable for these delay values, but it converges very slowly due to the placement of the eigenvalues. The dying cancer population, C_D, remains low and constant, which is the behavior induced by the zero eigenvalue. Despite the theoretical stability of these systems, the extremely slow rate of convergence may lead to numerical difficulties when trying to numerically evaluate the time evolution of the solutions, causing the solutions to appear periodic. This difficulty demonstrates that the resolution of using crossing curves for stability analysis exceeds that of directly evaluating the DDE system numerically.

Discussion

Surprisingly, the results indicate that the stability of the controlled state, FP-III, benefits from high values of $\tilde{\upsilon}$. This corresponds to T cells that have long turnaround times after killing cancer cells. These results imply that highly reactive T cells with low turnaround times may, in fact, be disadvantageous to the stability of the system. This phenomenon occurs because highly reactive T cells initially kill off cancer cells too rapidly, causing the T cell population to also decline rapidly due to lack of stimulus. The T cell population is then too low to prevent another cancer relapse resulting from the expansion of remaining cancer cells. More inert T cells, on the other hand, lead to more gradual declines of the cancer populations, which reduce the likelihood of a drastic rebound. Hence, in this formulation of the CML model, a more gradual reduction of the cancer population is favored over rapid elimination. However, we note that we did not have to deal with this issue in our previous CML work, [92], since there we studied the ruin problem; i.e., once the cancer population passes below a threshold, we consider the cancer to have been eliminated.

16.5 ▪ Notes and references

This chapter addressed the stability analysis in the delay-parameter space of time-delay systems encountered in modeling biochemical networks with delays, human respiration, and immune dynamics in chronic leukemia, respectively. Our choice of such different models was dictated by the particular structure of the linearized systems with respect to the delay presence; this structure helped us to derive some simple stability conditions and/or characterization by using relatively simple geometric arguments.

More precisely, in Section 16.2 of the chapter, we considered the stability analysis of two simplified systems including a single delay in their mathematical representation: biochemical networks with delay and delay models describing the dynamics of respiration, respectively. The analysis of the first class of systems closely follows [283] (see also [2] and the references therein for a deeper analysis of the nonlinear model).

In Section 16.3 (respiration), the analysis was derived by exploiting the rank one property of the delayed matrix. The crossing direction characterization was obtained straightforwardly from the approach proposed in Chapter 6. Another second-order model including one constant delay can be found in [18], while a third-order model was considered in [19]. The first two equations in both models describe the arterial partial pressures of O_2 and CO_2, and a peripheral controller. The third equation in the second system models the CO_2 level in the brain and a central controller. Both models possess the same particular structure (delayed matrix of rank one) leading to a very simple characteristic function of the form $p(\lambda;\, \tau) := Q(\lambda) + P(\lambda)e^{-\lambda\tau}$. In this chapter, we considered only a simplified second-order delay model from [394], but the ideas still apply to both models from [18, 19]. It is important to point out that, although the geometric argument was mentioned in [18], it was not fully exploited. The characterization of the crossing direction was developed in light of [80] and follows the same procedure as in [81], instead of taking advantage of the particular structure of the system.

Finally, a more general respiration model including five delay-differential equations [194] and including three independent delays can be found in [17]. The corresponding delays are as follows: τ_B is the lung to brain delay (equal to lung to tissue transport delay), τ_v is the venous side transport delay from tissue to lung, and finally, τ_a is the lung to carotid artery delay. We believe that the geometric idea proposed in the immune dynamics analysis can be applied also to the stability analysis of the corresponding linearization including three delays.

In Section 16.4 we addressed the characterization of stability boundaries in some delay-parameter space for a dynamical system including *four* (independent) *delays*, describing the post-transplantation dynamics of the immune response to CML. Such a model includes two small delays and two large delays [92]. Similarly to the previous case study, the corresponding delayed matrices are of rank one, which simplifies the analysis. More precisely, the stability analysis followed the general lines proposed in [299] and allowed us to briefly discuss two types of T/C cell interactions, weak and strong. Furthermore, a quantitative measure for the weak T/C interaction was introduced and explicitly computed. We also proved that the large delay values have a small influence on the stability properties in the weak cell interaction case. Next, the strong cell interaction case was analyzed in terms of stability crossing curves, and a classification of such crossing curves was given. Finally, an example was presented to illustrate the derived results. Several discussions completed the presentation.

Appendix

A.1 ▪ Rouché's theorem

The proofs of many continuity properties of the spectra of operators associated with linear time-delay systems rely on the celebrated Rouché's theorem (see, e.g. [199, section 5.3.]):

Theorem A.1. *Let f and g be analytic functions on an (open) domain $\mathscr{D} \subseteq \mathbb{C}$. Let $\mathscr{C} \subset D$ be a closed, simple contour, that is, without self-interactions. If*

$$\forall \lambda \in \mathscr{C} \ |g(\lambda)| < |f(\lambda)|,$$

then the functions f and $f+g$ have the same number of zeros inside $\mathscr{C}$, where each zeros is counted as many times as its multiplicity.

Corollary A.2 (see [242]). *Let f and the sequence $\{f_n\}_{n\geq 1}$ be analytic functions on a domain $\mathscr{D} \subseteq \mathbb{C}$. Assume that $\{f_n(\lambda)\}_{n\geq 1}$ converges uniformly to $f(\lambda)$ on the disc $D = \{\lambda : |\lambda - \lambda_0| \leq R\} \subset \mathscr{D}$ for some $R > 0$. Assume further that on this disc, λ_0 is the only zero of $f(\lambda)$, with multiplicity $k \geq 0$ ($k = 0$ means no zeros in D). Then there exists a number $N \in \mathbb{N}$ such that for all $n \geq N$, $f_n(\lambda)$ has exactly k zeros $\lambda_{n,1}, \ldots, \lambda_{n,k}$ in D and $\lim_{n\to\infty} \lambda_{n,j} = \lambda_0$ for all $j \in \{1, \ldots, k\}$.*

A.2 ▪ The structured singular value

We introduce the concept of structured singular values of matrices and outline the main principles behind the standard computational schemes. A more elaborate introduction can be found in the review paper [318], Chapter 11 of [414], and Chapter 4 of [164].

Let $G \in \mathbb{C}^{N\times M}$ and denote its singular values in decreasing order with $\sigma_1(G) \geq \sigma_2(G) \geq \ldots$. A classical result from linear algebra and robust control theory, which lays the basis for the celebrated small gain theorem, relates the largest singular value of G to the solutions of the equation

$$\det(I + G\Delta) = 0 \tag{A.1}$$

in the following way:

$$\sigma_1(G) = \begin{cases} 0 & \text{if } \det(I+G\Delta) \neq 0 \ \forall \Delta \in \mathbb{C}^{M\times N}, \\ \left(\min\left\{\sigma_1(\Delta) : \ \Delta \in \mathbb{C}^{M\times N} \text{ and } \det(I+G\Delta)=0\right\}\right)^{-1} & \text{otherwise.} \end{cases} \tag{A.2}$$

We refer to Δ as the "uncertainty," as in a robust control framework, (A.1) typically originates from a feedback interconnection of a nominal transfer function and an uncertainty block.

Next we reconsider the solutions of equation (A.1), where Δ is restricted to having a particular structure by imposing $\Delta \in \boldsymbol{\Delta}$, with $\boldsymbol{\Delta}$ a closed subset of $\mathbb{C}^{M\times N}$. In analogy with (A.2) one defines the *structured singular value* of the matrix G with respect to the uncertainty set $\boldsymbol{\Delta}$ as

$$\mu_{\boldsymbol{\Delta}}(G) := \begin{cases} 0 & \text{if } \det(I+G\Delta)\neq 0\ \forall \Delta\in\boldsymbol{\Delta}, \\ \left(\min\{\sigma_1(\Delta):\ \Delta\in\boldsymbol{\Delta} \text{ and } \det(I+G\Delta)=0\}\right)^{-1} & \text{otherwise.}\end{cases} \tag{A.3}$$

It directly follows from the definition that

$$\mu_{\boldsymbol{\Delta}}(G)\leq \sigma_1(G). \tag{A.4}$$

Furthermore, if $\mathbb{C}\boldsymbol{\Delta}=\boldsymbol{\Delta}$, then

$$\mu_{\boldsymbol{\Delta}}(G) = \max_{\Delta\in\boldsymbol{\Delta},\ \sigma_1(\Delta)=1} r_\sigma(G\Delta), \tag{A.5}$$

with $r_\sigma(\cdot)$ the spectral radius.

In what follows, for simplicity we restrict ourselves to an uncertainty set $\boldsymbol{\Delta}$ of the form

$$\begin{aligned}\boldsymbol{\Delta} := \Big\{\mathrm{diag}(\Delta_0,\ldots,\Delta_f,d_0 I_{m_0},\ldots,d_s I_{m_s}):\ \Delta_i\in\mathbb{C}^{k_i\times l_i},\ d_j\in\mathbb{C},\\ 0\leq i\leq f,\ 0\leq j\leq s\},\end{aligned} \tag{A.6}$$

where diag$(\cdot)$ represents a block diagonal matrix, $\sum_{i=0}^{f} k_i+\sum_{i=0}^{s} m_i = M$, and $\sum_{i=0}^{f} l_i + \sum_{i=0}^{s} m_i = N$. Such a set satisfies $\mathbb{C}\boldsymbol{\Delta}=\boldsymbol{\Delta}$. Furthermore, based on a slight generalization of [318, Lemma 6.3] to nonsquare block diagonal perturbations, the search space of the optimization in the right-hand side of (A.5) can be restricted. This results in

$$\mu_{\boldsymbol{\Delta}}(G) = \max_{U\in\mathcal{U}} r_\sigma(GU), \tag{A.7}$$

where $\mathcal{U}\subseteq\boldsymbol{\Delta}$ is defined as

$$\begin{aligned}\mathcal{U} := \Big\{\mathrm{diag}(U_0,\ldots,U_f,u_0 I_{m_0},\ldots,u_s I_{m_s}):\ U_i\in\mathbb{C}^{k_i\times l_i},\ u_j\in\mathbb{C},\\ \sigma_k(U_i)=1,\ 1\leq k\leq \min(k_i,l_i),\ |u_j|=1,\ 0\leq i\leq f,\ 0\leq j\leq s\Big\}.\end{aligned}$$

Note that the elements of $\mathcal{U}$ are *unitary* matrices if the uncertainty structure involves only square blocks, that is, $k_i=l_i$, $i=1,\ldots,f$.

Next, the following invariance property can easily be checked:

$$\mu_{\boldsymbol{\Delta}}(G) = \mu_{\boldsymbol{\Delta}}(D_2 G D_1^{-1})\quad \forall (D_1,D_2)\in\mathcal{D}, \tag{A.8}$$

where

$$\begin{aligned}\mathcal{D} := \Big\{(D_1,D_2):\ D_1=\mathrm{diag}(a_0 I_{k_1},\ldots,a_f I_{k_f},D_0,\ldots,D_s),\ D_2=\\ \mathrm{diag}(a_0 I_{l_1},\ldots,a_f I_{l_f},D_0,\ldots,D_s):\ a_i>0, D_i\in\mathbb{C}^{m_i\times m_i}, D_i^*=D_i>0\Big\}.\end{aligned}$$

From (A.7) and the combination of (A.8) and (A.4) we finally obtain

$$\max_{U\in\mathcal{U}} r_\sigma(GU) = \mu_{\boldsymbol{\Delta}}(G) \leq \min_{(D_1,D_2)\in\mathcal{D}} \sigma_1(D_2 G D_1^{-1}). \tag{A.9}$$

Therefore, *optimization* procedures are typically used to compute estimates for $\mu_\Delta(G)$. The function $U \in \mathcal{U} \to r_\sigma(GU)$ may have several local maxima and, for this, a local search for a maximum is not guaranteed to lead to $\mu_\Delta(G)$, but to lower bounds. An appropriate formulation of the optimality condition enables algorithms which resemble power algorithms for computing eigenvalues and singular values; see [319] for an example. Although the convergence of such algorithms to $\mu_\Delta(G)$ is not guaranteed either, and they may converge to values corresponding to lower bounds on $\mu_\Delta(G)$, they have proven their effectiveness in practice. The computation of the upper bound in (A.9) can be recast into a standard *convex optimization* problem. However, in general $\mu_\Delta(G)$ is not equal to the upper bound. An exception to this holds if the number of blocks in the matrices belonging to the uncertainty set Δ satisfies $f+2s \leq 3$ and all the blocks are square, $k_i = l_i$, $i = 0,\dots,f$.

A.3 ▪ Continuity properties

Throughout Chapter 6 (delay ratio analysis and delay-interference phenomenon), we encounter functions from $\mathcal{B}_+^m \subset \mathbb{R}^m$ to $\mathbb{R}$ and from $\mathcal{B}_+^m$ to $\mathcal{P}(\mathbb{C})$. We assume that $\mathbb{R}^m$ is equipped with the Euclidean norm and $\mathcal{P}(\mathbb{C})$ with the Hausdorff metric. A function $f : \mathcal{B}_+^m \to \mathbb{R}$ is lower semicontinuous, respectively upper semicontinuous, at $\vec{s}$ if and only if (see, for instance, [26])

$$\forall \epsilon > 0\ \exists \delta > 0\ \ \forall \vec{r} \in \mathcal{B}_+^m\ \ \|\vec{r}-\vec{s}\| < \delta \Rightarrow f(\vec{r})-f(\vec{s}) > -\epsilon,$$

respectively

$$\forall \epsilon > 0\ \exists \delta > 0\ \ \forall \vec{r} \in \mathcal{B}_+^m\ \ \|\vec{r}-\vec{s}\| < \delta \Rightarrow f(\vec{r})-f(\vec{s}) < \epsilon.$$

It is continuous at $\vec{s}$ when it is both upper and lower semicontinuous at $\vec{s}$.

A function $f : \mathcal{B}_+^m \to \mathcal{P}(\mathbb{C})$ is lower semicontinuous at $\vec{s}$ if and only if

$$\forall \epsilon > 0\ \exists \delta > 0\ \ \forall \vec{r} \in \mathcal{B}_+^m\ \ \|\vec{r}-\vec{s}\| < \delta \Rightarrow D(f(\vec{s}),f(\vec{r})) < \epsilon.$$

When replacing $D(f(\vec{s}),f(\vec{r}))$ with $D(f(\vec{r}),f(\vec{s}))$, respectively $D_h(f(\vec{s}),f(\vec{r}))$, we have upper semicontinuity, respectively continuity at $\vec{s}$.

A.4 ▪ Interdependency of numbers

The real numbers $r_1, r_2, \dots, r_m$ are *rationally independent* if and only if

$$\sum_{i=1}^{m} z_i r_i = 0, \quad z_i \in \mathbb{Z},$$

implies $z_i = 0$, $i = 1,\dots,m$. For example, two numbers are rationally independent if and only if their ratio is an irrational number.

If the real numbers $r_1,\dots,r_m$ are *rationally dependent* (that is, not rationally independent), then there always exists an integer $p < m$ and a matrix $\Gamma \in \mathbb{Z}^{m\times p}$ of full column rank such that

$$\begin{bmatrix} r_1 \\ \vdots \\ r_m \end{bmatrix} = \Gamma \begin{bmatrix} s_1 \\ \vdots \\ s_p \end{bmatrix},$$

with the numbers $s_1, \ldots, s_p$ rationally independent. Thus, rationally dependent numbers depend on a smaller number of rationally independent numbers. In the special case where $p = 1$, the numbers $r_1, \ldots, r_m$ are called *commensurate*, as they are all multiples of the same number.

For example, the numbers 1, π, and exp(1) are rationally independent, the numbers 1, 2, and 5/3 commensurate. The numbers 1, π, and $1+\pi$ are rationally dependent, yet not commensurate, as

$$\begin{bmatrix} 1 \\ \pi \\ 1+\pi \end{bmatrix} = \begin{bmatrix} 1 & 0 \\ 0 & 1 \\ 1 & 1 \end{bmatrix} \begin{bmatrix} 1 \\ \pi \end{bmatrix},$$

with 1 and π rationally independent.

A.5 ▪ Software

The following software packages were used for this book:

- Software for analysis and control of time-delay systems, available from

  http://twr.cs.kuleuven.be/research/software/delay-control/

  This software package in MATLAB includes routines for

  - computing all characteristic roots in a prescribed right half plane using a spectral method, following the approach of Chapter 2;
  - computing $\mathcal{H}_\infty$ norms using the method of Chapter 4, as well as computing pseudospectral abscissa [140];
  - computing stabilizing fixed-order controllers. These routines follow the approach of Chapter 7. As they are based on a DDAE description, they also allow us to treat neutral-type systems and to compute strongly stabilizing controllers;
  - solving the fixed-order $\mathcal{H}_\infty$ optimization problem, using the approach of Chapter 9;
  - evaluating and approximating FIR filters arising in the context of prediction-based controllers.

- DDE-BIFTOOL [100, 101]. This MATLAB package allows a numerical bifurcation and stability analysis of delay-differential equations with several fixed discrete and/or state-dependent delays. It contains routines for the computation, continuation, and stability analysis of steady state solutions, their Hopf and fold bifurcations, periodic solutions, and connecting orbits (but the latter only for the constant delay case). More information on the package can be found at

  http://twr.cs.kuleuven.be/research/software/delay/ddebiftool.shtml

  Unless stated otherwise, the plots of characteristic roots throughout the book were generated by means of this package.

The following packages are useful for problems addressed in this book:

- TRACE-DDE [46] is a MATLAB GUI (Graphic User Interface) package for numerical stability analysis of linear systems of DDEs with several discrete and/or distributed delays. It allows for the numerical computation of the characteristic roots and then it performs a two-parameter robust stability analysis producing the so-called stability chart, i.e., the set of asymptotically stable/unstable regions in the parameters plane. The package can be downloaded from

http://users.dimi.uniud.it/~dimitri.breda/software.html

- QPmR v.2 (Quasi-Polynomial Mapping Based Rootfinder, corresponding to [397]) is a MATLAB function for computing and analyzing the spectrum of characteristic quasi-polynomials corresponding to both retarded and neutral time-delay systems. For more details we refer the reader to

  http://www.cak.fs.cvut.cz/algorithms/qpmr

- The C++ package KNUT allows us to continue periodic solutions, analyze their stability properties, and continue periodic solution bifurcations of autonomous and periodically forced delay-differential equations with several pointwise delays. If one of the three common codimension-one bifurcations (fold, period doubling, Neimark–Sacker) is found along a branch of periodic solutions, the point can be used as a starting point for a continuation of the branch of bifurcation points in a two-parameter space. The package is available from

  https://gitorious.org/knut/pages/Home

- The MATLAB toolbox YALTA is dedicated to the study of classical and fractional systems with delay in the frequency-domain. Its objective is to provide basic but important information such as, for instance, the position of the neutral chains of poles and unstable poles, as well as the root locus with respect to the delay of the system. The toolbox can be obtained from

  http://yalta-toolbox.gforge.inria.fr/

- The MATLAB program HINFCON computes the $\mathscr{H}_\infty$ optimal performance and generates optimal and suboptimal controllers in mixed sensitivity minimization/reduction problems for SISO infinite-dimensional linear systems. It is available from

  http://www.ee.bilkent.edu.tr/~ozbay/

- The MATLAB package HIFOO [53] aims at solving fixed-order stabilization and local $\mathscr{H}_\infty$-$\mathscr{H}_2$ optimization problems for finite-dimensional systems. It depends on HANSO, a hybrid algorithm for nonsmooth, nonconvex optimization based on quasi-Newton updating, bundling, and gradient sampling. The HANSO routine can also be used for fixed-order control of time-delay systems, provided that it is coupled with routines for computing stability, performance, or robustness criteria and their derivatives with respect to controller parameters. More information on the package can be found at

  http://www.cs.nyu.edu/overton/software/hifoo/

For an updated overview of software packages for delay-differential equations, see the website

http://twr.cs.kuleuven.be/research/software/delay/software.shtml

Bibliography

[1] C. T. ABDALLAH, P. DORATO, J. BENITEZ-READ, AND R. BYRNE, *Delayed positive feedback can stabilize oscillatory systems*, in Proceedings of the 1993 American Control Conference, San Francisco, CA, 1993, pp. 3106–3107. (Cited on pp. 267, 284)

[2] M. AHSEN, H. OZBAY, AND S.-I. NICULESCU, *On the analysis of a dynamical model representing gene regulatory networks under negative feedback*, International Journal of Robust and Nonlinear Control, to appear. (Cited on p. 400)

[3] M. AIZERMAN, *Theory of Automatic Control*, Pergamon Press, Oxford, UK, 1st ed., 1963. (Cited on p. 276)

[4] M. AIZERMAN AND F. GANTMACHER, *Absolute Stability of Regulator Systems*, Holden-Day, San Francisco, CA, 1964. (Cited on p. 276)

[5] B. ANDERSON AND J. B. MOORE, *Optimal Control: Linear Quadratic Methods*, Prentice–Hall, Englewood Cliffs, NJ, 1990. (Cited on p. 93)

[6] D. ANGELI AND E. SONTAG, *Monotone controlled systems*, IEEE Transactions on Automatic Control, 48 (2003), pp. 1684–1698. (Cited on p. 376)

[7] W. ARNOLDI, *The principle of minimized iterations in the solution of the matrix eigenvalue problem*, Quarterly of Applied Mathematics, 9 (1951), pp. 17–29. (Cited on p. 59)

[8] Z. ARTSTEIN, *Linear systems with delayed control: A reduction*, IEEE Transactions on Automatic Control, 27 (1982), pp. 869–879. (Cited on p. 301)

[9] K. J. ASTRÖM, C. C. HANG, AND B. C. LIM, *A new Smith predictor for controlling a process with an integrator and long dead-time*, IEEE Transactions on Automatic Control, 39 (1994), pp. 343–345. (Cited on p. 285)

[10] K. J. ÅSTRÖM AND B. WITTENMARK, *Computer Controlled Systems: Theory and Design*, Prentice–Hall, Upper Saddle River, NJ, 3rd ed., 1997. (Cited on p. 275)

[11] F. ATAY, *Oscillator death in coupled functional differential equations near Hopf bifurcation*, Journal of Differential Equations, 221 (2006), pp. 190–209. (Cited on p. 354)

[12] K. ATKINSON, *An Introduction to Numerical Analysis*, John Wiley and Sons, New York, 2nd ed., 1989. (Cited on p. 58)

[13] C. E. AVELLAR AND J. K. HALE, *On the zeros of exponential polynomials*, Mathematical Analysis and Applications, 73 (1980), pp. 434–452. (Cited on pp. 19, 33, 287, 288, 289, 307)

[14] C. T. H. BAKER, G. A. BOCHAROV, AND F. A. RIHAN, *A Report on the Use of Delay Differential Equations in Numerical Modelling in the Biosciences*, Numerical Analysis Report No. 343, Manchester Centre for Computational Mathematics, Manchester, UK, 1999. (Cited on p. 375)

[15] M. Bando, K. Hasebe, K. Nakanishi, and A. Nakayama, *Analysis of optimal velocity model with explicit delay*, Physical Review E, 58 (1998), pp. 5429–5435. (Cited on p. 350)

[16] D. Bansal, H. Balakrishnan, S. Floyd, and S. Shenker, *Dynamic Behavior of Slowly-Responsive Congestion Control Algorithms*, in Proceedings of ACM SIGCOMM 2001, Aug. 2001, pp. 263–273. (Cited on p. 317)

[17] J. J. Batzel and T. H. Tran, *Modeling instability in the control system for human respiration: Applications to infant non-rem sleep*, Applied Mathematics and Computations, 110 (2000), pp. 1–51. (Cited on p. 400)

[18] ———, *Stability of the human respiratory control system. Part I: Analysis of a two-dimensional delay state-space model*, Journal of Mathematical Biology, 41 (2000), pp. 45–79. (Cited on pp. 375, 400)

[19] ———, *Stability of the human respiratory control system. Part II: Analysis of a three-dimensional delay state-space model*, Journal of Mathematical Biology, 41 (2000), pp. 80–102. (Cited on pp. 375, 400)

[20] N. Bekiaris-Liberis and M. Krstic, *Nonlinear Control Under Nonconstant Delays*, SIAM, Philadelphia, 2013. (Cited on p. 301)

[21] J. Bélair, *Stability in delayed neural networks*, in Ordinary and Delay Differential Equations, Pitman Research Notes in Mathematics Series 272, Longman Sci. Tech., Harlow, UK, 1992, pp. 6–9. (Cited on p. 376)

[22] A. Bellen and M. Zennaro, *Numerical Methods for Delay Differential Equations*, Oxford University Press, Oxford, UK, 2003. (Cited on p. 40)

[23] R. E. Bellman and K. L. Cooke, *Differential-Difference Equations*, Academic Press, New York, 1963. (Cited on pp. ix, xv, 3, 4, 5, 9, 37, 172)

[24] I. Belykh, M. Hasler, M. Lauret, and H. Nijmeijer, *Synchronization and graph topology*, International Journal of Bifurcation and Chaos, 15 (2005), pp. 3423–3433. (Cited on p. 354)

[25] A. Bensoussan, G. Da Prato, M. C. Defour, and S. K. Mitter, *Representation and Control of Infinite Dimensional Systems*, Birkhauser, Boston, Systems and Control: Foundation and Applications, 2 volumes, 1993. (Cited on pp. x, xvi)

[26] C. Berge, *Espaces topologiques. Fonctions multivoques*, Dunod, Paris, 1966. (Cited on p. 403)

[27] T. Betcke, N. J. Higham, V. Mehrmann, C. Schröder, and F. Tisseur, *NLEVP: A Collection of Nonlinear Eigenvalue Problems*, EPrint 2011.116, Manchester Institute for Mathematical Sciences, Manchester, UK, 2011. (Cited on p. x)

[28] P. A. Bliman, *LMI characterization of strong delay-independent stability of linear delay systems via quadratic Lyapunov-Krasovskii functionals*, Systems and Control Letters, 43 (2001), pp. 263–274. (Cited on p. 184)

[29] F. Boese, *Stability with respect to the delay: On the paper of K. L. Cooke and P. van den Driessche*, Journal of Mathematical Analysis and Applications, 228 (1998), pp. 293–321. (Cited on pp. 147, 260)

[30] J. C. Bolot and A. U. Shankat, *Analysis of a fluid control approximation to flow control dynamics*, in Proceedings of IEEE INFOCOM, Florence, Italy, 1992, pp. 2398–2407. (Cited on p. 315)

[31] N. BOSE, *Applied Multidimensional Systems Theory*, Van Nostrand Reinhold Co., New York, 1982. (Cited on p. 23)

[32] E.-K. BOUKAS AND Z.-K. LIU, *Deterministic and Stochastic Time Delay Systems*, Birkhäuser, Basel, 2002. (Cited on pp. x, xvi)

[33] I. BOUSSAADA, D. IROFTI, AND S.-I. NICULESCU, *Computing the codimension of the singularity at the origin for delay systems in the regular case: A Vandermonde-based approach*, in Proceedings of the 13th European Control Conference, Strasbourg, France, 2014. (Cited on p. 146)

[34] I. BOUSSAADA, I.-C. MORĂRESCU, AND S.-I. NICULESCU, *Inverted pendulum stabilization via a Pyragas-type controller: Revisiting the triple zero singularity*, in Proceedings of the 19th IFAC World Congress, Cape Town, South Africa, 2014. CD-ROM. (Cited on p. 224)

[35] S. BOYD AND V. BALAKRISHNAN, *A regularity result for the singular values of a transfer matrix and a quadratically convergent algorithm for computing its $\mathscr{L}_\infty$-norm*, Systems and Control Letters, 15 (1990), pp. 1–7. (Cited on p. 94)

[36] S. BOYD, V. BALAKRISHNAN, AND P. KABAMBA, *A bisection method for computing the $\mathscr{H}_\infty$ norm of a transfer matrix and related problems*, Mathematics of Control, Signals and Systems, 2 (1989), pp. 207–219. (Cited on pp. 93, 94)

[37] S. BOYD AND C. DESOER, *Subharmonic functions and performance bounds in linear time-invariant feedback systems*, IMA Journal on Mathematical Control and Information, 2 (1985), pp. 153–170. (Cited on p. 23)

[38] S. BOYD AND L. VANDENBERGHE, *Convex Optimization*, Cambridge University Press, Cambridge, UK, 2004. (Cited on p. 205)

[39] F. BRAUER, *Characteristic return times for harvested population models with time lag*, Mathematical Biosciences, 45 (1979), pp. 295–311. (Cited on p. 130)

[40] R. BRAYTON, *Small-signal stability criterion for electrical networks containing lossless transmission lines*, IBM Journal of Research and Development, 12 (1968), pp. 431–440. (Cited on p. 37)

[41] D. BREDA, *Solution operator approximation for delay differential equation characteristic roots computation via Runge-Kutta methods*, Applied Numerical Mathematics, 56 (2006), pp. 305–317. (Cited on pp. 39, 64)

[42] D. BREDA, S. MASET, AND R. VERMIGLIO, *Computing the characteristic roots for delay differential equations*, IMA Journal of Numerical Analysis, 24 (2004), pp. 1–19. (Cited on pp. 39, 40)

[43] ———, *Pseudospectral differencing methods for characteristic roots of delay differential equations*, SIAM Journal on Scientific Computing, 27 (2005), pp. 482–495. (Cited on pp. 39, 40, 41, 43, 54, 55, 98)

[44] ———, *Pseudospectral approximation of eigenvalues of derivative operators with non-local boundary conditions*, Applied Numerical Mathematics, 56 (2006), pp. 318–331. (Cited on pp. 56, 94, 98, 109)

[45] ———, *Numerical approximation of characteristic values of partial retarded functional differential equations*, Numerische Mathematik, 113 (2009), pp. 181–242. (Cited on p. 65)

[46] ———, *TRACE-DDE: A tool for robust analysis and characteristic equations for delay differential equations*, in Topics in Time Delay Systems: Analysis, Algorithms and Control, J. Loiseau et al., ed., Springer, Berlin, 2009, pp. 145–155. (Cited on pp. 40, 404)

[47] S. D. BRIERLEY, J. N. CHIASSON, E. B. LEE, AND S. H. ZAK, *On stability independent of delay for linear systems*, IEEE Transactions on Automatic Control, 27 (1982), pp. 252–254. (Cited on p. 189)

[48] J. W. BRUCE AND P. J. GIBLIN, *Curves and Singularities*, Cambridge University Press, Cambridge, UK, 1984. (Cited on pp. 137, 176)

[49] N. BRUINSMA AND M. STEINBUCH, *A fast algorithm to compute the $\mathcal{H}_\infty$-norm of a transfer function matrix*, Systems and Control Letters, 14 (1990), pp. 287–293. (Cited on pp. 94, 103, 235)

[50] W. E. BRUMLEY, *On the asymptotic behavior of solutions of differential-difference equations of neutral type*, Journal of Differential Equations, 7 (1970), pp. 175–188. (Cited on p. 18)

[51] S. BUNGAY AND S.-A. CAMPBELL, *Patterns of oscillation in a ring of identical cells with delayed coupling*, International Journal of Bifurcation and Chaos, 17 (2007), pp. 3109–3125. (Cited on p. 354)

[52] A. BUNSE-GERSTNER, D. KUBLIŃSKA, G. VOSSEN, AND D. WILCZEK, *h_2-norm optimal model reduction for large scale discrete dynamical MIMO systems*, Journal of Computational and Applied Mathematics, 233 (2010), pp. 1202–1216. (Cited on p. 93)

[53] J. V. BURKE, D. HENRION, A. S. LEWIS, AND M. L. OVERTON, *HIFOO - a* MATLAB *package for fixed-order controller design and H_∞ optimization*, in Proceedings of the IFAC Symposium on Robust Control Design (ROCOND), Toulouse, France, 2006. CD-ROM. (Cited on pp. 227, 405)

[54] ——, *Stabilization via nonsmooth, nonconvex optimization*, IEEE Transactions on Automatic Control, 51 (2006), pp. 1760–1769. (Cited on p. 195)

[55] J. V. BURKE, A. S. LEWIS, AND M. L. OVERTON, *Approximating subdifferentials by random sampling of gradients*, Mathematics of Operations Research, 27 (2002), pp. 567–584. (Cited on p. 198)

[56] ——, *A nonsmooth, nonconvex optimization approach to robust stabilization by static output feedback and low-order controllers*, in Proceedings of the IFAC Symposium on Robust Control Design (ROCOND), Milan, Italy, 2003. CD-ROM. (Cited on p. 194)

[57] ——, *Robust stability and a criss-cross algorithm for pseudospectra*, IMA Journal of Numerical Analysis, 23 (2003), pp. 359–375. (Cited on pp. 94, 103)

[58] ——, *A robust gradient sampling algorithm for nonsmooth, nonconvex optimization*, SIAM Journal on Optimization, 15 (2005), pp. 751–779. (Cited on pp. 194, 195, 197, 198, 209)

[59] E. BUTCHER AND O. BOBRENKOV, *On the Chebyshev spectral continuous time approximation for constant and periodic delay differential equations*, Communications in Nonlinear Science and Numerical Simulation, 16 (2011), pp. 1541–1554. (Cited on p. 39)

[60] R. BYERS, *A bisection method for measuring the distance of a stable matrix to the unstable matrices*, SIAM Journal on Scientific and Statistical Computing, 9 (1988), pp. 875–881. (Cited on pp. 93, 103)

[61] A. CALLENDER, D. R. HARTREE, AND A. PORTER, *Time lag in a control system*, Philosophical Transactions of the Royal Society of London, Series A, Mathematical and Physical Sciences, 235 (1936), pp. 415–444. (Cited on pp. ix, xv)

[62] A. CALLENDER AND A. G. STEVENSON, *The application of automatic control to a typical problem in chemical industry*, in the Proceedings of the Chemical Engineering Group, Society of Chemical Engineering, 18 (1936), pp. 108–117. (Cited on pp. ix, xv)

[63] S. Campbell and Y. Yuan, *Zero singularities of codimension two and three in delay differential equations*, Nonlinearity, 21 (2008), pp. 2671–2691. (Cited on p. 140)

[64] R. E. Chandler, R. Herman, and E. W. Montroll, *Traffic dynamics: Studies in car following*, Operations Research, 6 (1958), pp. 165–183. (Cited on p. 337)

[65] F. Chatté, B. Ducourthial, D. Nace, and S.-I. Niculescu, *Results on fluid modelling packet switched networks*, in Proceedings of the 3rd IFAC Workshop on Time-Delay Systems, Santa Fe, NM, 2001. CD-ROM. (Cited on p. 315)

[66] M. Chavez, D.-U. Hwang, A. Amann, H. Hentschel, and S. Boccaletti, *Synchronization is enhanced in weighted complex networks*, Physical Review Letters, 94 (2005), 218701. (Cited on p. 357)

[67] J. Chen, *Static output feedback stabilization for SISO systems and related problems: Solutions via generalized eigenvalues*, Control Theory and Advanced Technology, 10 (1995), pp. 2233–2244. (Cited on pp. 163, 255, 256, 279)

[68] J. Chen, P. Fu, and S.-I. Niculescu, *Asymptotic behavior of imaginary zeros of linear systems with commensurate delays*, in Proceedings of the 45th IEEE Conference on Decision and Control, San Diego, CA, 2006, pp. 1375–1380. (Cited on p. 158)

[69] J. Chen, P. Fu, S.-I. Niculescu, and Z. Guan, *An eigenvalue perturbation approach to stability analysis, part* ii: *When will zeros of time-delay systems cross imaginary axis?*, SIAM Journal on Control and Optimization, 48 (2010), pp. 5583–5605. (Cited on pp. 137, 152)

[70] J. Chen, G. Gu, and C. A. Nett, *A new method for computing delay margins for stability of linear delay systems*, Systems and Control Letters, 26 (1995), pp. 107–117. (Cited on pp. 155, 157, 166)

[71] J. Chen and H. A. Latchman, *Frequency sweeping tests for stability independent of delay*, Journal of Mathematical Analysis and Applications, 40 (1995), pp. 1640–1645. (Cited on p. 189)

[72] Y. Q. Chen and K. L. Moore, *Analytical stability bound for delayed second-order systems with repeating poles using Lambert function W*, Automatica, 38 (2002), pp. 891–895. (Cited on p. 148)

[73] J. Cheong, S.-I. Niculescu, and C. Kim, *Motion synchronization control of distributed multi-subsystems with invariant local natural dynamics*, IEEE Transactions on Robotics, 25 (2009), pp. 382–398. (Cited on p. 300)

[74] J. Cheong, J. Park, and S.-I. Niculescu, *Physically-based collaborative simulations under ring-like network configurations*, IEEE Transactions on Computer Graphics and Applications, 29 (2009), pp. 64–80. (Cited on p. 300)

[75] J. Chiasson and C. Abdallah, *Robust stability of time delay systems: Theory*, in Proceedings of the 3rd IFAC Workshop on Time-Delay Systems, Santa Fe, NM, 2001, pp. 125–130. (Cited on p. 284)

[76] J. N. Chiasson, S. D. Brierley, and E. B. Lee, *A simplified derivation of the Zeheb-Walach 2-d stability test with applications to time-delay systems*, IEEE Transactions on Automatic Control, 30 (1985), pp. 411–414. Corrections in IEEE Transactions on Automatic Control, 31 (1986), pp. 91–92. (Cited on p. 189)

[77] N. Chopra and M. Spong, *Output synchronization in strongly connected graphs*, in Proceedings of the 18th International Symposium on Mathematical Theory of Networks and Systems, Blacksburg, VA, 2008. (Cited on p. 370)

[78] K. L. COOKE, *Delay differential equations*, in Mathematics of Biology, M. Iannelli, ed., Liguori Editore, Napoli, 1981, pp. 5–80. (Cited on pp. 3, 6, 36)

[79] K. L. COOKE AND Z. GROSSMAN, *Discrete delay, distributed delay and stability switches*, Journal of Mathematical Analysis and Applications, 86 (1982), pp. 592–627. (Cited on pp. 163, 189, 331)

[80] K. L. COOKE AND J. TURI, *Stability, instability in delay equations modeling human respiration*, Journal of Mathematical Biology, 32 (1994), pp. 535–543. (Cited on p. 400)

[81] K. L. COOKE AND P. VAN DEN DRIESSCHE, *On zeroes of some transcendental equations*, Funkcialaj Ekvacioj, 29 (1986), pp. 77–90. (Cited on pp. 147, 163, 189, 400)

[82] R. M. CORLESS, G. H. GONNET, D. E. G. HARE, D. J. JEFFREY, AND D. E. KNUTH, *On the Lambert W function*, Advances in Computational Mathematics, 5 (1996), pp. 329–359. (Cited on p. 148)

[83] R. V. CULSHAW, S. RUAN, AND G. WEBB, *A mathematical model of cell-to-cell spread of HIV-1 that includes a time delay*, Journal of Mathematical Biology, 46 (2003), pp. 425–444. (Cited on p. 375)

[84] R. F. CURTAIN AND H. ZWART, *An Introduction to Infinite-Dimensional Linear Systems Theory*, vol. 21 of Texts in Applied Mathematics, Springer-Verlag, Berlin, 1995. (Cited on pp. x, 36, 114)

[85] R. DATKO, *A procedure for determination of the exponential stability of certain differential difference equations*, Quarterly of Applied Mathematics, 36 (1978), pp. 279–292. (Cited on p. 13)

[86] ——, *Remarks concerning the asymptotic stability and stabilization of linear delay differential equations*, Journal of Mathematical Analysis and Applications, 111 (1985), pp. 571–584. (Cited on pp. 181, 182)

[87] ——, *Not all feedback stabilized hyperbolic systems are robust with respect to small time delays in their feedbacks*, SIAM Journal on Control and Optimization, 26 (1988), pp. 697–713. (Cited on p. 308)

[88] ——, *Two examples of ill-posedness with respect to time delays revisited*, IEEE Transactions on Automatic Control, 42 (1997), pp. 434–452. (Cited on p. 308)

[89] R. DATKO, J. LAGNESE, AND M. P. POLIS, *An example on the effect of time delays in boundary feedback stabilization of wave equations*, SIAM Journal on Control and Optimization, 24 (1986), pp. 152–156. (Cited on p. 308)

[90] L. C. DAVIS, *Modifications of the optimal velocity traffic model to include delay due to driver reaction time*, Physica A, 319 (2003), pp. 557–567. (Cited on p. 350)

[91] L. DE CICCO, S. MASCOLO, AND S.-I. NICULESCU, *Robust stability analysis of Smith predictor-based congestion control algorithms for computer networks*, Automatica, 47 (2011), pp. 1685–1692. (Cited on pp. 300, 318, 319, 320, 321, 335)

[92] R. DECONDE, P. S. KIM, D. LEVY, AND P. P. LEE, *Post-transplantation dynamics of the immune response to chronic myelogenous leukemia*, Journal of Theoretical Biology, 236 (2005), pp. 39–59. (Cited on pp. 385, 387, 393, 399, 400)

[93] O. DIEKMANN, S. A. VAN GILS, S. M. VERDUYN LUNEL, AND H.-O. WALTHER, *Delay equations. Functional-, Complex-, and Nonlinear Analysis*, vol. 110 of Applied Mathematical Sciences, Springer-Verlag, New York, 1995. (Cited on pp. 3, 11, 36, 145, 189, 340, 342, 343)

[94] E. J. DOEDEL, A. R. CHAMPNEYS, T. J. FAIRGRIEVE, Y. A. KUZNETSOV, B. SANDSTEDE, AND X.-J. WANG, *AUTO97: Continuation and Bifurcation Software for Ordinary Differential Equations*, Technical Report, Department of Computer Science, Concordia University, Portland, OR, 1998. (Cited on p. 148)

[95] P. V. DOOREN, K. GALLIVAN, AND P.-A. ABSIL, *$\mathcal{H}_2$-optimal model reduction of MIMO systems*, Applied Mathematics Letters, 21 (2008), pp. 1267–1273. (Cited on p. 93)

[96] J. DOYLE, K. GLOVER, P. P. KHARGONEKAR, AND F. B. A. FRANCIS, *State-space solutions to standard $\mathcal{H}_2$ and $\mathcal{H}_\infty$ control problems*, IEEE Transactions on Automatic Control, 34 (1989), pp. 831–847. (Cited on p. 225)

[97] L. E. EL'SGOL'TS AND S. B. NORKIN, *Introduction to the theory and applications of differential equations with deviating arguments*, vol. 105 of Mathematics in Science and Engineering, Academic Press, New York, 1973. (Cited on pp. ix, xv, 36, 279)

[98] G. ENCISO, *On the asymptotic behavior of a cyclic biochemical system with delay*, in Proceedings of the 45th IEEE Conference on Decision and Control, San Diego, CA, 2006, pp. 2388–2393. (Cited on pp. 376, 377)

[99] K. ENGELBORGHS, M. DAMBRINE, AND D. ROOSE, *Limitations of a class of stabilization methods for delay equations*, IEEE Transactions on Automatic Control, 46 (2001), pp. 336–339. (Cited on p. 302)

[100] K. ENGELBORGHS, T. LUZYANINA, AND G. SAMAEY, *DDE-BIFTOOL v. 2.00: A MATLAB package for bifurcation analysis of delay differential equations*, TW Report 330, Department of Computer Science, Katholieke Universiteit Leuven, Belgium, October 2001. (Cited on pp. 40, 148, 326, 331, 333, 349, 361, 404)

[101] K. ENGELBORGHS AND D. ROOSE, *Numerical bifurcation analysis of delay differential equations using DDE-BIFTOOL*, ACM Transactions on Mathematical Software, 28 (2002), pp. 1–21. (Cited on pp. 40, 148, 331, 404)

[102] T. ERNEUX, *Applied Delay-Differential Equations*, vol. 3 of Surveys and Tutorials in the Applied Mathematical Sciences, Springer, New York, 2009. (Cited on p. ix)

[103] K. FALL AND K. VARADHAN, *The ns Manual*, 2003. Available from http://www.isi.edu/nsnam/ns/ns-documentation.html. (Cited on p. 333)

[104] A. FATTOUH, O. SENAME, AND J. DION, *Pulse controller design for linear time-delay systems*, in Proceedings of the First IFAC Symposium on Systems, Structure and Control, Prague, Czech Republic, 2001. (Cited on p. 302)

[105] F. FEINBERG, *Lectures on Chemical Reaction Networks*, Technical report, Mathematics Research Center, University of Wisconsin-Madison, Madison, WI, 1979. (Cited on p. 376)

[106] W. FENG, *On practical stability of linear multivariable feedback systems with time-delays*, Automatica, 27 (1991), pp. 389–394. (Cited on pp. 285, 286)

[107] A. FIORAVANTI, C. BONNET, S.-I. NICULESCU, AND J. GEROMEL, *Matrix norm approach for control of linear time-delay systems*, in Proceedings of the 50th IEEE Conference on Decision and Control and European Control Conference (CDC-ECC), Orlando, FL, 2011, pp. 4170–4175. (Cited on p. 36)

[108] E. FRIDMAN, *New Lyapunov-Krasovskii functionals for stability of linear retarded and neutral type systems*, Systems and Control Letters, 43 (2001), pp. 309–319. (Cited on p. 242)

[109] ——, *Stability of linear descriptor systems with delay: A Lyapunov-based approach*, Journal of Mathematical Analysis and Applications, 273 (2002), pp. 24–44. (Cited on pp. 32, 226)

[110] E. FRIDMAN AND U. SHAKED, *$\mathcal{H}_\infty$-state-feedback control of linear systems with small state delay*, Systems and Control Letters, 33 (1998), pp. 141–150. (Cited on pp. 241, 242)

[111] ——, *New bounded real lemma representations for time-delay systems and their applications*, IEEE Transactions on Automatic Control, 46 (2001), pp. 1973–1979. (Cited on p. 242)

[112] ——, *A descriptor system approach to $\mathcal{H}_\infty$ control of linear time-delay systems*, IEEE Transactions on Automatic Control, 47 (2002), pp. 253–270. (Cited on pp. 80, 241, 242)

[113] ——, *H-infinity control of linear state-delay descriptor systems: An LMI approach*, Linear Algebra and Its Applications, 351/352 (2002), pp. 271–302. (Cited on pp. 32, 226, 229, 241, 242)

[114] P. FU, J. CHEN, AND S.-I. NICULESCU, *Generalized eigenvalue-based stability tests for 2-D linear systems: Necessary and sufficient conditions*, Automatica, 42 (2006), pp. 1569–1576. (Cited on pp. 23, 24, 37)

[115] ——, *High-order analysis of critical stability properties of linear time-delay systems*, in Proceedings of the American Control Conference, New York, 2007, pp. 4921–4926. (Cited on p. 189)

[116] P. FU, S.-I. NICULESCU, AND J. CHEN, *Stability of linear neutral time-delay systems: Exact conditions via matrix pencil solutions*, IEEE Transactions on Automatic Control, 51 (2006), pp. 1063–1069. (Cited on pp. 145, 166, 167)

[117] P. GAHINET AND P. APKARIAN, *A linear matrix inequality approach to $\mathcal{H}_\infty$ control*, International Journal of Robust and Nonlinear Control, 4 (1994), pp. 421–448. (Cited on p. 225)

[118] Y. GENIN, R. STEFAN, AND P. VAN DOOREN, *Real and complex stability radii of polynomial matrices*, Linear Algebra and Its Applications, 351/352 (2002), pp. 381–410. (Cited on pp. 70, 72, 75, 76, 92, 235)

[119] T. T. GEORGIOU AND M. C. SMITH, *Graphs, causality and stabilizability: Linear, shift-invariant systems on $\mathcal{L}_2([0,\infty))$*, Mathematics of Control, Signals, and Systems, 6 (1993), pp. 195–223. (Cited on p. 305)

[120] H. GLUSING-LUERBEN, *First-order representations of delay-differential systems in a behavioural setting*, European Journal of Control, 3 (1997), pp. 137–149. (Cited on pp. x, xvi)

[121] A. GOLDBETER, *Biochemical Oscillations and Cellular Rhythms*, Cambridge University Press, Cambridge, UK, 1996. (Cited on p. 376)

[122] B. GOODWIN, *Oscillatory behaviour in enzymatic control processes*, Advances in Enzyme Regulation, 3 (1965), pp. 425–438. (Cited on pp. 376, 377)

[123] K. GOPALSAMY, *Stability and Oscillations in Delay Differential Equations of Population Dynamics*, vol. 74 of Mathematics and Its Applications, Kluwer Academic Publishers, Dordrecht, 1993. (Cited on pp. 124, 146, 258, 375)

[124] A. GRAHAM, *Kronecker Products and Matrix Calculus with Applications*, Ellis Horwood Ltd., Chichester, UK, 1981. (Cited on pp. 23, 155)

[125] K. GREEN AND T. WAGENKNECHT, *Pseudospectra and delay differential equations*, Journal of Computational and Applied Mathematics, 196 (2006), pp. 567–578. (Cited on pp. 70, 75, 78, 86)

[126] M. GREEN, *How long does it take to stop?*, Transportation Human Factors, 2 (2000), pp. 195–216. (Cited on p. 337)

[127] C. GROTTA-RAGAZZO, O. ARINO, J.-F. VIBERT, K. PAKDAMAN, AND C. MALTA, *Transient oscillations in continuous-time excitatory ring neural networks with delay*, Physical Review E, 55 (1997), pp. 3234–3248. (Cited on p. 376)

[128] E. GRYAZINA, *The D-decomposition theory*, Automation and Remote Control, 65 (2004), pp. 1872–1884. (Cited on p. 145)

[129] E. GRYAZINA, B. POLYAK, AND A. TREMBA, *D-decomposition technique state-of-the-art*, Automation and Remote Control, 69 (2008), pp. 1991–2026. (Cited on p. 145)

[130] K. GU, V. L. KHARITONOV, AND J. CHEN, *Stability of Time-Delay Systems*, Birkhäuser, Basel, 2003. (Cited on pp. ix, x, xv, xvi, 95, 154, 162, 189, 249)

[131] K. GU AND S.-I. NICULESCU, *Additional dynamics in transformed time-delay systems*, IEEE Transactions on Automatic Control, 45 (2000), pp. 572–575. (Cited on pp. 37, 208)

[132] K. GU, S.-I. NICULESCU, AND J. CHEN, *On stability of crossing curves for general systems with two delays*, Journal of Mathematical Analysis and Applications, 311 (2005), pp. 231–253. (Cited on pp. 37, 168, 169, 170, 171, 172, 173, 174, 176, 189, 277)

[133] ——, *Computing maximum delay deviation allowed to retain stability in systems with two delays*, in Application of Time Delay Systems, J. Chiasson and J. J. Loiseau, eds., vol. 352 of Lectures Notes in Control and Information Sciences, Springer, Berlin, 2007, pp. 157–164. (Cited on pp. 92, 178, 189)

[134] R. GUDIN AND L. MIRKIN, *On the delay margin of dead-time compensators*, International Journal of Control, 80 (2007), pp. 1316–1332. (Cited on p. 321)

[135] H. W. GUGGENHEIMER, *Differential Geometry*, Dover, New York, 1977. (Cited on pp. 128, 133)

[136] N. GUGLIELMI AND M. L. OVERTON, *Fast algorithms for the approximation of the pseudospectral abscissa and pseudospectral radius of a matrix*, SIAM Journal on Matrix Analysis and Applications, 32 (2011), pp. 1166–1192. (Cited on pp. 74, 118)

[137] S. GUMUSSOY, *On the relative stability of closed-loop SISO dead-time systems: Non-imaginary axis case*, Transactions of the Institute of Measurement and Control, 34 (2011), pp. 411–421. (Cited on p. 143)

[138] S. GUMUSSOY, D. HENRION, M. MILLSTONE, AND M. OVERTON, *Multiobjective robust control with HIFOO* 2.0, in Proceedings of the IFAC Symposium on Robust Control Design, Haifa, Israel, 2009. CD-ROM. (Cited on p. 193)

[139] S. GUMUSSOY AND W. MICHIELS, *Fixed-order H-infinity optimization of time-delay systems*, in Recent Advances in Optimization and Its Applications in Engineering, M. Diehl, F. Glineur, E. Jarlebring, and W. Michiels, eds., Springer, Berlin, 2010, pp. 103–112. (Cited on pp. 225, 239, 241, 243)

[140] ——, *A predictor-corrector type algorithm for the pseudospectral abscissa computation of time-delay systems*, Automatica, 46 (2010), pp. 657–664. (Cited on pp. 43, 74, 118, 404)

[141] ——, *Fixed-order H-infinity control for interconnected systems using delay differential algebraic equations*, SIAM Journal on Control and Optimization, 49 (2011), pp. 2212–2238. (Cited on pp. 30, 37, 231, 232, 235, 237, 238, 240, 242, 243, 245)

[142] ——, *Root locus for siso dead-time systems: A continuation based approach*, Automatica, 48 (2012), pp. 480–489. (Cited on pp. 142, 143, 144, 148)

[143] S. GUMUSSOY AND M. L. OVERTON, *Fixed-order H-infinity controller design via HIFOO, a specialized nonsmooth optimization package*, in Proceedings of the American Control Conference, Seattle, WA, 2008, pp. 2750–2754. (Cited on pp. 193, 225, 226, 239)

[144] K. P. HADELER, *Delay equations in biology*, in Functional Differential Equations and Approximation of Fixed Points, vol. 730 of Lecture Notes in Mathematics, Springer-Verlag, Berlin, 1979, pp. 136–157. (Cited on p. 375)

[145] A. HALANAY, *Differential Equations: Stability, Oscillations, Time Lags*, Academic Press, New York, 1966. (Cited on pp. ix, xv, 3, 5, 8, 36)

[146] A. HALANAY AND V. RĂSVAN, *General theory of linear hybrid control*, International Journal of Control, 26 (1977), pp. 621–634. (Cited on p. 275)

[147] A. HALANAY AND V. RĂSVAN, *Stability radii for some propagation models*, IMA Journal on Mathematical Control and Information, 14 (1997), pp. 95–107. (Cited on p. 37)

[148] J. K. HALE, *Theory of Functional Differential Equations*, vol. 3 of Applied Mathematical Sciences, Springer-Verlag, New York, 1977. (Cited on pp. ix, xv, 3, 5, 7, 36)

[149] ———, *Effects of delays on dynamics*, in Topological methods in differential equations and inclusions, A. Granas, M. Frigon, and G. Sabidussi, eds., Kluwer Academic Publishers, Dordrecht, 1995, pp. 191–238. (Cited on pp. 288, 290, 307)

[150] J. K. HALE, E. F. INFANTE, AND F. S.-P. TSEN, *Stability in linear delay equations*, Journal of Mathematical Analysis and Applications, 105 (1985), pp. 533–555. (Cited on pp. 127, 158, 258)

[151] J. K. HALE AND S. M. VERDUYN LUNEL, *Introduction to Functional Differential Equations*, vol. 99 of Applied Mathematical Sciences, Springer-Verlag, New York, 1993. (Cited on pp. ix, xv, 3, 21, 36, 37, 140, 186, 258, 288, 291, 305, 340, 342, 343)

[152] ———, *Strong stabilization of neutral functional differential equations*, IMA Journal of Mathematical Control and Information, 19 (2002), pp. 5–23. (Cited on pp. 21, 22, 35, 37, 226, 229, 307, 308)

[153] ———, *Stability and control of feedback systems with time delays*, International Journal of Systems Science, 34 (2003), pp. 497–504. (Cited on p. 37)

[154] K. B. HANNSGEN, Y. RENARDY, AND R. L. WHEELER, *Effectiveness and robustness with respect to time delays of boundary feedback stabilization in one-dimensional viscoelasticity*, SIAM Journal on Control and Optimization, 26 (1988), pp. 1200–1234. (Cited on p. 308)

[155] G. H. HARDY AND E. M. WRIGHT, *An Introduction to the Theory of Numbers*, Oxford University Press, Oxford, UK, 1968. (Cited on pp. 182, 289)

[156] B. D. HASSARD, *Counting roots of the characteristic equation for linear delay-differential systems*, Journal of Differential Equations, 136 (1997), pp. 222–235. (Cited on p. 146)

[157] D. HELBING, *Traffic and related self-driven many-particle systems*, Reviews of Modern Physics, 73 (2001), pp. 1067–1141. (Cited on p. 337)

[158] D. HENRION AND T. VYHLÍDAL, *Positive trigonometric polynomials for strong stability of difference equations*, Automatica, 48 (2012), pp. 2207–2212. (Cited on p. 59)

[159] D. HENRY, *Linear autonomous neutral functional differential equations*, Journal of Differential Equations, 15 (1974), pp. 106–128. (Cited on p. 21)

[160] D. HERTZ, E. I. JURY, AND E. ZEHEB, *Simplified analytical stability test for systems with commensurate time delays*, IEE Proceedings on Control Theory and Applications, Part D, 131 (1984), pp. 52–56. (Cited on p. 147)

[161] J. P. HESPANHA, P. NAGHSHTABRIZI, AND Y. XU, *A survey of recent results in Networked Control Systems*, Proceedings of IEEE, 95 (2007), pp. 138–162. (Cited on p. 284)

[162] N. J. HIGHAM, *Accuracy and Stability of Numerical Algorithms, 2nd ed.*, SIAM, Philadelphia, 2002. (Cited on p. 114)

[163] D. HINRICHSEN AND B. KELB, *Spectral value sets: A graphical tool for robustness analysis*, Systems and Control Letters, 21 (1993), pp. 127–136. (Cited on p. 78)

[164] D. HINRICHSEN AND A. J. PRITCHARD, *Mathematical Systems Theory* I. *Modelling, State Space Analysis, Stability and Robustness*, vol. 48 of Texts in Applied Mathematics, Springer-Verlag, Berlin, 2005. (Cited on pp. 69, 70, 75, 92, 401)

[165] C. V. HOLLOT AND Y. CHAIT, *Nonlinear stability analysis for a class of tcp/aqm networks*, in Proceeding of the 40th IEEE Conference on Decision and Control, Orlando, FL, 2001, pp. 2309–2314. (Cited on pp. 329, 330)

[166] C. V. HOLLOT, V. MISRA, D. TOWSLEY, AND W. B. GONG, *Analysis and design of controllers for AQM routers supporting TCP flow*, IEEE Transactions on Automatic Control, 47 (2002), pp. 945–956. (Cited on pp. 328, 329)

[167] F. C. HOPPENSTEADT AND C. S. PESKIN, *Mathematics in Medicine and Life Sciences*, vol. 10 of Texts in Applied Mathematics, Springer-Verlag, New York, 1992. (Cited on p. 375)

[168] R. A. HORN AND C. R. JOHNSON, *Matrix Analysis*, Cambridge University Press, Cambridge, UK, 1985. (Cited on p. 134)

[169] P. HÖVEL AND E. SCHÖLL, *Control of unstable steady states by time-delayed feedback methods*, Physical Review E, 72 (2005), 046203. (Cited on p. 354)

[170] R. HRYNIV AND P. LANCASTER, *On the perturbation of analytic matrix functions*, Integral Equations and Operator Theory, 34 (1999), pp. 325–338. (Cited on pp. 105, 106, 148, 359)

[171] H. HUIJBERTS, W. MICHIELS, AND H. NIJMEIJER, *Stabilizability via time-delayed feedback: An eigenvalue optimization approach*, SIAM Journal on Applied Dynamical Systems, 8 (2009), pp. 1–20. (Cited on pp. 222, 224)

[172] H. HUIJBERTS, H. NIJMEIJER, AND T. OGUCHI, *Anticipating synchronization of chaotic Lur'e systems*, Chaos, 17 (2007), 013117. (Cited on p. 354)

[173] R. IZMAILOV, *Adaptive feedback control algorithms for large data transfer in high-speed networks*, IEEE Transactions on Automatic Control, 40 (1995), pp. 1469–1471. (Cited on pp. 316, 322, 327, 335)

[174] ——, *Analysis and optimization of feedback control algorithms for data transfers in high-speed networks*, SIAM Journal on Control and Optimization, 34 (1996), pp. 1767–1780. (Cited on pp. 250, 316, 322, 323, 327, 335)

[175] V. JACOBSON, *Congestion avoidance and control*, ACM Communication Review, 18 (1988), pp. 314–329. (Cited on p. 315)

[176] E. JARLEBRING AND T. DAMM, *The Lambert W function and the spectrum of multidimensional time-delay systems*, Automatica, 43 (2007), pp. 2124–2128. (Cited on p. 148)

[177] E. JARLEBRING, K. MEERBERGEN, AND W. MICHIELS, *A Krylov method for the delay eigenvalue problem*, SIAM Journal on Scientific Computing, 32 (2010), pp. 3278–3300. (Cited on pp. 42, 60, 61, 64, 65, 66, 67)

[178] E. JARLEBRING AND W. MICHIELS, *Invariance properties in the root sensitivity of time-delay systems with double imaginary roots*, Automatica, 46 (2010), pp. 1112–1115. (Cited on pp. 152, 189)

[179] E. JARLEBRING, W. MICHIELS, AND K. MEERBERGEN, *A linear eigenvalue algorithm for the nonlinear eigenvalue problem*, Numerische Mathematik, 122 (2012), pp. 169–195. (Cited on pp. x, 64, 67)

[180] E. JARLEBRING, J. VANBIERVLIET, AND W. MICHIELS, *Characterizing and computing the H2 norm of time-delay systems by solving the delay Lyapunov equation*, IEEE Transactions on Automatic Control, 56 (2011), pp. 814–825. (Cited on pp. 111, 112, 113, 114, 118)

[181] W. JUST, T. BERNARD, M. OSTHEIMER, E. REIBOLD, AND H. BENNER, *Mechanism of time-delayed feedback control*, Physical Review Letters, 78 (1997), pp. 203–206. (Cited on p. 223)

[182] I. KABAKOV AND A. SOKOLOV, *Influence of the hydraulic shock on the process of steam turbine speed control*, Inzhenerov Sbornik, 2 (1946), pp. 61–76 (in Russian). (Cited on p. 37)

[183] T. KAILATH, *Linear Systems*, Prentice–Hall, Englewood Cliffs, NJ, 1990. (Cited on p. 297)

[184] E. W. KAMEN, *An operator theory of linear functional differential equations*, Journal of Differential Equations, 27 (1978), pp. 274–297. (Cited on pp. x, xvi)

[185] ——, *On the relationship between zero criteria for two-variable polynomials and asymptotic stability of delay differential equations*, IEEE Transactions on Automatic Control, 25 (1980), pp. 983–984. (Cited on p. 189)

[186] ——, *Linear systems with commensurate time delays: Stability and stabilization independent of delay*, IEEE Transactions on Automatic Control, 27 (1982), pp. 367–375. (Cited on p. 189)

[187] T. KATO, *Perturbation Theory for Linear Operations*, Springer-Verlag, Berlin, 1966. (Cited on pp. 134, 135, 139, 148)

[188] F. P. KELLY, *Mathematical modelling of the internet*, in Mathematics Unlimited–2001 and Beyond, B. Engquist and W. Schmid, eds., Springer-Verlag, Berlin, 2001, pp. 685–702. (Cited on pp. 250, 328, 329, 334)

[189] V. L. KHARITONOV, *Lyapunov functionals and Lyapunov matrices for neutral type time delay systems: A single delay case*, International Journal of Control, 78 (2005), pp. 783–800. (Cited on p. 95)

[190] ——, *Time-Delay Systems. Lyapunov Functionals and Matrices*, Birkhäuser, Basel, 2013. (Cited on p. 118)

[191] V. L. KHARITONOV, S.-I. NICULESCU, J. MORENO, AND W. MICHIELS, *Static output feedback stabilization: Necessary conditions for multiple delay controllers*, IEEE Transactions on Automatic Control, 50 (2005), pp. 82–86. (Cited on pp. 254, 268, 284, 379)

[192] V. L. KHARITONOV AND E. PLISCHKE, *Lyapunov matrices for time-delay systems*, System and Control Letters, 55 (2006), pp. 697–706. (Cited on pp. 95, 111, 113)

[193] V. L. KHARITONOV AND A. P. ZHABKO, *Lyapunov-Krasovskii approach to the robust stability analysis of time-delay systems*, Automatica J. IFAC, 39 (2003), pp. 15–20. (Cited on p. 95)

[194] M. C. KHOO, R. E. KRONAUER, K. P. STROHL, AND A. S. SLUTSKY, *Factors inducing periodic breathing in humans: A general model*, Journal of Applied Physiology, 53 (1982), pp. 644–659. (Cited on p. 400)

[195] H. KOKAME, K. HIRATA, K. KONISHI, AND T. MORI, *Difference feedback can stabilize unstable steady states*, IEEE Transactions on Automatic Control, 46 (2001), pp. 1908–1913. (Cited on pp. 268, 269)

[196] V. B. KOLMANOVSKII AND A. MYSHKIS, *Introduction to the Theory and Applications of Functional Differential Equations*, vol. 463 of Mathematics and Its Applications, Kluwer Academic Publishers, Dordrecht, 1999. (Cited on pp. 36, 37, 122, 123, 249, 331)

[197] V. B. KOLMANOVSKII AND V. R. NOSOV, *Stability of Functional Differential Equations*, vol. 180 of Mathematics in Science and Engineering, Academic Press, New York, 1986. (Cited on pp. 36, 123, 148)

[198] A. M. KRALL, *Stability criteria for feedback systems with a time lag*, SIAM Journal of Control, 2 (1965), pp. 160–170. (Cited on p. 126)

[199] S. G. KRANTZ, *Handbook of Complex Variables*, Birkhäuser, Boston, 1999. (Cited on p. 401)

[200] N. N. KRASOVSKII, *Stability and Motion: Applications of Lyapunov's Second Method to Differential Systems and Equations with Delay*, Stanford University Press, Stanford, CA, 1963. (Cited on pp. ix, xv, 36)

[201] M. KRSTIC, *Delay Compensation for Nonlinear, Adaptive and PDE Systems*, Birkhäuser, Basel, 2007. (Cited on p. 301)

[202] Y. KUANG, *Delay Differential Equations with Applications in Population Dynamics*, Academic Press, Boston, 1993. (Cited on pp. ix, xv, 147, 375)

[203] Y. A. KUZNETSOV, *Elements of Applied Bifurcation Theory*, vol. 112 of Applied Mathematical Sciences, Springer-Verlag, Berlin, 1998. (Cited on pp. 132, 148)

[204] W. H. KWON AND A. E. PEARSON, *Feedback stabilization of linear systems with delayed control*, IEEE Transactions on Automatic Control, 25 (1980), pp. 266–269. (Cited on p. 301)

[205] M. S. LEE AND C. S. HSU, *On the τ-decomposition method of stability analysis for retarded dynamical systems*, SIAM Journal of Control, 7 (1969), pp. 242–259. (Cited on pp. 147, 161)

[206] A. S. LEWIS AND M. L. OVERTON, *Nonsmooth optimization via quasi-Newton methods*, Mathematical Programming, 141 (2013), pp. 135–163. (Cited on pp. 198, 226, 239)

[207] X. LI, S.-I. NICULESCU, A. ÇELA, H.-H. WANG, AND T.-Y. CAI, *Invariance properties for a class of quasipolynomials*, Automatica, 50 (2014), pp. 890–895. (Cited on pp. 152, 153, 160, 161, 189)

[208] B. LIACU, I.-C. MORĂRESCU, S.-I. NICULESCU, C. ANDRIOT, D. DUMUR, F. COLLEDANI, AND P. BOUCHER, *Some remarks on Smith predictor-based control with distance feedback for a 3-dof haptic system with distributed delays*, in Proceedings of the 11th IFAC Workshop on Time-Delay Systems, Grenoble, France, 2013, pp. 214–219. (Cited on p. 300)

[209] F.-L. LIAN, *Analysis, Design, Modeling and Control of Networked Control Systems*, Ph.D. thesis, University of Michigan, Ann Arbor, MI, 2001. (Cited on p. 284)

[210] Z. LIN, *Low Gain Design*, vol. 240 of Lecture Notes in Control and Information Science, Springer-Verlag, Berlin, 1999. (Cited on p. 272)

[211] H. LOGEMANN, *Destabilizing effects of small time-delays on feedback-controlled descriptor systems*, Linear Algebra and Its Applications, 272 (1998), pp. 131–153. (Cited on pp. 226, 308)

[212] H. LOGEMANN AND R. REBARBER, *The effect of small time-delays on the closed-loop stability of boundary control systems*, Mathematics of Control, Signals and Systems, 9 (1996), pp. 123–151. (Cited on pp. 287, 290, 291, 308)

[213] H. LOGEMANN, R. REBARBER, AND G. WEISS, *Conditions for robustness and nonrobustness of the stability of feedback control systems with respect to small delays in the feedback loop*, SIAM Journal on Control and Optimization, 34 (1996), pp. 572–600. (Cited on p. 308)

[214] H. LOGEMANN AND S. TOWNLEY, *The effect of small delays in the feedback loop on the stability of neutral systems*, Systems and Control Letters, 27 (1996), pp. 267–274. (Cited on p. 308)

[215] J.-J. LOISEAU AND D. BRETHÉ, *The use of 2d systems theory for the control of time-delay systems*, Journal Européen de Systèmes Automatisés, 31 (1997), pp. 1043–1058. (Cited on pp. x, xvi)

[216] J. LOUISELL, *Absolute stability in linear delay-differential systems: Ill-posedness and robustness*, IEEE Transactions on Automatic Control, 40 (1995), pp. 1288–1291. (Cited on pp. 186, 187)

[217] S. LOW, F. PAGANINI, AND J. C. DOYLE, *Internet congestion control*, IEEE Control Systems Magazine, 22 (2002), pp. 28–43. (Cited on p. 329)

[218] T. LUZYANINA AND D. ROOSE, *Equations with distributed delays: Bifurcation analysis using computational tools for discrete delay equations*, Functional Differential Equations, 11 (2004), pp. 87–92. (Cited on pp. 37, 208, 349)

[219] N. MACDONALD, *Time lag in a model of a biochemical reaction sequence with end-product inhibition*, Journal of Theoretical Biology, 67 (1977), pp. 549–556. (Cited on p. 376)

[220] N. MACDONALD, *Comments on a simplified analytical test for systems with delay*, IEE Proceedings on Control Theory and Applications, Part D, 132 (1985), pp. 237–238. (Cited on p. 147)

[221] ——, *Two delays may not destabilize although either delay can*, Mathematical Biosciences, 82 (1986), pp. 127–140. (Cited on p. 180)

[222] ——, *An interference effect of independent delays*, IEE Proceeding on Control Theory and Applications, 134 (1987), pp. 38–42. (Cited on pp. 180, 187)

[223] ——, *Biological Delay Systems: Linear Stability Theory*, Cambridge University Press, Cambridge, UK, 1989. (Cited on pp. ix, xv, 148, 375, 376, 377)

[224] F. MAGHAMI ASL AND A. G. ULSOY, *Analytical solution of a system of homogeneous delay differential equations via the Lambert function*, in Proceedings of the 2000 American Control Conference, Chicago, IL, 2000, pp. 2496–2500. (Cited on p. 148)

[225] M. S. MAHMOUD, *Robust Control and Filtering for Time-Delay Systems*, vol. 5 of Control Engineering, Marcel Dekker, New York, 2000. (Cited on pp. x, xvi)

[226] M. MALEK-ZAVAREI AND M. JAMSHIDI, *Time Delay Systems: Analysis, Optimization and Applications*, vol. 9 of Systems and Control Series, North-Holland, Amsterdam, 1987. (Cited on p. 249)

[227] A. MANITIUS AND A. OLBROT, *Finite spectrum assignment problem for systems with delays*, IEEE Transactions on Automatic Control, 24 (1979), pp. 541–553. (Cited on pp. 301, 302, 303)

[228] M. MARDEN, *The Geometry of the Zeros of a Polynomial in the Complex Plane*, American Mathematical Society, Providence, RI, 1949. (Cited on p. 278)

[229] S. MASCOLO, *Classical control theory for congestion avoidance in high speed Internet*, in Proceedings of the 38th IEEE Conference on Decision and Control, Phoenix, AZ, 1999, pp. 2709–2714. (Cited on pp. 315, 335)

[230] ———, *Congestion control in high-speed communication networks using the Smith principle*, Special Issue on "Control Methods for Communication Networks," Automatica, 35 (1999), pp. 1921–1935. (Cited on pp. 315, 316, 317, 335)

[231] ———, *Modeling the Internet congestion control as a time delay system: A robust stability analysis*, in Proceedings of the IFAC Workshop on Time-Delay Systems, L'Aquila, Italy, September 2003. (Cited on p. 321)

[232] ———, *Modeling the Internet congestion control using a Smith controller with input shaping*, Control Engineering Practice, 14 (2006), pp. 425–435. (Cited on p. 317)

[233] M. R. MATAUŠEK AND A. D. MICIĆ, *On the modified Smith predictor for controlling a process with an integrator and long dead-time*, IEEE Transactions on Automatic Control, 44 (1999), pp. 1603–1607. (Cited on p. 285)

[234] F. MAZENC, S. MONDIÉ, AND R. FRANCISCO, *Global asymptotic stabilization of feedforward systems with delay in the input*, IEEE Transactions on Automatic Control, 49 (2004), pp. 844–850. (Cited on p. 269)

[235] F. MAZENC, S. MONDIÉ, AND S.-I. NICULESCU, *Global asymptotic stabilization for chains of integrators with a delay in the input*, IEEE Transactions on Automatic Control, 48 (2003), pp. 57–63. (Cited on pp. 269, 272, 284)

[236] F. MAZENC, S. MONDIÉ, AND S.-I. NICULESCU, *Global stabilization of oscillators with bounded delayed input*, Systems and Control Letters, 53 (2004), pp. 415–422. (Cited on p. 219)

[237] V. MEHRMANN AND H. VOSS, *Nonlinear eigenvalue problems: A challenge for modern eigenvalue methods*, GAMM Mitteilungen, 27 (2004), pp. 121–152. (Cited on pp. x, xvi)

[238] C. F. MÉNDEZ-BARRIOS, W. MICHIELS, AND S.-I. NICULESCU, *Further remarks on stabilizing chains of integrators by using network delays*, in Proceedings of the European Control Conference (ECC'09), August 23–26, 2009. (Cited on p. 276)

[239] F. MÉNDEZ-BARRIOS, S.-I. NICULESCU, J. CHEN, AND M. MAYA-MÉNDEZ, *Output feedback stabilization of SISO linear systems with I/O network-induced delays. An eigenvalue based-approach*, International Journal of Control, 87 (2014), pp. 346–362. (Cited on pp. 274, 277, 278, 280, 281, 282, 283, 284)

[240] W. MICHIELS, *Stability and Stabilization of Time-Delay Systems*, Ph.D. thesis, Katholieke Universiteit Leuven, Leuven, Belgium, 2002. (Cited on p. 224)

[241] ———, *Spectrum based stability analysis and stabilization of systems described by delay differential algebraic equations*, IET Control Theory and Applications, 5 (2011), pp. 1829–1842. (Cited on pp. 33, 36, 37, 54, 55, 57, 67, 206, 207, 208, 209)

[242] W. MICHIELS, K. ENGELBORGHS, D. ROOSE, AND D. DOCHAIN, *Sensitivity to infinitesimal delays in neutral equations*, SIAM Journal on Control and Optimization, 40 (2001), pp. 1134–1158. (Cited on pp. 20, 33, 37, 182, 226, 287, 289, 307, 401)

[243] W. MICHIELS, K. ENGELBORGHS, P. VANSEVENANT, AND D. ROOSE, *The continuous pole placement method for delay equations*, Automatica, 38 (2002), pp. 747–761. (Cited on pp. 208, 209)

[244] W. MICHIELS, E. FRIDMAN, AND S.-I. NICULESCU, *Robustness assessment via stability radii in delay parameters*, International Journal of Robust and Nonlinear Control, 19 (2009), pp. 1405–1426. (Cited on pp. 90, 91, 92, 93, 374)

[245] W. MICHIELS, K. GREEN, T. WAGENKNECHT, AND S.-I. NICULESCU, *Pseudospectra and stability radii for analytic matrix functions with application to time-delay systems*, Linear Algebra and Its Applications, 418 (2006), pp. 315–335. (Cited on pp. 72, 74, 75, 77, 85, 86, 92, 245)

[246] W. MICHIELS AND N. GUGLIELMI, *An iterative method for computing the pseudospectral abscissa for a class of nonlinear eigenvalue problems*, SIAM Journal on Scientific Computing, 34 (2012), pp. A2366–A2393. (Cited on p. 245)

[247] W. MICHIELS AND S. GUMUSSOY, *Characterization and computation of H_∞ norms for time-delay systems*, SIAM Journal on Matrix Analysis and Applications, 31 (2010), pp. 2093–2115. (Cited on pp. 95, 97, 99, 100, 101, 102, 104, 105, 108, 109, 110, 118, 235, 236)

[248] W. MICHIELS, E. JARLEBRING, AND K. MEERBERGEN, *Krylov-based model order reduction of time-delay systems*, SIAM Journal on Matrix Analysis and Applications, 32 (2011), pp. 1399–1421. (Cited on p. 67)

[249] W. MICHIELS, D. MELCHOR-AGUILLAR, AND S.-I. NICULESCU, *Stability analysis of some classes of TCP/AQM networks*, International Journal of Control, 79 (2005), pp. 1134–1144. (Cited on pp. 329, 330, 335)

[250] W. MICHIELS, S. MONDIÉ, AND D. ROOSE, *Robust Stabilization of Time-Delay Systems with Distributed Delay Control Laws: Necessary and Sufficient Conditions for a Safe Implementation*, TW Report 363, Department of Computer Science, KU Leuven, Leuven, Belgium, 2003. (Cited on pp. 302, 305, 307, 308, 309, 314)

[251] W. MICHIELS, S. MONDIÉ, D. ROOSE, AND M. DAMBRINE, *The effect of approximating distributed delay control laws on stability*, in Advances in Time-Delay Systems, vol. 38 of Lecture Notes in Computational Science and Engineering, Springer-Verlag, Berlin, 2004, pp. 207–222. (Cited on pp. 305, 306, 307, 309, 314)

[252] W. MICHIELS, C.-I. MORĂRESCU, AND S.-I. NICULESCU, *Consensus problems with distributed delays, with application to traffic flow models*, SIAM Journal on Control and Optimization, 48 (2009), pp. 77–101. (Cited on pp. 343, 348, 349, 350, 351)

[253] W. MICHIELS AND S.-I. NICULESCU, *On the delay sensitivity of Smith predictors*, International Journal of Systems Science, 34 (2003), pp. 543–552. (Cited on pp. 288, 289, 296, 300, 308)

[254] ——, *Stability analysis of a fluid flow model for TCP like behavior*, International Journal of Bifurcation and Chaos, 15 (2005), pp. 2277–2282. (Cited on pp. 332, 333, 334, 335)

[255] ——, *Characterization of delay-independent stability and delay-interference phenomena*, SIAM Journal on Control and Optimization, 45 (2007), pp. 2138–2155. (Cited on pp. 180, 182, 183, 184, 186, 187, 188, 190)

[256] ——, *Stability and Stabilization of Time-Delay Systems: An Eigenvalue Based Approach*, vol. 12 of Advances in Design and Control, SIAM, Philadelphia, 2007. (Cited on pp. 148, 190)

[257] W. MICHIELS, S.-I. NICULESCU, AND L. MOREAU, *Using delays and time-varying gains to improve the output feedback stabilizability of linear systems: A comparison*, IMA Journal of Mathematical Control and Information, 21 (2004), pp. 393–418. (Cited on pp. 251, 252, 253, 284)

[258] W. MICHIELS AND H. NIJMEIJER, *Synchronization of delay-coupled nonlinear oscillators: An approach based on the stability analysis of synchronized equilibria*, Chaos, 19 (2009), 033110. (Cited on pp. 362, 366, 367, 368, 369, 370, 371, 374)

[259] W. MICHIELS, T. PLOMTEUX, AND D. ROOSE, *Robust stabilization of linear time delay systems via the optimization of real stability radii*, in Proceedings of the 16th IFAC World Congress, Prague, Czech Republic, 2005, p. 639. (Cited on pp. 92, 245)

[260] W. MICHIELS AND D. ROOSE, *Global stabilization of multiple integrators with time delay and input constraints*, in Proceedings of the 3rd IFAC Workshop on Time-Delay Systems, Santa Fe, NM, 2001, pp. 266–271. (Cited on pp. 269, 271, 272, 284)

[261] ———, *Time-delay compensation in unstable plants using delayed state feedback*, in Proceedings of the 40th IEEE Conference on Decision and Control, Orlando, FL, 2001, pp. 1433–1437. (Cited on pp. 211, 286, 314)

[262] ———, *Limitations of delayed state feedback: A numerical study*, International Journal of Bifurcation and Chaos, 12 (2002), pp. 1309–1320. (Cited on pp. 214, 215, 216, 217, 218, 224)

[263] ———, *An eigenvalue based approach for the robust stabilization of linear time-delay systems*, International Journal of Control, 76 (2003), pp. 678–686. (Cited on pp. 70, 245)

[264] W. MICHIELS AND T. VYHLÍDAL, *An eigenvalue based approach to the robust stabilization of linear time-delay systems of neutral type*, Automatica, 41 (2005), pp. 991–998. (Cited on pp. 22, 33, 35, 37, 226)

[265] W. MICHIELS, T. VYHLÍDAL, AND P. ZÍTEK, *Control for time-delay systems based on quasi-direct pole placement*, Journal of Process Control, 20 (2010), pp. 337–343. (Cited on pp. 241, 243)

[266] W. MICHIELS, T. VYHLÍDAL, P. ZÍTEK, H. NIJMEIJER, AND D. HENRION, *Strong stability of neutral equations with an arbitrary delay dependency structure*, SIAM Journal on Control and Optimization, 48 (2009), pp. 763–786. (Cited on pp. 33, 37, 226, 229)

[267] L. MIRKIN AND Q.-C. ZHONG, *Are distributed delay control laws intrinsically unapproximable*, in Proceedings of the 4th IFAC Workshop on Time-Delay Systems, Paris, France, 2003. (Cited on pp. 302, 305)

[268] V. MISRA, W. B. GONG, AND D. TOWSLEY, *Fluid based analysis of a network of AQM routers supporting TCP flows with an application to RED*, in Proceeding ACM/SIGCOMM, Stockholm, Sweden, 2000, pp. 151–160. (Cited on pp. 316, 328, 329)

[269] C. MOLER AND C. VAN LOAN, *Nineteen dubious ways to compute the exponential of a matrix, twenty-five years later*, SIAM Review, 45 (2003), pp. 3–49. (Cited on p. 113)

[270] S. MONDIÉ, M. DAMBRINE, AND O. SANTOS, *Approximation of control laws with distributed delays: A necessary condition for stability*, Kybernetica, 38 (2002), pp. 541–551. (Cited on pp. 302, 306)

[271] S. MONDIÉ AND W. MICHIELS, *Finite spectrum assignment of unstable time-delay systems with a safe implementation*, IEEE Transactions on Automatic Control, 48 (2003), pp. 2207–2212. (Cited on pp. 302, 309, 314)

[272] S. MONDIÉ, S.-I. NICULESCU, AND J. J. LOISEAU, *Delay robustness of closed loop finite assignment for input delay systems*, in Proceedings of the 3rd IFAC Workshop on Time-Delay Systems, Santa-Fe, NM, 2001, pp. 2207–2212. (Cited on pp. 287, 297)

[273] M. MORARI AND E. ZAFIRIOU, *Robust Process Control*, Prentice-Hall, Englewood Cliffs, NJ, 1989. (Cited on p. 321)

[274] J. MORENO, *An extension of Lucas theorem to entire functions*, in Proceedings of the 1st IFAC Workshop on Linear Time-Delay Systems, Grenoble, France, 1998, pp. 159–163. (Cited on p. 254)

[275] Ö. MORGÜL, *On the stabilization and stability robustness against small delays of some damped wave equations*, IEEE Transactions on Automatic Control, 40 (1995), pp. 1626–1630. (Cited on p. 308)

[276] T. MORI AND H. KOKAME, *Stability of* $\dot{x}(t) = Ax(t) + Bx(t-\tau)$, IEEE Transactions on Automatic Control, 34 (1989), pp. 460–462. (Cited on pp. 36, 146)

[277] J. MORO, J. V. BURKE, AND M. L. OVERTON, *On the Lidskii–Vishik–Lyusternik perturbation theory for eigenvalues of matrices with arbitrary Jordan structure*, SIAM Journal on Matrix Analysis and Applications, 18 (1997), pp. 793–817. (Cited on p. 138)

[278] C.-I. MORĂRESCU, *Qualitative Analysis of Distributed Delay Systems. Methodology and Algorithms*, Ph.D. thesis, University of Bucharest, Romania/Université de Technologie de Compiègne, Compiègne, France, 2006. (Cited on pp. 37, 209, 284, 300, 338, 351)

[279] C.-I. MORĂRESCU AND S.-I. NICULESCU, *Stability crossing curves of SISO systems controlled by delayed output feedback*, Dynamics of Continuous, Discrete and Impulsive Systems, 14 (2007), pp. 659–678. (Cited on p. 284)

[280] C.-I. MORĂRESCU, S.-I. NICULESCU, AND K. GU, *Some remarks on Smith predictors: A geometric point of view*, in Proceedings of the 6th IFAC Workshop on Time-Delay Systems, L'Aquila, Italy, 2006, pp. 264–269. (Cited on p. 300)

[281] ——, *On the geometry of stability regions of Smith predictors subject to delay uncertainty*, IMA Journal on Mathematical Control and Information, 24 (2007), pp. 411–423. (Cited on pp. 295, 300)

[282] C. I. MORĂRESCU, S.-I. NICULESCU, AND W. MICHIELS, *Asymptotic stability of some distributed delay systems: An algebraic approach*, International Journal of Tomography and Statistics, 7 (2007), pp. 128–134. (Cited on p. 189)

[283] C.-I. MORĂRESCU AND S.-I. NICULESCU, *Some remarks on the delay effects on the stability of biochemical networks*, in Proceedings of the 16th IEEE Mediterranean Conference on Control and Automation, Ajaccio, France, 2008, pp. 25–27. (Cited on pp. 379, 381, 400)

[284] J. D. MURRAY, *Mathematical Biology*, 2nd ed., vol. 18 of BioMathematics Series, Springer-Verlag, Berlin, 1993. (Cited on p. 375)

[285] R. M. MURRAY, *Control in an Information Rich World*, Report of the Panel on Future Directions in Control, Dynamics, and Systems, Pasadena, CA, 2002. http://www.cds.caltech.edu/~murray/cdspanel. (Cited on p. 315)

[286] A. D. MYSHKIS, *General theory of differential equations with delay*, Uspehi, Matematičeskih Nauk, 4 (1949), pp. 99–141 (in Russian). American Mathematical Society Translations, 55 (1951), pp. 1–62. (Cited on pp. ix, xv)

[287] H. NAKAJIMA, *On analytical properties of delayed feedback control of chaos*, Physics Letters A, 232 (1997), pp. 207–210. (Cited on pp. 222, 223)

[288] J. NEIMARK, *D-subdivisions and spaces of quasi-polynomials*, Prikladnaja Matematika i Mehanika, 13 (1949), pp. 349–380. (Cited on pp. 121, 279)

[289] S.-I. NICULESCU, *Stability and hyperbolicity of linear systems with delayed state: A matrix pencil approach*, IMA Journal of Mathematical Control and Information, 15 (1998), pp. 331–347. (Cited on pp. 127, 155, 157, 158, 165, 166)

[290] ———, *Delay Effects on Stability. A Robust Control Approach*, vol. 269 of Lecture Notes in Control and Information Sciences, Springer-Verlag, 2001. (Cited on pp. ix, x, xv, xvi, 5, 36, 37, 48, 127, 128, 145, 147, 154, 166, 184, 189, 249, 258, 284, 326, 376, 378, 383)

[291] ———, *On delay robustness analysis of a simple control algorithm in high-speed networks*, Automatica, 38 (2002), pp. 858–889. (Cited on pp. 250, 284, 322, 326, 335)

[292] ———, *On interference phenomena in stability of linear systems with multiple delays*, in Proceeding of the 4th Asian Control Conference, Singapore, 2002. CD-ROM. (Cited on p. 187)

[293] ———, *On some stability regions of a multiple delays linear systems with applications*, in Proceedings of the 15th IFAC World Congress, Barcelona, Spain, 2002, p. 583. (Cited on pp. 322, 324, 325, 326, 335)

[294] S.-I. NICULESCU AND C. T. ABDALLAH, *Delay effects on static output feedback stabilization*, in Proceedings of the 39th IEEE Conference on Decision and Control, Sydney, Australia, 2000, pp. 2811–2816. (Cited on pp. 267, 284)

[295] S.-I. NICULESCU, J.-M. DION, AND L. DUGARD, *Robust stabilization for uncertain time-delay systems containing saturating actuators*, IEEE Transactions on Automatic Control, 41 (1996), pp. 742–747. (Cited on pp. 285, 292, 296)

[296] S.-I. NICULESCU, P. FU, AND J. CHEN, *Stability switches and reversals of linear systems with commensurate delays: A matrix pencil characterization*, in Proceedings of the 16th IFAC World Congress, Prague, Czech Republic, 2005, p. 637. (Cited on pp. 148, 155, 189)

[297] ———, *On the stability of linear delay-differential algebraic systems: Exact conditions via matrix pencil solutions*, in Proceedings of the 45th IEEE Conference on Decision and Control, San Diego, CA, 2006, pp. 834–839. (Cited on pp. 165, 189)

[298] S.-I. NICULESCU, K. GU, AND C. T. ABDALLAH, *Some remarks on the delay stabilizing effect in SISO systems*, in Proceedings of the 2003 American Control Conference, Denver, CO, 2003, pp. 2670–2675. (Cited on p. 219)

[299] S.-I. NICULESCU, P. KIM, K. GU, P. LEE, AND D. LEVY, *Stability crossing boundaries of delay systems modeling immune dynamics in leukemia*, Discrete and Continuous Dynamical Systems - Series B (DCDS-B), 15 (2010), pp. 129–156. (Cited on pp. 386, 395, 396, 397, 398, 399, 400)

[300] S.-I. NICULESCU AND W. MICHIELS, *Stabilizing a chain of integrators using multiple delays*, IEEE Transactions on Automatic Control, 49 (2004), pp. 802–807. (Cited on pp. 274, 284, 317)

[301] S.-I. NICULESCU, W. MICHIELS, K. GU, AND C. T. ABDALLAH, *Delay effects on output feedback control of dynamical systems*, in Complex Time-Delay Systems, F. M. Atay, ed., Understanding Complex Systems, Springer, Berlin, 2007, pp. 63–84. (Cited on pp. 162, 284)

[302] S.-I. NICULESCU, W. MICHIELS, D. MELCHOR-AGUILAR, T. LUZYANINA, F. MAZENC, K. GU, AND F. CHATTÉ, *Delay effects on the asymptotic stability of various fluid models in high-performance networks*, in Advances in Communication Control Networks, vol. 308 of Lecture Notes in Control and Information Science, Springer-Verlag, Heidelberg, 2004, pp. 87–110. (Cited on p. 335)

[303] S.-I. NICULESCU AND M. PUTINAR, *A toric positivstellensatz with applications to time delay systems*, Comptes Rendus Mathématique, 349 (2011), pp. 327–329. (Cited on p. 37)

[304] S.-I. NICULESCU, E. I. VERRIEST, L. DUGARD, AND J.-M. DION, *Stability of linear systems with delayed state: A guided tour*, in Stability and Control of Time-Delay Systems, L. Dugard and E. I. Verriest, eds., vol. 228 of Lecture Notes in Control and Information Sciences, Springer-Verlag, Berlin, 1998, pp. 1–71. (Cited on pp. ix, xv)

[305] J. NOCEDAL AND S. WRIGHT, *Numerical Optimization*, Springer Series in Operations Research and Financial Engineering, Springer, Berlin, 1999. (Cited on pp. 59, 233)

[306] J. E. NORMEY-RICO AND E. F. CAMACHO, *Robust tuning of dead-time compensators for processes with an integrator and long dead-time*, IEEE Transactions on Automatic Control, 44 (1999), pp. 1597–1603. (Cited on p. 285)

[307] R. NUSSBAUM, *Functional differential equations*, in Handbook of Dynamical Systems. vol. 2, B. Fiedler, ed., Elsevier, Amsterdam, 2002, pp. 461–499. (Cited on p. 145)

[308] G. OCHOA AND V. L. KHARITONOV, *Lyapunov matrices for neutral type time delay systems*, in Proceedings of the 2nd International Conference on Electrical and Electronics Engineering, Mexico City, Mexico, 2005, pp. 347–350. (Cited on p. 95)

[309] G. OCHOA, J. VELÁZQUEZ-VELÁZQUEZ, V. L. KHARITONOV, AND S. MONDIÉ, *Lyapunov matrices for neutral type time delay systems*, in Proceedings of the 7th IFAC workshop on time delay systems, Nantes, France, 2007. (Cited on pp. 95, 113)

[310] T. OGUCHI AND H. NIJMEIJER, *Prediction of chaotic behavior*, IEEE Transactions on Circuits and Systems I, 52 (2005), pp. 2464–2472. (Cited on pp. 212, 354)

[311] T. OGUCHI, H. NIJMEIJER, AND T. YAMAMOTO, *Synchronization in networks of chaotic systems with time-delay coupling*, Chaos, 18 (2008), 037108. (Cited on p. 354)

[312] R. OLFATI-SABER AND R. M. MURRAY, *Consensus problems in networks of agents with switching topology and time-delays*, IEEE Transactions on Automatic Control, 49 (2004), pp. 1520–1533. (Cited on pp. 339, 343, 345, 347)

[313] N. OLGAC AND R. SIPAHI, *An exact method for the stability analysis of time-delayed linear time-invariant (LTI) systems*, IEEE Transactions on Automatic Control, 47 (2002), pp. 793–797. (Cited on pp. 148, 164, 189, 260)

[314] A. V. OPPENHEIM AND R. W. SCHAFER, *Digital signal processing*, Prentice–Hall, Englewood Cliffs, NJ, 1975. (Cited on p. 148)

[315] G. OROSZ, B. KRAUSKOPF, AND R. E. WILSON, *Bifurcations and multiple traffic jams in a car-following model with reaction-time delay*, Physica D, 211 (2005), pp. 277–293. (Cited on p. 350)

[316] G. OROSZ AND G. STÉPÁN, *Subcritical Hopf bifurcations in a car-following model with reaction-time delay*, Proceedings of the Royal Society of London A, 462 (2006), pp. 2643–2670. (Cited on p. 350)

[317] M. OVERTON, *HANSO: Hybrid Algorithm for Nonsmooth Optimization*, 2009. http://cs.nyu.edu/overton/software/hanso/. (Cited on pp. 205, 226)

[318] A. PACKARD AND J. C. DOYLE, *The complex structured singular value*, Automatica, 29 (1993), pp. 71–109. (Cited on pp. x, 401, 402)

[319] A. PACKARD, M. FAN, AND J. C. DOYLE, *A power method for the structured singular value*, in Proceedings of the 27th IEEE Conference on Decision and Control, Austin, TX, 1988, pp. 2132–2137. (Cited on p. 403)

[320] Z. PALMOR AND M. BLAU, *An auto-tuner for Smith dead time compensator*, International Journal of Control, 60 (1994), pp. 117–135. (Cited on p. 317)

[321] Z. J. PALMOR, *Stability properties of Smith dead-time compensator controllers*, International Journal of Control, 32 (1980), pp. 937–949. (Cited on pp. 287, 290)

[322] ——, *Time-delay compensation—Smith predictor and its modifications*, in The Control Handbook, S. Levine, ed., CRC and IEEE Press, New York, 1996, Chap. 10, pp. 224–237. (Cited on pp. 286, 287, 308, 313, 314)

[323] L. PANDOLFI, *Stabilization of neutral functional differential equations*, Journal of Optimization Theory and Applications, 20 (1976), pp. 191–204. (Cited on p. 37)

[324] G. PAPPAS AND D. HINRICHSEN, *Robust stability of linear systems described by higher order dynamic equations*, IEEE Transactions on Automatic Control, 38 (1993), pp. 1430–1435. (Cited on pp. 70, 72, 75, 76, 80, 82)

[325] L. PECORA AND T. CARROLL, *Master stability function for synchronized coupled systems*, Physical Review Letters, 80 (1998), p. 2109. (Cited on pp. 354, 357)

[326] E. PINNEY, *Differential-Difference Equations*, University of California Press, Berkeley, CA, 1958. (Cited on pp. ix, xv, 145)

[327] E. PLISCHKE, *Transient Effects of Linear Dynamical Systems*, Ph.D. thesis, Universität Bremen, Bremen, Germany, 2005. (Cited on pp. 95, 111, 113)

[328] E. P. POPOV, *The dynamics of Automatic Control Systems*, Pergamon Press, New York, 1962. (Cited on p. 126)

[329] K. PYRAGAS, *Continuous control of chaos by self-controlling feedback*, Physics Letters A, 170 (1992), pp. 421–428. (Cited on pp. 222, 284, 354)

[330] L. QIU, B. BERNHARDSSON, A. RANTZER, E. DAVISON, P. YOUNG, AND J. DOYLE, *A formula for computation of the real stability radius*, Automatica, 31 (1995), pp. 879–890. (Cited on p. 92)

[331] Z. V. REKASIUS, *A stability test for systems with delays*, in Proceedings of 1980 Joint Automatic Control Conference, vol. TP9-A, San Francisco, CA, 1980. (Cited on pp. 147, 189)

[332] J.-P. RICHARD, *Time-delay systems: An overview of recent advances and open problems*, Automatica, 39 (2003), pp. 1667–1694. (Cited on pp. ix, xv)

[333] R. W. ROTHERY, *Car following models*, in Traffic Flow Theory. A State-of-the-Art Report, 2nd ed., H. Gartner, C. Messner, and A. Rathi, eds., 1998, pp. 4-1–4-42. (Cited on pp. 337, 338)

[334] M. R. ROUSSEL, *The use of delay differential equations in chemical kinetics*, Journal of Physical Chemistry, 100 (1996), pp. 8323–8330. (Cited on pp. ix, xv, 376)

[335] V. RĂSVAN, *Absolute stability of time lag control systems*, Ed. Academiei, Bucharest, 1975, (in Romanian). (Russian revised edition published by Nauka, Moscow, 1983). (Cited on pp. ix, xv, 37)

[336] V. RĂSVAN, *Dynamical systems with lossless propagation and neutral functional differential equations*, in Proceedings of MTNS, Padova, Italy, 1998, pp. 527–531. (Cited on p. 37)

[337] V. RĂSVAN AND D. POPESCU, *Control of systems with input delay: An elementary approach*, in Advances in Time-Delay Systems, Lecture Notes in Computational Science and Engineering, Springer-Verlag, Berlin, 2004, pp. 103–110. (Cited on pp. 302, 310)

[338] Y. SAAD, *Numerical Methods for Large Eigenvalue Problems*, Manchester University Press, Manchester, UK, 1992. (Cited on pp. 60, 65)

[339] A. SABERI, A. A. STOORVOGEL, AND P. SANNUTI, *Control of linear systems with regulation and input constraints*, Communication and Control Engineering, Springer-Verlag, London, 2000. (Cited on p. 272)

[340] C. SANTACESARIA AND R. SCATTOLINI, *Easy tuning of Smith predictor in presence of delay uncertainty*, Automatica, 29 (1993), pp. 1595–1597. (Cited on pp. 285, 286, 293)

[341] R. SEPULCHRE, *Slow peaking and low-gain design for global stabilization of nonlinear systems*, IEEE Transactions on Automatic Control, 45 (2000), pp. 453–461. (Cited on p. 269)

[342] R. SEYDEL, *Practical Bifurcation and Stability Analysis: From Equilibrium to Chaos*, vol. 5 of Interdisciplinary Applied Mathematics, Springer-Verlag, Berlin, 1994. (Cited on pp. 106, 132, 143, 144, 148, 199, 212, 217, 331)

[343] A. P. SEYRANIAN AND A. A. MAILYBAEV, *Interaction of eigenvalues in multi-parameter problems*, Journal of Sound and Vibration, 267 (2003), pp. 1047–1064. (Cited on p. 139)

[344] S. SHAKKOTTAI AND R. SRIKANT, *How good are deterministic fluid models of internet congestion control?*, in Proceedings IEEE INFOCOM 2002, New York, 2002, pp. 497–505. (Cited on p. 315)

[345] G. J. SILVA, A. DATTA, AND S. P. BHATTACHARYYA, *New results on the synthesis of PID controllers*, IEEE Transactions on Automatic Control, 47 (2000), pp. 241–252. (Cited on p. 284)

[346] G. J. SILVA, A. DATTA, AND S. P. BHATTACHARYYA, *PID Controllers for Time-Delay Systems*, Birkhäuser, Basel, 2005. (Cited on p. 148)

[347] J. G. SIMMONDS AND J. E. MAN, JR., *A First Look at Perturbation Theory*, Dover Books on Physics, Dover, Mineola, NY, 1998. (Cited on p. 148)

[348] R. SIPAHI, *Cluster Treatment of Characteristic Roots, CTCR, A Unique Methodology for the Complete Stability Robustness Analysis of Linear Time Invariant Multiple Time Delays Against Delay Uncertainties*, Ph.D. thesis, Mechanical Engineering Department, University of Connecticut, Storrs, CT, 2005. (Cited on pp. 148, 155, 164, 189)

[349] R. SIPAHI AND S.-I. NICULESCU, *Analytical stability study of a deterministic car following model under multiple delay interactions*, in Proceedings of the 6th IFAC Workshop on Time Delay Systems, L'Aquila, Italy, 2006, pp. 187–192. (Cited on p. 337)

[350] ——, *Deterministic time-delayed traffic flow models: A survey*, in Complex Time-Delay Systems, F. Atay, ed., vol. 2010 of Understanding Complex Systems, Springer, Berlin, 2010, pp. 297–322. (Cited on p. 338)

[351] H. SMITH, *An Introduction to Delay Differential Equations with Applications to the Life Sciences*, vol. 57 of Texts in Applied Mathematics, Springer, New York, 2011. (Cited on p. 124)

[352] H. L. SMITH AND P. WALTMAN, *The Theory of the Chemostat*, Cambridge University Press, Cambridge, UK, 1994. (Cited on p. 375)

[353] O. J. SMITH, *Closer control of loops with dead time*, Chemical Engineering Progress, 53 (1957), pp. 217–219. (Cited on pp. 285, 308)

[354] P. SMOLEN, D. BAXTER, AND J. BYRNE, *Frequency selectivity, multistability, and oscillations emerge from models of genetic regulatory systems*, American Journal of Physiology - Cell Physiology, 274 (1998), pp. C531–C542. (Cited on p. 377)

[355] J. SREEDHAR, P. VAN DOOREN, AND A. L. TITS, *A fast algorithm to compute the real structured stability radius*, in Stability Theory—Hurwitz Centenary Conference, R. Jeltsch and M. Mansour, eds., vol. 121 of International Series of Numerical Mathematics, Birkhäuser, Basel, 1996, pp. 219–230. (Cited on p. 92)

[356] R. SRIKANT, *Models and methods for analyzing Internet congestion control algorithms*, in Advances in Communication Control Networks, vol. 308 of Lectures Notes in Control and Information Sciences, Springer-Verlag, Heidelberg, 2005, pp. 65–84. (Cited on p. 335)

[357] E. STAFF, *The damping effect of time lag*, Engineer, 163 (1937), p. 439. (Cited on pp. ix, xv)

[358] N. STANKOVIC, S. OLARU, AND S.-I. NICULESCU, *On stability of discrete-time delay-difference equations for arbitrary delay variations*, in Proceedings of IFAC Joint Conference: 5th Symposium on System Structure and Control, 11th Workshop on Time-Delay Systems, and 6th Workshop on Fractional Differentiation and Its Applications, Grenoble, France, 2013, pp. 242–247. (Cited on p. 37)

[359] G. STÉPÁN, *Retarded Dynamical Systems: Stability and Characteristic Function*, vol. 210 of Research Notes in Mathematics, Longman Scientific, London, 1989. (Cited on pp. ix, xv, 8, 123, 146, 258)

[360] E. STEUR, W. MICHIELS, H. HUIJBERTS, AND H. NIJMEIJER, *Networks of diffusively time-delay coupled systems: Conditions for synchronization and its relation to the network topology*, Physica D, 277 (2014), pp. 22–39. Accepted. (Cited on p. 371)

[361] E. STEUR, T. OGUCHI, C. V. LEEUWEN, AND H. NIJMEIJER, *Partial synchronization in diffusively time-delay coupled oscillator networks*, Chaos, 22 (2012), 043144. (Cited on pp. 372, 373)

[362] S. STROGATZ, *Nonlinear Dynamics and Chaos: With Applications to Biology, Chemistry and Engineering*, Perseus Books Publishing, New York, 1994. (Cited on p. 362)

[363] ———, *Sync: The Emerging Science of Spontaneous Order*, Hyperion, New York, 2003. (Cited on p. 354)

[364] T. STYKEL, *On criteria for asymptotic stability of differential algebraic equations*, ZAMM Zeitschrift für Angewandte Mathematik und Mechanik, 82 (2002), pp. 147–158. (Cited on p. 229)

[365] H. SUSSMANN AND P. KOKOTOVIC, *The peaking phenomenon and the global stabilization of nonlinear systems*, IEEE Transactions on Automatic Control, 36 (1991), pp. 424–440. (Cited on p. 269)

[366] V. L. SYRMOS, C. T. ABDALLAH, P. DORATO, AND K. GRIGORIADIS, *Static output feedback—A survey*, Automatica, 33 (1997), pp. 125–137. (Cited on p. 255)

[367] M. SZYDLIWSKI AND A. KRAWIEC, *Differential delay equations in chemical kinetics: Some simple linear model systems*, Journal of Chemical Physics, 92 (1990), pp. 1702–1712. (Cited on pp. ix, xv, 376)

[368] ———, *The Kaldor-Kalecki model of business cycles as a two-dimensional dynamical system*, Journal of Nonlinear Mathematical Physics, 8 (2001), pp. 266–271. (Cited on pp. ix, xv)

[369] A. R. TEEL, *Global stabilization and restricted tracking for multiple integrators with bounded controls*, Systems and Control Letters, 18 (1992), pp. 165–171. (Cited on p. 284)

[370] A. THOWSEN, *An analytical stability test for a class of time-delay systems*, IEEE Transactions on Automatic Control, 25 (1981), pp. 735–736. (Cited on pp. 147, 155, 189)

[371] ———, *The Routh-Hurwitz method for stability test for a class of time-delay systems*, International Journal of Control, 33 (1981), pp. 991–995. (Cited on pp. 147, 189)

[372] Y. TIAN AND F. GAO, *Adaptive control of chaotic continuous-time systems with delay*, Physica D, 117 (1998), pp. 1–12. (Cited on p. 334)

[373] F. TISSEUR AND N. J. HIGHAM, *Structured pseudospectra for polynomial eigenvalue problems, with applications*, SIAM Journal on Matrix Analysis and Applications, 23 (2001), pp. 187–208. (Cited on pp. 70, 75, 76, 82)

[374] F. TISSEUR AND K. MEERBERGEN, *The quadratic eigenvalue problem*, SIAM Review, 43 (2001), pp. 235–286. (Cited on p. 70)

[375] O. TOKER AND H. ÖZBAY, *On the $\mathcal{NP}$-hardness of the purely complex μ computation, analysis/synthesis, and some related problems in multidimensional systems*, in Proceedings of the 1995 American Control Conference, Seattle, WA, 1995, pp. 447–451. (Cited on p. 81)

[376] L. N. TREFETHEN, *Pseudospectra of linear operators*, SIAM Review, 39 (1997), pp. 383–406. (Cited on p. 69)

[377] ——, *Spectral methods in MATLAB*, vol. 10 of Software, Environments, and Tools, SIAM, 2000. (Cited on pp. 41, 54, 63, 65, 94, 114)

[378] L. N. TREFETHEN AND M. EMBREE, *Spectra and Pseudospectra: The Behavior of Nonnormal Matrices and Operators*, Princeton University Press, Princeton, NJ, 2005. (Cited on pp. 69, 78)

[379] H. UNAL AND W. MICHIELS, *Prediction of partial synchronization in delay-coupled nonlinear oscillators, with application to hindmarsh-rose neurons*, Nonlinearity, 26 (2013), pp. 3101–3126. (Cited on pp. 373, 374)

[380] V. VAN ASSCHE, M. DAMBRINE, J.-F. LAFAY, AND J.-P. RICHARD, *Implementation of a distributed control law for a class of systems with delay*, in Proceedings of the 3rd IFAC Workshop on Time-Delay Systems, Santa Fe, NM, 2001, pp. 266–271. (Cited on pp. 302, 310)

[381] V. VAN ASSCHE, A. GANGULI, AND W. MICHIELS, *Practical stability analysis of systems with delay and periodic coefficients*, in Proceedings of the 5th IFAC Workshop on Time-Delay Systems, Leuven, Belgium, 2004, pp. 191–196. (Cited on pp. 302, 304)

[382] A. VAN DER SCHAFT, *L_2-Gain and Passivity Techniques in Nonlinear Control*, 2nd ed., Communications and Control Engineering Series, Springer-Verlag, London, 2000. (Cited on pp. 92, 370)

[383] P. VAN DOOREN AND V. VERMAUT, *On stability radii of generalized eigenvalue problems*, in Proceedings of the 1997 European Control Conference (ECC'97), Brussels, Belgium, 1997. (Cited on p. 75)

[384] G. H. M. VAN TARTWIJK AND D. LENSTRA, *Semiconductor lasers with optical injection and feedback*, Journal of Optics B: Quantum and Semiclassical Optics, (1995), pp. 87–143. (Cited on p. 87)

[385] J. VANBIERVLIET, W. MICHIELS, AND E. JARLEBRING, *Using spectral discretization for the optimal h2 design of time-delay systems*, International Journal of Control, 84 (2011), pp. 228–241. (Cited on pp. 114, 115, 118, 245)

[386] J. VANBIERVLIET, B. VANDEREYCKEN, W. MICHIELS, S. VANDEWALLE, AND M. DIEHL, *The smoothed spectral abscissa for robust stability optimization*, SIAM Journal on Optimization, 20 (2009), pp. 156–171. (Cited on p. 93)

[387] J. VANBIERVLIET, K. VERHEYDEN, W. MICHIELS, AND S. VANDEWALLE, *A nonsmooth optimization approach for the stabilization of linear time-delay systems*, ESAIM: Control, Optimisation and Calculus of Variations, 14 (2008), pp. 478–493. (Cited on pp. 195, 196, 197, 198, 201, 202, 203, 209)

[388] J. VELÁZQUEZ-VELÁZQUEZ AND V. L. KHARITONOV, *Lyapunov-Krasovskii functionals for scalar neutral type time delay equation*, Systems and Control Letters, 58 (2009), pp. 17–25. (Cited on pp. 95, 113)

[389] A. VERES AND M. BODA, *The chaotic nature of TCP-AQM congestion control*, in Proceedings of IEEE INFOCOM 2000, Tel Aviv, Israel, 2000, pp. 1715–1723. (Cited on p. 333)

[390] K. VERHEYDEN, K. GREEN, AND D. ROOSE, *Numerical stability analysis of a large-scale delay system modeling a lateral semiconductor laser subject to optical feedback*, Physical Review E, 69 (2004), 036702. (Cited on pp. 200, 201, 202)

[391] K. VERHEYDEN, T. LUZYANINA, AND D. ROOSE, *Efficient computation of characteristic roots of delay differential equations using LMS methods*, Journal of Computational and Applied Mathematics, 214 (2008), pp. 209–226. (Cited on pp. 40, 51)

[392] K. VERHEYDEN AND D. ROOSE, *Efficient numerical stability analysis of delay equations: A spectral method*, in Proceedings of the 5th IFAC Workshop on Time-Delay Systems, Leuven, Belgium, 2004. (Cited on p. 39)

[393] E. VERRIEST AND W. MICHIELS, *Routh criterion and stability of neutral systems*, Systems and Control Letters, 55 (2006), pp. 711–718. (Cited on p. 18)

[394] B. VIELLE AND G. CHAUVET, *Delay equation analysis of human respiratory stability*, Mathematical Biosciences, 152 (1998), pp. 105–122. (Cited on pp. 375, 381, 382, 383, 385, 400)

[395] M. I. VISHIK AND L. A. LYUSTERNIK, *The solution of some perturbation problems for matrices and selfadjoint or non-selfadjoint differential equations*, Russian Mathematical Surveys, 15 (1960), pp. 1–74. (Cited on p. 138)

[396] T. VYHLÍDAL, W. MICHIELS, AND P. MCGAHAN, *Synthesis of strongly stable state-derivative controllers for a time delay system using constrained non-smooth optimization*, IMA Journal of Mathematical Control and Information, 27 (2010), pp. 437–455. (Cited on p. 205)

[397] T. VYHLÏDAL AND P. ZÏTEK, *Mapping based algorithm for large-scale computation of quasi-polynomial zeros*, IEEE Transactions on Automatic Control, 54 (2009), pp. 171–177. (Cited on pp. 66, 405)

[398] T. VYHLÍDAL, P. ZÍTEK, AND K. PAULŮ, *Design, modelling and control of the experimental heat transfer set-up*, in Topics in Time Delay Systems. Analysis, Algorithms, and Control, J. J. L. et al., eds., vol. 308 of Lecture Notes in Control and Information Sciences, Springer-Verlag, Berlin, 2009, pp. 303–314. (Cited on p. 208)

[399] T. WAGENKNECHT, K. GREEN, S. ADHIKARI, AND W. MICHIELS, *Structured pseudospectra and random eigenvalue problems in vibrating systems*, AIAA Journal, 44 (2006), pp. 2404–2414. (Cited on pp. 81, 92)

[400] T. WAGENKNECHT AND J. AGARWAL, *Structured pseudospectra in structural engineering*, International Journal for Numerical Methods in Engineering, 64 (2005), pp. 1735–1751. (Cited on p. 78)

[401] T. WAGENKNECHT, W. MICHIELS, AND K. GREEN, *Structured pseudospectra for nonlinear eigenvalue problems*, Journal of Computational and Applied Mathematics, 212 (2006), pp. 245–259. (Cited on pp. 71, 81, 82, 92, 245)

[402] K. WALTON AND J. MARSHALL, *Direct method for TDS stability analysis*, IEE Proceedings, Part D, 134 (1987), pp. 101–107. (Cited on pp. 147, 189)

[403] Q. G. WANG, T. H. LEE, AND K. K. TAN, *Finite Spectrum Assignment for Time-Delay Systems*, vol. 239 of Lecture Notes in Control and Information Sciences, Springer-Verlag, Berlin, 1999. (Cited on p. 301)

[404] S. S. WANG, *Further results on stability of* $\dot{x}(t) = Ax(t) + Bx(t-\tau)$, Systems and Control Letters, 19 (1992), pp. 165–168. (Cited on p. 36)

[405] K. WATANABE, *Finite spectrum assignment and observer for multivariable systems with commensurate delays*, IEEE Transactions on Automatic Control, 31 (1986), pp. 543–550. (Cited on p. 301)

[406] K. WATANABE AND M. ITO, *A process-model control for linear systems with delay*, IEEE Transactions on Automatic Control, 26 (1981), pp. 1261–1269. (Cited on p. 285)

[407] C. WU, *Synchronization in coupled chaotic circuits and systems*, World Scientific, Singapore, 2002. (Cited on p. 354)

[408] J. WU, *Theory and Applications of Partial Functional Differential Equations*, vol. 119 of Applied Mathematical Sciences, Springer-Verlag, New York, 1996. (Cited on p. 65)

[409] Z. WU AND W. MICHIELS, *Reliably computing all characteristic roots of delay differential equations in a given right half plane*, Journal of Computational and Applied Mathematics, 236 (2012), pp. 2499–2514. (Cited on pp. 44, 45, 46, 47, 48, 51, 52, 53, 67)

[410] K. YAMANAKA AND E. SHIMEMURA, *Effects of mismatched Smith controller on stability in systems with time-delay*, Automatica, 23 (1987), pp. 787–791. (Cited on pp. 285, 286)

[411] W. ZHANG, M. S. BRANICKY, AND S. M. PHILLIP, *Stability of networked control systems*, IEEE Control Systems Magazine, 21 (2001), pp. 84–99. (Cited on pp. 275, 276, 284)

[412] Q.-C. ZHONG, *Robust stability analysis of simple systems controlled over communication networks*, Automatica, 39 (2003), pp. 1309–1312. (Cited on p. 335)

[413] ———, *Robust Control of Time-Delay Systems*, Springer-Verlag, London, 2006. (Cited on p. 313)

[414] K. ZHOU, J. C. DOYLE, AND K. GLOVER, *Robust and Optimal Control*, Prentice–Hall, Englewood Cliffs, NJ, 1995. (Cited on pp. 93, 95, 111, 225, 401)

Index